Peptide Science

Peptide Science

Chemical Ligation, Lead Generation, and Therapeutic Advances

Edited by

Fa Liu

Focus-X Therapeutics
Watchung, USA

WILEY

Published by John Wiley & Sons, Inc., Hoboken, New Jersey.

Published simultaneously in Canada.

Library of Congress Cataloging-in-Publication Data

Hardback: 9781119824671
ePDF: 9781119824688
epub: 9781119824695
oBook: 9781119824701

Cover Image: Courtesy of Fa Liu
Cover design by: Wiley

Contents

List of Contributors

Yvonne M. Angell
WuXi TIDES
San Francisco, CA, USA

Zachary T. Ball
Department of Chemistry
Rice University
Houston, TX, USA

Gaurav Bhardwaj
Institute for Protein Design
University of Washington
Seattle, Washington, D.C., USA

and

Department of Medicinal Chemistry
University of Washington
Seattle, Washington, D.C., USA

Yihui Cao
Department of Chemistry
University of Hong Kong
Hong Kong SAR, P.R. China

Sunay V. Chankeshwara
Medicinal Chemistry, Research and
Early Development, Respiratory and
Immunology, BioPharmaceuticals R&D
AstraZeneca
Gothenburg, Sweden

Nick Cox
Novo Nordisk, Chemical Biology
Global Research Technologies
Lexington, MA, USA

Werngard Czechtizky
Medicinal Chemistry, Research and
Early Development, Respiratory and
Immunology, BioPharmaceuticals
R&D
AstraZeneca
Gothenburg, Sweden

Ratmir Derda
Department of Chemistry
University of Alberta
Edmonton, AB, Canada

Richard D. DiMarchi
Department of Chemistry
Indiana University
Bloomington, IN, USA

Yuxuan Ding
Department of Chemistry
Rice University
Houston, TX, USA

Arunika Ekanayake
Department of Chemistry
University of Alberta
Edmonton, AB, Canada

Wendy J. Hartsock
Aralez Bio
Berkeley, CA, USA

Rongjun He
Neurocrine Biosciences, Inc.
San Diego, CA, USA

Michael T. Jacobsen
Novo Nordisk, Chemical Biology
Global Research Technologies
Lexington, MA, USA

Ming-Jin Jheng
Full-Life Technologies Limited
Shanghai, China

David Juergens
Institute for Protein Design
University of Washington
Seattle, Washington, D.C., USA

Anna Koijen
EnzyTag BV
Nuth, the Netherlands

Katharina L. Kopp
Novo Nordisk, Functional Assays & Screening
Global Research Technologies
Måløv, Denmark

Christin Kossmann
Zealand Pharma A/S
Søborg, Denmark

Pengyun Li
Full-Life Technologies Limited
Shanghai, China

Xiang Li
Entrada Therapeutics, Inc.
Boston, MA, USA

Xuechen Li
Department of Chemistry
University of Hong Kong
Hong Kong SAR, P.R. China

and

Laboratory for Marine Drugs and Bioproducts
Qingdao National Laboratory for Marine Science and Technology
Qingdao, P.R. China

Fa Liu
Full-Life Technologies Limited
Shanghai, China

John P. Mayer
Department of Molecular, Cellular and Developmental Biology
University of Colorado
Boulder, CO, USA

Hans Melo
Menten AI
San Francisco, CA, USA

Adam Mezo
Neurocrine Biosciences, Inc.
San Diego, CA, USA

Ethan Mickelson
Department of Bioengineering
University of Washington
Seattle, WA, USA

Jørgen Olsen
Novo Nordisk, Discovery ADME
Global Research Technologies
Måløv, Denmark

Christian Poulsen
Novo Nordisk, Biophysics and
Injectable Formulation
Global Research Technologies
Måløv, Denmark

Ziqing Qian
Entrada Therapeutics, Inc.
Boston, MA, USA

Timothy M. Reichart
Department of Chemistry
Hampden-Sydney College
Hampden-Sydney, VA, USA

Stephen Rettie
Molecular and Cell Biology Program
University of Washington
Seattle, Washington, D.C., USA

and

Institute for Protein Design
University of Washington
Seattle, Washington, D.C., USA

Ashweta Sahni
Entrada Therapeutics, Inc.
Boston, MA, USA

Weijun Shen
Center for Translational Research
(CTR)
Shenzhen Bay Laboratory
Shenzhen, China

Xiulong Shen
Entrada Therapeutics, Inc.
Boston, MA, USA

Sumeet Singh
Neurocrine Biosciences, Inc.
San Diego, CA, USA

Michael H.B. Stowell
Department of Molecular, Cellular and
Developmental Biology
University of Colorado
Boulder, CO, USA

Wu Su
Medicinal Chemistry, Research and
Early Development, Respiratory and
Immunology, BioPharmaceuticals
R&D
AstraZeneca
Gothenburg, Sweden

Xiao-xuan Su
Center for Translational Research
(CTR)
Shenzhen Bay Laboratory
Shenzhen, China

Kishore Thalluri
Department of Chemistry
Indiana University
Bloomington, IN, USA

Leendert van den Bos
EnzyTag BV
Nuth, the Netherlands

Xiaozhu Wang
Full-Life Technologies Limited
Shanghai, China

Jack Xu
Elixtide LLC
San Marcos, CA, USA

Shengping Zhang
Center for Translational Research (CTR)
Shenzhen Bay Laboratory
Shenzhen, China

Nan Zheng
Center for Translational Research (CTR)
Shenzhen Bay Laboratory
Shenzhen, China

Preface

With great excitement, the past 10 years have witnessed significant advancements in the broad peptide science field. These include the development of the fundamentally novel peptide ligation and protein modification chemistries, the re-engineering of automated peptide synthesizers enabling fast and high-quality protein synthesis, and numerous milestone achievements in peptide drug discovery and development. On the chemistry front, novel chemical methods have resulted in promising progress in enhancing the permeability of peptides to reach the targets inside cells or behind the blood-brain barrier (BBB). Relatedly but differently, new formulation approaches have enabled the oral administration of peptide drugs, where an incredible milestone was marked by the 2019 approval of Rybelsus®, the oral tablet formulation of semaglutide, a 31-mer glucagon-like peptide 1 (GLP-1) analog. On the therapeutics front, the huge success of peptide drugs in metabolic diseases, such as semaglutide, tirzepatide, liraglutide, and multiple modern insulin analogs, has now been expanded into oncology and other therapeutic areas. Two notable recent successes are the approval of Lutathera® in 2018 and Pluvicto® in 2022. Lutathera® is a peptide-radionuclide conjugate consisting of a somatostatin receptor agonist as the homing vector and a beta-emitter Lutethium-177 as the warhead to treat gastroenteropancreatic neuroendocrine tumors. Pluvicto® shares the same mechanism of action as Lutathera®, while targeting prostate-specific membrane antigen for the treatment of metastatic castration-resistant prostate cancer. Further in the metabolic disease areas, the 2022 approval of tirzepatide (Mounjaro®), a dual-acting peptide that can activate both GLP-1 receptor and gastric inhibitory polypeptide receptor, opens the era of clinical application of unimolecular polypharmacy where numerous promising multi-acting peptides are racing in the late development phases. Finally, it needs to be noted that there are many other equally exciting successes that were also developed within the same time frame. However, these are not mentioned above, only due to the limited space in this Preface.

These important advancements in general peptide chemistry and broad peptide therapeutics areas prompted us to assemble this book: *Peptide Science: Chemical*

Ligation, Lead Generation, and Therapeutic Advances. The motivation was to provide the readers with a collective set of quality reviews and case studies that cover the representative developments within the past 10 years or so in a logical and systematic fashion. With a total of five sections, it is intended to offer a highlight of modern peptide chemistries and comprehensive coverage of peptide therapeutics, including discovery, development, and regulatory.

Section I, New Developments in Peptide Chemistry, highlights recent developments in chemical and enzymatic peptide ligation technologies and modern peptide and protein modification chemistries. Please note that this section is not arranged to provide lecture chapters on the fundamentals of peptide chemistry, but rather focuses on selected advancements.

Section II, Peptide Drug Hit Identification, aims to review the approaches and technologies that have been frequently practiced for peptide hit identification, an important early step of the peptide drug discovery process. The classical source is naturally existing peptides, which are still one of the most reliable starting places as these peptides are already well-optimized via evolution to effectively execute their endogenous biological tasks. In addition to the nature source, artificial libraries, including phage display library and its chemically modified versions, and mRNA peptide display libraries, have become increasingly important as quality hit sources. More recently, computational-based approaches have also been rapidly developed, and impressive outcomes on peptide hit generation have been frequently achieved.

Section III, Peptide Drug Lead Generation and Optimization, includes two chapters focusing on lead generation and lead optimization, respectively. There is some overlapping between Chapter 6 and the chapters in Section II. The intention is for Chapter 6 to provide a thorough review of the hit and lead generation while Section II allows readers to dive deeply into the individual specific approaches. Chapter 7, contributed by a group of Novo Nordisk scientists, serves as a great reference for peptide therapeutics lead optimization with a focus on long-acting peptide hormone analogs, particularly the fatty acid derivatization approach.

Section IV, Peptide Drug Case Studies, dives deeply into the established peptide therapeutics and emerging developments in selected therapeutic areas. It consists of six chapters covering several therapeutic areas, including diabetes, obesity, NASH, cardiovascular diseases, cancer, inflammation, and autoimmune. It also involves various peptide drug molecular formats such as insulin (a mini protein with two chains and three pairs of disulfide bonds), incretin (a linear peptide with 30–40 residues), peptide-radionuclide conjugate (radioligand therapy, also called RDC), peptide–cytotoxin conjugate, and finally, peptide–oligonucleotides conjugates.

Section V, Peptide CMC and Regulatory, discusses peptide CMC processes and the related regulatory requirements.

What I hoped to assemble while missing in this book is the alternative formulation approach, such as oral tablet formulation of peptides, the methods enabling targeting intracellular receptors and targets beyond BBB, and another chapter in Section V to cover more regulatory side of peptide drug development.

In any case, I expect this book can serve as a valuable resource for the entire peptide science community, from junior researchers to university professors and seasoned industrial veterans, whoever is interested in the highlight of recent developments in peptide chemistry and a deep dive into the discovery and development of peptide therapeutics.

I thank all chapter authors for their contributions. The time they devoted to this book is highly appreciated. I also thank all Wiley editors and friends involved in this project for their long-lasting support. Finally, I thank my daughter Doris (fifth grade), who designed the cover image for the book.

Section I

New Developments in Peptide Chemistry

1

Chemical Peptide Ligations

Yihui Cao and Xuechen Li

Department of Chemistry, the University of Hong Kong, Hong Kong SAR, P. R. China

1.1 Introduction

Proteins play crucial roles in basic physiological processes and are responsible for a variety of biochemical functions, including signaling transduction, energy utilization, and immune response. Correlating protein structure with function has always been a charming topic among researchers. Although recombinant expression from bacteria or cell lines is a convenient means to produce proteins, it is still difficult to control specific post-translational modifications such as glycosylations, incorporate any uncanonical amino acid, or introduce unnatural reporters such as fluorescent tags, using the natural cellular machinery [1]. Herein, chemical protein synthesis that assembles protein sequence through atom-by-atom control provides a solution for generating site-specific natural or unnatural modification(s), and mirror-image proteins.

The solid-phase peptide synthesis (SPPS) by Merrifield provides an efficient peptide synthesis approach [2]. Utilization of SPPS methodology, along with the condensation of protected peptide fragments, has significantly expanded the range of polypeptide lengths that can be achieved via chemical synthesis. However, SPPS is limited by the peptide length. Due to the statistical reasons for linear stepwise coupling, each step during SPPS is incomplete, which causes byproducts to accumulate with peptide chain elongation. The peptide length from SPPS mostly remains within 50 amino acids to maintain good synthesis quality. Besides, the limited solubility of protected peptides in organic solvents hampers the ability of this method to meet the increasing synthesis demand for

Peptide Science: Chemical Ligation, Lead Generation, and Therapeutic Advances, First Edition. Fa Liu.

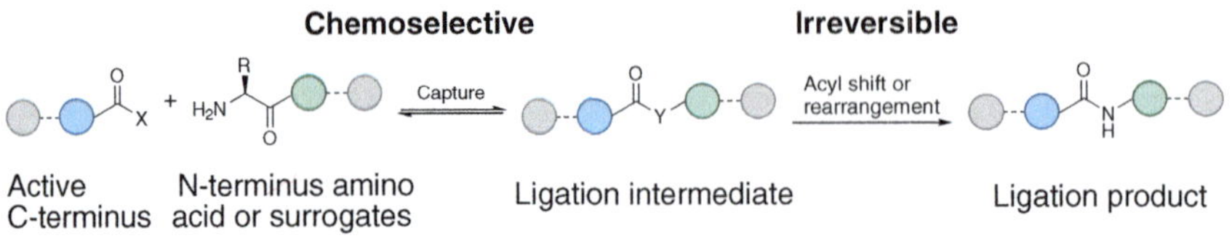

Figure 1.1 Generic chemical peptide ligation.

complex protein structures [3]. Consequently, novel synthetic approaches that can be conducted in aqueous buffers for handling unprotected peptide segments are strongly demanded.

The concept of peptide ligation, which allows condensation of unprotected peptide segments, was first proposed in the 1980s [4,5]. It involves a weakly activated peptide C-terminus to chemoselectively react with the N-terminus of the second peptide, resulting in a ligation intermediate that links the two fragments together, followed by an irreversible rearrangement step to form a natural peptide linkage (Figure 1.1). In this chapter, the driving forces of ligation, chemoselectivity details, and their applications in protein synthesis are discussed.

1.2 Ligation Driven by *Trans*-esterification

1.2.1 Native Chemical Ligation

Native chemical ligation (NCL) (Figure 1.2a), developed by Kent et al. in 1994, is the most widely applied ligation method [6]. NCL requires one peptide with a C-terminal thioester and the second peptide with an N-terminal cysteine (Cys). Ligation occurs when two fragments are mixed in a neutral or slightly basic aqueous buffer. The thiol group of N-terminal Cys undergoes reversible *trans*-thioesterification, replacing the thioester at the C-terminal of the first fragment. After that, a rapid [1,4] *S*-to-*N* acyl transfer converts the thioester intermediate into a native Xaa-Cys peptide (Xaa represents any amino acid) and generates the desired ligation product. Chemoselectivity can be considered to originate from "soft base-soft acid" interaction between the thioester (soft acid) and the thiol group (soft base) from free Cys. Other nucleophiles present on unprotected peptides, such as amines, are "hard" bases that do not have the same reactivity as the thiol group. Next, the irreversible and rapid acyl transfer drives the equilibrium. Even though the internal Cys could be involved in the reversible *trans*-thioesterification, the reaction equilibrium cannot move forward and does not produce a stable product. The ligating C-terminal residue and the thioester type highly affect the NCL reaction rate. For instance, β-branched amino acids (Val, Ile, Thr) significantly decrease ligation rates due to their bulky side chains [7]. Additionally, because of carbonyl oxygen interference,

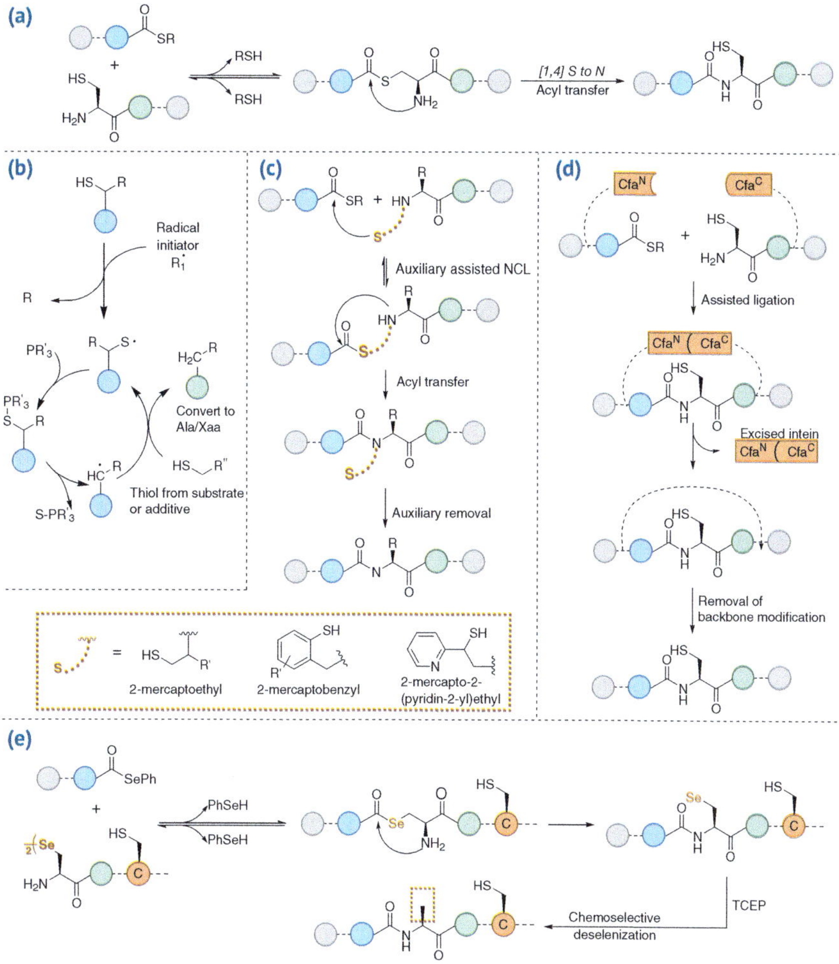

Figure 1.2 Ligation through *trans*-esterification.

proline carbonyl is less electrophilic, which restricts *trans*-esterification [6]. For the formation of intermolecular thioesters, 4-mercaptophenylacetic acid (MPAA) is widely used as an additive due to its low pKa, good water solubility in NCL buffer, and odorless nature. Other thiols with higher pKa values, such as trifluoroethanethiol (TFET), can be also used [8].

1.2.1.1 Desulfurization

Even though NCL is a powerful technique, the demand for one N-terminal Cys restricts its potential applications in protein chemical synthesis due to its low natural abundance (1.8%). The situation has changed after the invention of post-ligation desulfurization. Desulfurization, first reported in 2001 by Dawson, took advantage of metal-catalyzed reduction under a hydrogen atmosphere [9]. Nevertheless, the requirement of excess metal could occasionally cause side reactions and result in low yields. In addition, utilization of metal catalysts potentially induced epimerization of secondary alcohols and caused reduction of thiols and thioesters [10]. Later, the establishment of free radical-based desulfurization by Danishefsky et al. provided a milder and more reliable means for chemoselective peptide desulfurization (Figure 1.2b) [10]. Radical-induced desulfurization requires a radical initiator, phosphine compound, and hydrogen source. The reaction is initiated by water-soluble radical initiator 2,2α′-azobis[2-(2-imidazolin-2-yl)propane]dihydrochloride (VA-044) at 37 °C. The initiator radical attacks the Cys on the peptide, generating a thiyl radical. Subsequently, the thiyl radical rapidly reacts with tris(2-carboxyethyl)phosphine (TCEP) to generate the phosphonium radical. After that, the phosphine sulfide TCEP=S is cleaved from the complex, driving the alkyl radical formation. The desulfurized product is eventually generated after the application of hydrogen atom transfer (HAT) from thiol additives such as *tert*-butylthiol (tBuSH) or glutathione (GSH). Other thiyl radicals generated continue the radical chain reaction until its full conversion.

Based on the radical desulfurization strategy, photon-induced radical initiation has been developed. Pyane's group designed a flow chemistry system allowing NCL and UV-induced desulfurization to take place sequentially [11]. Later, visible light–induced metal complexes [12] and peroxide [13] radical initiator were reported, expanding the scope of radical initiator. Lately, Li et al. applied a novel radical generator, tetra-organylborate, to peptide desulfurization [14]. This strategy significantly increases the desulfurization reaction rate. It can be accomplished through a simple add-and-done procedure to finish within 30 seconds. Sodium tetraethylborate effectively serves as a radical initiator in the presence of atmospheric oxygen, inducing peptide desulfurization. Besides, the byproduct triethylborane from the initiation step serves as a hydrogen donor, demonstrating comparable efficiency to thiol additives such as tBuSH. Therefore, odorous thiol additives are not necessary for this strategy. In addition, the expeditious production of radicals can surpass the inhibitory impact of traces of MPAA residue (<100 μM), which allows NCL and desulfurization to perform in one pot. Furthermore, the mild conditions are compatible with some reductive functional groups, such as serotonylated substrates.

With the effective desulfurization technique established, the development of thiolated amino acids has drawn wide attention. The first thiolated amino acid to be used in NCL-desulfurization was β-thiolated phenylalanine [15]. NCL-desulfurization via both the nickel method and the free radical VA-044 method has been successfully applied to it [16,17]. Remarkably, commercially available penicillamine (Pen) has been used as the valine surrogate to expand the ligation site to one of the most abundant amino acids (6.8%) [18]. So far, a number of research groups have contributed to developing 13 Fmoc-SPPS-compatible thiolated amino acids, including β-thiolated, γ-thiolated, and δ-thiolated amino acids [1,19,20]. Although desulfurization may be difficult to some steric hindrance residue [18], NCL surrogate has been significantly expanded.

1.2.1.2 Auxiliary-assisted NCL

In parallel with the development of desulfurization, another strategy to overcome the low Cys abundance problem is the auxiliary-assisted NCL (Figure 1.2c). It introduces a thiol-containing handle at or near the peptide N-terminus, which mediates *trans*-thioesterification and *N*-to-*S* acyl transfer in a similar manner as Cys. After ligation, the auxiliary thiol hand can be removed. For glycopeptides, the branching sugar can serve as the handle through a thiol substitution. Its ligation site can be 1–6 amino acids away from the sugar-substituted amino acid [21,22]. In such cases, the acyl transfer via a large ring transition state significantly affects the reaction rate and efficiency, as compared to NCL. On the other hand, a variety of N^{α} substitutions, including N^{α}-2-mer-captoethyl type and N^{α}-2-mercaptobenzyl type [23], have been developed. They are designed to allow for acyl transfer through a five- or six-member ring for a rapid transfer rate. However, the secondary N-terminal amine carrying an auxiliary has more steric hindrance than a typical N-terminus. Thus, the *N*-to-*S* acyl transfer rate is much slower than that of NCL and becomes a rate-determining step. Due to the low ligation rate, the significant hydrolysis of the peptide thioesters and C-terminal residues epimerization problem may occur. Although this strategy could be applied to any ligation site in theory, a less hindered ligation site such as Gly-Gly or Gly-Ala is preferred. Recently, another auxiliary, 2-mercapto-2-(pyridine-2-yl) ethyl (MPyE), has been reported to have a significant improvement in the acyl transfer efficiency [24]. Its pyridine structure on auxiliary handle serves as an internal base catalyst that accepts hydrogen from the amine group, providing an extra driving force for acyl transfer.

1.2.1.3 Backbone-installed Split Intein-assisted Ligation

NCLs usually require millimolar (mM) concentration to proceed. Otherwise, their conversion yields could be dramatically low. However, such concentration cannot be achieved for low-solubility peptides, especially membrane protein fragments, due to their high aggregation potency. Recently, Liu's group developed backbone-installed split intein-assisted ligation (BISIAL) to overcome this problem (Figure 1.2d) [25]. The technique combines NCL and the consensus-fast (Cfa) split intein, which has a unique sequence allowing condensation and further engaging fusion of N- and C-extein without the presence of transpeptidases under denaturing conditions [26,27]. In this strategy, some short peptide sequences were designed as "exteins" (Exts) and installed on the desired peptide segment as a branch of removable backbone modification (RBM). Then, Cfas were linked to the extein containing peptides through NCL. BISIAL occurred once the Cfa^N-containing peptide and Cfa^C-containing peptide are mixed together. The reaction maintained a high reaction rate (within 2 hours) and high conversion yield at low concentration (50 μM). After NCL, Cfas was automatically split, and only connected Exts remained on the ligated product through RMB. Finally, RMBs were cleaved by acidic cleavage cocktails, providing a native peptide structure. Moreover, BISAL showed high efficiency on a variety of C-terminal amino acids, even those with steric hindrance, including Val (88% yield) and Pro (79% yield).

1.2.2 Diselenide/Selenoester Ligation

The 21st amino acid selenocysteine (Sec) was also investigated for peptide ligation with a similar mechanism and chemoselectivity with NCL (Figure 1.2e) [28]. Compared with Cys, Sec is more prone to undergo air oxidation. Thus, reducing reagents are required under ligation conditions to maintain the existence of selenolate monomer [29]. Due to Sec's lower pKa [1], the ligation can occur under slightly lower pH, under which less hydrolysis happens on the thioester. In the meantime, peptide selenoesters exhibit fast *trans*-selenoesterification even under high diluted concentrations due to its liability. Moreover, the removal of a selenol group is milder and more accessible than desulfurization due to a weaker C—Se bond. Deselenization does not require a radical initiator, and it can be achieved with a reducing agent TCEP and hydrogen donor present [30]. The production of a phosphine selenide, which is highly thermodynamically favorable, further facilitates the cleavage of the C—Se bond. The produced β-carbon-centered radical can then abstract a hydrogen atom to form the alanine residue [1]. In-chain native Cys residues do not participate in the deselenization conditions, so specific Cys protection is not necessarily required. Similar to NCL, researchers have contributed to the synthesis of selenylated amino acids, thereby synthesizing five β- or γ-substituted selenylated amino acids.

1.3 Ligation Driven by Imine Capture

1.3.1 Serine/Threonine Ligation

Chemoselective imine formation between the N-terminus and a C-terminus aldehyde has been also used to realize chemoselective condensation. Kemp et al. first demonstrated the possibility of applying imine capture and *O*-to-*N* acyl transfer to the amide bond formation with benzaldehyde [31]. Later, unprotected peptides with

C-terminal glycolaldehyde esters were found to chemoselectively react unprotected peptides with N-terminal Cys/Ser/Ter to form a pseudoproline-like thiazolidine/oxazolidine ring structure at the ligation site by Tam et al. in 1994 [32]. However, the obtained thiazolidine/oxazolidine ring structure could not be restored to the native peptide bonds, and they could be used to mimic the proline residue. In 2010, Li et al. reported serine/threonine ligation (STL), which utilizes unprotected peptides with C-terminal salicylaldehyde (SAL) esters to react with N-terminal serine/threonine of another peptide, generating native Xaa-Ser/Thr linkage at the ligation (Figure 1.3a).

STL occurs in a pyridine/acetic acid solvent mixture, which can dissolve most unprotected peptides. After the reversible imine capture between the N-terminal amine and the C-terminal SAL ester, the β-hydroxyl group participates in generating an oxazolidine structure via ring-chain tautomerism, which is then followed by an irreversible [1,5] *O*-to-*N* acyl transfer via a fused [5,6] transition to generate an *N,O*-benzylidene acetal structure as the ligation site. Nucleophilic functional groups from other amino acid side chains, including the amine group of lysine and the thiol group of cysteine, can also react with the SAL. However, the hemiaminal, imine, or hemithioacetal formation is reversible and unproductive. The reaction will finally be driven to the irreversible ligation pathway. Next, the resultant *N,O*-benzylidene acetal at the ligation site can be readily restored to the native peptide linkage by Trifluoroacetic acid (TFA)-mediated acidolysis.

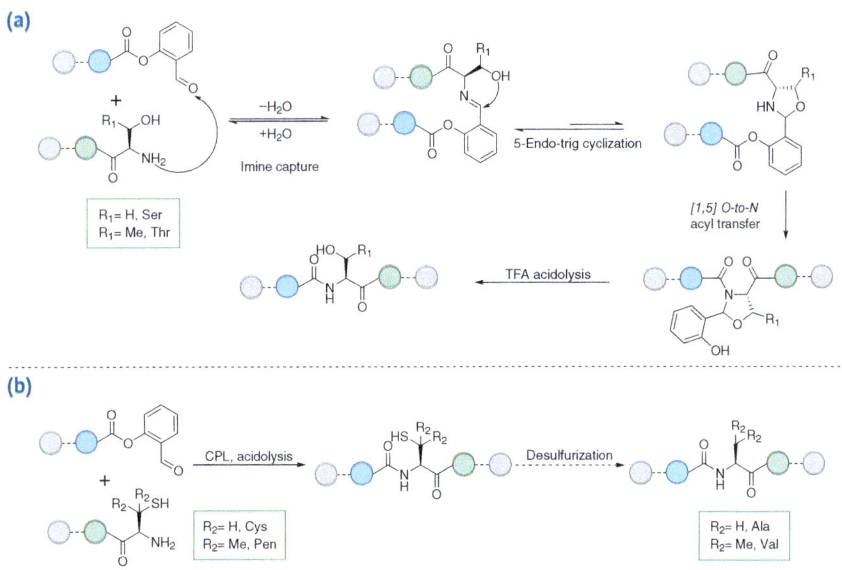

Figure 1.3 Ligation through imine capture.

1.3.2 Cysteine/Penicillamine Ligation

Cysteine/penicillamine ligation (CPL) shares a similar mechanism as STL, replacing the N-terminal Ser/Thr with Cys/Pen (Figure 1.3b) [33]. The penicillamine is selected because it is commercially available and serves as a representative of thiolated amino acids. In this case, the ring-closing becomes driven by thiazolidine formation instead of oxazolidine, which results in a significantly more rapid reaction kinetic. Similarly, the *N,S*-benzylidene acetal ligation intermediate can be removed by TFA-mediated acidolysis. However, due to the raised stability, TFA conditions with ethane-1,2-dithiol (EDT) or strong Lewis acid such as trimethylsilyl trifluoromethanesulfonate (TMSOTf) are typically required. This problem has been recently solved by using peptide MeO-SAL esters [34], for which the *N,S*-MeO-benzylidene acetal can be removed within minutes.

Both NCL and CPL use N-terminal Cys to mediate peptide ligations. Two features of CPL have been found as compared to NCL. Firstly, the CPL rate is affected by the steric hindrance of the C-terminal ligating amino acid, such as Val, Ile, Thr, and Pro. Secondly, CPL uses nonreductive acidic conditions for reaction, including pyridine/acetic acid buffer and pH 3.0 aqueous buffer. These conditions tolerate disulfide bonds and base-sensitive modifications, such as *S*-palmitoylation and *O*-acetylation. Similar to NCL-desulfurization, desulfurization can also be combined with CPL to expand the ligation site to other thiolated amino acids. For instance, penicillamine after ligation can be converted to a Val residue [33].

1.4 Ligation Initiated by Hydroxylamine

α-Ketoacid-hydroxylamine (KAHA) ligation was developed by Bode et al. in 2006 [35]. As the name indicates, the reaction occurs through condensing a C-terminal ketoacid and an N-terminal hydroxylamine or its derivatives. Unlike amines whose nucleophilicity is affected by protonation, hydroxylamines remain nucleophilic under acidic conditions. The hydroxylamine reacting with the keto group of the ketoacid, followed by dehydration and irreversible decarboxylation, provides the driving force to the peptide ligation.

KAHA ligation is categorized into two types [36]. Both type reactions are initiated by N-terminal hydroxylamine attacking C-terminal α-keto to afford an intermediate nitrone through dehydration equilibrium. For Type I (Figure 1.4a) with free hydroxylamine, the nitrone is attacked by the adjacent carboxylate, resulting in an α-lactone. Then, it converts into a desired amide product through rearrangement and irreversible decarboxylation. Mechanism of Type II (Figure 1.4b) proposes the presence of a nitrilium ion after dehydration and decarboxylation [36]. Additional water will drive

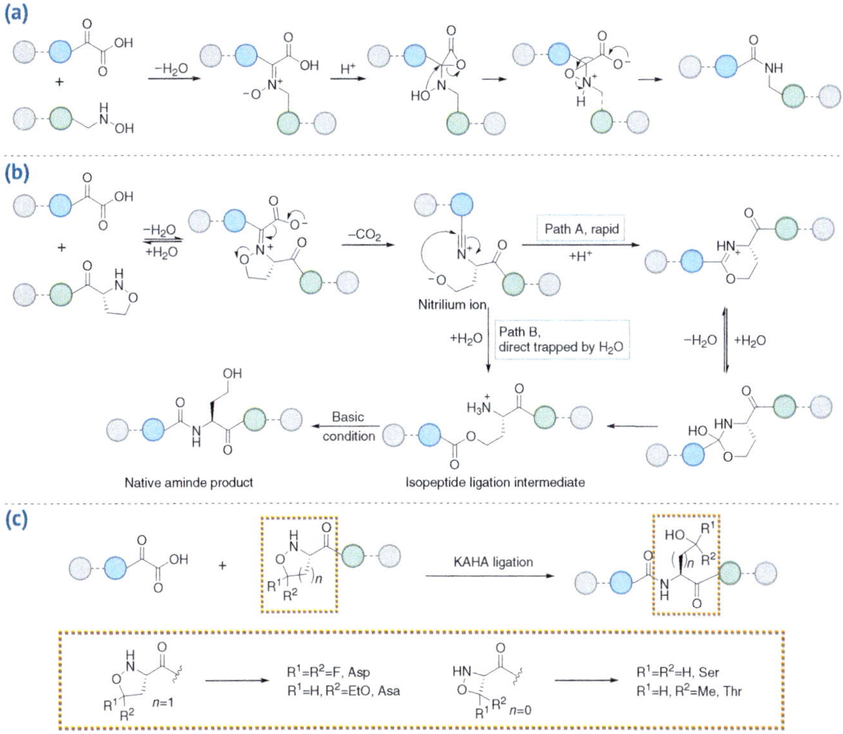

Figure 1.4 KAHA ligation.

the reaction to produce an amine product. When the hydroxylamine is 5-oxaproline, the nitrilium ion is more ready to undergo 6-endo-trig cyclization, go through iminium ether, and finally give rise to a depsipeptide intermediate. Afterward, the ester intermediate undergoes *O*-to-*N* acyl transfer under basic conditions to form an amide bond with a homoserine residue at the ligation site.

Suitable substitution at the 5-oxaproline ring can be used to generate other amino acids rather than homoserine at the ligation site, broadening the range of substrates for KAHA ligation (Figure 1.4c). By substituting the fourth position of the oxaproline ring, aspartic acid (Asp) and aspartyl aldehyde (Asa) can be produced after KAHA ligation [37,38]. In addition, the N-terminal cyclic five-member oxaproline ring can be replaced with a four-member ring oxazetidine to produce a native Ser residue at the ligation site. Compared with 5-oxaproline-mediated KAHA ligations, the oxazetidine-mediated ones are faster due to spontaneous condensation between C-terminal keto acid and N-terminal hydroxyl amine facilitated by higher ring strain [39]. Besides, introducing an additional methyl group to the side chain of oxazetidine leads to a threonine-forming KAHA ligation, which is also accompanied with enhanced stability of the oxazetidine structure as compared to the unsubstituted one [40].

1.5 Ligation in Protein Synthesis

1.5.1 Sequential Ligation and Chemical Protein Synthesis

1.5.1.1 Terminal Latent and Activation

In chemical protein synthesis, proteins are usually obtained from multiple ligations with several segments to maintain good SPPS quality, as peptides from SPPS are usually less than 50 amino acids. Therefore, the middle peptide fragments need to carry both reactive C-terminus and N-terminus. To avoid self-cyclization of the middle fragment, it is essential to accurately distinguish between required and irrelevant active sites in different fragments to direct fragment condensations properly. Techniques for temporarily masking the reactive terminus enable controllable ligation direction on demand (Figure 1.5a).

For N-terminal Cys (as in NCL and CPL) and Ser/Thr (as in STL), either protecting the side-chain functional nucleophile or masking the amine can serve the purpose. For instance, for thiol-substituted amino acids or selenol amino acids involved in ligations, the N-terminal thiol or selenol group is protected. Thiolthiazoline (Thz) and acetamidomethyl (Acm) are typical N-terminal thio-protecting groups for NCL. They can be removed by methoxamine and metal ions (usually Ag or Pd), respectively. Selenazolidine (Sez), which has a similar ring structure to Thz, is used to protect Sec with the same deprotecting conditions [41]. Such a protecting group can also be applied to CPL. Particularly, the nonreducing and acidic conditions of CPL allow disulfide masking groups to be used on Cys. The thiol-protecting group can either be aryl thiol or thiol-substituted solubilizing tags. Both can be easily removed with TCEP treatment [42]. In addition, selenol can be protected by *p*-methoxybenzyl [43]. On the other hand, Fmoc group to temporarily protect the N-terminal amine has been used for NCL, STL, and KAHA ligations. It can be easily removed under basic conditions. Moreover, KAHA ligation has a unique photocaged protecting group to be deprotected using 365 nm UV radiation [44].

A frequently utilized C-terminus variant for *trans*-esterification-driven ligation is the peptidyl hydrazide. Peptide hydrazines can be activated either through oxidation by $NaNO_2$ via azido intermediate [45] or condensation with

Figure 1.5 Sequential ligation and protein synthesis. *Source*: (a, b) Ref. [58–65]/ American Chemical Society.

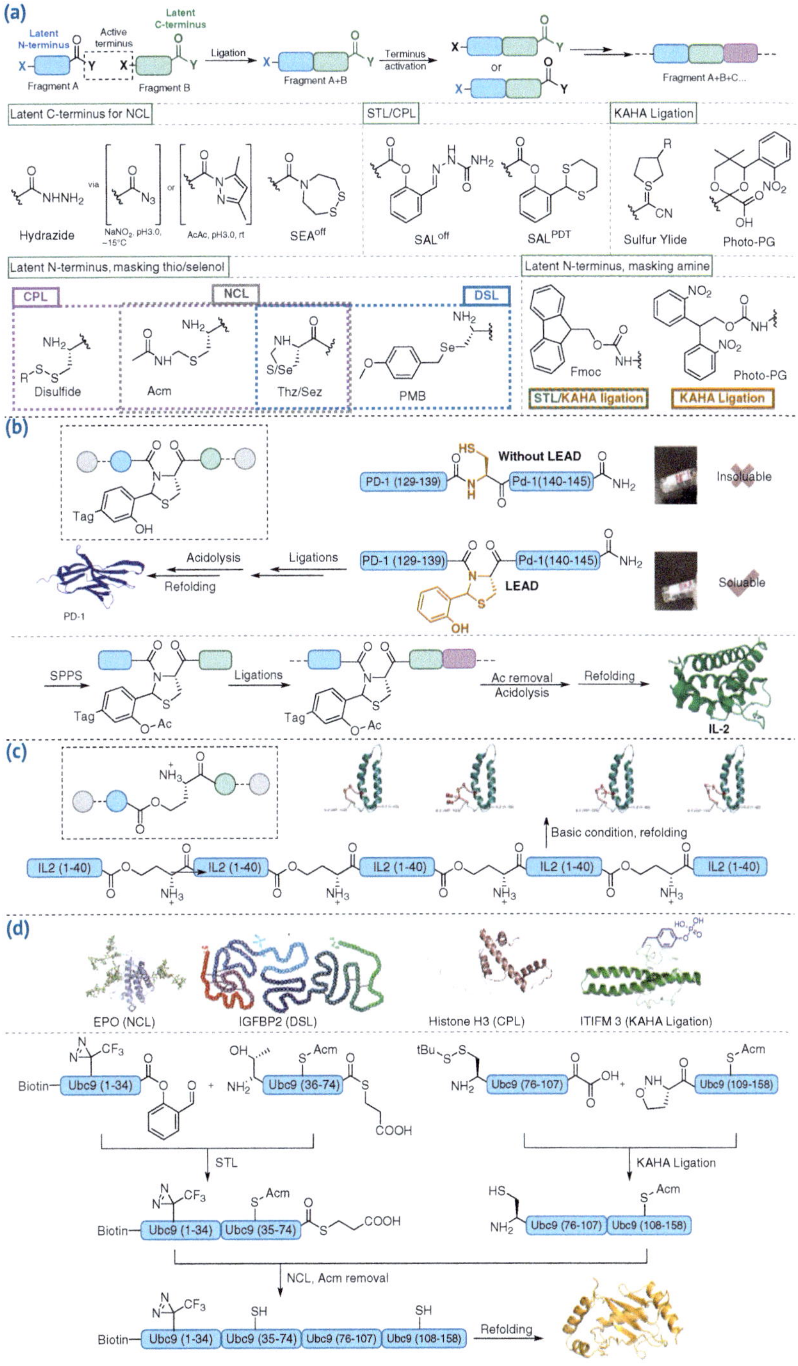
(a)
Latent N-terminus
Active terminus
Latent C-terminus
Fragment A
Fragment B
Ligation
Fragment A+B
Terminus activation
or
Fragment A+B+C...
Latent C-terminus for NCL
STL/CPL
KAHA Ligation
Hydrazide
via
or
SEA^off
SAL^off
SAL^PDT
Sulfur Ylide
Photo-PG
Latent N-terminus, masking thio/selenol
Latent N-terminus, masking amine
CPL
NCL
DSL
Disulfide
Acm
Thz/Sez
PMB
Fmoc
Photo-PG
STL/KAHA ligation
KAHA Ligation
(b)
Tag
Without LEAD
PD-1 (129-139)
Pd-1(140-145)
Insoluable
Soluable
LEAD
Acidolysis
Ligations
Refolding
PD-1
SPPS
Ac removal
Acidolysis
IL-2
(c)
Basic condition, refolding
IL2 (1-40)
(d)
EPO (NCL)
IGFBP2 (DSL)
Histone H3 (CPL)
ITIFM 3 (KAHA Ligation)
Biotin
Ubc9 (1-34)
Ubc9 (36-74)
Ubc9 (76-107)
Ubc9 (109-158)
COOH
STL
KAHA Ligation
Ubc9 (35-74)
Ubc9 (108-158)
NCL, Acm removal
Refolding

acetylacetone (AcAc) via Knorr pyrazole [46]. The intermediates can be rapidly transformed into active thioesters by employing thioesterification with MPAA or other thiol derivatives. Another example is the bis(2-sulfanylethyl)amido (SEA) system [47]. SEA in a cyclic disulfide form (called SEA^{off}) exhibits no reactivity toward Cys peptides with MPAA present. After being reduced by TCEP in a slightly acidic buffer, thiols are released by ring opening and undergo an *N*-to-*S* acyl shift, which generates the C-terminal alkyl thioester, causing the transition from SEA^{off} to SEA^{on}. This species is ready to undergo further conversion into a more reactive form, such as an MPAA thioester. The entire process, which involves the NCL reaction and the subsequent exchange of the SEA group, can be carried out in one pot [48].

For STL/CPL using peptide SAL esters, it is expected to employ protective groups on SAL. The latent C-terminus in STL/CPL has been established by employing the semicarbazone (SAL^{off}) to protect the aldehyde. The reactivation of SAL^{on} can be achieved through pyruvic acid treatment in TFA [49]. In addition, propanedithiol (PDT) has recently been discovered as a SAL protecting group, which provides stability under both CPL and NCL conditions, allowing *N*-to-*C* sequential ligation between NCL and STL. The PDT-protected SAL ester can be reactivated using NCS/$AgNO_3$. However, it is essential to note that unprotected Cys residues are not compatible with these conditions [50].

Cyanosulfurylide serves as an α-ketoacid precursor in KAHA ligations, which can be activated by Oxone or DMDO generating α-ketoacid [51,52]. However, the activation conditions are incompatible with reducible amino acid residues such as Cys and methionine (Met) because of the use of oxidative reagents. Subsequently, a photocaging group was devised for the α-ketoacid to enhance its compatibility with delicate amino acids [53].

1.5.1.2 Ligation Intermediates Facilitate Difficult Peptide Synthesis

Sequences that potentially adopt β-sheet or α-helical conformations often result in high aggregation tendency and low solubility, especially for membrane proteins [54]. These "difficult sequences" become a big challenge for both chemical synthesis and recombinant expression [55]. For chemical synthesis, aggregation highly restricts peptide synthesis, purification, and interfragment ligation efficiency. To overcome such difficulties, backbone protecting groups including pseudoproline, 2-hydroxy-4-methoxybenzyl, and *O*-acyl isopeptides are developed for disrupting peptide aggregation during SPPS for better synthesis quality [56–58]. Although these techniques are widely adopted in protein synthesis, they require extra building block synthesis, and some of them would be removed during global deprotection. Herein, some ligation intermediates can provide a convenient way to insert backbone modification to unprotected peptides and maintain such benefit till the end of synthesis.

1.5.1.2.1 STL/CPL Ligation Intermediate as Aggregation Disruptor

The ligation *N,O/S*-benzylidene acetal intermediate after STL/CPL was found to prevent aggregation (Figure 1.5b). This strategy, named ligation embedding aggregation disruptor (LEAD), was applied in the synthesis of the extracellular domain of programmed cell death protein 1 (PD-1) [59]. The ligation *N,O/S*-benzylidene acetal intermediate could improve fragments' solubility and reactivity, allowing for smooth subsequent ligations, desulfurization, and purification. Moreover, through removable acylation on the phenol group, *N/S*-benzylidene acetal can be maintained from acidolysis. Thus, CPL intermediate could be installed by on-resin ligation, serving as tunable backbone modification (TBM). The temporary acyl "capping" can be removed via hydrazine treatment. The successful synthesis of interleukin-2 (IL-2) demonstrated that TBM is a powerful tool for difficult sequence synthesis [60].

1.5.1.2.2 Depsipeptide Increases Solubility of Difficult Sequence

As a KAHA intermediate generated by Type II oxaproline ligation, the depsipeptide intermediate significantly prevents peptide fragments from aggregation, increases solubility, and improves separation (Figure 1.5c). By controlling buffer $pH < 3$, the *O*-acyl isopeptide can be maintained and remains stable to be isolated. Besides, the ligation solvent DMSO/H_2O is also a powerful solvent to denature aggregated peptides and dissolve hydrophobic sequences. The IL-2 sequence was assembled through convergent 5-oxaproline KAHA ligation, incorporating multiple isoacyl modifications [61]. After ligation and Acm deprotection, *O*-to-*N* acyl transfer and folding were performed in one pot to obtain bioactive versions of IL-2. The KAHA ligation method successfully constructs an efficient synthesis routine for IL-2 with various distinct disulfide moieties.

1.5.1.3 Chemical Protein Synthesis

Over the past decades, with the development of ligation technology, a variety of important protein targets have been chemically synthesized (Figure 1.5d). Remarkably, the glycosylated erythropoietin synthesis was reported in 2013 [62]. This work maintained a uniform pattern of attached sugars, which was helpful in the analysis of how variation in the sugar components of erythropoietin impacts its function. In addition, various proteins have been synthesized by NCL, diselenide-selenoester ligation (DSL), STL/CPL, and KAHA ligations [49,63,64].

Moreover, different types of ligation could be effectively combined to one protein synthesis. Bode et al. demonstrated using NCL, STL, and KAHA ligations in a convergent synthesis of SUMO E2 conjugating enzymes Ubc9 [65]. Fragment 1, Ubc9 (1–34), with an N-terminal biotin and photo-Phe22, was ligated with fragment 2, Ubc9 (35–74), with an N-terminal Thr and a C-terminal thioester. The N-terminal Cys of ketoacid fragment 3, Ubc9 (75–107), was protected by disulfide.

After KAHA ligation with fragment 4, Ubc9 (108–158), the disulfide protection was removed with pH adjustment in one pot. The final NCL was pressed to join Ubc9 (1–74) and Ubc9 (75–158) together. Then, after Acm deprotection and folding, Ubc9 protein was obtained.

1.5.2 Protein Semi-synthesis

Semi-synthesis takes advantage of both chemical synthesis and recombinant expression. Instead of synthesizing an entire protein, only a small portion of fragments that contain the required modifications needs to be synthesized. The unmodified protein domain can be biologically expressed, avoiding tedious ligations and purifications. The expressed fragment is then ligated with the peptide fragment containing the desired modifications, producing protein modification precisely and efficiently (Figure 1.6). Protein semi-synthesis has been developed for both NCL and STL.

Expressed protein ligation was first reported in 1998 [66]. This technique is based on protein splicing of inteins, a post-translational process that removes an internal segment from a precursor protein through intramolecular *N*-to-*S* or *N*-to-*O* acyl transfer and *trans*-esterification. The internal segment is intein, and the joint segments are N-/C- exteins. The whole progress is catalyzed by intein sequence and does not have specific requirements of exteins. By intercepting the thioester formation during protein splicing with thiol-derivatives or hydrazine, it becomes possible to conveniently express the truncated protein fragment containing a thioester [66,67] or hydrazine surrogate [68] at the C-terminus. Thus, the expressed peptide thioesters are then ligated with a synthetic peptide containing N-terminal cysteine via NCL to produce semi-synthetic proteins.

On the other hand, peptides with N-terminal Cys can be expressed, which is then ligated with synthetic peptide thioesters via NCL. Mutating the conserved intein N-terminal to Ala can suspend splicing driven by succinimide formation.

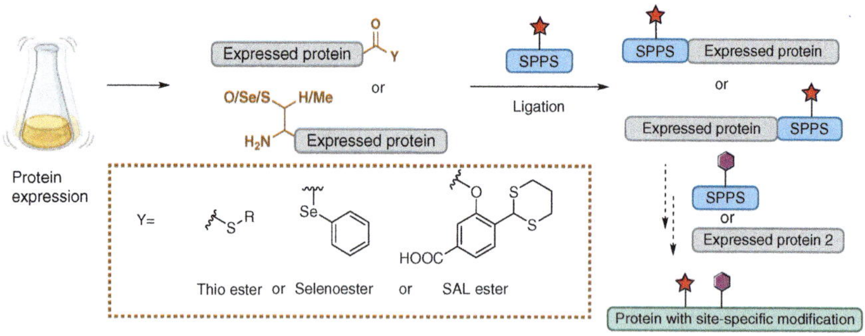

Figure 1.6 Semi-ligation.

The desired protein was expressed as C-extein. Its N-terminal Cys allows intein splicing between the C-extein and N-extein-intein complex.

A protein with N-terminal Sec could be expressed by replacing Cys to Sec in substrate amino acids during expression, allowing EPL to be carried out under a diselenide-selenoester ligation strategy. Although all Cys residues in the protein must be mutated, protein behavior is not supposed to be influenced significantly due to low Cys abundance [69]. Furthermore, AcAc hydrazine activation breaks the barrier of expression C-terminal selenoester, allowing DSL to occur at the C-terminal of the expressed part [70].

STL/CPL semi-synthesis has been also advanced for using either expressed peptide SAL esters or expressed peptides with N-terminal Ser/Thr. N-terminal Ser/Thr allows intein splicing as well. Peptides with N-terminal Ser/Thr can be obtained via similar intein mutation. These expressed proteins were ligated with synthetic peptides with SAL esters to produce proteins with specific modifications such as glycosylated IL-2 [71] and acetylated HMGB1 [42,72]. Recently, the peptide C-terminal SAL ester expression technique was developed after the introduction of the PDT protecting group [73]. In a similar manner to EPL for NCL, the intein system was used. After proteins with inteins were expressed, followed by hydrazinolysis and $NaNO_2$ activation, protein phenolysis with 3-(1,3-dithian-2-yl)-4-hydroxybenzoic acid ($SAL(-COOH)^{PDT}$) followed by PDT removal produces proteins/peptides with C-terminal SAL(—COOH) ester.

1.6 Ligations in Drug Development

1.6.1 Ligation in Macrocyclic Peptide Synthesis

Macrocyclic peptides have been isolated from a variety of species and are widely recognized as a promising modality for drug discovery because of their unique biological activities [74]. Compared with linear peptides, cyclic peptides encounter more conformational constraints. The constrained structure renders them with better specificity, selectivity, and efficacy for targeting [75,76]. Meanwhile, they are more resistant to peptidase degradation [51]. Although cyclization through direction coupling of the fully protected peptide could be carried out, the strategy may suffer from underlying risks, including low solubility, epimer generation, and oligomerization from intermolecular reactions [77]. Therefore, the effective synthesis of macrocyclic peptides is still an essential but challenging goal. Herein, peptide ligations provide effective, highly chemoselective, and epimerization-free methods for nonprotected peptide cyclization.

Not long after the invention of NCL, Tam et al. first applied NCL into peptide cyclization in 1997 [78]. In the initial instances, they successfully synthesized cyclic

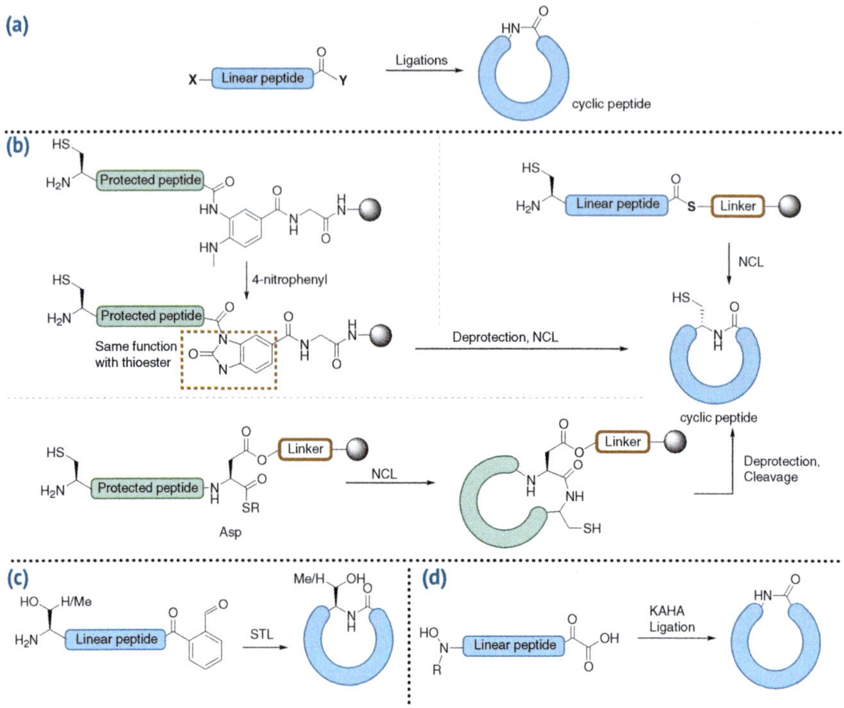

Figure 1.7 Ligation and macrocyclic peptides.

derivatives of enkephalins ranging in size from five to eight amino acids (Figure 1.7a). Additionally, they synthesized two larger cyclic peptides by utilizing sequences obtained from human parathyroid hormone (14 mer) and salmon calcitonin (16 mer). It has been demonstrated that oligomer formation did not occur even when the cyclization was performed at relatively high concentrations (mM level). In this scenario, employing high dilution techniques for conventional peptide cyclization is unnecessary, showing advantages over the conventional peptide cyclization using fully protected peptides.

Furthermore, two types of on-resin NCL-mediated cyclization have been reported (Figure 1.7b). Muir's group developed on-resin free amino acid NCL. They anchor peptides through thiol esters. After the completion of Boc SPPS, all protecting groups were removed, and then the resin was immersed in an NCL buffer. Thus, after cyclic peptides were formed, they were spontaneously cleaved from the resin [79]. Later, a similar strategy was applied to Fmoc-SPPS. In this case, the peptide was linked to the resin through *O*-amino(methyl)aniline (MeDbz), which served as the thioester surrogate after activation with 4-nitrophenyl [80]. If a sequence contained Asp, it could be anchored through a carboxylic acid group of the Asp side chain, allowing the installation of an amino acid thioester at last [81]. Besides, the Cys protecting group was chosen to be highly acid-labile so that it was orthogonal to others. After treatment with dilute acid, the resin was added to the NCL buffer for cyclization. A final global deprotection and cleavage step was then required.

In 2013, STL was first applied to peptide cyclization on daptomycin (Figure 1.7c), a lipocyclic peptide antibiotic drug with 13 amino acids [82]. The linear peptide was assembled via a solution phase and solid-phase hybrid approach. After the preparation of peptide SAL ester, ligation occurred between Gly10 and D-Ser11 smoothly to complete the total synthesis of daptomycin. Furthermore, several cyclic tetrapeptides without any turn-inducing residues (such as Pro, Gly, and D-amino acid) with 4–10 amino acid ring sizes have been successfully synthesized using STL-mediated peptide cyclization [83]. Without any discernible epimerization, all of these syntheses produced a high monomer-to-dimer ratio at a concentration of 1 mM [77].

Type I KAHA ligation was applied to the synthesis of cyclic peptides in 2012 (Figure 1.7d). The cyclic peptides were prepared using sulfur ylide resin through Fmoc-based SPPS. After cleavage from the resin, dimethoxyldioxirane (DMDO) instead of ozone was selected to oxidize the sulfur ylide into α-ketoacids to protect N-terminal nitrone from hydrolysis. Excess oxidant was quenched by dimethylsulphide (DMS). The precursor linear peptide was dissolved in a reaction solution DMF/H_2O (50 : 1) with an excess amount of oxalic acid under 40 °C. Under such conditions, the nitrone-protecting group was hydrolyzed by a trace amount of water, thus steadily releasing an active N-terminus. However, 5% of epimerization has been detected at the C-terminal α-ketoacid [51].

The Type II KAHA strategy has been applied to the synthesis of AS-48, a cyclic antibacterial protein consisting of 70 amino acids [84]. The protein is produced from two-fragment condensation instead of single-segment cyclization. One significant challenge of AS-48 for chemical synthesis was its low solubility due to the extreme hydrophobic sequence. The ligations were successfully carried out using DMSO/H_2O (7 : 3) as the solvent. The N-terminus and C-terminus of the two fragments were protected with a photosensitive protecting group, respectively. After the first ligation, both the protecting groups were released by UV radiation in one pot. Then it proceeded for the second ligation for a cyclic structure, followed by *O*-to-*N* acyl transfer.

1.6.2 Mirror-image Phage Display

Phage display is a high-throughput affinity screening technique for peptide drug development, demonstrated in 1985 (Figure 1.8a) [85]. It is used to discover peptides that selectively bind to the target. The procedure is as follows: First, a phage clone library is created by cloning the combined DNA sequences. The immobilized

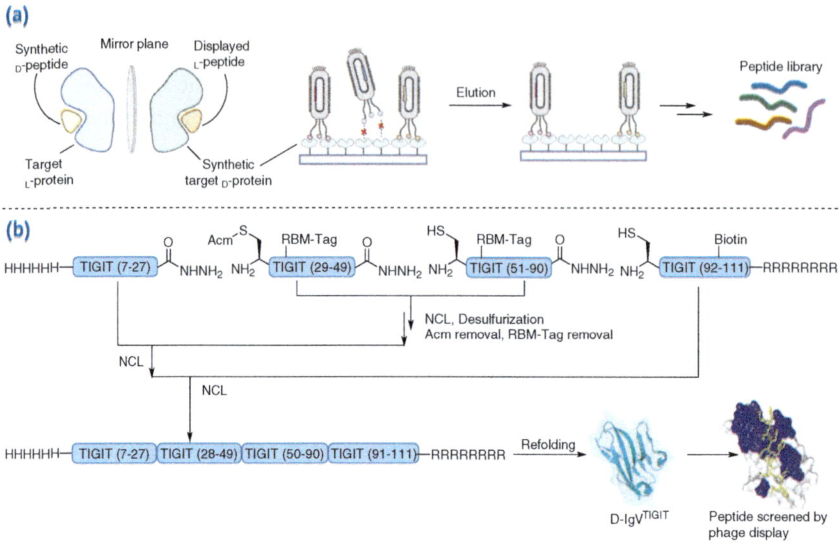

Figure 1.8 Phage display.

protein to be screened is allowed to incubate with phage library. Subsequently, phages that do not bind selectively to targets or lack binding affinity can be washed off. Then, "elution" occurs, in which phages bound to the target get washed off from the immobilized target through either competitive elution or incubation in a low pH buffer. Finally, infection with *Escherichia coli* amplifies the eluted phage, creating a more selective phage library for the next round of screening. Three to five screening rounds are typically necessary to isolate reliable, high-affinity peptide binders. Every time, by decreasing the amount of targets and increasing the number of washes, a stricter screening is reached. Finally, DNA sequencing and ELISA are used to determine individual phages with high binding affinity to the target and decode the peptide sequence [86].

The Kent group demonstrated that the L-form and D-form HIV protease could only cleave substrates with the same chirality [87], that is, the L-form enzyme only worked on the L-substrate and did not harm the D-version. The finding proves that D-peptides are resistant to proteolysis. When compared to L-peptides, D-peptides are significantly superior. Proteases are unable to identify and break them down since their chirality differs from that of native compounds [88]. The protease resistance significantly increases the D-peptide's *in vivo* lifetime [89]. Furthermore, the stability of D-peptides offers a more versatile approach to administration. D-peptides can remain stable and be absorbed orally, in contrast to L-enantiomeric peptides, which require injections to remain active [90]. In 1999, Kim reported D-peptide inhibitors targeting the gp41 coiled-coiled pocket [91].

Mirror-image phage display was first reported by Schumacher in 1996 [92]. This technique, based on the laws of symmetry, provides an approach for identifying D-peptide ligands to target proteins. It shares the basic concept and experimental procedure with standard phage display. However, the target protein is changed into D-form proteins. After screening, amplification, and elution, a set of L-form peptide ligands can be discovered. Because the complexes formed by L-ligands and D-proteins are mirror images of D-ligands and L-proteins, a D-peptide with the same sequence as the selected L-peptide would be an exemplary ligand for the natural L-protein.

With the capability of protein chemical synthesis, D-form proteins can be chemically synthesized, which dramatically expands the scope of mirror-image phage display. There have been several attempts to use mirror-image phage display screening for drug discovery. An inhibitor of the immune checkpoint TIGIT (T cell immunoreceptor with Ig and ITIM domains) was obtained through this approach (Figure 1.8b) [93]. In the context, the D-IgV domain of TIGIT was synthesized using the peptide ligation approach. After folding, the D-IgV^{TIGIT} had an exact mirror image with the expressed native form. Six clones were selected after five rounds of screening against synthesized D-IgV^{TIGIT}. Although the obtained peptides showed low membrane permeability, they were active, as the TIGIT IgV domain is located outside the cell membrane.

1.7 Summary and Outlook

After decades of development, the ligation chemistry scope has been largely extended. More diverse fragment assembly approaches allow the total synthesis of more complicated and challenging protein sequences. Chemical protein synthesis provides a more precise way to generate proteins with natural and unnatural modification(s) as compared to traditional recombinant protein expression technology. Due to novel developments in ligation techniques, cyclic peptide synthesis has become more available and predictable, which has expanded the possibility of peptide drug production. Moreover, large proteins with hundreds of residues can possibly be synthesized through ligation technology. The possibilities for modifications through chemical synthesis are unlimited. Thus, chemical protein synthesis enables the creation of larger and increasingly tricky mirror-image proteins, which is expected to result in the identification of several novel therapeutic possibilities using mirror-image display technologies in the future. Ligations offer more opportunities to explore protein function and develop novel therapeutics.

References

1 Kulkarni, S.S., Sayers, J., Premdjee, B., and Payne, R.J. Rapid and efficient protein synthesis through expansion of the native chemical ligation concept. *Nat Rev Chem*. 2018;2(4):0122.
2 Merrifield, R.B. Automated synthesis of peptides. *Science*. 1965;150(3693): 178–185.
3 Kent, S.B.H. Chemical synthesis of peptides and proteins. *Annu Rev Biochem*. 1988;57(1):957–989.
4 Blake, J. and Li, C.H. New segment-coupling method for peptide synthesis in aqueous solution: application to synthesis of human [Gly17]-beta-endorphin. *Proc Natl Acad Sci*. 1981;78(7):4055–4058.
5 Yamashiro, D. and Li, C.H. New segment synthesis of α-inhibin-92 by the acyl disulfide method. *Int J Pept Protein Res*. 1988;31(3):322–334.
6 Agouridas, V., El Mahdi, O., Diemer, V., Cargoët, M., Monbaliu, J.-C.M., and Melnyk, O. Native chemical ligation and extended methods: mechanisms, catalysis, scope, and limitations. *Chem Rev*. 2019;119(12):7328–7443.
7 Hackeng, T.M., Griffin, J.H., and Dawson, P.E. Protein synthesis by native chemical ligation: expanded scope by using straightforward methodology. *Proc Natl Acad Sci*. 1999;96(18):10068–10073.
8 Diemer, V., Firstova, O., Agouridas, V., and Melnyk, O. Pedal to the metal: the homogeneous catalysis of the native chemical ligation reaction. *Chem Eur J*. 2022;28(16):e202104229.
9 Yan, L.Z. and Dawson, P.E. Synthesis of peptides and proteins without cysteine residues by native chemical ligation combined with desulfurization. *J Am Chem Soc*. 2001;123(4):526–533.
10 Wan, Q. and Danishefsky, S.J. Free-radical-based, specific desulfurization of cysteine: a powerful advance in the synthesis of polypeptides and glycopolypeptides. *Angew Chem Int Ed*. 2007;46(48):9248–9252.
11 Chisholm, T.S., Clayton, D., Dowman, L.J., Sayers, J., and Payne, R.J. Native chemical ligation–photodesulfurization in flow. *J Am Chem Soc*. 2018;140(29):9020–9024.
12 Gao, X.-F., Du, J.-J., Liu, Z., and Guo, J. Visible-light-induced specific desulfurization of cysteinyl peptide and glycopeptide in aqueous solution. *Org Lett*. 2016;18(5):1166–1169.
13 Qiu, W., Shi, S., Li, R., Lin, X., Rao, L., and Sun, Z. A mild, general, metal-free method for desulfurization of thiols and disulfides induced by visible-light. *Chin J Chem*. 2021;39(5):1255–1258.
14 Sun, Z., Ma, W., Cao, Y., et al. Superfast desulfurization for protein chemical synthesis and modification. *Chem*. 2022;8(9):2542–2557.

15 Botti, P. and Tchertchian, S. Side-chain extended ligation. WO2006133962 A1. 21 December 2006;42.
16 Crich, D. and Banerjee, A. Native chemical ligation at phenylalanine. *J Am Chem Soc*. 2007;129(33):10064–10065.
17 Malins, L.R., Giltrap, A.M., Dowman, L.J., and Payne, R.J. Synthesis of β-thiol phenylalanine for applications in one-pot ligation–desulfurization chemistry. *Org Lett*. 2015;17(9):2070–2073.
18 Haase, C., Rohde, H., and Seitz, O. Native chemical ligation at valine. *Angew Chem Int Ed*. 2008;47(36):6807–6810.
19 Malins, L.R. and Payne, R.J. Synthetic amino acids for applications in peptide ligation–desulfurization chemistry. *Aust J Chem*. 2015;68(4):521.
20 Guan, I., Williams, K., Liu, J.S.T., and Liu, X. Synthetic thiol and selenol derived amino acids for expanding the scope of chemical protein synthesis. *Front Chem*. 2022;9:826764.
21 Brik, A., Yang, Y.-Y., Ficht, S., and Wong, C.-H. Sugar-assisted glycopeptide ligation. *J Am Chem Soc*. 2006;128(17):5626–5627.
22 Payne, R.J., Ficht, S., Tang, S., et al. Extended sugar-assisted glycopeptide ligations: development, scope, and applications. *J Am Chem Soc*. 2007;129(44):13527–13536.
23 Offer, J. Native chemical ligation with Nα acyl transfer auxiliaries. *Pept Sci*. 2010;94(4):530–541.
24 Fuchs, O., Trunschke, S., Hanebrink, H., Reimann, M., and Seitz, O. Enabling cysteine-free native chemical ligation at challenging junctions with a ligation auxiliary capable of base catalysis. *Angew Chem Int Ed*. 2021;60(35):19483–19490.
25 Zhang, B., Zheng, Y., Chu, G., et al. Backbone-installed split intein-assisted ligation for the chemical synthesis of mirror-image proteins. *Angew Chem Int Ed*. 2023;62(33):e202306270.
26 Stevens, A.J., Brown, Z.Z., Shah, N.H., Sekar, G., Cowburn, D., and Muir, T.W. Design of a split intein with exceptional protein splicing activity. *J Am Chem Soc*. 2016;138(7):2162–2165.
27 Thompson, R.E., Stevens, A.J., and Muir, T.W. Protein engineering through tandem transamidation. *Nat Chem*. 2019;11(8):737–743.
28 Mitchell, N.J., Malins, L.R., Liu, X., et al. Rapid additive-free selenocystine–selenoester peptide ligation. *J Am Chem Soc*. 2015;137(44):14011–14014.
29 Chisholm, T.S., Kulkarni, S.S., Hossain, K.R., Cornelius, F., Clarke, R.J., and Payne, R.J. Peptide ligation at high dilution via reductive diselenide-selenoester ligation. *J Am Chem Soc*. 2020;142(2):1090–1100.
30 Metanis, N., Keinan, E., and Dawson, P.E. Traceless ligation of cysteine peptides using selective deselenization. *Angew Chem Int Ed*. 2010;49(39):7049–7053.

31 Kemp, D.S., Grattan, J.A., and Reczek, J. Peptide bond formation by the prior amine capture principle. *J Org Chem.* 1975;40(23):3465–3466.

32 Liu, C.-F. and Tam, J.P. Chemical ligation approach to form a peptide bond between unprotected peptide segments: concept and model study. *J Am Chem Soc.* 1994;116(10):4149–4153.

33 Tan, Y., Li, J., Jin, K., et al. Cysteine/penicillamine ligation independent of terminal steric demands for chemical protein synthesis. *Angew Chem Int Ed.* 2020;59(31):12741–12745.

34 Wu, H., Sun, Z., and Li, X. Selective peptide cysteine manipulation on demand and difficult protein chemical synthesis enabled by controllable acidolysis of N,S-benzylidene thioacetals. *Angew Chem Int Ed.* 2024;136(19):e202403396.

35 Bode, J.W., Fox, R.M., and Baucom, K.D. Chemoselective amide ligations by decarboxylative condensations of N-alkylhydroxylamines and α-ketoacids. *Angew Chem Int Ed.* 2006;45(8):1248–1252.

36 Rohrbacher, F., Wucherpfennig, T.G., and Bode, J.W. Chemical protein synthesis with the KAHA ligation. In: *Protein Ligation and Total Synthesis II* (ed. L. Liu). Cham: Springer International Publishing; 2015. pp. 1–31.

37 Baldauf, S., Ogunkoya, A.O., Boross, G.N., and Bode, J.W. Aspartic acid forming α-ketoacid–hydroxylamine (KAHA) ligations with (S)-4,4-difluoro-5-oxaproline. *J Org Chem.* 2020;85(3):1352–1364.

38 Murar, C.E., Thuaud, F., and Bode, J.W. KAHA ligations that form aspartyl aldehyde residues as synthetic handles for protein modification and purification. *J Am Chem Soc.* 2014;136(52):18140–18148.

39 Pusterla, I. and Bode, J.W. An oxazetidine amino acid for chemical protein synthesis by rapid, serine-forming ligations. *Nat Chem.* 2015;7(8):668–672.

40 Baldauf, S., Schauenburg, D., and Bode, J.W. A threonine-forming oxazetidine amino acid for the chemical synthesis of proteins through KAHA ligation. *Angew Chem Int Ed.* 2019;58(36):12599–12603.

41 Reddy, P.S., Dery, S., and Metanis, N. Chemical synthesis of proteins with non-strategically placed cysteines using selenazolidine and selective deselenization. *Angew Chem Int Ed.* 2016;55(3):992–995.

42 Liu, J., Wei, T., Tan, Y., Liu, H., and Li, X. Enabling chemical protein (semi)synthesis via reducible solubilizing tags (RSTs). *Chem Sci.* 2022;13(5):1367–1374.

43 Mitchell, N.J., Sayers, J., Kulkarni, S.S., et al. Accelerated protein synthesis via one-pot ligation-deselenization chemistry. *Chem.* 2017;2(5): 703–715.

44 Bode, J.W. Chemical protein synthesis with the α-ketoacid–hydroxylamine ligation. *Acc Chem Res.* 2017;50(9):2104–2115.

45 Fang, G.-M., Li, Y.-M., Shen, F., et al. Protein chemical synthesis by ligation of peptide hydrazides. *Angew Chem Int Ed.* 2011;50(33):7645–7649.

46 Flood, D.T., Hintzen, J.C.J., Bird, M.J., Cistrone, P.A., Chen, J.S., and Dawson, P.E. Leveraging the Knorr pyrazole synthesis for the facile generation of thioester surrogates for use in native chemical ligation. *Angew Chem Int Ed*. 2018;57(36):11634–11639.

47 Agouridas, V., Ollivier, N., Vicogne, J., Diemer, V., and Melnyk, O. Redox-controlled chemical protein synthesis: sundry shades of latency. *Acc Chem Res*. 2022;55(18):2685–2697.

48 Raibaut, L., Cargoët, M., Ollivier, N., et al. Accelerating chemoselective peptide bond formation using bis(2-selenylethyl)amido peptide selenoester surrogates. *Chem Sci*. 2016;7(4):2657–2665.

49 Lee, C.L., Liu, H., Wong, C.T.T., Chow, H.Y., and Li, X. Enabling N-to-C Ser/Thr ligation for convergent protein synthesis via combining chemical ligation approaches. *J Am Chem Soc*. 2016;138(33):10477–10484.

50 Huang, D.-L., Li, Y., Liang, J., et al. The new salicylaldehyde S,S-propanedithioacetal ester enables N-to-C sequential native chemical ligation and Ser/Thr ligation for chemical protein synthesis. *J Am Chem Soc*. 2020;142(19):8790–8799.

51 Fukuzumi, T., Ju, L., and Bode, J.W. Chemoselective cyclization of unprotected linear peptides by α-ketoacid–hydroxylamine amide-ligation. *Org Biomol Chem*. 2012;10(30):5837–5844.

52 Ju, L., Lippert, A.R., and Bode, J.W. Stereoretentive synthesis and chemoselective amide-forming ligations of C-terminal peptide α-ketoacids. *J Am Chem Soc*. 2008;130(13):4253–4255.

53 Thuaud, F., Rohrbacher, F., Zwicky, A., and Bode, J.W. Photoprotected peptide α-ketoacids and hydroxylamines for iterative and one-pot KAHA ligations: synthesis of NEDD8. *Helv Chim Acta*. 2016;99(11):868–894.

54 Mueller, L.K., Baumruck, A.C., Zhdanova, H., and Tietze, A.A. Challenges and perspectives in chemical synthesis of highly hydrophobic peptides. *Front Bioeng Biotechnol*. 2020;8:162.

55 Shen, F., Huang, Y.-C., Tang, S., Chen, Y.-X., and Liu, L. Chemical synthesis of integral membrane proteins: methods and applications. *Isr J Chem*. 2011; 51(8–9):940–952.

56 Haack, T. and Mutter, M. Serine derived oxazolidines as secondary structure disrupting, solubilizing building blocks in peptide synthesis. *Tetrahedron Lett*. 1992;33(12):1589–1592.

57 Johnson, T., Quibell, M., Owen, D., and Sheppard, R.C. A Reversible protecting group for the amide bond in peptides. Use in the synthesis of 'difficult sequences'. *J Chem Soc, Chem Commun*. 1993;(4):369–372.

58 Sohma, Y., Sasaki, M., Hayashi, Y., Kimura, T., and Kiso, Y. Novel and efficient synthesis of difficult sequence-containing peptides through O–N intramolecular acyl migration reaction of O-acyl isopeptides. *Chem Commun*. 2004;(1):124–125.

59 Wu, H., Wei, T., Ngai, W.L., Zhou, H., and Li, X. Ligation embedding aggregation disruptor strategy enables the chemical synthesis of PD-1 immunoglobulin and extracellular domains. *J Am Chem Soc*. 2022;144(32):14748–14757.

60 Wu, H., Tan, Y., Ngai, W.L., and Li, X. Total synthesis of interleukin-2 via a tunable backbone modification strategy. *Chem Sci*. 2023;14(6):1582–1589.

61 Murar, C.E., Ninomiya, M., Shimura, S., Karakus, U., Boyman, O., and Bode, J.W. Chemical synthesis of interleukin-2 and disulfide stabilizing analogues. *Angew Chem Int Ed*. 2020;59(22):8425–8429.

62 Wang, P., Dong, S., Shieh, J.-H., et al. Erythropoietin derived by chemical synthesis. *Science*. 2013;342(6164):1357–1360.

63 Premdjee, B., Andersen, A.S., Larance, M., Conde-Frieboes, K.W., and Payne, R.J. Chemical synthesis of phosphorylated insulin-like growth factor binding protein 2. *J Am Chem Soc*. 2021;143(14):5336–5342.

64 Harmand, T.J., Pattabiraman, V.R., and Bode, J.W. Chemical synthesis of the highly hydrophobic antiviral membrane-associated protein IFITM3 and modified variants. *Angew Chem Int Ed*. 2017;56(41):12639–12643.

65 Zhang, Y., Hirota, T., Kuwata, K., Oishi, S., Gramani, S.G., and Bode, J.W. Chemical synthesis of atomically tailored SUMO E2 conjugating enzymes for the formation of covalently linked SUMO–E2–E3 ligase ternary complexes. *J Am Chem Soc*. 2019;141(37):14742–14751.

66 Muir, T.W., Sondhi, D., and Cole, P.A. expressed protein ligation: A general method for protein engineering. *Proc Natl Acad Sci*. 1998;95(12):6705–6710.

67 Flavell, R.R. and Muir, T.W. Expressed protein ligation (EPL) in the study of signal transduction, ion conduction, and chromatin biology. *Acc Chem Res*. 2009;42(1):107–116.

68 Thom, J., Anderson, D., McGregor, J., and Cotton, G. Recombinant Protein Hydrazides: Application to Site-Specific Protein PEGylation. *Bioconjugate Chem*. 2011;22(6):1017–1020.

69 Liu, J., Chen, Q., and Rozovsky, S. Utilizing selenocysteine for expressed protein ligation and bioconjugations. *J Am Chem Soc*. 2017;139(9):3430–3437.

70 Kulkarni, S.S., Watson, E.E., Maxwell, J.W.C., et al. Expressed protein selenoester ligation. *Angew Chem Int Ed*. 2022;61(20):e202200163.

71 Zhao, J., Liu, J., Liu, X., et al. Revealing functional significance of interleukin-2 glycoproteoforms enabled by expressed serine ligation. *Chin J Chem*. 2022;40(7):787–793.

72 Wei, T., Liu, J., Tan, Y., et al. Revealing the function of HMGB1 N-terminal acetylation by a protein semi-synthesis approach. 2023; 10.1101/2021.10.05.463167.

73 Lin, S., Mo, Z., Wang, P., and He, C. Oxidation and phenolysis of peptide/protein C-terminal hydrazides afford salicylaldehyde ester surrogates for chemical protein synthesis. *J Am Chem Soc*. 2023;145(30):16843–16851.

74 Tapeinou, A., Matsoukas, M.-T., Simal, C., and Tselios, T. Review cyclic peptides on a merry-go-round; towards drug design. *Pept Sci.* 2015;104(5):453–461.
75 Davies, J.S. The cyclization of peptides and depsipeptides. *J Pept Sci.* 2003;9(8):471–501.
76 White, A.M. and Craik, D.J. Discovery and optimization of peptide macrocycles. *Expert Opin Drug Discov.* 2016;11(12):1151–1163.
77 Chow, H.Y., Zhang, Y., Matheson, E., and Li, X. Ligation technologies for the synthesis of cyclic peptides. *Chem Rev.* 2019;119(17):9971–10001.
78 Zhang, L. and Tam, J.P. Synthesis and application of unprotected cyclic peptides as building blocks for peptide dendrimers. *J Am Chem Soc.* 1997;119(10):2363–2370.
79 Camarero, J.A., Cotton, G.J., Adeva, A., and Muir, T.W. Chemical ligation of unprotected peptides directly from a solid support. *J Pept Res.* 1998;51(4):303–316.
80 Gless, B.H. and Olsen, C.A. Direct peptide cyclization and one-pot modification using the MeDbz linker. *J Org Chem.* 2018;83(17):10525–10534.
81 Tulla-Puche, J. and Barany, G. On-resin native chemical ligation for cyclic peptide synthesis1,2. *J Org Chem.* 2004;69(12):4101–4107.
82 Lam, H.Y., Zhang, Y., Liu, H., et al. Total synthesis of daptomycin by cyclization via a chemoselective serine ligation. *J Am Chem Soc.* 2013;135(16): 6272–6279.
83 Lee, C.L. and Li, X. Serine/threonine ligation for the chemical synthesis of proteins. *Curr Opin Chem Biol.* 2014;22:108–114.
84 Rohrbacher, F., Zwicky, A., and Bode, J.W. Chemical synthesis of a homoserine-mutant of the antibacterial, head-to-tail cyclized protein AS-48 by α-ketoacid–hydroxylamine (KAHA) ligation. *Chem Sci.* 2017;8(5):4051–4055.
85 Smith, G.P. Filamentous fusion phage: novel expression vectors that display cloned antigens on the virion surface. *Science.* 1985;228(4705):1315–1317.
86 Tan, Y., Tian, T., Liu, W., Zhu, Z., and Yang, C.J. Advance in phage display technology for bioanalysis. *Biotechnol J.* 2016;11(6):732–745.
87 Milton RC deL., Milton, S.C.F., and Kent, S.B.H. Total chemical synthesis of a D-enzyme: the enantiomers of HIV-1 protease show demonstration of reciprocal chiral substrate specificity. *Science.* 1992;256(5062):1445–1448.
88 Chalifour, R.J., McLaughlin, R.W., Lavoie, L., et al. Stereoselective interactions of peptide inhibitors with the β-amyloid peptide *. *J Biol Chem.* 2003;278(37):34874–34881.
89 Wei, G.-X. and Bobek, L.A. Human salivary mucin MUC7 12-Mer-l and 12-Mer-d peptides: antifungal activity in saliva, enhancement of activity with protease inhibitor cocktail or EDTA, and cytotoxicity to human cells. *Antimicrob Agents Chemother.* 2005;49(6):2336–2342.

90 Pappenheimer, J.R., Dahl, C.E., Karnovsky, M.L., and Maggio, J.E. Intestinal absorption and excretion of octapeptides composed of D amino acids. *Proc Natl Acad Sci*. 1994;91(5):1942–1945.

91 Eckert, D.M., Malashkevich, V.N., Hong, L.H., Carr, P.A., and Kim, P.S. Inhibiting HIV-1 entry: discovery of D-peptide inhibitors that target the gp41 coiled-coil pocket. *Cell*. 1999;99(1):103–115.

92 Schumacher, T.N.M., Mayr, L.M., Minor, D.L., Milhollen, M.A., Burgess, M.W., and Kim, P.S. Identification of d-peptide ligands through mirror-image phage display. *Science*. 1996;271(5257):1854–1857.

93 Zhou, X., Zuo, C., Li, W., et al. A novel d-peptide identified by mirror-image phage display blocks TIGIT/PVR for cancer immunotherapy. *Angew Chem Int Ed*. 2020;59(35):15114–15118.

2

Protein and Peptide Ligation Using Peptide Ligases

Anna Koijen and Leendert J. van den Bos

EnzyTag BV, Nuth, the Netherlands

2.1 Introduction

Thousands of naturally occurring peptides have been discovered, many of which play vital roles in human physiology as hormones, neurotransmitters, growth factors, ion channel ligands, or anti-infectives. Peptides are highly selective and effective signaling molecules that interact with specific cell surface receptors, such as G protein-coupled receptors or ion channels, to initiate intracellular responses. Due to their favorable pharmacological profiles and inherent properties, peptides serve as excellent candidates for the development of new therapeutics. Their high specificity often results in exceptional safety, tolerability, and efficacy in humans, distinguishing them from traditional small molecules. In many respects, peptides represent an optimal middle ground between small molecules and biopharmaceuticals. Long peptides, typically defined as peptides exceeding 30 amino acids, offer a vast array of structural diversity and functional versatility, rendering them promising candidates for tackling complex diseases such as cancer, autoimmune disorders, and infectious diseases [1,2].

Today, solid-phase peptide synthesis (SPPS), developed by Merrifield in 1963 [3], stands out as the primary method employed for the preparation of synthetic peptides. This technique allows for peptide synthesis on a solid phase with precisely controlled amino acid sequence. As opposed to recombinant techniques, it allows for limited peptide length, up to 40–50 amino acids [4]. Nevertheless, a major advantage of this approach is the possibility of designing peptides with a specific sequence, including the incorporation of D-amino acids, noncanonical amino acids, and specific backbone or side chain modifications, expanding the scope for novel drug design [5]. SPPS has proven to be a reliable method for efficiently synthesizing complex peptide

Peptide Science: Chemical Ligation, Lead Generation, and Therapeutic Advances, First Edition. Fa Liu.

targets, thereby opening up a wide chemical space for peptide therapeutics. However, SPPS and most current peptide synthesis technologies require large amounts of hazardous reagents and solvents, highlighting the need for improvements to reduce their environmental impact [6–8].

Proteins are another class of biomolecules, serving as catalysts, signaling molecules, molecular and ion transporters, structural scaffolds, receptors, and more. In medicinal applications, proteins can perform these roles and offer several advantages over small molecule drugs. They are less likely to cause side effects by interfering with normal biological processes because they have evolved to carry out highly specific functions. Furthermore, protein therapeutics generally exhibit high potency and are capable of executing complex functions due to their intricate three-dimensional structures. Proteins' inherent tendencies to aggregate, degrade, denature, and lose activity presented significant challenges. Furthermore, rapid clearance from the body, nonspecific distribution, immunogenicity, and toxicity were also critical issues. Advances in rational design and the strategic introduction of chemical and structural modifications have led to a significant shift in how these properties can be optimized [9]. Chemical modifications to protein structures introduce numerous advantageous properties and these can also alter the activity or potency of the drugs. Typically, high activity or potency is a desirable property as it can lower the therapeutic dose. Established chemical design strategies to achieve this balance include site-specific mutagenesis, antibody-drug conjugate, PEGylation, Fc fusion, fusion to other proteins, glycosylation, lipidation, and computational modeling. Emerging chemical design strategies include supercharging, attachment of non-PEG, fusion to other partners, and use of unnatural amino acids (UAAs).

Recombinant expression is a powerful technique widely used in biotechnology for producing proteins, including enzymes, hormones, and antibodies. It offers several advantages, such as high yield, scalability, and the ability to produce proteins with complex post-translational modifications when using eukaryotic systems. Additionally, it allows for precise genetic manipulation to optimize protein expression and function. However, recombinant expression also has drawbacks. It can be costly and time-consuming, particularly in optimizing expression conditions. Expression in prokaryotic systems often results in the formation of inclusion bodies or improper folding, while eukaryotic systems may require complex culture conditions and have lower yields. Chemical modification, as described above, can be challenging to implement in an expression system. For example, the susceptibility of UAAs for inefficient incorporation often results in low protein yield after recombinant expression [9]. Despite these challenges, recombinant expression remains a crucial tool in modern biotechnology for the production of diverse and complex proteins [10,11].

Over the past decades, significant efforts have been dedicated to developing chemoselective and widely applicable peptide ligation strategies for the efficient synthesis of complex peptides, (chemically modified) proteins, and protein conjugates. Driven by the increasing significance of the peptide and protein therapeutics market and the use of peptides in nutritional and cosmetic products, enzyme-mediated ligation methods have gained substantial attention. Enzymes function under mild, physiological reaction conditions and are exquisitely regio- and stereoselective. Peptide ligases, enzymes that form peptide bonds with site and substrate selectivity, are promising tools for protein engineering, chemoselective labeling, and macrocyclization of peptides and proteins. An efficient ligase with precise site specificity is highly desirable for protein modifications, while a ligase with a broad substrate scope, enabling "traceless" ligation, would greatly benefit peptide semisynthesis. Most notably, the vast amounts of peptide needed for treating metabolic disorders call for this approach. Although only a few naturally occurring enzymes are known to catalyze peptide ligations, approaches such as genome mining and protein engineering have successfully led to the discovery and enhancement of various ligases.

The lack of natural peptide ligases has intrigued many researchers to investigate the use of proteases in reverse to catalyze peptide bond formation. In general, proteases obey the principle of reversibility and therefore enhance the rates of peptide bond hydrolysis and peptide bond formation by the same amount. However, because enzymes alter the rates of these reactions but not the thermodynamic stabilities of the starting materials and product, whether a peptide bond is predominantly formed or hydrolyzed is determined by the position of the equilibrium. Under physiological conditions, the position of the equilibrium lies strongly in favor of peptide bond hydrolysis. Therefore, a key limitation in the application of proteases to catalyze direct reversal of peptide bond hydrolysis is the high thermodynamic stability of the proteolytic products compared to the ligation product. The position of the equilibrium can be shifted by altering reaction conditions, including solvent polarity, temperature, concentration of water, and pH. Many practical limitations to this approach remain, including the need to individually optimize the conditions for each reaction or step and the decreased stability and solubility of proteases in organic solvents [12].

An alternative approach for using proteases in reverse is to run the reaction under kinetically controlled conditions, in which the kinetic properties of the enzyme, rather than the thermodynamic stabilities of the substrates and products, determine the reaction yield. The kinetic control reaction can be achieved by the use of an activated ester substrate. This approach has been applied in the context of serine and cysteine proteases, whose reaction mechanisms involve the formation of an acyl-enzyme intermediate. In the context of these enzymes, the protease-catalyzed hydrolysis of both esters and amides proceeds through a similar chemical mechanism as shown in Figure 2.1 [12].

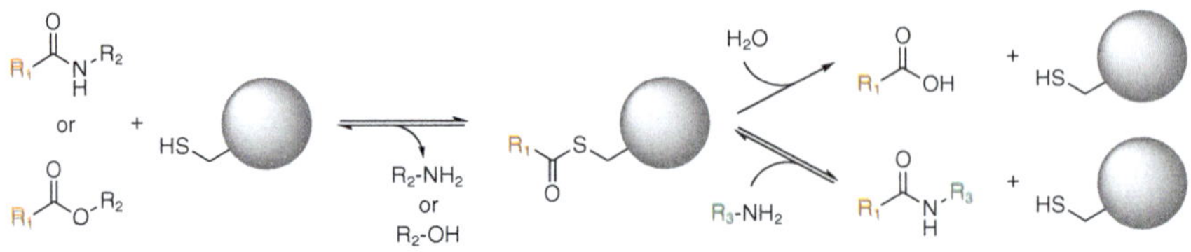

Figure 2.1 General reaction mechanism of peptide ligases. An acyl intermediate is formed via cleavage of an amide or ester bond. Nucleophilic attack by a water molecule leads to hydrolysis, while nucleophilic attack by an N-terminal amine leads to aminolysis, resulting in the formation of a new peptide bond.

2.2 Peptide Ligases Mechanism of Action

Enzyme-mediated amide bond formation involves a series of steps catalyzed by enzymes known as ligases or synthetases. This process is essential in biological systems for the synthesis of proteins, peptides, and other important biomolecules.

A simplified overview of the common mechanism:

1. Substrate binding: The enzyme binds to its substrate molecules, typically an amino acid or a peptide fragment containing a carboxyl group (—COOH) and an amino group (—NH_2).
2. Activation of the carboxyl group: The carboxyl group of one substrate molecule (acyl donor) is activated through the transfer of an activating group.
3. Nucleophilic attack: The activated carboxyl group undergoes nucleophilic attack by the amino group of another substrate molecule (acyl acceptor). This leads to the formation of a tetrahedral intermediate.
4. Amide bond formation: The tetrahedral intermediate collapses, resulting in the formation of a new amide bond (—CONH—) between the carboxyl carbon of one substrate and the amino nitrogen of the other substrate.
5. Release of the product: The newly formed amide bond is stabilized, and the product molecule is released from the enzyme active site.
6. Regeneration of the enzyme: In some cases, the enzyme may require additional steps to regenerate its active form, such as the release of any cofactors or the removal of any inhibitory groups.

This mechanism allows for the specific and efficient formation of amide bonds under mild conditions, crucial for the synthesis of complex biological molecules within living organisms. Enzymes play a vital role in catalyzing these reactions, speeding up the rate of amide bond formation and ensuring high selectivity and fidelity in the resulting products.

Despite the aforementioned advantages of the use of peptide ligases for fragment condensation, there are also limitations to the use of enzymes. Natural evolution has shaped enzymes as highly efficient and specialized catalysts that operate best under natural, aforementioned mild reaction conditions. However, evolution has not optimized enzymes for the conditions of an industrial process, which may vary greatly from the conditions found in nature. As a consequence, they do not always meet the process demands regarding their activity, stability, and selectivity.

In our view, an ideal enzyme "possesses" the following properties: the enzyme should be highly stable under a wide range of conditions, including variations in temperature (~75 °C), pH (2–10), and solvent composition (chaotropic agents and organic solvents). It should not require any cofactors for its catalytic activity, simplifying its use and reducing costs associated with cofactor synthesis or supplementation. The enzyme should be easy and cost-effective to produce on a large scale, preferably using recombinant DNA technology in a microbial host system. The enzyme should exhibit exceptionally high catalytic activity, enabling rapid conversion of substrate to product, thereby increasing process efficiency. It should have minimal environmental impact, with no toxic byproducts or waste generated during its use. Ideally, the enzyme should possess both broad substrate specificity and high selectivity, allowing it to catalyze a wide range of reactions while maintaining high efficiency and specificity for each substrate. The enzyme should promote atom-efficient reactions, minimizing the use of excess reagents and maximizing the conversion of reactants into desired products. The enzyme should be adaptable to protein engineering techniques, allowing for the optimization of its properties such as stability, activity, specificity, and substrate range. This could involve methods such as directed evolution or rational design. By meeting these criteria, the ideal enzyme would serve as a versatile and sustainable catalyst for a wide range of industrial and biotechnological applications, facilitating efficient and environmentally friendly chemical transformations.

While obviously the ideal enzyme that meets all the aforementioned criteria still needs to be discovered/developed, considerable efforts have been undertaken to optimize peptide ligases. This review will highlight key recent findings, including the discovery, engineering, and application of versatile ligases such as sortase, asparaginyl ligases, subtilisin-derived ligases, and trypsiligase.

2.3 Sortase

Only a few enzymes are known to catalyze the formation of peptide bonds in nature. Among them, sortase A from *Staphylococcus aureus* is one of the most prominent examples. The natural function of sortase A – the covalent anchoring of proteins to the cell wall of gram-positive bacteria – was first described in 1999 [13,14]. Only a few months later, it was shown that the N-terminal membrane anchor domain can be removed without loss of the transpeptidation activity. Replacing the

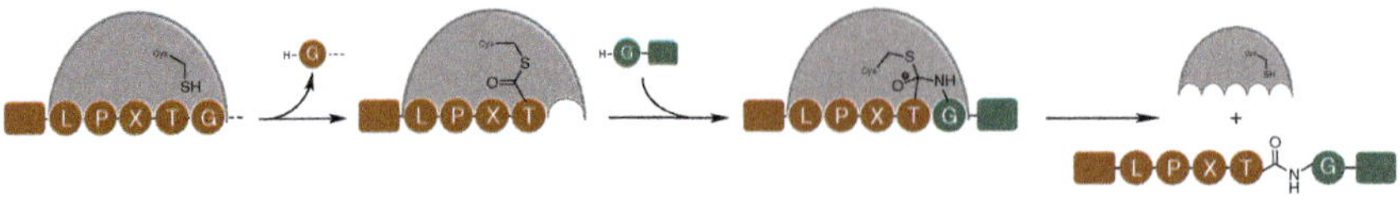

Figure 2.2 Catalytic mechanism of sortase A.

domain with a hexahistidine tag led to an enzyme that can be produced solubly in *Escherichia coli* and easily purified via affinity chromatography [15]. Since then, sortase A has been utilized for a wide range of applications, including site-selective bioconjugation for peptide and protein modifications, cyclization, and immobilization of proteins on surfaces.

Staphylococcus aureus sortase A requires an LPXTG recognition motif, in which X denotes any amino acid, but most commonly glutamic acid is used. Upon binding to the active site, the motif is cleaved between the threonine and the glycine residues, and the acyl enzyme intermediate is formed via a thioester (Figure 2.2). The requirement for the acyl acceptor is a sterically available N-terminal glycine residue. Here, the acyl acceptor can also be a single glycine residue under the premise that its C-terminus is amidated; otherwise, a second amino acid is required [16]. The comparably large recognition motif of sortase A limits its application for peptide- and protein semisynthesis. Besides sortase A, sortases from five other classes (B–F) have been identified, which differ in specificity, for example, *Staphylococcus aureus* sortase B has a NPXTG motif. However, most sortases of these other classes have low *in vitro* activity and are therefore not commonly used in ligation reactions [17]. In addition, sortases from other organisms show different specificities. For example, sortases from *Streptococci* prefer a leucine in place of a threonine in the P1 position and are more promiscuous for the N-terminus of the acyl acceptor, where they accept not only glycine but also alanine and serine [18,19]. Even branched primary amines have been reported as nucleophiles for sortase A from *Streptococcus pyogenes* [20]. Potentially exploring the natural sequence space of sortases will lead to the identification of new sortases with altered recognition motifs. But the substrate scope of sortase A has also been broadened by enzyme engineering. Directed evolution approaches, using phage display or yeast surface display as a screening system, have been successful in identifying variants with different or lower specificity [21–24]. These variants have been described in detail in extensive reviews [14,16,17,25].

As the LPXTG recognition motif is also present in the ligation product and the byproduct was cleaved off from the acyl donor during the ligation reaction can act as an acyl acceptor, sortase-mediated ligation reactions are reversible. This means that the ligation reaction will eventually reach equilibrium, which is dependent on the concentration of the substrates/products involved. As shifting the equilibrium to one side by using an excess of one substrate is not preferred from the viewpoint of atom efficiency, several strategies have been developed to circumvent the reverse reaction (Figure 2.3). One approach is to create a secondary structure in the ligation product that prevents enzyme binding and thereby makes the product unreactive. This has been achieved by introducing

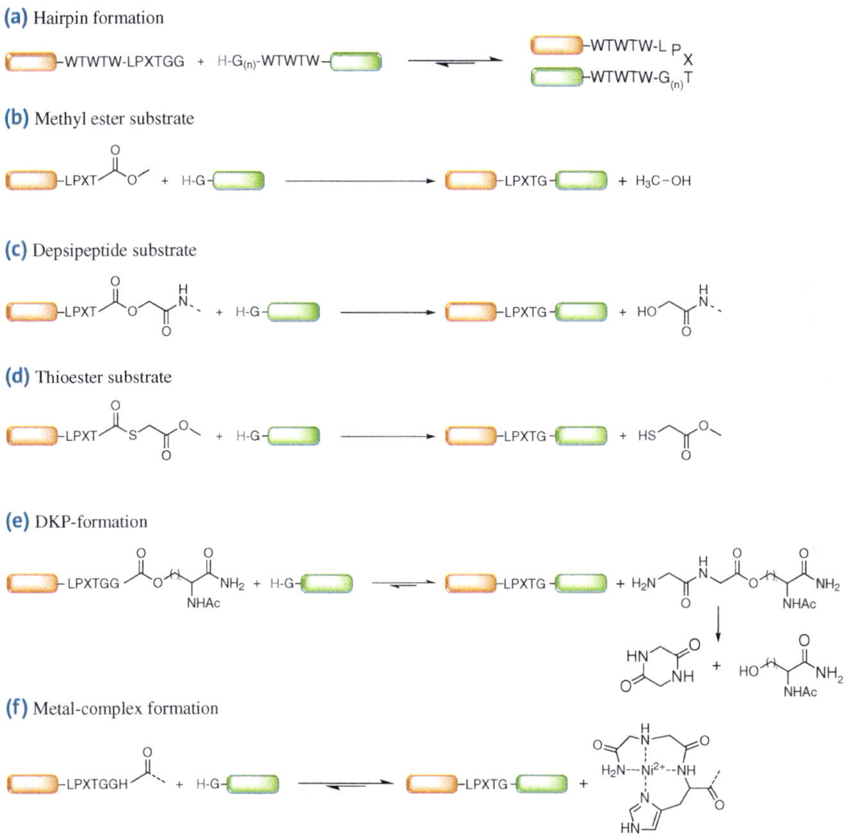

Figure 2.3 Strategies to circumvent the reversibility of sortase-mediated ligation reactions. (a) The ligation product forms a hairpin that prevents binding of sortase. *Source*: Ref. [26]/The Royal Society of Chemistry. (b) The use of a methyl ester substrate leads to an unreactive alcohol byproduct. *Source*: Ref. [27]/The Royal Society of Chemistry. (c) The use of a depsipeptide substrate leads to an unreactive alcohol byproduct. *Source*: Ref. [28]/John Wiley & Sons. (d) The use of a thioester substrate leads to an unreactive thiol byproduct. *Source*: Ref. [29]/John Wiley & Sons. (e) The byproduct is cyclized as a result of DKP formation. *Source*: Ref. [30]/American Chemical Society. (f) The byproduct is removed from the reaction via chelation with metal ions. *Source*: Refs. [31,32].

hairpin-forming sequences adjacent to the ligation site [26]. However, the requirement to insert additional peptide sequences that form the hairpin makes this strategy not suitable for peptide- and protein semisynthesis purposes. Other strategies focus on the removal or inactivation of the byproduct to render the reaction irreversible. The use of ester substrate is one approach to achieve this, as the byproduct will contain an unreactive alcohol in place of the primary amine. Initially, this was attempted with methyl-ester substrates, but their poor recognition by sortase limited their application [27]. A better alternative is the use of depsipeptides, namely peptides containing a carboxyamidomethyl (Cam) ester, which is essentially an ester derivative of a glycine residue. Using Cam esters, quantitative ligation yields can be reached with the use of a small excess of the ester substrate [28]. Recently, the use of thioesters for sortase-mediated ligations has been shown. Here, close to 100% conversion was reached in ligation reactions with equimolar substrate concentrations [29]. Notably, in thioester-assisted ligations, sortase A tolerates a greater diversity of sequences at the ligation site, which is a striking example of how substrate engineering can affect the enzyme specificity. An alternative approach using ester substrates places the ester not in the sortase recognition motif but two amino acids C-terminal of the cleave site of the LPXT-G motif. The resulting byproduct of the ligation reaction is prone to spontaneous diketopiperazine (DKP) formation, which removes it from the reaction equilibrium [30]. Furthermore, extension of the sortase motif of the acyl donor with a glycine and one or six histidine residues (LPXT-GGH_x) leads to a GGH_x byproduct that can be removed from the reaction by chelation with metal ions, such as Ni^{2+} [31,32].

A major drawback of sortase A is their low catalytic efficiency – typical k_{cat}/K_M values are around 200 M^{-1} s^{-1} [16]. This is due to a low affinity of sortase A to their substrates, which is likely because in nature, sortase A and its substrates are bound to the cell wall, resulting in a high local substrate concentration [17]. In practice, the low catalytic efficiency means that high enzyme concentrations and/or prolonged reaction times are required for *in vitro* applications of sortase A. Several directed evolution approaches have been used to improve the catalytic efficiency of sortase A, including the use of yeast display, FRET-based assays, selection systems based on complementation of a protein that confers antibiotic resistance, and a high-throughput assay named "SortEvolve" that combines screening and semi-purification in one microtiter plate [33–36]. In these studies, variants with more than a 100-fold improvement in k_{cat}/K_M values have been identified, which allows to decrease the reaction times and/or the enzyme concentration. For further details, the reader is again referred to excellent reviews [14,16,17,25]. A downside of the increased catalytic efficiency is that it often goes hand in hand with an increased irreversible hydrolysis of the enzyme-acyl complex. Typically, proximity-based approaches are taken to circumvent this, for example, covalent linking of one substrate to the sortase or use of the traceless spy catcher technology [16,37].

Furthermore, sortase variants with improved catalytic efficiency commonly suffer from decreased thermostability [17]. Therefore, enzyme engineering has been used to improve the robustness of sortase A and its variants with enhanced catalytic efficiency. The aforementioned SortEvolve assay was used to screen a set of site-saturation mutagenesis libraries [38]. Recombination of beneficial mutations led to a 6 °C increase in melting temperature. The same assay was used to improve the tolerance for organic solvents such as dimethyl sulfoxide (DMSO) and dimethylformamide (DMF), which are frequently added to aid the solubility of the peptide substrates but drastically decrease the activity of wildtype sortase [39]. Other strategies to stabilize sortase A do not rely on mutagenesis but rather on cyclization. It was shown that head-to-tail backbone cyclization increases the stability and tolerance against chaotropic agents of sortase A without reduction of the enzyme activity [38,40]. Furthermore, *in situ* cyclization of proteins (INCYPRO) has been applied to stabilize *Staphylococcus aureus* sortase A as well as one of its variants with enhanced catalytic efficiency [41,42]. This method introduces three spatially aligned, solvent-exposed cysteines in the protein sequence that are subsequently cross-linked. The resulting bicyclic enzymes showed improved thermostability with a melting temperature increase of more than 10 °C as well as tolerance against chaotropic agents.

Last but not least, the calcium dependency of *Staphylococcus aureus* sortase A has also been addressed by enzyme engineering. It should be noted that many sortases from other organisms are not calcium dependent, but are less attractive for ligations due to their lower activity. The calcium dependency limits the application of sortase A under conditions that require low Ca^{2+} concentrations, such as in living cells, as well as in phosphate buffers or buffers containing chelators. Calcium-independent variants were rationally designed through replacing the Ca^{2+} interaction by ionic interactions between a Lys- and an Asp-residue, which were introduced in the Ca^{2+} binding site [43,44].

In summary, sortase is a well-established enzyme for protein modifications and bioconjugations. It can be easily produced by recombinant expression. The catalyzed transpeptidation reaction is reversible, but several strategies can be applied to drive the reaction to completion without the use of an extensive excess of one reaction partner. The use of sortase for peptide and protein semisynthesis is limited by its comparably large recognition motif and low catalytic efficiency. Both of these issues have been addressed by enzyme engineering, although all sortase variants described to date will leave a clear footprint in the ligation product.

2.4 Peptide Asparaginyl Ligases

Another example of peptide ligases that can be found in nature are peptide asparaginyl ligases (PALs). Being closely related to asparaginyl endoproteases (AEPs), they recognize the peptide motif P1-P1′-P2′, in which P1 is an Asn, or less commonly an Asp; P1′ is typically a small amino acid and P2′ a hydrophobic amino acid [16]. PALs cleave the peptide motif C-terminal of the Asn/Asp, but in contrast to AEPs catalyze aminolysis instead of hydrolysis of the peptide substrate acyl complex (Figure 2.4). All PALs described to date originate from plants, where they catalyze the backbone cyclization of peptides, for example, in the synthesis of cyclotides such as KalataB1 or MCoTI-II [45]. In 2014 Nguyen et al. were the first to describe the ligating activity of a PAL isolated from *Clitoria ternatea* (butterfly pea), which they named butelase-1 after the local name of the plant (**bu**nga **tel**ang lig**ase**) [46–48]. In contrast to sortases, butelase-1 shows very high catalytic efficiency – k_{cat}/K_M values up to 1.3×10^6 M^{-1} s^{-1} have been reported [47]. This and the shorter recognition motif make it attractive for peptide ligations. After butelase-1, several other PALs were identified, of which especially *Oa*AEP1 [49] and *Vy*PAL2 [50,51] showed potential for peptide ligations. However, none of the later identified PALs matched the catalytic activity of butelase-1. Notably, the catalytic kinetics of *Oa*AEP-1 could be improved significantly by a single amino acid substitution C247A [52]. This residue was later termed the ligase activity determinant 1 (LAD1) because hydrolyzing AEPs usually have a glycine residue in this position, while ligating PALs often have a hydrophobic, aliphatic residue [50]. In addition, a second ligase activity determinant (LAD2) was identified, in which an alanine residue is preferred for ligating activity [53].

Different PALs differ slightly in their specificity for the amino acids in the P1′ and P2′ positions. So are the preferred sequences for butelase-1, *Oa*AEP1, and *Vy*PAL2 N-HV, N-GL, and N-SL, respectively. The sequence requirements of the prime site are overall less strict, especially for the acyl acceptor. PALs therefore only have a minimal and comparably promiscuous recognition motif.

As for sortases, the ligation reaction catalyzed by PALs is reversible, as the P1′-P2′ dipeptide is a competing nucleophile. In the natural cyclization reaction, this poses

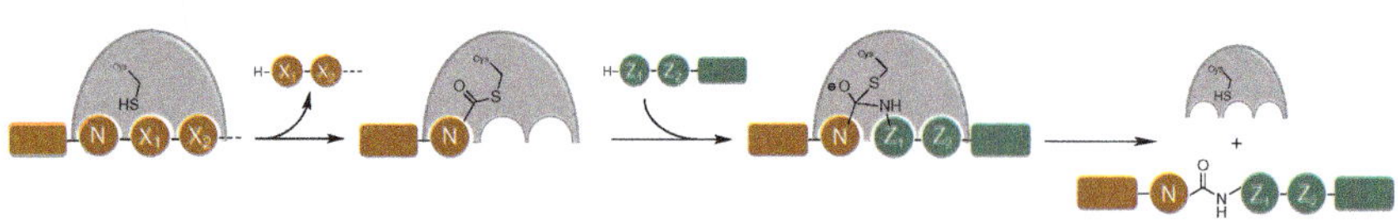

Figure 2.4 Catalytic mechanism of peptide asparaginyl ligases. In nature, PALs catalyze the backbone cyclization of peptides. In this case, the N-terminus of the acyl donor peptide serves as an acyl acceptor.

less of a problem, as the structurally constrained confirmation of the cyclic peptide reduces the chance of the product docking to the active site of the enzyme [54]. For ligation of two peptides toward a linear product, an excess of the acyl acceptor is required to compete with the reverse reaction. Various strategies have been employed to address this challenge (Figure 2.5). One option is to make use of the fact that the specificity for the acyl donor is stricter than for the acyl acceptor. While *Oa*AEP1 requires the N//GL motif for the acyl donor, the enzyme is promiscuous for the acyl acceptor. Rehm et al. showed that the use of an acyl acceptor

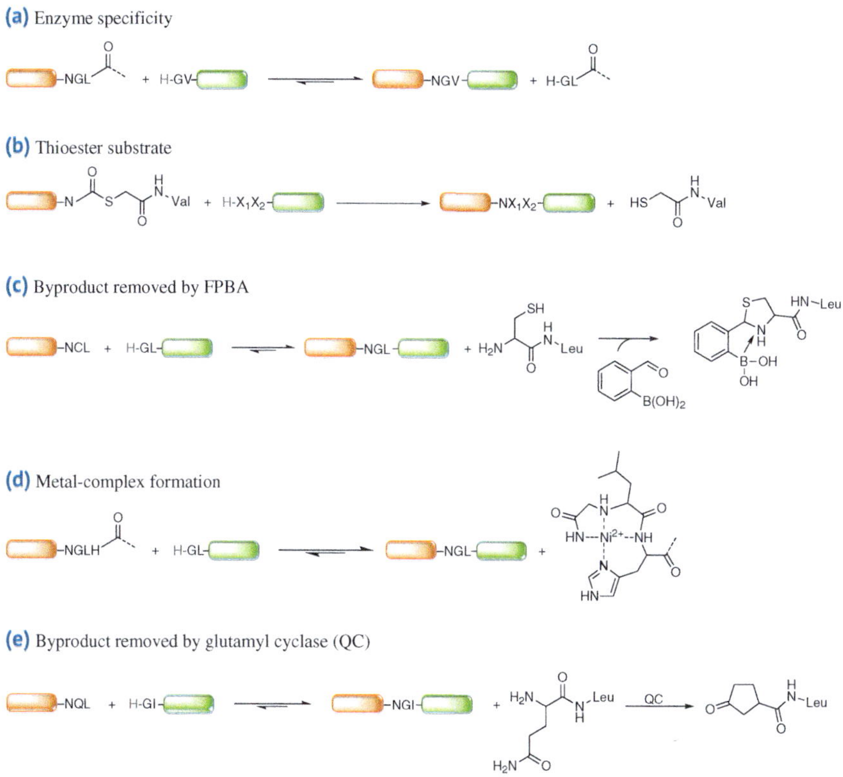

Figure 2.5 Strategies to circumvent the reversibility of PAL-mediated ligation reactions. (a) The sequence is designed in such a way that the ligation product is a poor substrate. *Source*: Ref. [55]/American Chemical Society. (b) The use of a thioester substrate leads to an unreactive thiol byproduct. *Source*: Ref. [48]/John Wiley & Sons. (c) The byproduct is removed from the reaction by a reaction with FPBA. *Source*: Ref. [56]/The Royal Society of Chemistry. (d) The byproduct is removed from the reaction via chelation with metal ions. *Source*: Ref. [45]/The Royal Society of Chemistry. (e) The byproduct removed from the reaction by cyclization catalyzes by a glutamyl cyclase (QC). *Source*: Ref. [57]/American Chemical Society.

with the N-terminal sequence GV creates a ligation product that is resistant to the reverse reaction [55]. Another approach is the use of thiodepsipeptides as acyl donors. The resulting leaving group is an α-thiol, which cannot serve as a nucleophile anymore [48]. A disadvantage of this strategy is that the required thioester is prone to hydrolysis. As a consequence of this instability, an excess of the acyl donor rather than the acyl acceptor is required for an efficient ligation reaction. Several strategies aim to use reaction cascades to remove the leaving group from the reaction mixture. *Oa*AEP1 C247A accepts the acyl donor motif N//CL, resulting in a leaving group with a N-terminal Cys. Here, 2-formyl phenyl boronic acid (FPBA) can be used as a scavenger that reacts with the N-terminal Cys, forming a nonreactive thiazolidine and thus effectively removing the nucleophile and rendering the reaction irreversible [56]. Similarly, a His can be introduced in the P3′ position, which enables the quenching of the nucleophile via complex formation with divalent cations such as Ni^{2+} [45]. Additionally, an enzymatic cascade has been reported, in which a Gln in the P1′ position is utilized. Here, a glutamyl cyclase removes the leaving group by cyclization of the N-terminal Gln of the leaving group into a lactam [57].

A long-standing challenge in the use of butelase-1 was the requirement to isolate it from plant material, as in the first instance, the attempt of a recombinant production of butelase-1 led to the formation of inclusion bodies. Follow-up studies that optimized the expression by using optimized strains, substitution of the signal peptides by a hexahistidine tag, and truncation of the N-terminal pro-domain enabled the recombinant production with *E. coli* [58,59]. The process requires the activation of the enzyme, during which the C-terminal cap-domain and the residual part of the pro-domain are autocatalytically cleaved at low pH. This mimics the maturation of PALs and AEPs in nature, where the zymogen is activated in the acidic environment of the plant vacuole. It has been reported that replacement of the cap-domain with ubiquitin and a hexahistidine tag permits the purification of *Oa*AEP1 in its active form without the acidic activation step [56]. Other strategies include the expression as a fusion protein with MBP [60] and the use of *Pichia pastoris* as the expression host [61]. For *Vy*PAL2, expression protocols in insect cells have been described [50]. Recently, a *de novo* PAL variant that was engineered for efficient recombinant production has been developed [62]. In the first step, a consensus sequence was derived from 1500 AEP and PAL sequences that gave a 20-fold higher yield than natural AEPs and PALs when recombinantly produced in *E. coli*. Secondly, the resulting enzyme was converted from a protease to a ligase by substituting the ligase activity-determining residues LAD1 and LAD2 with amino acids that are typically found in PALS, namely G155A and G225V. Despite these significant improvements, the recombinant production of PALs still required relatively many purification steps and results in low yields [45].

In summary, PALSs are ligases with a very high catalytic efficiency that can be used for cyclization, as well as N- and C-terminal modifications of peptides and proteins. Their minimal recognition sequence makes them applicable for the semisynthesis of native proteins and peptides. Various strategies have been developed to circumvent the reverse reaction and reduce the excess of acyl acceptor that is required to reach full conversion. A drawback of PALs is their limited availability by means of recombinant production, which has been addressed in various studies but could still be further improved.

2.5 Subtilisin-derived Peptide Ligases

While sortases and PALs are some of the few enzymes that have a natural peptide ligase activity, which is leveraged in *in vitro* applications, efforts have also been undertaken to engineer ligases from proteases. One prominent example is the engineering of subtilisin variants. As early as the 1960s, so-called thiol-subtilisin was prepared by the chemical transformation of the active site serine into a cysteine. A few years later, it was shown that thiol-subtilisin barely acts as an amidase, but has remaining esterase activity [63]. In combination with the fact that thioesters are more prone to aminolysis than hydrolysis, this held great potential for the engineering of a ligase. In 1991, Wells et al. created a subtilisin BPN′ variant that was genetically modified with the S221C substitution. They also found that the ligation activity could be further improved by the substitutions of proline 225 with alanine, as it counteracts steric crowding of the sulfur that replaced the oxygen of serin 221. The resulting enzyme was named subtiligase [64]. Subtiligase, just as other later-developed subtilisin-derived peptide ligases, can be produced in high yields by recombinant expression in *Bacillus subtilis* and have high catalytic efficiency, so that only sub-stoichiometric amounts of enzyme are required.

As the S221C substitution diminishes the amidase function but not the esterase function, ligation reactions using subtilisin-derived ligases do not rely on transpeptidation but instead use activated (thio-) esters as substrates (Figure 2.6). As the ester-to-amide conversion is not reversible and the loss of the amidase function prevents secondary hydrolysis of the ligation product, in theory, a yield of 100% can be obtained without the need of using one reaction partner in excess. In practice, however, a 10-fold excess of the acyl acceptor was required in subtiligase-catalyzed ligation reactions to suppress the hydrolytic reaction [64].

Different esters can be used for ligation reactions that are catalyzed by subtilisin-derived ligases, but typically, a carboxyamidomethyl (Cam) ester or the ester derived from 4-hydroxymethylbenzoic acid (HMBA) is used. These esters can be easily introduced during Fmoc-based SPPS, using Fmoc-glycolic acid or HMBA as a building block, respectively [65,66]. One disadvantage of the requirement

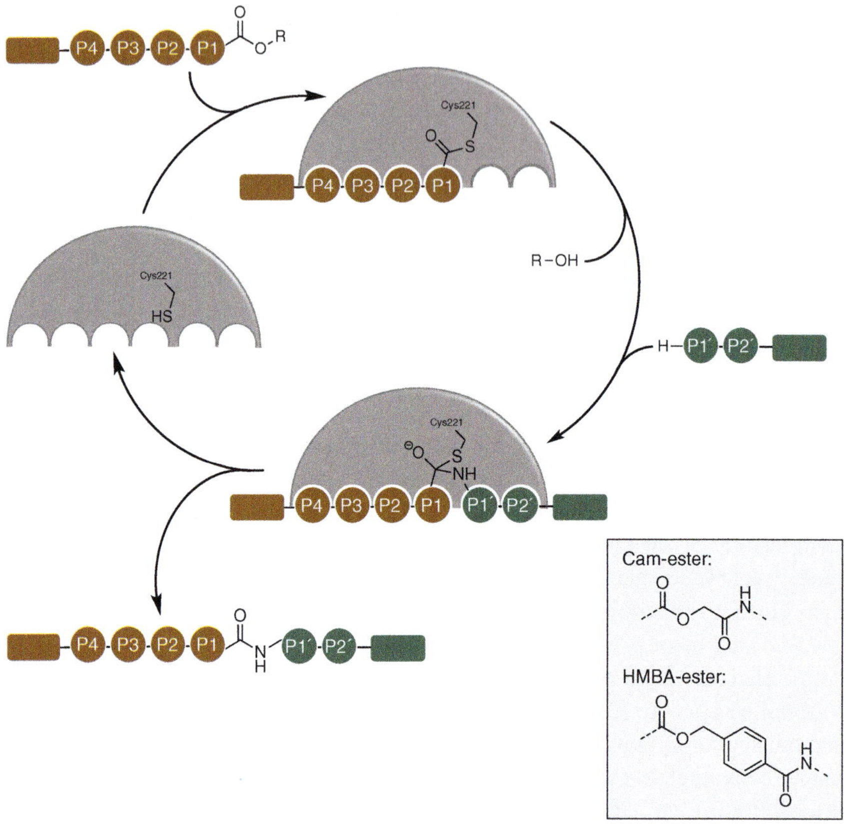

Figure 2.6 Catalytic mechanism of ligations catalyzed by subtilisin-derived peptide ligases.

for an activated ester is that it limits C-terminal modification to synthetic peptides, while for the modification of proteins, only the N-terminus is accessible with this technology.

A major advantage of subtilisin-derived peptide ligases is that no specific recognition sequence is required for these enzymes. This is due to the fact that also the parent protease subtilisin has no recognition sequence and can bind to a range of different amino acid sequences. Subtilisin and all related peptide ligases have six binding pockets, of which four (S4–S1) bind amino acids N-terminal to the hydrolysis/ligation site and two (S1′–S2′) bind C-terminal to the hydrolysis/ligation site. As subtilisin-derived ligases are not specific for one N-terminal sequence, undesired cyclization of the ester-containing fragment needs to be prevented by an N-terminal protection group. While there is no specific recognition sequence, there are still preferences for binding certain amino acids in specific pockets. For example, a hydrophobic amino acid in the P4 position is required for proper recognition in the S4 pocket. Different studies have been conducted to further broaden the specificity of subtilisin-derived ligases [12,67]. These studies often focused on the pockets that bind the amino acids in positions P1′ and P2′, as good recognition of the acyl acceptor is decisive to prevent hydrolysis of the ester as a side reaction.

In order to characterize the specificity of the S1′ and S2′ pockets of subtiligase, among others, the proteomic identification of ligation sites (PILS) was developed [68]. In this method *E. coli* protein extract was digested with two orthogonal proteases (i.e. trypsin and Glu-C) to create a peptide library that contained almost all of the 400 possible P1′–P2′ combinations. This library was used as an acyl acceptor in subtiligase-catalyzed ligations to a peptide containing an N-terminal biotin, a TEV-protease cleavage site, and a C-terminal ester. After ligation, the biotinylated peptides were enriched on streptavidin resin, eluted by TEV cleavage, and then sequenced by LC-MS/MS. This way the frequency of each amino acid in each binding pocket could be determined, showing that small amino acids are preferred in the S1′ pocket and aromatic or large hydrophobic amino acids are preferred in the S2′ pocket. In addition, the PILS assay was combined with an alanine screening of residues within 7 Å of the catalytic triad to identify key residues for the recognition of the P1′–P2′ amino acids. It was found that P1′ and P2′ specificity is strongly dependent on positions 217 and 189, respectively. A total of 72 subtiligase variants with substitutions in one or each of these positions were created and again screened with the PILS assay. This resulted in a toolbox of subtiligase variants with varying but known specificity for the prime sequence, thereby broadening the scope of possible target sequences for subtiligase. An online tool named ALPINE was made available that enables identification of the best subtiligase variants for ligation to a given N-terminal sequence [12,68].

In order to improve the stability of subtiligase, five substitutions (M50F, N76D, N109S, K213R, and N218S) that had previously been described to enhanced the stability toward heat, basic conditions, and organic solvents have been combined in an enzyme named stabiligase [12,69]. It was shown that stabiligase remains active in the presence of chaotropic agents and detergents. Another approach to engineer a more stable subtilisin-derived peptide ligase focused on the calcium-binding domain of subtilisin, which is required for folding and makes subtilisin and its derivatives relatively unstable. However, hyperstable variants of subtilisin BPN′, in which the calcium-binding domain was deleted and a set of stabilizing substitutions were introduced, have been described [70]. Introducing the S221C and P225A substitutions into one of these subtilisin variants created the calcium-independent peptide ligase "peptiligase" [71]. This ligase proved to be highly stable and also retained activity in the presence of organic solvent and chaotropic agents. Most remarkably, unlike subtiligase, peptiligase does not require a large excess of the acyl acceptor but gives high conversion using only 1–1.5 equivalents of the acyl acceptor.

For peptiligase, especially the S1′ pocket is very discriminatory, allowing only small amino acids such as Gly, Ala, and Ser in the P1′ position. This problem was addressed through rational design and mutagenesis of residues lining the S1′ pocket, namely M222, L217, and A225. In addition, F189 of the S2′ pocket was also investigated, and an I107V substitution in the S4 pocket was introduced. The resulting variant (I107, F189W, L217H, M222P, A225N) showed a significantly broadened substrate range and was named omniligase-1 [72,73]. Omniligase-1 has been used, among others, for the synthesis of GLP-1 analogs like exenatide, cyclotides like kalata B1, and affibody molecules [66,67,74–76]. Another example shows the possibility to design substrate-tailored peptiligase variants for the synthesis of specific peptides, in this case, thymosin-α_1. In addition to the S1′ and S2′ residues described above, the S1 residues S156 and S166 were also targeted in mutagenesis studies. The resulting enzyme "thymoligase" accommodates positively charged amino acids in the S1 pocket and negatively charged amino acids in the S1′ pocket and can be used in the chemoenzymatic synthesis of thymosin-α_1 with more than a 5-fold higher catalytic efficiency and a more than 20-fold increased S/H ratio compared to its parent enzyme, peptiligase [77]. Advantages of substrate-tailored ligase variants such as thymoligase include not only increased catalytic efficiency for the specific product but also that no N-terminal protection of the acyl donor is required.

In summary, subtilisin-derived peptide ligases like subtiligase and peptiligase are very versatile and powerful tools for the semisynthesis of peptides and proteins, as they are readily available by recombinant production, show high catalytic efficiency, and most importantly do not have a specific recognition sequence, meaning that ligations are footprint-free. Enzyme engineering has been used to improve the stability of subtilisin-derived ligases, to broaden their specificity, and to create substrate-tailored ligases. The use of an activated ester substrate not only renders the ligation reaction irreversible but also limits the application for protein modifications to the N-terminus.

2.6 Trypsiligase

Trypsin is another example for a protease whose use in the catalysis of amide-bond formation has been described from early on [78]. However, secondary hydrolysis of the product limits the application of trypsin for peptide synthesis. In 2014, Bordusa et al. reported a highly specific variant of anionic trypsin 2 from rats, named trypsiligase. By introducing the substitutions D189K in the S1-pocket, K60E in the S1′ pocket, and N143H and E151H in the S2′ pocket, they engineered specificity for the tripeptide sequence YRH, in which the Y—R bond is cleaved (Figure 2.7a) [79].

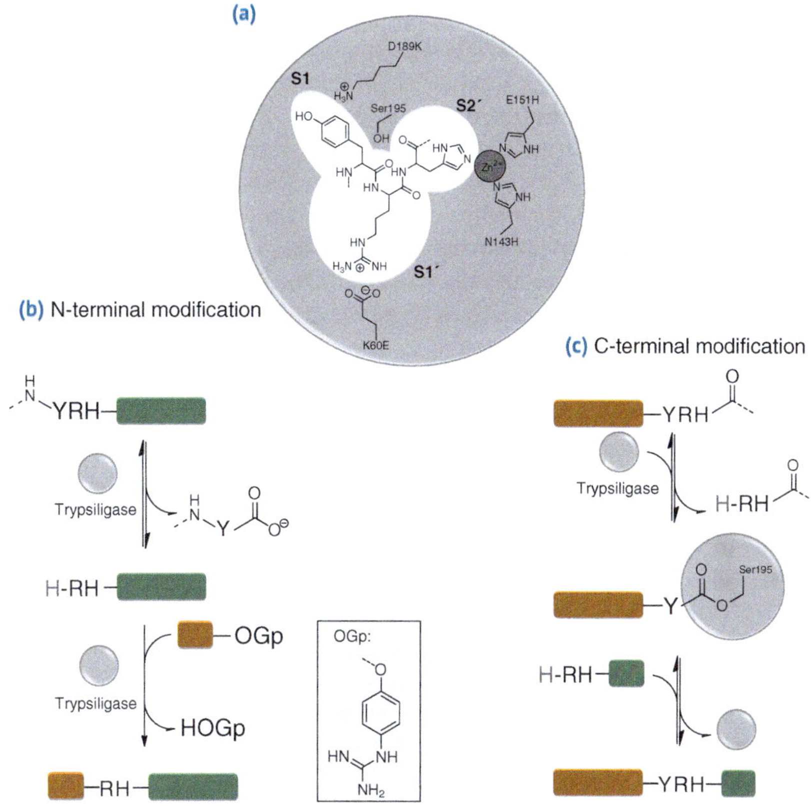

Figure 2.7 (a) Substrate binding site of trypsiligase including amino acid substitutions required for the recognition of the YRH-motif. (b) Catalytic mechanism for N-terminal modification of peptides and proteins using peptidyl-4-guanidinophenyl (OGp) esters. (c) Catalytic mechanism for C-terminal modification of peptides and proteins through a transpeptidation reaction. *Source*: Ref. [80]/Springer Nature.

The proteolytic activity of trypsiligase toward basic residues outside of the recognition sequence is negligible. Notably, this could be assigned to a partially disordered, inactive, zymogen-like confirmation that trypsiligase adopts in the absence of a substrate. Only upon binding of high-affinity substrates containing the YRH-motif, the equilibrium is shifted toward a trypsin-like active conformation, providing an example of substrate-activated catalysis [79]. For this process, Zn^{2+} is required as a cofactor, as the binding of histidine in the S2-pocket is Zn dependent. Trypsiligase can be produced by recombinant expression in *Pichia pastoris* and, in contrast to trypsin, is highly robust against pH and additives with high long-term stability, as it does not possess autoproteolytic activity [80].

Trypsiligase has an increased aminolysis-over-hydrolysis ratio when the activated substrate mimetics, such as peptidyl-4-guanidinophenyl esters (OGp), are used as the acyl donor. This can be used for N-terminal modification of peptides and proteins under kinetic control, as shown in Figure 2.7b. In the first reaction step, trypsiligase binds and cleaves the YRH-motif. The resulting N-terminus on the RH-sequence serves as the acyl acceptor in a selective bioconjugation reaction with the OGp-ester. The resulting product does not contain the YRH-sequence and is therefore not suspectable to hydrolysis by trypsiligase.

In addition, trypsiligase catalyzes a transpeptidation reaction between an acyl donor with the YRH-motif and an acyl acceptor with an N-terminal RH-sequence (Figure 2.7c). This reaction has been used for site-specific modification at the C-terminus of proteins, for example, of antibody Fab fragments [81]. However, as the recognition motif YRH is still present in the reaction product, the reaction is reversible, thus limiting the reaction yield. Recently, an enhanced trypsiligase with five additional substitutions (H40P, A55S, K97D, L99F, and Q192E) was published, which has an increased synthesis potential and reaches product yields close to the thermodynamic maximum of the reaction type catalyzed [82,83].

A drawback of trypsiligase is that, for complete conversion of one reaction partner, typically a 5- to 10-fold excess of the other reaction partner is required [80]. Furthermore, the YRH recognition motif of trypsiligase is only found in 0.5% of all known peptide sequences [79]. While this is an advantage for site-specific protein labeling (given that it is possible to engineer the YRH-sequence into the sequence of the target protein), it limits the application of trypsiligase for the semisynthesis of native peptides and proteins.

In summary, trypsiligase can be used for N- and C-terminal modifications and is highly selective for its recognition sequence, YRH, making it a very efficient enzyme for protein labeling. N-terminal modification proceeds through an activated ester and is not reversible. C-terminal modification relies on transpeptidation and its reversibility has been addressed through enzyme engineering.

2.7 Conclusion

The past decades have seen a shift in the nature of new drugs entering the market. The portion of small molecules is declining, and more antibodies, antibody-drug conjugates, (functionalized) peptides, and oligonucleotides are being approved. The level of functionalization of these antibodies and other proteins and peptides requires sophisticated tools for modification. This review highlights the use of sortase, peptide asparaginyl ligases, subtilisin-derived peptide ligases, and trypsiligase as summarized in Table 2.1. Enzymes enable the manipulation of completely

Table 2.1 Summary of the presented peptide ligases.

Type	Enzyme	Source	Cofactor	Recognition sequence	Other requirements	Reversible reaction	Application
Sortase	Sortase A[a]	*Staphylococcus aureus*[a]	Ca^{2+}	LPXT//GX[a]	–	Yes	• N-terminal modification • C-terminal modification
PAL	Butelase	*Clitoria ternatea*	–	B//HV[b]	–	Yes	• Peptide/protein semisynthesis (limited, due to recognition sequence) • Peptide cyclization • N-terminal modification • C-terminal modification
	*Oa*AEP1	*Oldenlandia affinis*		B//GL[b]			
	*Vy*PAL2	*Viola yedoensis*		B//SL[b]			
Subtilisin derived ligases	Subtiligase Stabiligase	*Bacillus amyloliquefaciens*	Ca^{2+}	None[c]	Activated ester required for the acyl donor	No	• Peptide/protein semisynthesis • Peptide cyclization • N-terminal modification
	Peptiligase Omniligase		–				
Engineered protease	Trypsiligase	*Rattus norvegicus*	Zn^{2+}	Y//RH	Activated ester required for the acyl donor in case of N-terminal modifications	N-terminal modifications: No C-terminal modifications: Yes	• N-terminal modification • C-terminal modification

[a]Several sortases from other classes, organisms, and engineered sortases have been described that differ slightly in specificity and recognition sequence; see Refs. [14,16,17,25].
[b]Promiscuous sequence requirements for the prime site, especially for the acyl acceptor. Typically, a hydrophobic amino acid is preferred in the P2′ position.
[c]Hydrophobic and slightly polar amino acids preferred in the P4 position; proline is not allowed in the P3, P1, P1′, and P2′ positions.

deprotected proteins and peptides without the risk of undesired modifications at other sites. The enzymes described do not require any (expensive) cofactors either. However, some of these enzymes require a recognition motif, leaving a scar in the product, which may limit their applications. Subtilisin-derived peptiligase is a so-called broad-specificity enzyme and does not require a specific sequence of amino acids. A typical peptide sequence always provides sufficient recognition motifs for the enzyme, making it particularly useful for the fragment-based assembly of peptides.

References

1 Fosgerau, K. and Hoffmann, T. Peptide therapeutics: current status and future directions. *Drug Discov Today*. 2015;20(1):122–128.

2 Otvos, L. and Wade, J.D. Big peptide drugs in a small molecule world. *Front Chem*. 2023; 11(December):1302169.

3 Merrifield, R.B. Solid phase peptide synthesis. I. The synthesis of a tetrapeptide. *J Am Chem Soc*. 1963;85(14):2149–154.

4 Jensen, K.J., Tofteng Shelton, P., and Pedersen, S.L. (eds.) *Peptide Synthesis and Applications* [Internet]. Totowa, NJ: Humana Press; 2013. (*Methods in Molecular Biology*; vol. 1047).

5 Leisle, L., Valiyaveetil, F., Mehl, R.A., and Ahern, C.A. Incorporation of non-canonical amino acids. *Adv Exp Med Biol*. 2015;869:119–151.

6 Kekessie, I., Wegner, K., Martinez, I., et al. Process mass intensity (PMI): A holistic analysis of current peptide manufacturing processes informs sustainability in peptide synthesis. *J Org Chem*. 2024;89(7):4261–4282.

7 Isidro-Llobet, A., Kenworthy, M.N., Mukherjee, S., et al. Sustainability challenges in peptide synthesis and purification: from R&D to production. *J Org Chem*. 2019;84(8):4615–4628.

8 Martin, V., Egelund, P.H.G., Johansson, H., Thordal Le Quement, S., Wojcik, F., and Sejer Pedersen, D. Greening the synthesis of peptide therapeutics: an industrial perspective. *RSC Adv*. 2020;10(69):42457–42492.

9 Ebrahimi, S.B. and Samanta, D. Engineering protein-based therapeutics through structural and chemical design. *Nat Commun*. 2023;14(1):1–11.

10 Sharma, S.K. Key issues in the purification and characterization of recombinant proteins for therapeutic use. *Adv Drug Deliv Rev*. 1989;4(2):87–111.

11 Assenberg, R., Wan, P.T., Geisse, S., and Mayr, L.M. Advances in recombinant protein expression for use in pharmaceutical research. *Curr Opin Struct Biol*. 2013;23(3):393–402.

12 Weeks, A.M. and Wells, J.A. Subtiligase-catalyzed peptide ligation. *Chem Rev*. 2020;120(6):3127–3160.

13 Mazmanian, S.K., Liu, G., Ton-That, H., and Schneewind, O. Staphylococcus aureus sortase, an enzyme that anchors surface proteins to the cell wall. *Science (80)*. 1999;285(5428):760–763.

14 Pishesha, N., Ingram, J.R., and Ploegh, H.L. Sortase A: a model for transpeptidation and its biological applications. *Annu Rev Cell Dev Biol*. 2018;34(1):163–188.

15 Ton-That, H., Liu, G., Mazmanian, S.K., Faull, K.F., and Schneewind, O. Purification and characterization of sortase, the transpeptidase that cleaves surface proteins of Staphylococcus aureus at the LPXTG motif. *Proc Natl Acad Sci*. 1999;96(22):12424–12429.

16 Morgan, H.E., Turnbull, W.B., and Webb, M.E. Challenges in the use of sortase and other peptide ligases for site-specific protein modification. *Chem Soc Rev*. 2022;51(10):4121–4145.

17 Freund, C. and Schwarzer, D. Engineered sortases in peptide and protein chemistry. *ChemBioChem*. 2021;22(8):1347–1356.

18 Nikghalb, K.D., Horvath, N.M., Prelesnik, J.L., et al. Expanding the scope of sortase-mediated ligations by using sortase homologues. *ChemBioChem*. 2018;19(2):185–195.

19 Schmohl, L., Bierlmeier, J., von Kügelgen, N., et al. Identification of sortase substrates by specificity profiling. *Bioorg Med Chem*. 2017;25(18):5002–5007.

20 Zou, Z., Nöth, M., Jakob, F., and Schwaneberg, U. Designed *Streptococcus pyogenes* sortase A accepts branched amines as nucleophiles in sortagging. *Bioconjug Chem*. 2020;31(11):2476–2481.

21 Podracky, C.J., An, C., DeSousa, A., Dorr, B.M., Walsh, D.M., and Liu, D.R. Laboratory evolution of a sortase enzyme that modifies amyloid-β protein. *Nat Chem Biol*. 2021;17(3):317–325.

22 Piotukh, K., Geltinger, B., Heinrich, N., et al. Directed evolution of sortase A mutants with altered substrate selectivity profiles. *J Am Chem Soc*. 2011;133(44):17536–17539.

23 Schmohl, L., Bierlmeier, J., Gerth, F., Freund, C., and Schwarzer, D. Engineering sortase A by screening a second-generation library using phage display. *J Pept Sci*. 2017;23(7–8):631–635.

24 Dorr, B.M., Ham, H.O., An, C., Chaikof, E.L., and Liu, D.R. Reprogramming the specificity of sortase enzymes. *Proc Natl Acad Sci U S A*. 2014;111(37):13343–13348.

25 Zou, Z., Ji, Y., and Schwaneberg, U. Empowering site-specific bioconjugations in vitro and in vivo: advances in sortase engineering and sortase-mediated ligation. *Angew Chemie*. 2024;136(12):e202310910.

26 Yamamura, Y., Hirakawa, H., Yamaguchi, S., and Nagamune, T. Enhancement of sortase A-mediated protein ligation by inducing a β-hairpin structure around the ligation site. *Chem Commun*. 2011;47(16):4742–4744.

27 Antos, J.M., Chew, G.L., Guimaraes, C.P., et al. Site-specific N- and C-terminal labeling of a single polypeptide using sortases of different specificity. *J Am Chem Soc.* 2009;131(31):10800–10801.

28 Williamson, D.J., Fascione, M.A., Webb, M.E., and Turnbull, W.B. Efficient N-terminal labeling of proteins by use of sortase. *Angew Chem Int Ed Engl.* 2012;51(37):9377–9380.

29 Zuo, C., Ding, R., Wu, X., et al. Thioester-assisted sortase-A-mediated ligation. *Angew Chemie Int Ed.* 2022;61(28):e202201887.

30 Liu, F., Luo, E.Y., Flora, D.B., and Mezo, A.R. Irreversible sortase a-mediated ligation driven by diketopiperazine formation. *J Org Chem.* 2014;79(2):487–492.

31 Reed, S.A., Brzovic, D.A., Takasaki, S.S., Boyko, K.V., and Antos, J.M. Efficient sortase-mediated ligation using a common C-terminal fusion tag. *Bioconjug Chem.* 2020;31(5):1463–1473.

32 David Row, R., Roark, T.J., Philip, M.C., Perkins, L.L., and Antos, J.M. Enhancing the efficiency of sortase–mediated ligations through nickel–peptide complex formation. *Chem Commun.* 2015;51(63):12548–12551.

33 Chen, I., Dorr, B.M., and Liu, D.R. A general strategy for the evolution of bond-forming enzymes using yeast display. *Proc Natl Acad Sci U S A.* 2011;108(28):11399–11404.

34 Chen, L., Cohen, J., Song, X., et al. Improved variants of SrtA for site-specific conjugation on antibodies and proteins with high efficiency. *Sci Rep.* 2016;6:31899.

35 Suliman, M., Santosh, V., Seegar, T.C.M., et al. Directed evolution provides insight into conformational substrate sampling by SrtA. *PLoS One.* 2017;12(8):e0184271.

36 Zou, Z., Mate, D.M., Rübsam, K., Jakob, F., and Schwaneberg, U. Sortase-mediated high-throughput screening platform for directed enzyme evolution. *ACS Comb Sci.* 2018;20(4):203–211.

37 Wang, H.H., Altun, B., Nwe, K., and Tsourkas, A. Proximity-based sortase-mediated ligation. *Angew Chemie Int Ed.* 2017;56(19):5349–5352.

38 Zou, Z., Mate, D.M., Nöth, M., Jakob, F., and Schwaneberg, U. Enhancing robustness of sortase A by loop engineering and backbone cyclization. *Chem – A Eur J.* 2020;26(60):13537.

39 Zou, Z., Alibiglou, H., Mate, D.M., Davari, M.D., Jakob, F., and Schwaneberg, U. Directed sortase A evolution for efficient site-specific bioconjugations in organic co-solvents. *Chem Commun.* 2018;54(81):11467–11470.

40 Zhulenkovs, D., Jaudzems, K., Zajakina, A., and Leonchiks, A. Enzymatic activity of circular sortase A under denaturing conditions: An advanced tool for protein ligation. *Biochem Eng J.* 2014;82:200–209.

41 Kiehstaller, S., Hutchins, G.H., Amore, A., et al. Bicyclic engineered sortase A performs transpeptidation under denaturing conditions. *Bioconjug Chem.* 2023;34(6):1114–1121.

42 Pelay-Gimeno, M., Bange, T., Hennig, S., and Grossmann, T.N. In situ cyclization of native proteins: structure-based design of a bicyclic enzyme. *Angew Chem Int Ed*. 2018;57(35):11164–11170.

43 Hirakawa, H., Ishikawa, S., and Nagamune, T. Design of Ca2+-independent Staphylococcus aureus sortase A mutants. *Biotechnol Bioeng*. 2012;109(12): 2955–2961.

44 Wuethrich, I., Peeters, J.G.C., Blom, A.E.M., et al. Site-specific chemoenzymatic labeling of aerolysin enables the identification of new aerolysin receptors. *PLoS One*. 2014;9(10):e109883.

45 Tang, T.M.S. and Luk, L.Y.P. Asparaginyl endopeptidases: enzymology, applications and limitations. *Org Biomol Chem*. 2021;19(23):5048–5062.

46 Nguyen, G.K.T., Wang, S., Qiu, Y., Hemu, X., Lian, Y., and Tam, J.P. Butelase 1 is an Asx-specific ligase enabling peptide macrocyclization and synthesis. *Nat Chem Biol*. 2014;10(9):732–738.

47 Nguyen, G.K.T., Kam, A., Loo, S., Jansson, A.E., Pan, L.X., and Tam, J.P. Butelase 1: a versatile ligase for peptide and protein macrocyclization. *J Am Chem Soc*. 2015;137(49):15398–15401.

48 Nguyen, G.K.T., Cao, Y., Wang, W., Liu, C.F., and Tam, J.P. Site-specific N-terminal labeling of peptides and proteins using Butelase 1 and Thiodepsipeptide. *Angew Chem Int Ed*. 2015;54(52):15694–15698.

49 Harris, K.S., Durek, T., Kaas, Q., et al. Efficient backbone cyclization of linear peptides by a recombinant asparaginyl endopeptidase. *Nat Commun*. 2015;6:1–10.

50 Hemu, X., Sahili A El, Hu, S., et al. Structural determinants for peptide-bond formation by asparaginyl ligases. *Proc Natl Acad Sci U S A*. 2019;116(24):11737–11746.

51 Zhang, D., Wang, Z., Hu, S., Lescar, J., Tam, J.P., and Liu, C.F. Vypal2: a versatile peptide ligase for precision tailoring of proteins. *Int J Mol Sci*. 2021;23(1):458.

52 Yang, R., Wong, Y.H., Nguyen, G.K.T., Tam, J.P., Lescar, J., and Wu, B. Engineering a catalytically efficient recombinant protein ligase. *J Am Chem Soc*. 2017;139(15):5351–5358.

53 Hemu, X., El Sahili, A., Hu, S., et al. Turning an asparaginyl endopeptidase into a peptide ligase. *ACS Catal*. 2020;10(15):8825–8834.

54 Rehm, F.B.H., Tyler, T.J., Xie, J., Yap, K., Durek, T., and Craik, D.J. Asparaginyl ligases: new enzymes for the protein engineer's toolbox. *ChemBioChem*. 2021;22(12):2079–2086.

55 Rehm, F.B.H., Harmand, T.J., Yap, K., Durek, T., Craik, D.J., and Ploegh, H.L. Site-specific sequential protein labeling catalyzed by a single recombinant ligase. *J Am Chem Soc*. 2019;141(43):17388–17393.

56 Tang, T.M.S., Cardella, D., Lander, A.J., et al. Use of an asparaginyl endopeptidase for chemo-enzymatic peptide and protein labeling. *Chem Sci*. 2020;11(23):5881–5888.

57 Xia, Y., Li, F., Zhang, X., et al. A cascade enzymatic reaction scheme for irreversible transpeptidative protein ligation. *J Am Chem Soc.* 2023;145(12):6838–6844.
58 James, A.M., Haywood, J., Leroux, J., et al. The macrocyclizing protease butelase 1 remains autocatalytic and reveals the structural basis for ligase activity. *Plant J.* 2019;98(6):988–999.
59 Hemu, X., Zhang, X., Nguyen, G.K.T., et al. Characterization and application of natural and recombinant butelase-1 to improve industrial enzymes by end-to-end circularization. *RSC Adv.* 2021;11(37):23105–23112.
60 Zhao, J., Ge, G., Huang, Y., Hou, Y., and Hu, S.Q. Study on activation mechanism and cleavage sites of recombinant butelase-1 zymogen derived from Clitoria ternatea. *Biochimie.* 2022;199:12–22.
61 Pi, N., Gao, M., Cheng, X., et al. Recombinant butelase-mediated cyclization of the p53-binding domain of the oncoprotein MdmX-stabilized protein conformation as a promising model for structural investigation. *Biochemistry.* 2019;58(27):3005–3015.
62 Hemu, X., Zhang, X., Chang, H.Y., Poh, J.E., and Tam, J.P. Consensus design and engineering of an efficient and high-yield peptide asparaginyl ligase for protein cyclization and ligation. *J Biol Chem.* 2023;299(3):102997.
63 Bordusa, F. Proteases in organic synthesis †. *Chem Rev.* 2002;102(12):4817–4868.
64 Abrahmsen, L., Tom, J., Burnier, J., Butcher, K.A., Kossiakoff, A., and Wells, J.A. Engineering subtilisin and its substrates for efficient ligation of peptide bonds in aqueous solution. *Biochemistry.* 1991;30(17):4151–4159.
65 Nuijens, T., Toplak, A., van de Meulenreek MBAC, Schmidt, M., Goldbach, M., and Quaedflieg, P.J.L.M. Improved solid phase synthesis of peptide carboxyamidomethyl (Cam) esters for enzymatic segment condensation. *Tetrahedron Lett.* 2016;57(32):3635–3638.
66 Pawlas, J., Nuijens, T., Persson, J., et al. Sustainable, cost-efficient manufacturing of therapeutic peptides using chemo-enzymatic peptide synthesis (CEPS). *Green Chem.* 2019;21(23):6451–6467.
67 Nuijens, T., Toplak, A., Quaedflieg PJLM, Drenth, J., Wu, B., and Janssen, D.B. Engineering a diverse ligase toolbox for peptide segment condensation. *Adv Synth Catal.* 2016;358(24):4041–4048.
68 Weeks, A.M. and Wells, J.A. Engineering peptide ligase specificity by proteomic identification of ligation sites. *Nat Chem Biol.* 2018;14(1):50–57.
69 Chang, T.K., Jackson, D.Y., Burnier, J.P., and Wells, J.A. Subtiligase: a tool for semisynthesis of proteins. *Proc Natl Acad Sci.* 1994;91(26):12544–12548.
70 Ruan, B., London, V., Fisher, K.E., Gallagher, D.T., and Bryan, P.N. Engineering substrate preference in subtilisin: structural and kinetic analysis of a specificity mutant. *Biochemistry.* 2008;47(25):6628–6636.

71 Toplak, A., Nuijens, T., Quaedflieg PJLM, Wu, B., and Janssen, D.B. Peptiligase, an enzyme for efficient chemoenzymatic peptide synthesis and cyclization in water. *Adv Synth Catal.* 2016;358(13):2140–2147.

72 Toplak, A., Teixeira de Oliveira, E.F., Schmidt, M., et al. From thiol-subtilisin to omniligase: design and structure of a broadly applicable peptide ligase. *Comput Struct Biotechnol J.* 2021;19:1277–1287.

73 Schmidt, M., Toplak, A., Quaedflieg PJLM, et al. Omniligase-1: a powerful tool for peptide head-to-tail cyclization. *Adv Synth Catal.* 2017;359(12):2050–2055.

74 Schmidt, M., Huang, Y.H., Texeira de Oliveira, E.F., et al. Efficient enzymatic cyclization of disulfide-rich peptides by using peptide ligases. *ChemBioChem.* 2019;20(12):1524–1529.

75 Liu, Y., Güler, R., Liao, Y., et al. Biologic Evaluation of a heterodimeric HER2-albumin targeted affibody molecule produced by chemo-enzymatic peptide synthesis. *Pharmaceutics.* 2022;14(11):2519.

76 Chandrashekar, C., Nishiuchi, Y., White, B.F., et al. Glycosylation improves the proteolytic stability of exenatide. *Bioconjug Chem.* 2023;34(6):1014–1018.

77 Schmidt, M., Toplak, A., Rozeboom, H.J., et al. Design of a substrate-tailored peptiligase variant for the efficient synthesis of thymosin-α 1. *Org Biomol Chem.* 2018;16(4):609–618.

78 Tatsushi, O. and Morihara, K. Trypsin as a catalyst for peptide synthesis. *J Biochem.* 1977;82(4):1055–1062.

79 Liebscher, S., Schöpfel, M., Aumüller, T., et al. N-Terminal protein modification by substrate-activated reverse proteolysis. *Angew Chem Int Ed.* 2014;53(11):3024–3028.

80 Liebscher, S. and Bordusa, F. Trypsiligase-catalyzed peptide and protein ligation. In: *Enzyme-Mediated Ligation Methods. Methods in Molecular Biology* (ed. T. Nuijens and M. Schmidt). Humana; 2019. pp. 111–133.

81 Liebscher, S., Kornberger, P., Fink, G., et al. Derivatization of antibody fab fragments: a designer enzyme for native protein modification. *ChemBioChem.* 2014;15(8):1096–1100.

82 Voigt, J., Meyer, C., and Bordusa, F. Synthesis of multiple bispecific antibody formats with only one single enzyme based on enhanced trypsiligase. *Int J Mol Sci.* 2022;23(6):3144.

83 Wartner, R., Boehme, M., Bordusa, F., Simon, A.H., and Richter, T. Trypsin variants with improved enzymatic properties. WO/2020/127808. 2020.

3

Modern Methods for Late-stage Peptide Modification

Yuxuan Ding and Zachary T. Ball

Department of Chemistry, Rice University, Houston, Texas, USA

3.1 Introduction

Peptide chemistry is a dynamic, interdisciplinary field at the nexus of organic, biological, and medicinal chemistry. Peptides play crucial roles in diverse applications, including drug development [1–3], biomaterials [4–6], and diagnostic imaging [7–9]. Historically, the term "peptide" often implied structures made of naturally occurring amino acids or fairly modest variations thereof. This situation has changed dramatically in recent years. While secondary metabolites such as vancomycin have always demonstrated the possibilities of radically diverse peptide structures, advances in biosynthetic engineering, synthetic methodology, combinatorial screening approaches, and analytical methods have allowed more routine access to a remarkable diversity of peptide structures, including structures that diverge markedly from natural structures. Incorporating unnatural amino acids via chemical or biochemical synthesis has become commonplace, and methods exist to create even more profound diversification of the basic polypeptide amino acid structure. Indeed, the very definition of "peptide" is less clear today, with blurred lines separating peptides from peptoids [10], peptidomimetics [11], stapled structures [12,13], foldamers [14,15], hybrid conjugates, and other diverse structural types.

The advent of modern peptide chemistry has revolutionized drug discovery, diagnostics, and materials science, offering tailored solutions to complex biomedical challenges with structures that present complex synthetic challenges [16].

Peptide Science: Chemical Ligation, Lead Generation, and Therapeutic Advances, First Edition. Fa Liu.

Over 60 peptide drugs have been approved by the FDA, and more than 600 are currently in clinical or preclinical trials [1,17,18]. To get improved drug properties, such as cellular uptake, *in vivo* stability, and selective localization, noncanonical modification of natural peptides is increasingly employed [17,19,20].

There are many ways to gain access to noncanonical peptide structures. Incorporating unnatural structures and linkages from the beginning of peptide synthesis is a straightforward solution in many contexts. A complementary approach, late-stage modification of natural polypeptide structures, has several advantages in some contexts. Divergent synthesis of diverse target structures from a single polypeptide scaffold derived from routine, inexpensive amino acid monomers becomes possible. Additionally, "lead" structures in medicinal contexts are often biological in origin, and thus late-stage manipulation provides a path to SAR studies directly from the initial hit compound.

Late-stage manipulation of canonical peptides is also an important fundamental challenge that creates opportunities to uncover new selectivity paradigms necessary for performing reactions in a complex polyfunctional environment [21,22]. Classical methods center on electrophiles or acylation chemistries that target nucleophilic side chains [23]. While some side chains – i.e. lysine, cysteine, and aspartate/glutamate – may serve as uniquely reactive sites, chemoselectivity challenges are common, especially in larger structures. The past decades have seen an explosion in approaches to late-stage functionalization, with the promise of dramatically increasing the types of substrates and product structures amenable to late-stage manipulation. Modern synthetic concepts, such as transition-metal catalysis [24–26], photoredox catalysis [27–29], organocatalysis [30], and electrochemical synthesis [31], have been essential to this effort.

This chapter aims to provide a summary of important approaches to side-chain modification methods developed in the past two decades, focusing on fundamentally enabling mechanistic, selectivity, and reactivity ideas. This chapter prioritizes methods compatible with complex, unprotected polypeptides, although several fundamentally interesting reactions of small/protected peptides are highlighted. Some important and related ideas, including peptide stapling [32,33], ligation [34,35], macrocyclization [33,36], C- and N-terminus reactivity [37–40], and enzymatic methods [41,42], are well reviewed elsewhere and are generally outside the focus of this chapter. Methods that utilize biorthogonal or other unique, unnatural handles for late-stage manipulation fall outside the scope of this review, but very many examples of these concepts have appeared in recent years and provide fascinating opportunities to build complex structures [43–56]. This chapter is organized by residue type, with sp^3 C—H functionalization methods combined into one section.

3.2 Methionine

Methionine is one of only two sulfur-containing proteinogenic amino acids and plays a crucial role in metal binding and redox regulation [57–59]. In addition to its unique biochemical properties, methionine is a rare amino acid in proteins, and it is often possible to find or engineer a uniquely reactive methionine residue in a protein of interest. There are two general strategies in the current methionine chemical modification toolbox. The first is nucleophilicity-based modification. Remarkably, methionine is the only native amino acid that can be alkylated at strongly acidic pH (below 3) because the thioether group is highly resistant to protonation. As early as 1940s, alkylation of methionine with alkyl halides was reported to give sulfonium salts [60,61]. In the following decades, this reaction was successfully applied to methionine-containing peptides and proteins [62–64]. Besides alkyl halides, the cyanogen bromide reaction with methionine residues is a classic approach to peptide degradation and sequencing [65]. In 2013, Deming showed that alkylation to a sulfonium salt could be made reversible in the presence of added nucleophiles (Figure 3.1a) [66]. Residue compatibility is generally excellent,

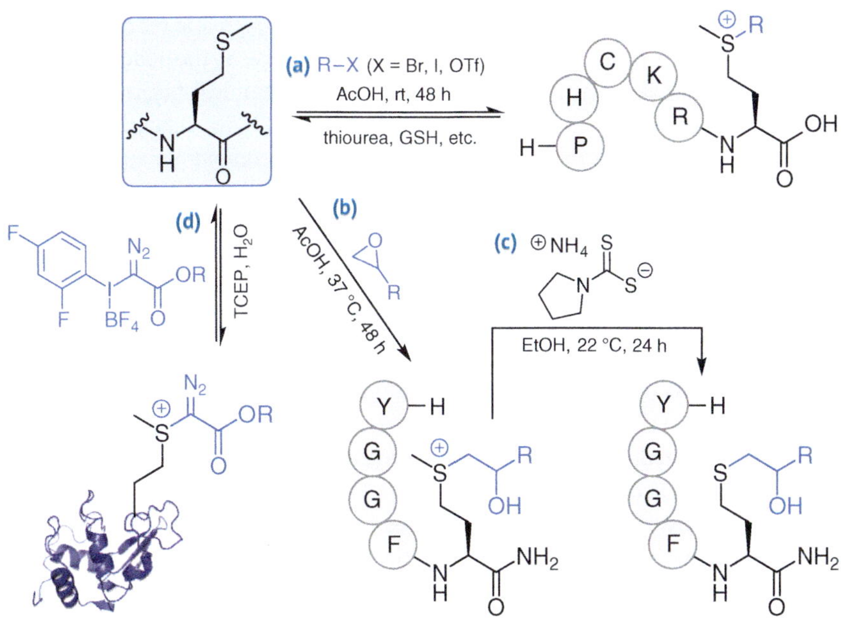

Figure 3.1 Nucleophilicity-based methionine modification with (a) alkyl halides, (b–c) epoxides, and (d) iodonium reagents.

and subsequent work demonstrated compatibility with other sensitive groups, such as functionalized epoxides (Figure 3.1b) [67]. Demethylation of the sulfonium salt products using ammonium pyrrolidinedithiocarbamate (APDC) was subsequently reported as a novel method for the synthesis of functional homocysteine residues (Figure 3.1c) [68]. Hypervalent iodine reagents have seen enormous development in recent years, and Gaunt recently developed a hypervalent iodine moiety as a distinct class of electrophile for methionine functionalization (Figure 3.1d) [69]. The mild aqueous reaction conditions allow modification of methionine within whole proteins. Cysteine is not tolerated in this method, and there is little evidence that any of these electrophilic reagents afford reactivity selective for methionine over cysteine. Interestingly, the product methionine sulfonium with a diazo group can undergo some photochemical processes for further functionalization.

The second strategy is redox-based modification: methionine features a relatively low oxidation potential (1.48 V) [70], allowing redox selectivity over most other residues. The thioether in methionine is readily oxidized to sulfoxide or sulfone oxidation states in a process that can be either chemical or enzymatic [71]. In living systems, methionine sulfoxide can be reduced by methionine sulfoxide reductases to regenerate a methionine residue [72,73]. This reversible process plays a role in cellular redox signaling [74–76]. While enzymatic redox processes have been known for decades, chemists took advantage of the redox potential of methionine only recently, with a redox-activated chemical tagging (ReACT) strategy for methionine bioconjugation (Figure 3.2a) [77,78]. Oxaziridines react with methionine to give sulfimide products under biocompatible aqueous conditions. Strikingly, free cysteine is tolerated in this reaction. Recently, selective oxidation of methionine residues with hypervalent iodine and ammonium carbamate was shown to produce NH-sulfoximines, which are useful in medicinal chemistry, including as protease transition-state mimics (Figure 3.2b) [79]. For late-stage diversification, the resulting methionine sulfoximine is a uniquely reactive handle for further elaboration by copper-catalyzed cross-coupling with arylboronic acids (Figure 3.2c). The oxidation/coupling sequence is compatible with all canonical amino acids except cysteine. In addition to traditional chemical oxidation, photoredox methods have also begun to appear. A lumiflavin photocatalyst provides access to functionalization of the *S*-methyl group with Michael acceptors (Figure 3.2d) [80]. The method is compatible with not only polypeptides, but also with proteins as large as antibodies.

Electrochemical activation is an additional pathway that has been recently exploited. Electrochemical acyloxylation of methionine with a carboxylic acid coupling partner was achieved at C/Pt electrodes (Figure 3.2e) [81]. Although this initial report included only protected dipeptides, it represents an unusual electrochemical approach for the functionalization of methionine methyl

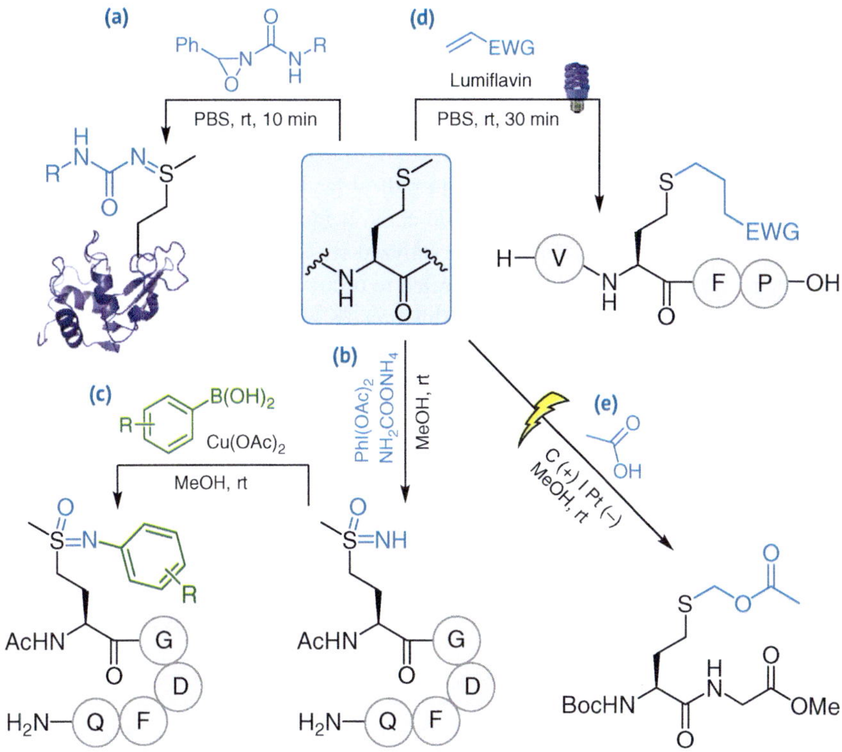

Figure 3.2 Redox-based methionine modification via (a–c) nitrene transfer (d) Michael acceptors, or (e) electrochemical C—H activation.

group and may indicate significant future potential for selective structure manipulation.

Overall, methionine functionalization methods remain fairly limited in number. However, these methods demonstrate generally very broad amino acid tolerance and reliability. One particular benefit is the ability to access a broad range of functional group structures, including sulfimines, sulfoximines, trialkyl sulfonium salts, and alkylation or acetoxylation of S-Me C—H bonds.

3.3 Aspartate and Glutamate

Carboxylic acid side chains (aspartate, glutamate) are among the most straightforward sites of facile structural manipulation by classical amidation, esterification, and thioesterification reactions [82,83]. These acylation products can suffer from proteolytic instability, and completely new reaction manifolds probably have significant value for peptide chemists, given the commonplace nature of carboxylic acid side chains as sites of selective manipulation. Decarboxylative functionalization [84,85] could provide dramatically different product structures, but presents obvious selectivity challenges in a polyfunctional peptide environment. Decarboxylative functionalization of free acids is established at the C-terminus [39], where the unique stabilization of an intermediate α-amino radical facilitates reactivity. Methods for decarboxylative functionalization of side chain acids require pre-activation as an activated ester. After activation, nickel-catalyzed cross-coupling with sp^3, sp^2, or sp organozinc reagents has been demonstrated to occur with decarboxylation (Figure 3.3a) [86]. A photocatalytic version of this process has been reported for on-resin decarboxylative coupling with aryl halides under purple light illumination (Figure 3.3b), using the Hantzsch ester as a stoichiometric reductant [87]. This method avoids handling and chemoselectivity issues associated with organozinc reagents,

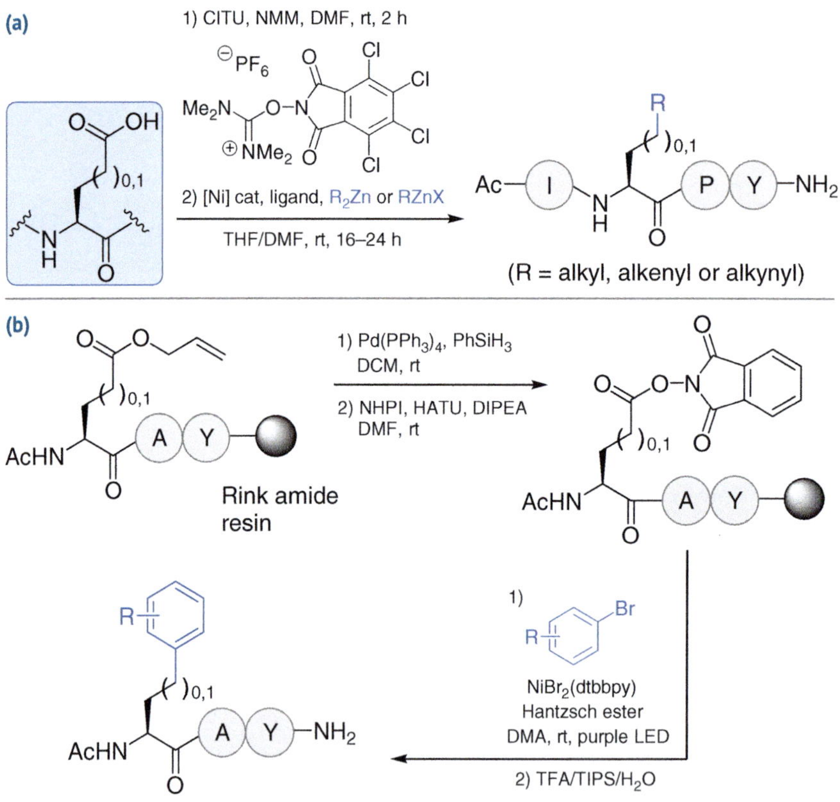

Figure 3.3 Decarboxylative functionalization of Asp/Glu side chains (a) in solution or (b) on resin.

and demonstrates good tolerance for a variety of (protected) amino acids; however, orthogonal protecting groups are needed in solid-phase peptide synthesis (SPPS), and the yields obtained in the activation/coupling sequence are rather modest (typically <20%).

3.4 Arginine

Arginine-selective modifications are quite limited, due in part to the high basicity and low nucleophilicity of the guanidine side chain. For bioconjugation applications, condensation with 1,2-dicarbonyl compounds in water has shown promise (Figure 3.4a) [88–94]. In addition, a single report demonstrated arginine acylation in DMF at mild temperatures (Figure 3.4b) [95].

Recently, a palladium-catalyzed stereospecific glycosylation of guanidine side chains was achieved on protected di- and tripeptides. The resulting glycosylated products could be converted to glycopeptide analogs subsequently (Figure 3.5) [96]. Although aprotic solvent (CH_2Cl_2) was used, this method represents an interesting approach for selective arginine functionalization with transition-metal catalysis. All current arginine modification methods have limited investigations into the scope

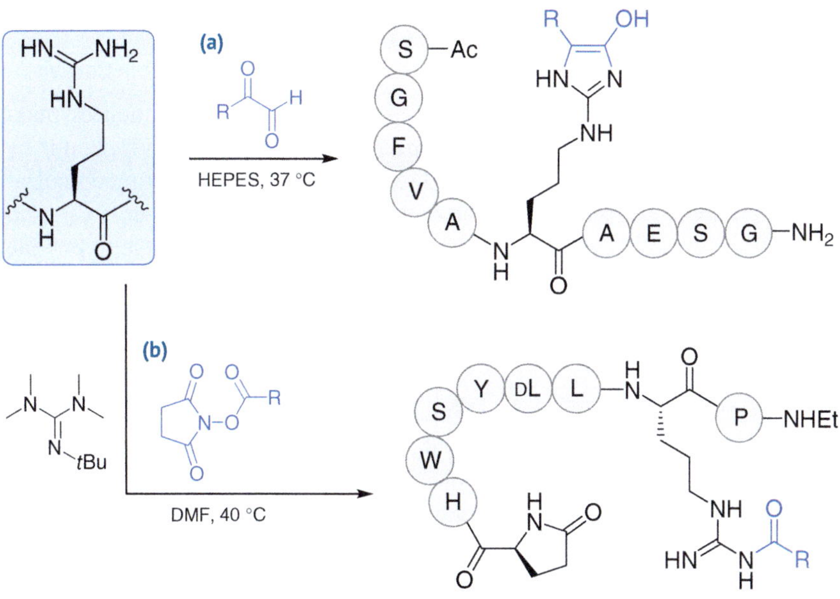

Figure 3.4 Arginine-selective peptide modification with (a) 1,2-dicarbonyl reagents and (b) activated esters.

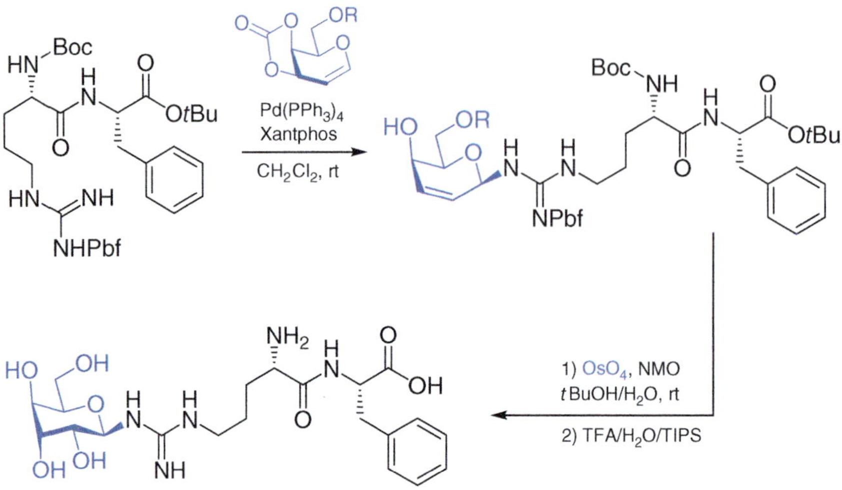

Figure 3.5 Pd-catalyzed arginine glycosylation.

for late-stage peptide diversification. The amino acid tolerance/specificity is not fully described for other nucleophilic residues (e.g. Lys, Cys).

3.5 Tryptophan

Tryptophan has unique reactivity and low abundance in natural sequences, making it appealing as a target for residue-specific manipulation. Tryptophan is by far the most successful and well-studied residue for peptide C—H functionalization, primarily due to the ease of electrophilic metalation of the indole C_2— and C_3—H bonds [97,98]. The field is large enough that it is not possible to provide a comprehensive discussion here. Recent reviews of this topic exist [99], and we instead focus on broad mechanistic classes of tryptophan reactivity. Several Pd-mediated C_2—H arylation methods have been demonstrated with a plethora of arylation reagents, including aryl iodides [100], arylboronic acids [101], aryldiazonium salts [102,103], and aryliodonium reagents [104,105] (Figure 3.6a) [99]. The reaction substrates vary from small, protected peptides all the way to full proteins. It is noteworthy the intramolecular C_2—H arylation of tryptophan with 4-iodo-phenylalanine provided a simple method for peptide macrocyclization (Figure 3.6b) [106]. Gold catalysis has also proven useful in tryptophan functionalization. Alkynylation at C_2 was achieved with TIPS-EBX and Au catalysts (Figure 3.6c) [107]. While many of these methods have only been reported in the context of short, protected peptides, it is noteworthy that recent examples of

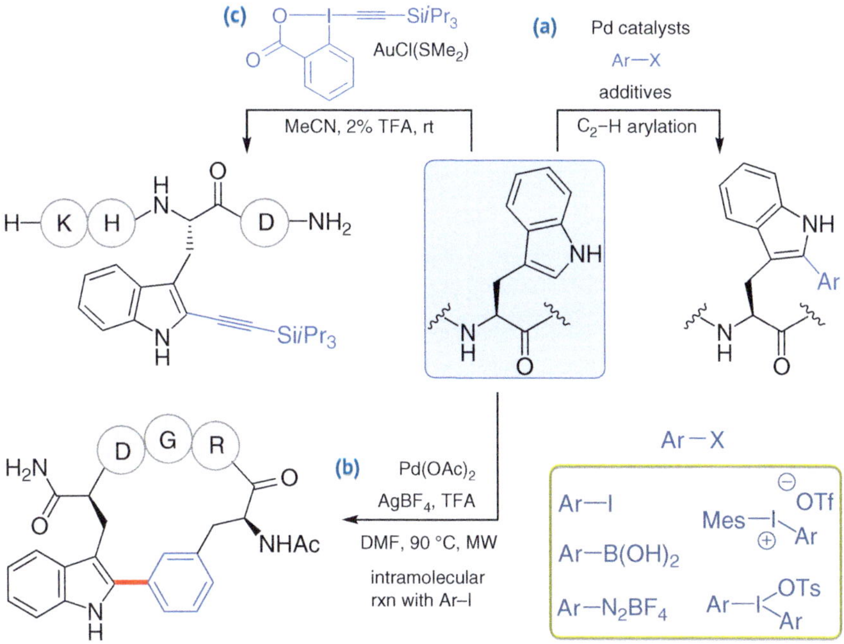

Figure 3.6 C_2—H functionalization at tryptophan residues with (a, b) Pd or (c) Au catalysts.

aqueous conditions and entire protein substrates have been reported with both Au [107] and Pd [105], indicating that further advances in substrate scope may be feasible [99].

While functionalization of C_2 is well established, functionalization of other positions on the indole ring is largely unknown, despite the importance of accessing diverse substitution patterns. Recently, C—H functionalizations targeting the C_7 position were reported, which serve as a complementary strategy for tryptophan modification. Ackermann et al. reported C_7 amidation with 3-phenyl-1,4,2-dioxazol-5-ones as the amidating reagent under rhodium(III) catalysis (Figure 3.7a) [108]. The method can be applied to fairly large peptides. A range of unprotected amino acids, including tyrosine, serine, and cysteine disulfides, are well tolerated. Similarly, Rh-catalyzed C_7—H maleimidation of tryptophan was also achieved (Figure 3.7b) [109]. Although the reaction requires elevated temperatures in aprotic solvent, the pivaloyl directing group on Trp can be easily removed. A cyclic RGD peptide was synthesized through an intramolecular maleimidation of tryptophan, followed by TFA removal of all protecting groups (Figure 3.7b). Introducing maleimide, a highly thiol-reactive moiety, offers new opportunities for secondary modification or peptide ligation. An interesting

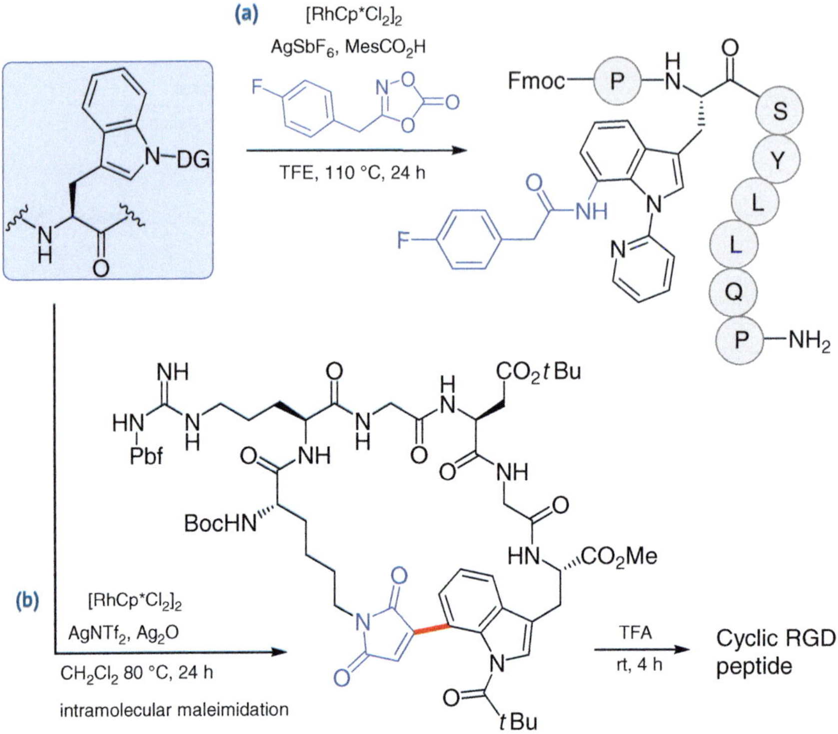

Figure 3.7 C_7—H functionalization at tryptophan residue in (a) intermolecular or (b) intramolecular coupling reactions.

rhodium-catalyzed C_7—H alkylation of tryptophan recently appeared, employing *N*-P^tBu_2 as a directing group for alkylation with olefins [110].

There are also a few reports of C_4—H functionalization of tryptophan residues [111,112]. Although currently limited in scope and functional group tolerance, these reports hint that broad access to other substitution patterns may be possible.

The electron-rich indole side chain is readily oxidized, which makes redox-based chemistry an attractive strategy for tryptophan modification. Several tryptophan oxidation reactions have been reported since the 1960s, employing oxidants such as cerium(IV) ammonium nitrate, chlorine dioxide, Fenton's reagent, and AAPH [113]. Early methods tend to exhibit limited chemoselectivity, but the past decade has brought new redox-based tryptophan modification methodologies, benefiting from the development of milder redox methods, including photo- and electrochemistry.

Nitrite ($NaNO_2$) and a nitroxyl radical also mediate a highly tryptophan-selective bioconjugation [114]. Unprotected polypeptides up to full-length proteins are tolerated (Figure 3.8a). Shi used a photoredox strategy for C_β carbofunctionalization with olefin reagents, in a strategy that can be envisioned via an initial photoinduced indole radical cation (Figure 3.8b) [115]. The reaction is compatible with complex native polypeptides and represents a rare modification method selective for the Trp β-carbon, although diastereomeric mixtures are formed. Tryptophan modification at C_2 is also observed upon treatment with a variety of radical precursors under oxidative conditions (Figure 3.8c). Several groups reported tryptophan trifluoromethylation with trifluoromethanesulfinate [113]. Notably, Gouverneu and Davis introduced ^{18}F to tryptophan-containing peptides through a ^{18}F-trifluoromethylation process with $[^{18}F]CF_2SO_2NH_4$, achieving *in vivo* positron emission tomography (PET) imaging [116]. A light-induced perfluoroalkylation of Trp with perfluoroalkyl iodides has also recently appeared [117]. Photocatalytic tryptophan C_2—H alkylation was demonstrated using α-bromo

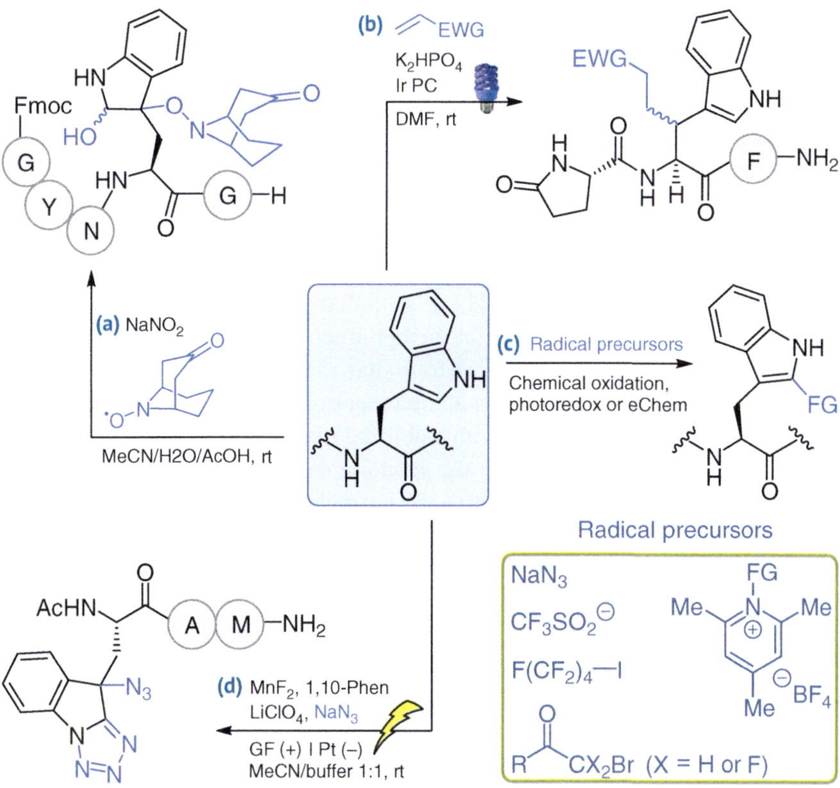

Figure 3.8 Redox-based tryptophan modification with (a) nitroxide, (b) Michael acceptors, (c) radical intermediates, or (d) electrochemical oxidation.

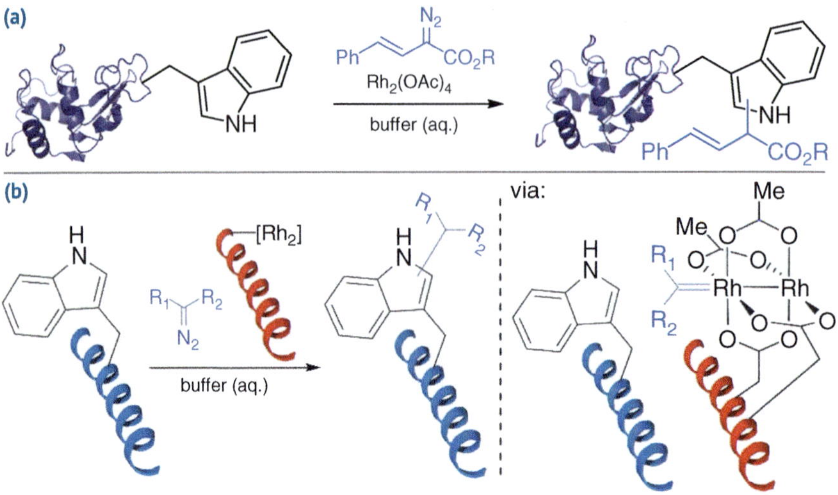

Figure 3.9 Tryptophan modification with rhodium carbenoids (a), which can also be directed by molecular recognition (b).

carbonyl compounds [118,119]. Besides, functionalized pyridinium salts have been identified as an emerging class of radical precursors for tryptophan modification under photocatalysis [120–122]. Remarkably, Weng reported an electrochemical approach for manganese-catalyzed tandem radical azidation/heterocyclization of tryptophan residues in unprotected peptides using NaN_3 as a radical precursor (Figure 3.8d) [123]. Modified tryptophan residues with an azide moiety are effective handles for further functionalizations.

Rhodium metallocarbenes, derived from diazo reagents, also react selectively with tryptophan residues. Francis et al. first reported a $Rh_2(OAc)_4$-catalyzed tryptophan modification in proteins with stabilized diazo compounds (Figure 3.9a) [124,125]. By appending ligands to the rhodium catalyst, it was also possible to demonstrate selective modification based on molecular recognition of peptide [126], protein [127,128], and glycopeptide [129] substrates (Figure 3.9b), for applications including the discovery [130] of new drug–protein interactions [131,132].

3.6 Functionalization of sp^2 C—H Bonds: Tyr, Phe, and His

3.6.1 Tyrosine

Tyrosine, containing an electron-rich and nucleophilic phenol side chain, could undergo a variety of $C(sp^2)$—H functionalization reactions via diverse mechanisms

[99]. S_EAr reaction is one of the most common strategies for *ortho* C—H functionalization, including halogenation [133], Mannich-type reactions [134], and coupling with diazonium salts [135], among others. An ene-type *ortho* C—H amination of tyrosine with triazolinediones was also developed by Ban et al. [136]. In a mechanistically different approach, the phenol group is readily oxidized to form a tyrosyl radical, which can then react with tyrosyl radical trapping agents. For example, photoredox catalysis promotes the generation of tyrosyl radicals, which can be trapped by functionalized *N′*-acyl-*N,N*-dimethyl-1,4-phenylenediamines (Figure 3.10a) [137]. The reaction conditions are mild and compatible with protein substrates. Electrochemical tyrosine oxidation provides a tyrosyl radical that reacts with phenothiazine derivatives to form *ortho* C—N bonds (Figure 3.10b) [138]. Large, unprotected polypeptides and small proteins are tolerated. Phenoxazine dialdehyde tags have been developed based on a similar photoredox approach [139].

In addition to S_EAr, ene-type, and radical-mediated reactions, there are scattered reports of transition-metal-catalyzed C—H functionalization of tyrosine [99]. However, most methods require harsh conditions and external directing groups, primarily involving single amino acid substrates. Xiong used an O-linked silanol as a directing group and achieved Pd-catalyzed *ortho* C—H alkenylation

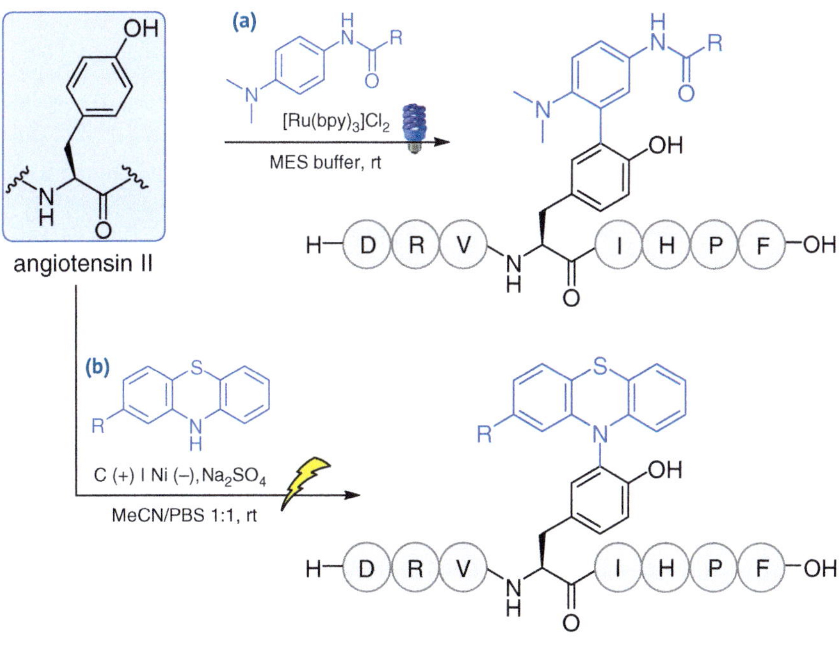

Figure 3.10 Radical-mediated C(sp^2)—H functionalization of tyrosine under photochemical (a) or electrochemical (b) conditions.

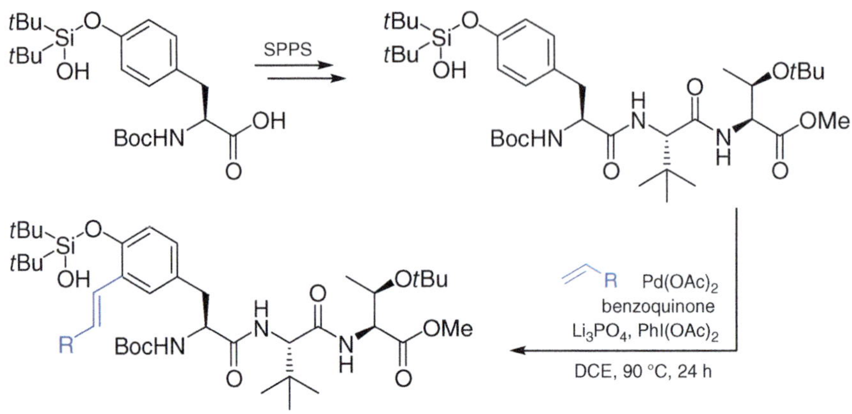

Figure 3.11 Silanol-directed C(sp^2)—H alkenylation of tyrosine in peptides.

of tyrosine in short peptides (Figure 3.11) [140]. Notably, silanol can also be used as a protecting group for standard SPPS technologies, which can be readily removed by TBAF.

3.6.2 Phenylalanine

The C—H functionalization methods of phenylalanine are rare because of the inert nature of the electron-neutral phenyl group. Lectka reported a visible light-induced benzylic C—H fluorination at the β-position of phenylalanine with Selectfluor and dibenzosuberenone [141].

One of the most exciting applications of phenylalanine C—H functionalization was reported by a group at AureoGen Biosciences [142]. Elhammer et al. developed an Ir-catalyzed, undirected C—H borylation of phenylalanine in aureobasidin A (Figure 3.12), which is a potent antifungal agent that covers several fungi except *Aspergillus fumigatus*, an important human pathogen. Interestingly, only one of the two phenylalanine residues has good accessibility in solvent. A mixture of C_3—H and C_4—H borylation products was obtained, which can be further functionalized by halogenation or Suzuki coupling. The resulting aureobasidin A derivatives showed great activity against *A. fumigatus*, which may serve as broad-spectrum antifungal drug candidates.

Another practical phenylalanine C—H functionalization reaction is directed by the peptide backbone amide. A Pd-catalyzed C(sp^2)—H olefination and alkynylation of phenylalanine residues in protected peptides (Figure 3.13a) was recently reported [143], and a similar strategy with olefin reagents for intramolecular C—H olefination has also appeared (Figure 3.13b) [144]. Both reactions likely go

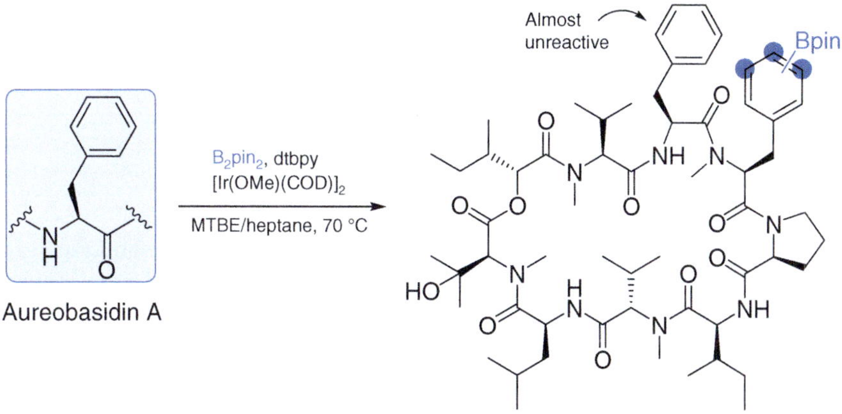

Figure 3.12 Ir-catalyzed undirected C(sp^2)—H borylation of phenylalanine.

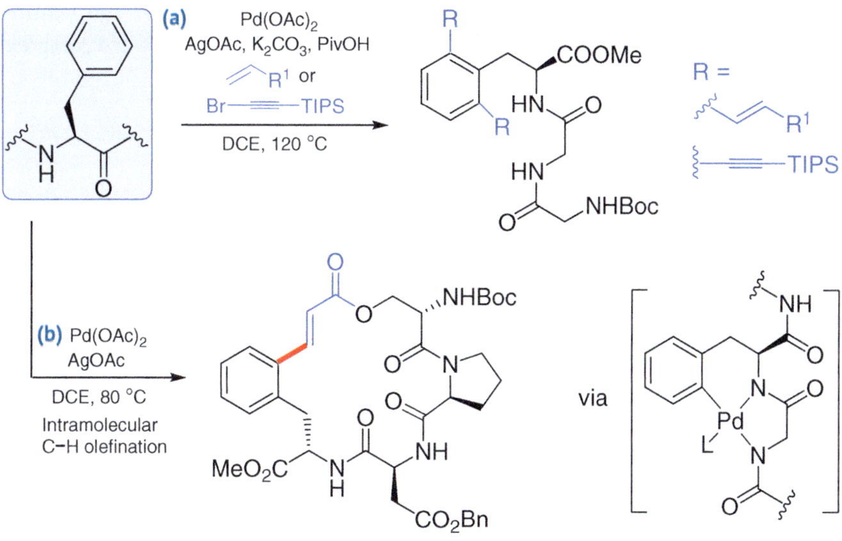

Figure 3.13 Backbone-directed C(sp^2)—H olefination and alkynylation of phenylalanine via (a) intermolecular or (b) intramolecular coupling.

through a six-membered palladacycle intermediate chelated by backbone amide groups.

A few Pd-catalyzed C—H functionalization reactions of phenylalanine using external directing groups have also been developed [99]. For instance, Correa reported a picolinamide (PA)-directed C_2—H acylation of N-terminal phenylalanine in small peptides with functionalized aldehydes [145]. A PA-directed

C_2—H functionalization of phenylalanine can also incorporate tandem functionalization, including C—C coupling with benzoquinone and intramolecular C—N cyclization [146]. This approach usually requires harsh reaction conditions, and peptide scope/complexity is limited to date.

3.6.3 Histidine

The imidazole side chain of histidine could also be modified through C(sp^2)—H functionalization reactions. The C_2—H bond is the most acidic one on imidazole ring [147], and radical-mediated C—H functionalization reactions targeting histidine C_2 position in unprotected peptides have been explored [148,149]. Histidine C_2—H alkylation is also feasible using aliphatic sulfinate salts as radical precursors and *t*-butyl hydroperoxide (TBHP) as an oxidant (Figure 3.14a) [148]. A photochemical version of this method for C_2—H alkylation of histidine in polypeptides employed substituted Hantzsch esters as alkylating reagents with blue LED irradiation (Figure 3.14b) [149]. Both reactions are transition-metal-free and feature relatively mild conditions. A variety of alkylating reagents and unprotected complex

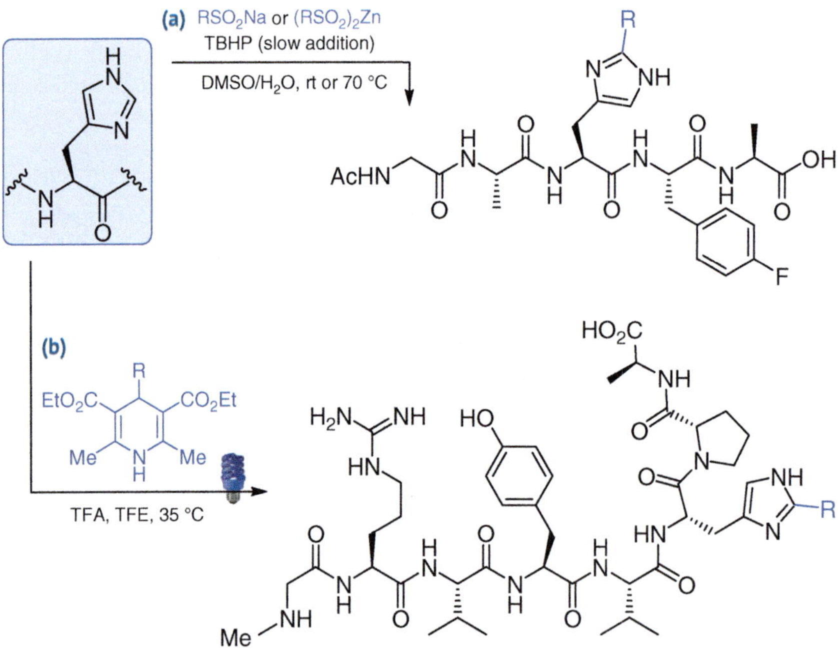

Figure 3.14 C_2—H functionalization at histidine with radicals generated from (a) sulfinate or (b) Hantzsch esters.

peptides are compatible with the reactions. Other redox-sensitive residues, such as Tyr, Trp, and Met, are well tolerated, although the tolerance of Cys has not been demonstrated, and yields reported can be modest.

3.7 Functionalization of sp^3 C—H Bonds

Photoredox catalysis has been significantly developed for a host of unique transformations in recent years, and the fundamental ideas of this field have begun to affect peptide chemistry as well. In the context of peptides, photoredox catalysis offers distinct advantages: radical intermediates are generally compatible with the Lewis basic sites common in peptides and—at least in some circumstances—are compatible with polar solvents, up to and including water [150,151].

Radicals at the α-position can be reliably generated by a two-step mechanistic process involving photooxidation of the amine, forming a radical cation, followed by deprotonation of the α-carbon, producing a neutral C-centered radical, which can be alkylated with a variety of agents, including acyl *N*-hydroxyphthalimides [152] and pyridinium salts (Figure 3.15) [153].

Metal-catalyzed functionalization of unactivated C—H bonds driven by directing groups is a general approach to C—H activation generally. Similar ideas have been used within peptide substrates, and the topic has been reviewed in recent years [102,154]. In most cases, substrates are limited to short peptides (1–4 amino acids). Reactions require nonaqueous conditions, and tolerance of polar or charged side chains is limited.

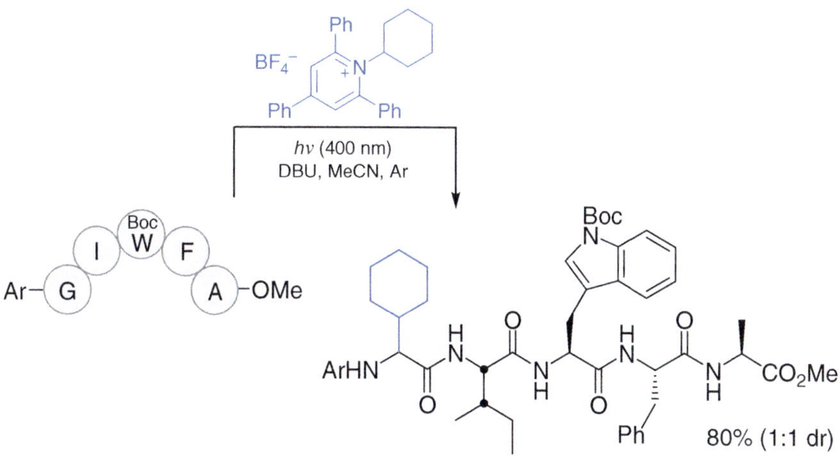

Figure 3.15 Photoredox alkylation of *N*-arylglycine residues at the N-terminus.

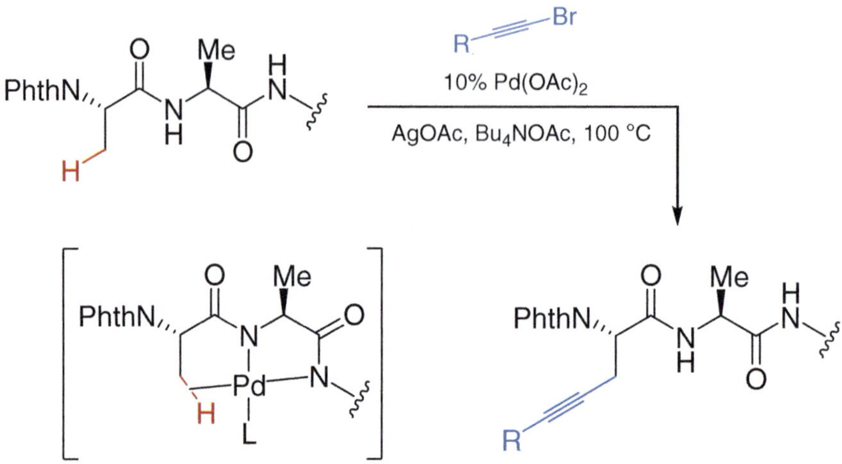

Figure 3.16 Alanine alkynylation catalyzed by palladium(0).

Several reported transformations derive from the observation that the amide backbone itself can template the modification of N-terminal alanine residues (Figure 3.16) [155,156]. In one important study, alkynylation of N-terminal alanine residues, via C—H activation of the methyl group, was demonstrated for di- and tripeptides (Figure 3.16) [155]. Some substantial residue tolerance was noted: protected tryptophan and tyrosine residues were tolerated, as was an aspartate ester. The fundamental reactivity at work here – directed N-terminal C—H functionalization at alanine – has been used for coupling with aryl iodides as well [156]. A related carbonylative reaction of N-terminal leucine methyl groups indicates that other coordination modes facilitate γ-activation, as opposed to the β-methyl of alanine.

These fundamental ideas for methyl-group functionalization may be poised to profoundly change the way we think about late-stage diversification. However, much work remains to be done. The complexity of bioactive sequences of interest is typically far beyond the fairly simple di-, tri-, and occasionally tetrapeptides employed in these studies. Functional group tolerance, too, is limited to a few robust side chains, and protecting groups are typically required. It will take substantial new developments in our understanding of the rules governing selective activation and reaction efficiency in order to predictably design late-stage manipulation of complex peptides based on these ideas.

Perhaps the most developed system for methyl C—H functionalization is a methionine-directed approach, which enabled arylation of macrocyclic peptides of up to nine residues (Figure 3.17) [157]. Building on earlier work with simpler systems [158], the work showed impressive reaction efficiencies, but tolerance

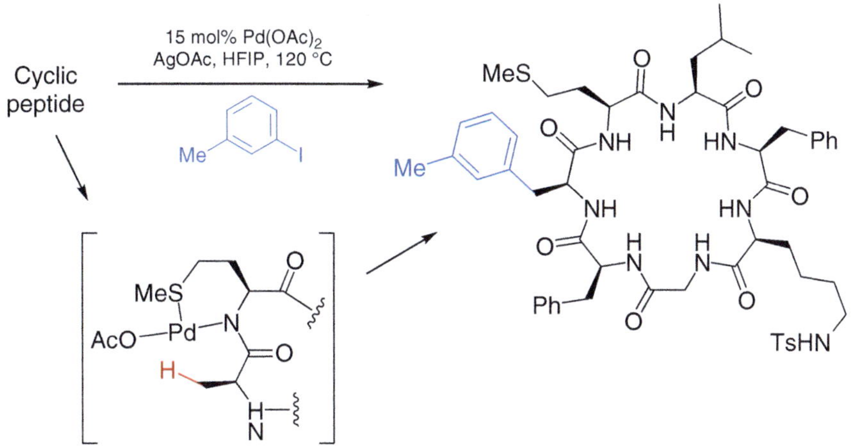

Figure 3.17 Methionine-directed methyl-group arylation.

of polar residues remains modest. Only a few successes, such as tosyl-protected lysine side chains, were reported. Nonetheless, the work conveys a significant advance in the complexity tolerated in such late-stage sp^3 C—H coupling reactions.

3.8 Other X—H Functionalizations

In addition to C—H activations, several fundamentally useful functionalizations of X—H (such as N—H and O—H) bonds have been reported. A gold-catalyzed, site-selective N—H arylation of carboxamides (Asn, Gln, and C-terminus) was developed. The reaction is compatible with unprotected polypeptides, although the conversions are moderate [159]. A solid-phase N—H arylation of tryptophan indoles was achieved under nickel/photoredox dual catalysis [160]. A free N-terminal amine is well tolerated under the reaction conditions, enabling stepwise arylation with subsequent backbone elongation. Similarly, N_3—H alkylation of histidine imidazole groups was also possible with bioinspired thiophosphorodichloridate reagents [161] or was enabled by a photoredox-mediated thioacetal activation process [162]. Both methods are under biocompatible conditions and tolerate unprotected peptides and proteins. Compared to N—H functionalization, reactions targeting O—H bonds are limited. Serine-selective bioconjugation was achieved with phosphorothioate reagents that mimic natural P(V)-based phosphorylation of serine [163]. The reaction shows great tolerance of all canonical amino acids, including Cys, Thr, and Lys, although moderate conversions were observed for complex polypeptides and protein substrates.

3.9 Conclusion

The proliferation of outstanding methodological advances in this area in recent years reflects the importance of biologics and related structures in medicine, chemical biology, and biomaterials research. While peptide modification was once a tiny field, it has grown quickly into a robust and mature field that allows exquisitely selective transformations on complex polypeptides with a significant degree of predictable selectivity. The field has become far too big to cover in a single book chapter, and this is reflected in the difficult choices about topics to cover here. There are certainly important, interesting, and useful methods that are not given justice here, due simply to space limitations.

At the same time, the work described here reflects the remarkable chemoselectivity insights that enable selective manipulation of complex polyfunctional peptides. Although traditional organic approaches still represent fertile grounds for the discovery of new reagents and selectivity paradigms, modern concepts of metal catalysis, photocatalysis, and redox-mediated reaction design have significantly expanded the toolbox to facilitate new reaction types.

Despite these efforts, considerable challenges remain. Ideally, residue-selective methods would exist for each natural amino acid, but efficient methods with complex substrates are still primarily limited to a few aromatic or protic residues. Even this is changing rapidly. Activation of hydrocarbon side chains via C—H activation remains in its infancy, with most studies focusing on small sequences with limited functional group complexity. But here, too, a few recent examples point to the potential for application of aliphatic C—H activation ideas within truly complex substrates.

References

1 Muttenthaler, M., King, G.F., Adams, D.J., and Alewood, P.F. Trends in peptide drug discovery. *Nat Rev Drug Discov*. 2021;20(4):309–325.

2 Henninot, A., Collins, J.C., and Nuss, J.M. The current state of peptide drug discovery: back to the future? *J Med Chem*. 2018;61(4):1382–1414.

3 Fosgerau, K. and Hoffmann, T. Peptide therapeutics: current status and future directions. *Drug Discov Today*. 2015;20(1):122–128.

4 Varanko, A., Saha, S., and Chilkoti, A. Recent trends in protein and peptide-based biomaterials for advanced drug delivery. *Adv Drug Deliv Rev*. 2020;156:133–187.

5 Chen, J. and Zou, X. Self-assemble peptide biomaterials and their biomedical applications. *Bioact Mater*. 2019;4:120–131.

6 Hamley, I.W. Small bioactive peptides for biomaterials design and therapeutics. *Chem Rev*. 2017;117(24):14015–14041.

7 Qi, G., Gao, Y., Wang, L., and Wang, H. Self-assembled peptide-based nanomaterials for biomedical imaging and therapy. *Adv Mater.* 2018;30(22):1703444.
8 Lee, S., Xie, J., and Chen, X. Peptides and peptide hormones for molecular imaging and disease diagnosis. *Chem Rev.* 2010;110(5):3087–3111.
9 Okarvi, S.M. Peptide-based radiopharmaceuticals: future tools for diagnostic imaging of cancers and other diseases. *Med Res Rev.* 2004;24(3):357–397.
10 Clapperton, A.M., Babi, J., and Tran, H. A field guide to optimizing peptoid synthesis. *ACS Polym Au.* 2022;2(6):417–429.
11 Li Petri, G., Di Martino, S., and De Rosa, M. Peptidomimetics: an overview of recent medicinal chemistry efforts toward the discovery of novel small molecule inhibitors. *J Med Chem.* 2022;65(11):7438–7475.
12 Moiola, M., Memeo, M.G., and Quadrelli, P. Stapled peptides—a useful improvement for peptide-based drugs. *Molecules.* 2019;24(20):3654.
13 Walensky, L.D. and Bird, G.H. Hydrocarbon-stapled peptides: principles, practice, and progress. *J Med Chem.* 2014;57(15):6275–6288.
14 Hill, D.J., Mio, M.J., Prince, R.B., Hughes, T.S., and Moore, J.S. A field guide to foldamers. *Chem Rev.* 2001;101(12):3893–4012.
15 Horne, W.S. and Gellman, S.H. Foldamers with heterogeneous backbones. *Acc Chem Res.* 2008;41(10):1399–1408.
16 Alleyne, C., Amin, R.P., Bhatt, B., et al. Series of novel and highly potent cyclic peptide PCSK9 inhibitors derived from an mRNA display screen and optimized via structure-based design. *J Med Chem.* 2020;63(22):13796–13824.
17 Wang, L., Wang, N., Zhang, W., et al. Therapeutic peptides: current applications and future directions. *Sig Transduct Target Ther.* 2022;7(1):48.
18 Rossino, G., Marchese, E., Galli, G., et al. Peptides as therapeutic agents: challenges and opportunities in the green transition era. *Molecules.* 2023;28(20):7165.
19 Barman, P., Joshi, S., Sharma, S., Preet, S., Sharma, S., and Saini, A. Strategic approaches to improvise peptide drugs as next generation therapeutics. *Int J Pept Res Ther.* 2023;29(4):61.
20 Al Musaimi, O., Lombardi, L., Williams, D.R., and Albericio, F. Strategies for improving peptide stability and delivery. *Pharmaceuticals.* 2022;15(10):1283.
21 Fischer, N.H., Oliveira, M.T., and Diness, F. Chemical modification of proteins – challenges and trends at the start of the 2020s. *Biomater Sci.* 2023;11(3):719–748.
22 deGruyter, J.N., Malins, L.R., and Baran, P.S. Residue-specific peptide modification: a chemist's guide. *Biochemistry.* 2017;56(30):3863–3873.
23 Boutureira, O. and Bernardes, G.J.L. Advances in chemical protein modification. *Chem Rev.* 2015;115(5):2174–2195.
24 Rodríguez, J. and Martínez-Calvo, M. Transition-metal-mediated modification of biomolecules. *Chem - Eur J.* 2020;26(44):9792–9813.

25 Ohata, J., Martin, S.C. and Ball, Z.T. Metal-mediated functionalization of natural peptides and proteins: panning for bioconjugation gold. *Angew Chem Int Ed*. 2019;58(19):6176–6199.
26 Malins, L.R. Peptide modification and cyclization via transition-metal catalysis. *Curr Opin Chem Biol*. 2018;46:25–32.
27 Lechner, V.M., Nappi, M., Deneny, P.J., Folliet, S., Chu, J.C.K., and Gaunt, M.J. Visible-light-mediated modification and manipulation of biomacromolecules. *Chem Rev*. 2022;122(2):1752–1829.
28 King, T.A., Mandrup Kandemir, J., Walsh, S.J., and Spring, D.R. Photocatalytic methods for amino acid modification. *Chem Soc Rev*. 2021;50(1):39–57.
29 Bottecchia, C. and Noël, T. Photocatalytic modification of amino acids, peptides, and proteins. *Chem - Eur J*. 2019;25(1):26–42.
30 Dalko, P.I. and Moisan, L. Enantioselective organocatalysis. *Angew Chem Int Ed*. 2001;40(20):3726–3748.
31 Mackay, A.S., Payne, R.J., and Malins, L.R. Electrochemistry for the chemoselective modification of peptides and proteins. *J Am Chem Soc*. 2022;144(1):23–41.
32 Lau, Y.H., de, Andrade, P., Wu, Y., and Spring, D.R. Peptide stapling techniques based on different macrocyclisation chemistries. *Chem Soc Rev*. 2014;44(1):91–102.
33 Reguera, L. and Rivera, D.G. Multicomponent reaction toolbox for peptide macrocyclization and stapling. *Chem Rev*. 2019;119(17):9836–9860.
34 Chow, H.Y., Zhang, Y., Matheson, E., and Li, X. Ligation technologies for the synthesis of cyclic peptides. *Chem Rev*. 2019;119(17):9971–10001.
35 Hackenberger, C.P.R. and Schwarzer, D. Chemoselective ligation and modification strategies for peptides and proteins. *Angew Chem Int Ed*. 2008;47(52):10030–10074.
36 White, C.J. and Yudin, A.K. Contemporary strategies for peptide macrocyclization. *Nat Chem*. 2011;3(7):509–524.
37 Tantipanjaporn, A. and Wong, M.K. Development and recent advances in lysine and N-terminal bioconjugation for peptides and proteins. *Molecules*. 2023;28(3):1083.
38 Jiang, H., Chen, W., Wang, J., and Zhang, R. Selective N-terminal modification of peptides and proteins: Recent progresses and applications. *Chin Chem Lett*. 2022;33(1):80–88.
39 Malins, L.R. Decarboxylative couplings as versatile tools for late-stage peptide modifications. *Pept Sci*. 2018;110(3):e24049.
40 Rosen, C.B. and Francis, M.B. Targeting the N terminus for site-selective protein modification. *Nat Chem Biol*. 2017;13(7):697–705.
41 Debon, A., Siirola, E., and Snajdrova, R. Enzymatic bioconjugation: a perspective from the pharmaceutical industry. *JACS Au*. 2023;3(5):1267–1283.

42 Rashidian, M., Dozier, J.K., and Distefano, M.D. Enzymatic labeling of proteins: techniques and approaches. *Bioconjugate Chem.* 2013;24(8):1277–1294.
43 Zielke, F.M. and Rutjes, F.P.J.T. Recent advances in bioorthogonal ligation and bioconjugation. *Top Curr Chem (Z).* 2023;381(6):35.
44 Bird, R.E., Lemmel, S.A., Yu, X., and Zhou, Q.A. Bioorthogonal chemistry and its applications. *Bioconjugate Chem.* 2021;32(12):2457–2479.
45 Scinto, S.L., Bilodeau, D.A., Hincapie, R., et al. Bioorthogonal chemistry. *Nat Rev Methods Primers.* 2021;1(1):30.
46 Salmain, M., Fischer-Durand, N., and Rudolf, B. Bioorthogonal conjugation of transition organometallic complexes to peptides and proteins: strategies and applications. *Eur J Inorg Chem.* 2020;2020(1):21–35.
47 Devaraj, N.K. The future of bioorthogonal chemistry. *ACS Cent Sci.* 2018;4(8):952–959.
48 Akgun, B. and Hall, D.G. Boronic acids as bioorthogonal probes for site-selective labeling of proteins. *Angew Chem, Int Ed.* 2018;57(40):13028–13044.
49 Qin, L.H., Hu, W., and Long, Y.Q. Bioorthogonal chemistry: optimization and application updates during 2013–2017. *Tetrahedron Lett.* 2018;59(23):2214–2228.
50 Lang, K. and Chin, J.W. Cellular incorporation of unnatural amino acids and bioorthogonal labeling of proteins. *Chem Rev.* 2014;114(9):4764–4806.
51 Yang, M., Li, J., and Chen, P.R. Transition metal-mediated bioorthogonal protein chemistry in living cells. *Chem Soc Rev.* 2014;43(18):6511–6526.
52 Ramil, C.P. and Lin, Q. Bioorthogonal chemistry: strategies and recent developments. *Chem Commun.* 2013;49(94):11007.
53 Schumacher, D. and Hackenberger, C.P. More than add-on: chemoselective reactions for the synthesis of functional peptides and proteins. *Curr Opin Chem Biol.* 2014;22:62–69.
54 Debets, M.F., van Hest, J.C.M., and Rutjes, F.P.J.T. Bioorthogonal labelling of biomolecules: new functional handles and ligation methods. *Org Biomol Chem.* 2013;11(38):6439–6455.
55 Best, M.D. Click chemistry and bioorthogonal reactions: unprecedented selectivity in the labeling of biological molecules. *Biochemistry.* 2009;48(28):6571–6584.
56 Van Swieten, P.F., Leeuwenburgh, M.A., Kessler, B.M., and Overkleeft, H.S. Bioorthogonal organic chemistry in living cells: novel strategies for labeling biomolecules. *Org Biomol Chem.* 2005;3(1):20.
57 Levine, R.L., Mosoni, L., Berlett, B.S., and Stadtman, E.R. Methionine residues as endogenous antioxidants in proteins. *Proc Natl Acad Sci USA.* 1996;93(26):15036–15040.
58 Levine, R.L., Moskovitz, J., and Stadtman, E.R. Oxidation of methionine in proteins: roles in antioxidant defense and cellular regulation. *IUBMB Life.* 2000;50(4–5):301–307.

59 Battin, E.E. and Brumaghim, J.L. Antioxidant activity of sulfur and selenium: a review of reactive oxygen species scavenging, glutathione peroxidase, and metal-binding antioxidant mechanisms. *Cell Biochem Biophys*. 2009;55(1):1–23.

60 Toennies, G. Sulfonium reactions of methionine and their possible metabolic significance. *J Biol Chem*. 1940;132(1):455–456.

61 Toennies, G. and Kolb, J.J. Methionine studies. VII. Sulfonium Derivatives[1]. *J Am Chem Soc*. 1945;67(5):849–851.

62 Gundlach, H.G., Stein, W.H., and Moore, S. The nature of the amino acid residues involved in the inactivation of ribonuclease by iodoacetate. *J Biol Chem*. 1959;234(7):1754–1760.

63 Vithayathil, P.J. and Richards, F.M. Modification of the methionine residue in the peptide component of ribonuclease-S. *J Biol Chem*. 1960;235(8):2343–2351.

64 Kleanthous, C. and Coggins, J.R. Reversible alkylation of an active site methionine residue in dehydroquinase. *J Biol Chem*. 1990;265(19):10935–10939.

65 Yeung, C.W.T., Carpenter, F.H., and Busse, W.D. Cyanogen bromide treatment of methionine-containing compounds. *Biochemistry*. 1977;16(8):1635–1641.

66 Kramer, J.R. and Deming, T.J. Reversible chemoselective tagging and functionalization of methionine containing peptides. *Chem Commun*. 2013;49(45):5144–5146.

67 Gharakhanian, E.G. and Deming, T.J. Versatile synthesis of stable, functional polypeptides via reaction with epoxides. *Biomacromolecules*. 2015;16(6):1802–1806.

68 Gharakhanian, E.G. and Deming, T.J. Chemoselective synthesis of functional homocysteine residues in polypeptides and peptides. *Chem Commun*. 2016;52(30):5336–5339.

69 Taylor, M.T., Nelson, J.E., Suero, M.G., and Gaunt, M.J. A protein functionalization platform based on selective reactions at methionine residues. *Nature*. 2018;562(7728):563–568.

70 Brunelle, P. and Rauk, A. One-electron oxidation of methionine in peptide environments: the effect of three-electron bonding on the reduction potential of the radical cation. *J Phys Chem A*. 2004;108(50):11032–11041.

71 Brot, N. and Weissbach, H. The biochemistry of methionine sulfoxide residues in proteins. *Trends Biochem Sci*. 1982;7(4):137–139.

72 Brot, N., Weissbach, L., Werth, J., and Weissbach, H. Enzymatic reduction of protein-bound methionine sulfoxide. *Proc Natl Acad Sci USA*. 1981;78(4):2155–2158.

73 Boschi-Muller, S., Olry, A., Antoine, M., and Branlant, G. The enzymology and biochemistry of methionine sulfoxide reductases. *Biochim Biophys Acta, Proteins Proteomics*. 2005;1703(2):231–238.

74 Lee, B.C., Péterfi, Z., Hoffmann, F.W., et al. MsrB1 and MICALs regulate actin assembly and macrophage function via reversible stereoselective methionine oxidation. *Mol Cell*. 2013;51(3):397–404.

75 Drazic, A. and Winter, J. The physiological role of reversible methionine oxidation. *Biochim Biophys Acta, Proteins Proteomics*. 2014;1844(8): 1367–1382.
76 Manta, B. and Gladyshev, V.N. Regulated methionine oxidation by monooxygenases. *Free Radic Biol Med*. 2017;109:141–155.
77 Lin, S., Yang, X., Jia, S., et al. Redox-based reagents for chemoselective methionine bioconjugation. *Science*. 2017;355(6325):597–602.
78 Christian, A.H., Jia, S., Cao, W., et al. A physical organic approach to tuning reagents for selective and stable methionine bioconjugation. *J Am Chem Soc*. 2019;141(32):12657–12662.
79 Ding, Y., Pedersen, S.S., Lin, A., Qian, R., and Ball, Z.T. Direct formation and site-selective elaboration of methionine sulfoximine in polypeptides. *Chem Sci*. 2022;13(47):14101–14105.
80 Kim, J., Li, B.X., Huang, R.Y.C., Qiao, J.X., Ewing, W.R., and MacMillan, D.W.C. Site-selective functionalization of methionine residues via photoredox catalysis. *J Am Chem Soc*. 2020;142(51):21260–21266.
81 Wang, H., He, M., Li, Y., et al. Electrochemical oxidation enables regioselective and scalable α-C(sp^3)-H acyloxylation of sulfides. *J Am Chem Soc*. 2021;143(9):3628–3637.
82 Giri, R.S., Manne, S.R., Dolai, G., Paul, A., Kalita, T., and Mandal, B. $FeCl_3$-mediated side chain modification of aspartic acid- and glutamic acid-containing peptides on a solid support. *ACS Omega*. 2017;2(10):6586–6597.
83 Barnes, N.G., Nyandoro, K., Jin, H., and Macmillan, D. Rapid access to Asp/Glu sidechain hydrazides as thioester precursors for peptide cyclization and glycosylation. *Chem Commun*. 2021;57(8):1006–1009.
84 Qin, T., Cornella, J., Li, C., et al. A general alkyl-alkyl cross-coupling enabled by redox-active esters and alkylzinc reagents. *Science*. 2016;352(6287):801–805.
85 Edwards, J.T., Merchant, R.R., McClymont, K.S., et al. Decarboxylative alkenylation. *Nature*. 2017;545(7653):213–218.
86 deGruyter, J.N., Malins, L.R., Wimmer, L., et al. CITU: A peptide and decarboxylative coupling reagent. *Org Lett*. 2017;19(22):6196–6199.
87 Pal, S., Openy, J., Krzyzanowski, A., Noisier, A., and 't Hart, P. On-resin photochemical decarboxylative arylation of peptides. *Org Lett*. 2024;26(14):2795–2799.
88 Gauthier, M.A. and Klok, H.A. Arginine-specific modification of proteins with polyethylene glycol. *Biomacromolecules*. 2011;12(2):482–493.
89 Conibear, A.C., Farbiarz, K., Mayer, R.L., Matveenko, M., Kählig, H., and Becker, C.F.W. Arginine side-chain modification that occurs during copper-catalysed azide–alkyne click reactions resembles an advanced glycation end product. *Org Biomol Chem*. 2016;14(26):6205–6211.
90 Thompson, D.A., Ng, R., and Dawson, P.E. Arginine selective reagents for ligation to peptides and proteins. *J Pept Sci*. 2016;22(5):311–319.

91 Gong, Y., Andina, D., Nahar, S., Leroux, J.C., and Gauthier, M.A. Releasable and traceless PEGylation of arginine-rich antimicrobial peptides. *Chem Sci.* 2017;8(5):4082–4086.

92 Dovgan, I., Erb, S., Hessmann, S., et al. Arginine-selective bioconjugation with 4-azidophenyl glyoxal: application to the single and dual functionalisation of native antibodies. *Org Biomol Chem.* 2018;16(8):1305–1311.

93 Zhang, Z., Morstein, J., Ecker, A.K., Guiley, K.Z., and Shokat, K.M. Chemoselective covalent modification of K-Ras(G12R) with a small molecule electrophile. *J Am Chem Soc.* 2022;144(35):15916–15921.

94 Leone, D., Pohl, R., Hubálek, M., et al. Glyoxal-linked nucleotides and DNA for bioconjugations and crosslinking with arginine-containing peptides and proteins. *Chem - Eur J.* 2022;28(14):e202104208.

95 Grundler, V. and Gademann, K. Direct arginine modification in native peptides and application to chemical probe development. *ACS Med Chem Lett.* 2014;5(12):1290–1295.

96 Yang, J., Dai, Y., Bartlett, R. and Zhang, Q. Convergent palladium-catalyzed stereospecific arginine glycosylation using glycals. *Org Lett.* 2021;23(10):4008–4012.

97 Sandtorv, A.H. Transition metal-catalyzed C—H activation of indoles. *Adv Synth Catal.* 2015;357(11):2403–2435.

98 Leitch, J.A., Bhonoah, Y., and Frost, C.G. Beyond C2 and C3: transition-metal-catalyzed C—H functionalization of indole. *ACS Catal.* 2017;7(9):5618–5627.

99 Tong, H.R., Li, B., Li, G., He, G., and Chen, G. Postassembly modifications of peptides via metal-catalyzed C—H functionalization. *CCS Chem.* 2021;3(3):1797–1820.

100 Ruiz-Rodríguez, J., Albericio, F., and Lavilla, R. Postsynthetic modification of peptides: chemoselective C-arylation of tryptophan residues. *Chem - Eur J.* 2010;16(4):1124–1127.

101 Williams, T.J., Reay, A.J., Whitwood, A.C., and Fairlamb, I.J.S. A mild and selective Pd-mediated methodology for the synthesis of highly fluorescent 2-arylated tryptophans and tryptophan-containing peptides: a catalytic role for Pd^0 nanoparticles? *Chem Commun.* 2014;50(23):3052–3054.

102 Zhu, Y., Bauer, M., and Ackermann, L. Late-stage peptide diversification by bioorthogonal catalytic C—H arylation at 23 °C in H_2O. *Chem - Eur J.* 2015;21(28):9980–9983.

103 Reay, A.J., Williams, T.J., and Fairlamb, I.J.S. Unified mild reaction conditions for C2-selective Pd-catalysed tryptophan arylation, including tryptophan-containing peptides. *Org Biomol Chem.* 2015;13(30):8298–8309.

104 Reay, A.J., Hammarback, L.A., Bray, J.T.W., Sheridan, T., Turnbull, D., Whitwood, A.C., et al. Mild and regioselective $Pd(OAc)_2$-catalyzed C—H

arylation of tryptophans by [ArN_2]X, promoted by tosic acid. *ACS Catal.* 2017;7(8):5174–5179.

105 Perez-Rizquez, C., Abian, O., and Palomo, J.M. Site-selective modification of tryptophan and protein tryptophan residues through PdNP bionanohybrid-catalysed C—H activation in aqueous media. *Chem Commun.* 2019;55(86):12928–12931.

106 Dong, H., Limberakis, C., Liras, S., Price, D., and James, K. Peptidic macrocyclization via palladium-catalyzed chemoselective indole C-2 arylation. *Chem Commun.* 2012;48(95):11644.

107 Hansen, M.B., Hubálek, F., Skrydstrup, T., and Hoeg-Jensen, T. Chemo- and regioselective ethynylation of tryptophan-containing peptides and proteins. *Chem - Eur J.* 2016;22(5):1572–1576.

108 Wang, W., Wu, J., Kuniyil, R., Kopp, A., Lima, R.N., and Ackermann, L. Peptide late-stage diversifications by rhodium-catalyzed tryptophan C7 amidation. *Chem.* 2020;6(12):3428–3439.

109 Wang, P., Liu, J., Zhu, X., et al. Modular synthesis of clickable peptides via late-stage maleimidation on C(7)-H tryptophan. *Nat Commun.* 2023;14(1):3973.

110 Liu, L., Fan, X., Wang, B., et al. P^{III}-Directed late-stage ligation and macrocyclization of peptides with olefins by rhodium catalysis. *Angew Chem Int Ed.* 2022;61(31):e202206177.

111 Zhang, Q., Xie, X., Peng, J., et al. Direct C4-acetoxylation of tryptophan and tryptophan-containing peptides *via* palladium(II)-catalyzed C—H activation. *Org Lett.* 2021;23(12):4699–4704.

112 Bai, Z., Cai, C., Sheng, W., Ren, Y., and Wang, H. Late-stage peptide macrocyclization by palladium-catalyzed site-selective C—H olefination of tryptophan. *Angew Chem Int Ed.* 2020;132(34):14794–14800.

113 Hu, J., He, P., and Li, Y. Chemical modifications of tryptophan residues in peptides and proteins. *J Pept Sci.* 2021;27(1):e3286.

114 Seki, Y., Ishiyama, T., Sasaki, D., et al. Transition metal-free tryptophan-selective bioconjugation of proteins. *J Am Chem Soc.* 2016;138(34):10798–10801.

115 Yu, Y., Zhang, L.K., Buevich, A.V., et al. Chemoselective peptide modification via photocatalytic tryptophan β-position conjugation. *J Am Chem Soc.* 2018;140(22):6797–6800.

116 Kee, C.W., Tack, O., Guibbal, F., et al. ^{18}F-trifluoromethanesulfinate enables direct C—H ^{18}F-trifluoromethylation of native aromatic residues in peptides. *J Am Chem Soc.* 2020;142(3):1180–1185.

117 Wang, Y., Wang, J., Li, G.X., He, G., and Chen, G. Halogen-bond-promoted photoactivation of perfluoroalkyl iodides: a photochemical protocol for perfluoroalkylation reactions. *Org Lett.* 2017;19(6):1442–1445.

118 Lima, R.N., Delgado, J.A.C., Bernardi, D.I., et al. Post-synthetic functionalization of tryptophan protected peptide sequences through indole (C-2) photocatalytic alkylation. *Chem Commun*. 2021;57(47):5758–5761.

119 Lee, J.C., Cuthbertson, J.D., and Mitchell, N.J. Chemoselective late-stage functionalization of peptides via photocatalytic C2-alkylation of tryptophan. *Org Lett*. 2023;25(29):5459–5464.

120 Tower, S.J., Hetcher, W.J., Myers, T.E., Kuehl, N.J., and Taylor, M.T. Selective modification of tryptophan residues in peptides and proteins using a biomimetic electron transfer process. *J Am Chem Soc*. 2020;142(20):9112–9118.

121 Hoopes, C.R., Garcia, F.J., Sarkar, A.M., et al. Donor–acceptor pyridinium salts for photo-induced electron-transfer-driven modification of tryptophan in peptides, proteins, and proteomes using visible light. *J Am Chem Soc*. 2022;144(14):6227–6236.

122 Laroche, B., Tang, X., Archer, G., Di Sanza, R., and Melchiorre, P. Photochemical chemoselective alkylation of tryptophan-containing peptides. *Org Lett*. 2021;23(2):285–289.

123 Weng, Y., Xu, X., Chen, H., Zhang, Y., and Zhuo, X. Tandem electrochemical oxidative azidation/heterocyclization of tryptophan-containing peptides under buffer conditions. *Angew Chem Int Ed*. 2022;61(41):e202206308.

124 Antos, J.M. and Francis, M.B. Selective tryptophan modification with rhodium carbenoids in aqueous solution. *J Am Chem Soc*. 2004;126(33):10256–10257.

125 Antos, J.M., McFarland, J.M., Iavarone, A.T., and Francis, M.B. Chemoselective tryptophan labeling with rhodium carbenoids at mild pH. *J Am Chem Soc*. 2009;131(17):6301–6308.

126 Popp, B.V. and Ball, Z.T. Structure-selective modification of aromatic side chains with dirhodium metallopeptide catalysts. *J Am Chem Soc*. 2010;132(19):6660–6662.

127 Vohidov, F., Coughlin, J.M., and Ball, Z.T. Rhodium(II) metallopeptide catalyst design enables fine control in selective functionalization of natural SH3 domains. *Angew Chem, Int Ed*. 2015;54(15):4587–4591.

128 Ohata, J. and Ball, Z.T. A Hexa-rhodium metallopeptide catalyst for site-specific functionalization of natural antibodies. *J Am Chem Soc*. 2017;139(36):12617–12622.

129 Vargas, R.D., Ding, Y., Trial, H.O., Qian, R., and Ball, Z.T. Polyol recognition in catalysis: toward selective modification of glycosylated polypeptides with boronic acid–rhodium(II) catalysts. *Chem Commun*. 2023;59(87):13030–13033.

130 Minus, M.B., Liu, W., Vohidov, F., et al. Rhodium(II) proximity-labeling identifies a novel target site on STAT3 for inhibitors with potent anti-leukemia activity. *Angew Chem, Int Ed*. 2015;54:13085–13089.

131 Ball, Z.T. Designing enzyme-like catalysts: A rhodium(II) metallopeptide case study. *Acc Chem Res*. 2013;46(2):560–570.

132 Ball, Z.T. Molecular recognition in protein modification with rhodium metallopeptides. *Curr Opin Chem Biol.* 2015;25(0):98–102.

133 Vilaró M, Arsequell, G., Valencia, G., Ballesteros, A., and Barluenga, J. Arylation of Phe and Tyr side chains of unprotected peptides by a Suzuki–Miyaura reaction in water. *Org Lett.* 2008;10(15):3243–3245.

134 Joshi, N.S., Whitaker, L.R., and Francis, M.B. A three-component Mannich-type reaction for selective tyrosine bioconjugation. *J Am Chem Soc.* 2004;126(49):15942–15943.

135 Jones, M.W., Mantovani, G., Blindauer, C.A., et al. Direct peptide bioconjugation/PEGylation at tyrosine with linear and branched polymeric diazonium salts. *J Am Chem Soc.* 2012;134(17):7406–7413.

136 Ban, H., Gavrilyuk, J., and Barbas, C.F. Tyrosine bioconjugation through aqueous ene-type reactions: a click-like reaction for tyrosine. *J Am Chem Soc.* 2010;132(5):1523–1525.

137 Sato, S. and Nakamura, H. Ligand-directed selective protein modification based on local single-electron-transfer catalysis. *Angew Chem, Int Ed.* 2013;52(33):8681–8684.

138 Song, C., Liu, K., Wang, Z., et al. Electrochemical oxidation induced selective tyrosine bioconjugation for the modification of biomolecules. *Chem Sci.* 2019;10(34):7982–7987.

139 Li, B.X., Kim, D.K., Bloom, S., et al. Site-selective tyrosine bioconjugation via photoredox catalysis for native-to-bioorthogonal protein transformation. *Nat Chem.* 2021;13(9):902–908.

140 Hu, Q.L., Hou, K.Q., Li, J., et al. Silanol: a bifunctional group for peptide synthesis and late-stage functionalization. *Chem Sci.* 2020;11(23):6070–6074.

141 Bume, D.D., Pitts, C.R., Jokhai, R.T., and Lectka, T. Direct, visible light-sensitized benzylic C—H fluorination of peptides using dibenzosuberenone: selectivity for phenylalanine-like residues. *Tetrahedron.* 2016;72(40):6031–6036.

142 Wuts, P.G.M., Simons, L.J., Metzger, B.P., Sterling, R.C., Slightom, J.L., and Elhammer, A.P. Generation of broad-spectrum antifungal drug candidates from the natural product compound aureobasidin A. *ACS Med Chem Lett.* 2015;6(6):645–649.

143 Zheng, Y. and Song, W. Pd-catalyzed site-selective C(sp^2)—H olefination and alkynylation of phenylalanine residues in peptides. *Org Lett.* 2019;21(9):3257–3260.

144 Bai, Z., Cai, C., Yu, Z., and Wang, H. Backbone-enabled directional peptide macrocyclization through late-stage palladium-catalyzed δ-C(sp^2)—H olefination. *Angew Chem Int Ed.* 2018;130(42):14108–14112.

145 San Segundo, M. and Correa, A. Pd-catalyzed site-selective C(sp^2)—H radical acylation of phenylalanine containing peptides with aldehydes. *Chem Sci.* 2019;10(38):8872–8879.

146 Tang, J., Lu, F., Geng, Y., Liu, Y., and Zhang, E. Site-selective modification of peptides via late-stage Pd-catalyzed tandem reaction of phenylalanine with benzoquinone. *Org Lett*. 2023;25(28):5378–5382.

147 Shen, K., Fu, Y., Li, J.N., Liu, L., and Guo, Q.X. What are the pKa values of C—H bonds in aromatic heterocyclic compounds in DMSO? *Tetrahedron*. 2007;63(7):1568–1576.

148 Noisier, A.F.M., Johansson, M.J., Knerr, L., et al. Late-stage functionalization of histidine in unprotected peptides. *Angew Chem Int Ed*. 2019;131(52):19272–19278.

149 Chen, X., Ye, F., Luo, X., et al. Histidine-specific peptide modification via visible-light-promoted C—H alkylation. *J Am Chem Soc*. 2019;141(45):18230–18237.

150 Bu M.-J. Cai, C., Gallou, F., and Lipshutz, B.H. PQS-enabled visible-light iridium photoredox catalysis in water at room temperature. *Green Chem*. 2018;20(6):1233–1237.

151 Russo, C., Brunelli, F., Tron, G.C., and Giustiniano, M. Visible-light photoredox catalysis in water. *J Org Chem*. 2023;88(10):6284–6293.

152 Wang, C., Guo, M., Qi, R., et al. Visible-light-driven, copper-catalyzed decarboxylative $C(sp^3)$—H alkylation of glycine and peptides. *Angew Chem Int Ed*. 2018;130(48):16067–16072.

153 Wang, C., Qi, R., Xue, H., et al. Visible-light-promoted $C(sp^3)$—H alkylation by intermolecular charge transfer: preparation of unnatural α-amino acids and late-stage modification of peptides. *Angew Chem Int Ed*. 2020;132(19):7531–7536.

154 Sengupta, S. and Mehta, G. Late stage modification of peptides via C—H activation reactions. *Tetrahedron Lett*. 2017;58(14):1357–1372.

155 Liu, T., Qiao, J.X., Poss, M.A., and Yu, J.Q. Palladium(II)-catalyzed site-selective $C(sp^3)$—H alkynylation of oligopeptides: a Linchpin approach for oligopeptide–drug conjugation. *Angew Chem, Int Ed*. 2017;56(36):10924–10927.

156 Gong, W., Zhang, G., Liu, T., Giri, R., and Yu, J.Q. Site-Selective $C(sp^3)$—H functionalization of di-, tri-, and tetrapeptides at the N-terminus. *J Am Chem Soc*. 2014;136(48):16940–16946.

157 Li, G., Yuan, F., and Yao, B. Post-assembly modification of head-to-tail cyclic peptides by methionine-directed β-$C(sp^3)$—H arylation. *Org Lett*. 2022;24(31):5767–5771.

158 Chen, X., Li, B., Tong, H., Qi, L., He, G., and Chen, G. Palladium-catalyzed methionine-facilitated β and γ $C(sp^3)$—H arylation of N-terminal aliphatic amino acids of peptides. *Chin J Chem*. 2022;40(21):2502–2506.

159 Li, W., Chen, Y., Chen, Y., et al. Site-selective arylation of carboxamides from unprotected peptides. *J Am Chem Soc*. 2023;145(27):14865–14873.

160 Delgado, J.A.C., Tian, Y.M., Marcon, M., König, B., and Paixão, M.W. Side-selective solid-phase metallaphotoredox $N_{(\text{in})}$-arylation of peptides. *J Am Chem Soc*. 2023;145(48):26452–26462.

161 Jia, S., He, D., and Chang, C.J. Bioinspired thiophosphorodichloridate reagents for chemoselective histidine bioconjugation. *J Am Chem Soc*. 2019;141(18):7294–7301.

162 Wan, C., Wang, Y., Lian, C., et al. Histidine-specific bioconjugation *via* visible-light-promoted thioacetal activation. *Chem Sci*. 2022;13(28):8289–8296.

163 Vantourout, J.C., Adusumalli, S.R., Knouse, K.W., et al. Serine-selective bioconjugation. *J Am Chem Soc*. 2020;142(41):17236–17242.

Section II

Peptide Drug Hit Identification

4

Generating Phage Libraries Bearing Unnatural Pharmacophores

Christin Kossmann[1], Arunika Ekanayake[2] and Ratmir Derda[2]

[1]*Zealand Pharma A/S, Søborg, Denmark*
[2]*Protomer Technologies, A Wholly Owned Subsidiary of Eli Lilly and Company. Department of Chemistry, University of Alberta, Edmonton, AB, Canada*

4.1 Introduction

Since the initial development of peptide phage display in the 1980s [1], it has been used to discover peptide ligands for diverse molecular targets. As canonical phage display systems are limited to 20 naturally occurring amino acids, there is a need to increase the diversity of libraries to include chemotypes other than those present in the naturally occurring residues. Diversification can be approached genetically through the incorporation of unnatural amino acids (UAAs) or through posttranslational modification (PTM) of translated peptide chains. Early reports of UAA incorporation into phage libraries [2] often gave rise to libraries with diversity bias against phage that contains UAA residues [3]. Selection from such libraries was possible when the selection pressure during library panning maintained the presence of the UAA in the library: one example is the selection of phage-displayed libraries that contain sulfated tyrosines against receptors that recognize sulfated tyrosines [3]. Significant advances have been made in the UAA mutagenesis field in the last 5–10 years, making a new iteration of UAA phage libraries more diverse, less prone to biases, and more practical [3–7]. PTM and chemical PTM (cPTM) are approaches orthogonal to UAA mutagenesis; both approaches can be used in concert. For example, most publications that describe UAA incorporation into phage libraries employed further cPTM of the unnatural side chains [5–7]. This chapter will not discuss the incorporation of UAA, rather, it will describe architecturally diverse examples of cPTM of canonical peptides displayed on phage.

Peptide Science: Chemical Ligation, Lead Generation, and Therapeutic Advances, First Edition. Fa Liu.

Such cPTM multiplies the complexity of these libraries using site-specific chemical modification to thiols of Cys or N-terminal aldehydes generated from Ser/Thr residues, and the logic of design and monitoring of the modification of these residues can be readily expanded to encompass modifications of UAA side chains. There exists a diverse palette of biocompatible chemical reactions for modification of proteins and peptides in water: many can be applied to phage display libraries and increase the diversity of the phage-displayed libraries by converting millions of peptides ("starting materials") into millions of peptides bearing unnatural modifications ("products"). Chemical diversification of peptide libraries can be viewed agnostically as a chemical reaction that acts on a mixture of starting materials producing a mixture of products. As with any reaction, cPTM of phage-displayed libraries of peptides should be characterized by key factors such as reaction rate, reaction conversion or yield, regioselectivity of reaction, and presence of unreacted substrates or side products. Any cPTM of phage-displayed libraries of peptides, whether using a well-established chemical reaction or a new class of chemical transformation, needs companion characterization tools to characterize the rate, conversion, and chemoselectivity.

The reactions described in this chapter can be divided topologically into three classes (Figure 4.1):

1. Linear peptides → linear peptides with unnatural N-terminal modification
2. Peptide disulfide macrocycle → peptide macrocycle with unnatural cross-linker
3. Combination of 1 and 2: simultaneous modification of the N-terminus and cross-linking

Conversion of peptide disulfide to macrocycle with unnatural cross-linker has been practiced by researchers in peptide science since the early 1990s to modulate degradation of peptide sequence by peptidases by distorting the conformation around the amide bond [8]. Macrocyclization introduces other beneficial properties, such as depleted ground state conformational ensemble [9]. Conformational depletion (or "rigidity") can give rise to increased potency of preorganized macrocyclic peptides when compared to lesser-structured linear peptides because the rigidity provided by macrocyclization reduces the entropic cost of binding of such macrocycle to a protein surface [10–14]. Grafting unnatural chemical moieties onto linear peptides or macrocycles described in examples 1–3 can be used to incorporate pharmacophores known to bind to a specific site of a protein (Figure 4.1). Grafting of unnatural fragments has been described as "genetically encoded fragment-based discovery (GE-FBD)," which can be viewed as a variant of traditional fragment-based discovery (FBD) of pharmaceutical leads (Figure 4.2).

In developing small molecule drugs, FBD starts from chemical fragments that bind to a target protein with low affinity and high promiscuity. Combining two such fragments through a linker can increase the affinity and specificity of the

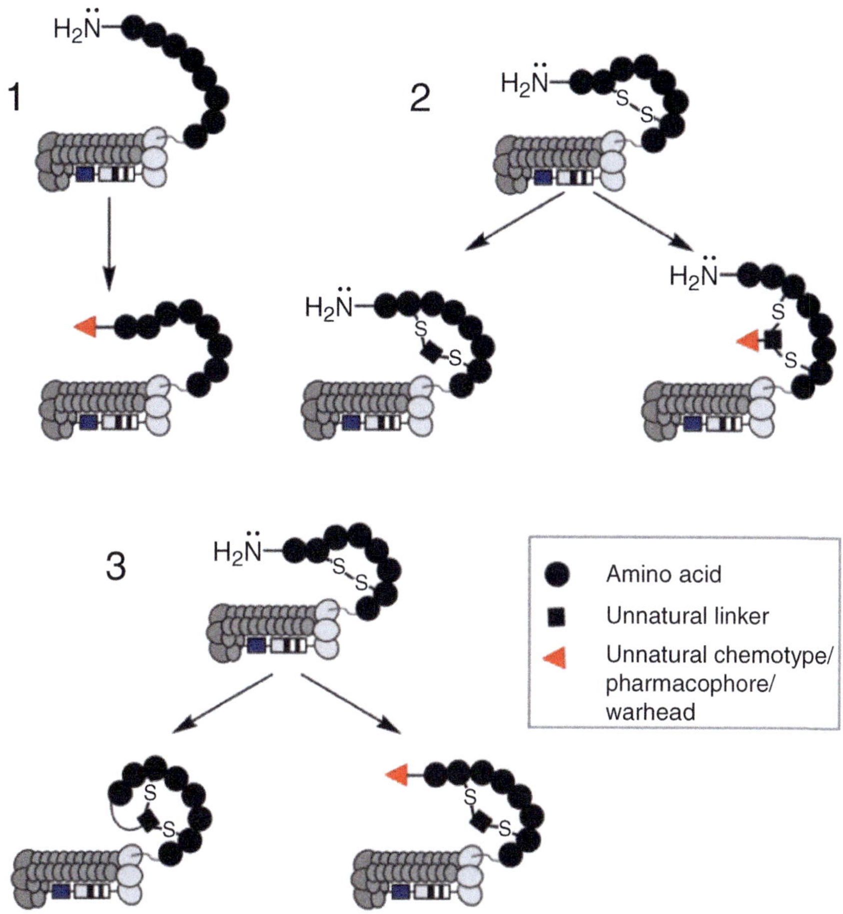

Figure 4.1 Chemical posttranslational modification reactions on phage can be divided into three classes. (1) Linear peptides with unnatural N-terminal modification. (2) Disulfide peptides converted to unnatural pharmacophore-bearing macrocycle. (3) Combination of 1 and 2: simultaneous modification of the N-terminus and cross-linking (not described in this chapter).

fragment heterodimer to the target of interest [15]. The same FBD concept can be applied to peptide libraries bearing an unnatural pharmacophore that binds to a target protein with low affinity and high promiscuity. Selection of such libraries against a target protein of interest can identify a combination of peptide "fragment" and unnatural fragment that can bind to the protein with increased affinity and selectivity [16]. cPTMs described in examples 1–3 can also be used to incorporate chemical moieties that improve pharmacokinetic (PK) properties of peptides.

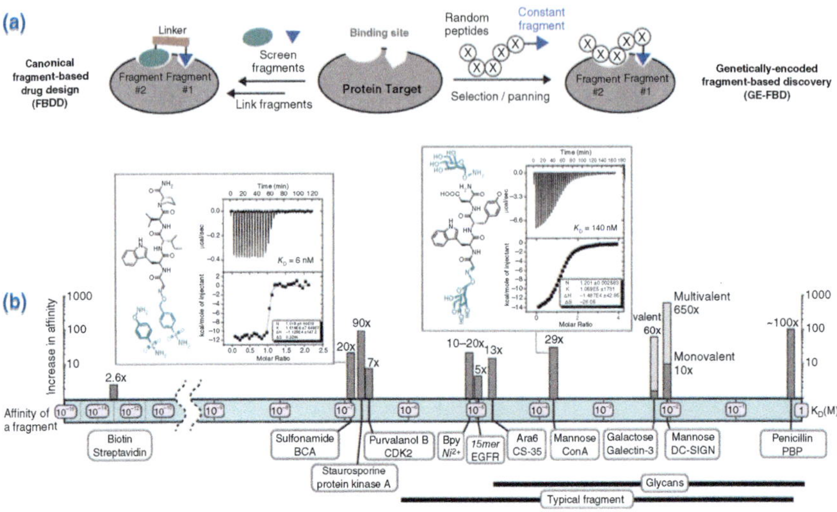

Figure 4.2 Concept of GE-FBD. (a) Comparison between canonical FBD and GE-FBD. (b) Grafting unnatural pharmacophores onto naturally occurring peptides helps to increase the affinity of the fragment toward the protein target of interest.

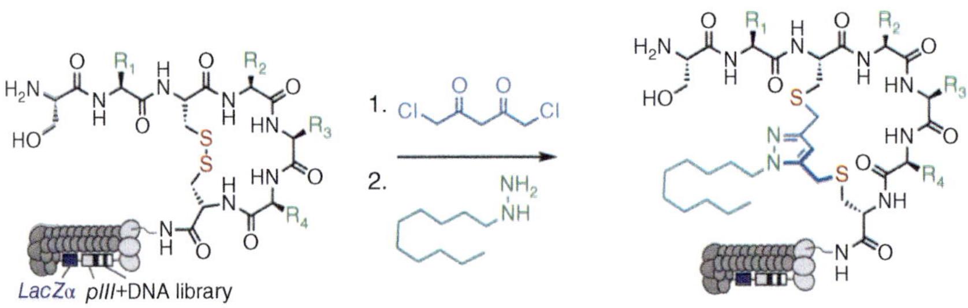

Figure 4.3 Lipidation of phage-displayed peptide libraries. Pegylation can be done similarly. *Source*: Ref. [17]/American Chemical Society.

Known examples are lipidation and pegylation of peptides; both can improve the PK profile of peptides *in vivo*. Strategies that incorporate such chemical modifications directly into the naïve phage-displayed library make it possible to discover ligands that are already bearing these modifications (i.e. pegylated or lipidated peptides) (Figure 4.3).

Panning with chemically modified libraries follows the same procedure as unmodified libraries but with an additional step to incorporate the chemical modification and a quality control step to measure the yield of modification. If the chemical fragment improves binding of phage libraries to the target, increasing the stringency of washes will be required to deplete populations of weaker binders and enrich the desired stronger binders. Weak wash can result in selection of a random sub-sample of library, in which all binders interact with the target via the unnatural fragment rather than the peptide sequence. If the library bears high-affinity fragments that can interact strongly with the target protein or reactive fragments (covalent warheads), incorporation of washes at mildly denaturing conditions such as in the presence of high concentrations of guanidium chloride can aid in effectively washing away nonspecific binders.

4.2 Biotin Capture Assay to Detect Chemical Modification of Phage Libraries

This protocol employs biotin–avidin interactions to monitor any cPTM of phage-displayed peptide libraries. The biotin–avidin interaction is one of the strongest known noncovalent interactions with a pK_d value of 15 and it allows detecting biotinylated phage particles with single molecule (single particle) resolution and determine the percentage of modification of chemical reactions on phage-displayed peptides present in concentrations as low as 100 attomolar.

cPTM of phage-displayed peptides by a biotin derivative with a reactive group X (Biotin-X) incorporates biotin handles onto the phage particles that bear a complementary reactive group in the displayed peptide sequence (Figure 4.4). In turn, biotinylation allows to immobilize or "capture" the chemically modified phage particles onto para-magnetic beads coated with streptavidin that can be separated from the unmodified phage particles remaining in the solution (Figure 4.5). Therefore, the plaque-forming unit numbers before capture (PFU^{BC}) and after capture (PFU^{AC}) depict the total number of phage particles and unmodified number of phage particles in the reaction mixture. Their difference ($PFU^{BC} - PFU^{AC}$) indicates the number of phage particles that were not modified by biotin derivative. Therefore, capture rate $CapR = (PFU^{BC} - PFU^{AC})/PFU^{BC}$ will indicate the fraction of modified phage particles within a library of phage-displayed peptides (Figure 4.5a). An increase of CapR as a function of time or concentration of (Biotin-X) indicates consumption of reactive moieties on phage and it can be used to extrapolate the kinetics of reactions on phage-displayed peptides.

When using chemical biotinylation as a strategy to monitor chemical reactions, a "pulse-chase" approach allows the detection of modification by chemical fragments other than biotin. In the "pulse-chase" approach, the phage-displayed peptides are first "pulsed" with the chemical fragment R, with reactive group X (R-X) to graft the desired chemical functionality R onto phage particles. The same sample can then be "chased" by (exposed to) a biotin derivative with the same reactive handle X and the Biotin-X reagent should cap any unreacted groups on phage-displayed peptides. In this case, a decrease of CapR as a function of time or

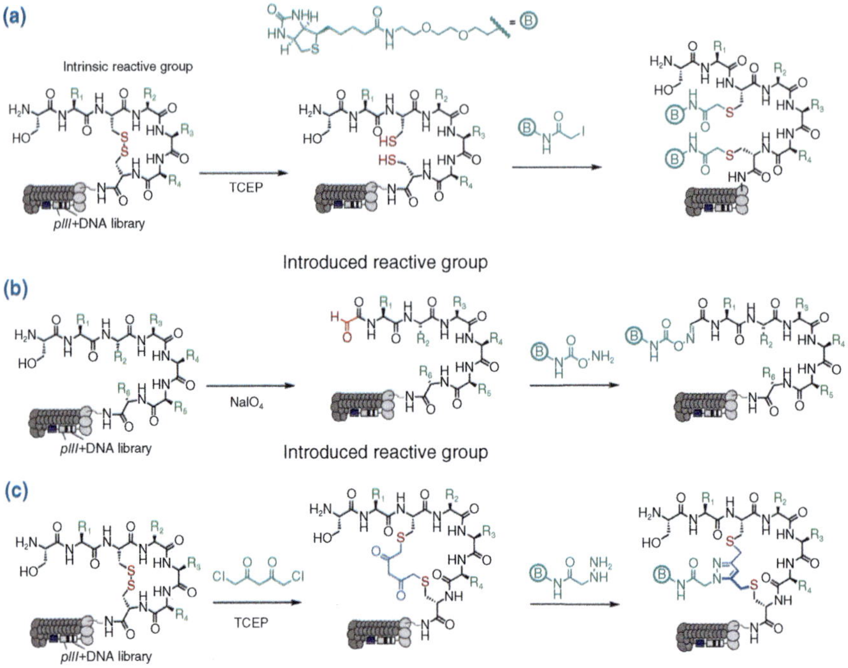

Figure 4.4 Biotinylation of phage-displayed peptide libraries. (a) Grafting biotins onto phage-displayed peptide libraries using the intrinsic reactivity of cysteine residues. (b) Grafting biotins onto phage-displayed peptide libraries through introduced reactive groups such as aldehydes and (c) Grafting biotin onto phage displayed libraries though introduced 1,3-diketone functionality.

concentration of (R-X) indicates consumption of reactive moieties on phage and it can be used to extrapolate the kinetics of reactions with the "pulse" reagent on phage-displayed peptides (Figure 4.5b).

We will describe three reactions to graft chemical moieties onto peptides displayed on phage. Each has a designated biotin fragment with corresponding functional groups to detect the chemical modification by the desired fragments. When using cysteine reactive chemical groups, biotin-iodoacetamide (BIA) that bears an electrophilic reagent for nucleophilic reactions with the thiol groups can be used as the "chase" reagent. Aldehyde-reactive biotin probes can be used in monitoring reactions with oxidized N-terminal serine. A biotin hydrazine compound allows quantifying the percentage of 1,3-diketone handles grafted onto phage-displayed peptides.

In cPTM of phage-displayed peptides, one must always question the regioselectivity: is the reaction taking place on peptides, or does it also occur on other phage proteins? To validate the regioselectivity of the chemical reactions, it is beneficial

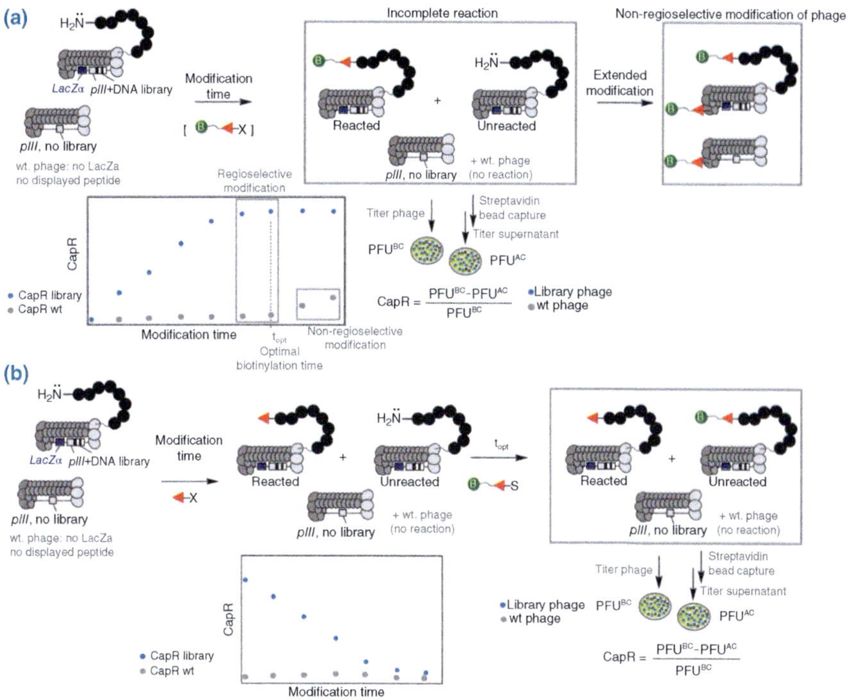

Figure 4.5 Biotinylation and "pulse-chase" monitoring of chemical modification of phage-displayed peptide libraries and control phage particles that display no peptide. (a) Exposure of the mixture of phage that displays the peptide library and phage that displays no peptide to biotinylation reagent (Biotin-X) for specific duration and capture of the biotinylated phage determines the ratio of modified (biotinylated) phage particles in the phage-displayed peptide library. Such experiment determines the optimal concentrations of Biotin-X reagent and reaction time that yield the highest regioselective conversion. Capture of phage particles that display no peptide library wild-type (WT) serves as an indication of non-regioselective reaction. (b) "Pulse" with a pharmacophore-X, where X is a reactive group followed by "chase" with Biotin-X to determine the ratio of reacted and unreacted peptides in the phage-displayed peptide library. Decrease of biotinylation over the course of the "pulse" reaction indicates disappearance of the reactive group and such progressive decrease indicates progressive modification of the phage-displayed peptide with pharmacophore-X reagent.

to include a phage that does not bear any peptides in the reaction mixture (we refer to such phage as "wildtype" (WT)) and does not express a genetic reporter, such as *LacZα*. A benefit of this strategy is the possibility to easily distinguish the WT and library phage particles in a blue-white PFU assay in agar overlay supplemented with colorimetric substrate X-gal. Here, phage particles that displayed a library of peptides and transduced *LacZα*, a reporter, produce PFU of blue color, while WT phage, which displayed no peptides and transduced no reporter, produce PFU of white color. Manifestation of regioselective reactions is different capture of blue and white phage particles: The "white PFU" phage particles that do not display any peptides should not be modified in the regioselective chemical reactions. The white PFU should not decrease after biotin capture. In contrast, the capture of the blue-colored particles indicates the percentage of regioselective modification (with a biotin tag) that occurs on phage-displayed peptides.

Protocol 1: Measurement of phage concentration by titer (PFUs) in agar overlay

Materials

1. *E. coli* ER2738 host strain: tet-plate. Streak an LB-tetracycline plate with *E. coli* from supplied culture. Invert and incubate the plate at 37 °C overnight. Wrap the plate with parafilm and store at 4 °C in the dark for a maximum of 1 month.
2. Lysogeny broth (LB): 10 g/L tryptone, 5 g/L yeast extract, 5 g/L NaCl. Autoclave the medium and store it at room temperature.
3. Falcon® 14 mL round bottom high-clarity polypropylene test tube, graduated, with snap cap (corning product number 352059) or similar tube (purpose: overnight culture).
4. *E. coli* ER2738 host strain: overnight culture. Inoculate 3 mL of LB in a 14-mL round bottom polypropylene tube with a colony of *E. coli* ER2738 from a tet-plate. Shake the tube at 37 °C, 250–300 rpm, overnight.
5. Top agar for titering: 10 g/L tryptone, 5 g/L yeast extract, 5 g/L NaCl, 7 g/L agar. Autoclave and dispense into 50–200 mL aliquots. Store solid at room temperature (up to 6 months) and melt in a microwave as needed.
6. LB/IPTG/X-Gal Plate (1 L LB + 15 g/L agar). Autoclave and cool to <70 °C. Add a solution of 50 mg IPTG (isopropyl β-D-1-thiogalactopyranoside) and 40 mg X-Gal (5-bromo-4-chloro-3-indolyl β-D-galactopyranoside) in 1 mL of DMF. Mix well and pour onto a plate. Store plates at 4 °C in the dark.
7. Phage dilution buffer 0.1% (w/v) BSA in phosphate-buffered saline (PBS). Sterile filter and store at 4 °C.
8. Autoclaved 0.5 and 1.5 mL microcentrifuge tube.
9. 5 mL polycarbonate round bottom tubes with snap caps.
10. Filter/barrier pipette tips (do not use non-barrier tips with any phage samples!).
11. Water bath or heat block set to 50 °C. We recommend heat block over water bath to minimize carryover of liquids during the procedure.

12. 0.5–1 L beaker filled with 50–100 mL of bleach solution. We recommend all tips and small tubes that come in contact with phage to be discarded into this beaker.
13. Time required: 1–2 hours.

Procedure

1. PFU assay will detect any M13 phages (clones, libraries, modified, unmodified) that have ability to infect *E. coli* ER2738 host strain. If cPTM impairs infectivity of M13 phage, PFU assay will not detect such "non-infective" or "dead" phage. Still, such detection may be detected by qPCR (see Protocol 2). "Dead"/"non-infective" phage, if they are intact, are suitable for panning, PCR and analysis by NGS, but they should not be used for amplification in *E. coli* host strain. A discrepancy between PFU and qPCR assay can be used to assess the fraction of infective vs. non-infective types of phages after the reaction.
2. Throughput: n = 30–60 instances of titering can be performed per day. We recommend performing no more than 10 titering procedures in parallel. Every titer requires one 0.5 mL reaction tube with 190 μL of overnight culture and one X-Gal/IPTG plate and one tube of molten top agar.
3. Melt top agar in microwave. Use serological pipette to dispense ~3 mL of molten agar into the (n + 2) 5 mL polycarbonate tubes (n = number of samples) and keep the tubes in 50 °C bath or heat block for no more than one day. Agar that remains molten at 50 °C for >24 hours does not form functional overlay culture and should be discarded.
4. Distribute 190 μL of overnight *E. coli* ER2738 culture to 0.5 mL tubes (prepare n + 2 tubes). These tubes are ok for up to two days at room temperature but we recommend preparing them fresh daily.
5. Label IPTG/X-Gal plates and, if they are cold, let them adjust to room temperature (prepare n + 2 plates). Cold plates can be used as well; they form perfectly normal agar overlay, but they require agility because top agar solidifies on these plates very fast.
6. Dilute a phage sample to ~10^4 PFU/mL concentration. If concentration is unknown, perform a serial 1 : 100 dilution by sampling 10 μL and mixing it with 990 μL of phage dilution buffer.
7. Dispense 10 μL of diluted phage sample into a tube that contains 190 μL of overnight culture (inoculation of bacteria). Vortex briefly and place the tube on top of the designated agar plate. Wait for 1–2 minutes or, if other

(Continued)

(Continued)

phage samples are available, inoculate 5–10 other overnight culture tubes with 5–10 other phage samples. All inoculated tubes should be processed within 5–7 minutes of inoculation.

8. [Approximately 2–5 minutes later] Aspirate 200 μL of inoculated phage + bacteria mixture and dispense 200 μL of inoculum into tube with 3 mL molten Agar. Vortex the tube briefly, uncap, and immediately invert the tube onto the X-gal/IPTG plate. Immediately after pouring, gently tilt/rock the plate for 2–3 seconds to spread the agar. Close the plate lid and let the plate solidify for 15 minutes. Note: we recommend practicing this step and setting up an ergonomic arrangement of all the necessary tools on the bench for this step: a comfortable mechanical dexterity in the inoculation and agar overlay steps can greatly increase the throughput of titering and robustness of all downstream analyses.
9. Repeat Step 8 for all inoculated tubes. Repeat Steps 8 and 9 for all phage samples (batch of 5-10 per Step 8).
10. [Optional but highly recommended] Mix a leftover, non-inoculated overnight bacteria sample with agar and pour on the plate. This is a "negative control," which must remain blank if all solutions are sterile and free of phage contaminants. Use the other overnight bacteria, top agar and plate to titer one well-defined reference sample with known phage concentration. This is a "positive control," which will show you whether your reagents are fine and bacteria is capable of being infected by phage.
11. Discard all plasticware and reagent that came in touch with phage into the "bleach beaker"; best practice is to discard them immediately after use. Cover the "bleach beaker" with aluminum foil and let it sit overnight before discarding it. Wipe the area with 10% bleach, then 95 or 70% ethanol and let it dry. Do not discard any diluted phage samples yet: keep them until the next day at 4 °C because some samples may need re-titering at a different dilution.
12. Don't forget to place all plates into the 37 °C incubator within ~1 hour. Plates that remain at room temperature for multiple hours (or overnight) and are then placed into the 37 °C incubator do not form plaques anymore.
13. Next day: examine the positive and negative control plates and count plaques on all plates. Discard the content of the bleach beaker, discard all diluted samples or re-titer samples as needed, and discard the plates. Most contaminations and problems occur when plates, tubes, and waste accumulate over 1–2 weeks.

Protocol 2: Detection of phage by qPCR

Materials

1. Bio-Rad CFX96 qPCR C1000 Thermal cycler (or similar).
2. Universal qPCR mix (2×; contains Taq hot start DNA polymerase, dNTP, SYBR Green, and buffers). Store in dark at 4 °C.
3. 10 μM qPCR primers mix (forward + reverse primers, IDT); store at −20 °C (short term) or −80 °C (long term); thaw before each use.
4. 1.5 mL microcentrifuge tube.
5. PCR reaction tubes or plates.
6. Time required: 2 hours.

Procedure

1. Prepare master mix for *n* + 2 samples (*n* = number of samples) by combining the following components:
   a. qPCR primers mix (10 μM): 3 μL * (*n* + 2)
   b. Nuclease free water: 5 μL * (*n* + 2)
   c. Universal qPCR mix (2×): 10 μL * (*n* + 2)
2. Mix the master mix by pipetting up and down and spin down in a microcentrifuge. Cover with aluminum foil and keep on ice.
3. Aliquot 18 μL of master mix into the 8-strip PCR tubes; use the same tip and electronic repeater pipette or multichannel pipette if available.
4. Add 2 μL of phage samples. Cap the tubes with flat PCR caps, mix by brief vortexing and centrifuge briefly to level the liquid at the bottom of the tube.
5. Start qPCR on the Thermal cycler using the following cycle:

98 °C	3 min	
98 °C	10 s	39×
60 °C	20 s	
72 °C	10 s	
Melting curve point step 65–95 °C		

6. Record the Cq values for each sample and use the difference between Cq values, as is or with calibration curve, to extrapolate the differences in phage concentrations between the samples (see Protocol 4).

Protocol 3: Measurement of capture of biotinylated phage: evaluation by titer

Materials

1. Phage of phage library biotinylated by any of the methods described in this chapter.
2. Wild-type phage that transduces no α-galactosidase as a background control.
3. Reagents needed for titer of phage by agar overlay (see Protocol 1).
4. 10X PBS (pH 7.4): 0.119 M phosphates, 1.37 M sodium chloride, and 0.027 M potassium chloride.
5. Bead blocking solution: 2% (w/v) BSA in PBS. Sterile filter and store at 4 °C.
6. Phage blocking/dilution buffer 0.1% (w/v) BSA in PBS. Sterile filter and store at 4 °C.
7. Streptavidin Magnasphere™ paramagnetic particles (binding capacity: 2 pmol/μL) (Promega, catalog no. Z5481).
8. Magnetic rack (New England Biolabs, catalog no. S1509S).
9. Orbital shaker.
10. 360° rotator.
11. Zeba desalting columns (ThermoFisher Scientific, catalog no. 89882).

Procedure

1. This procedure uses a phage library, or a phage clone previously modified by any biotinylation reaction. Examples of chemical modifications are described in Sections 4.3–4.5. If the solution of biotinylated phage does not contain WT phage (control), we recommend supplementing it to the biotinylated phage solution in ~1 : 1 ratio. Excess of biotinylation reagent and other reagents should be removed using dialysis or Zeba desalting column (see modification protocol). The concentration of the biotinylated phage should be known at least approximately because the phage should be diluted to 10^5–10^6 PFU/mL for the capture procedure described below.
2. Briefly vortex or mix by inversion Promega streptavidin magnetic beads to create a uniform slurry. Transfer 50 μL of this slurry to a 1.5 mL microcentrifuge tube.
3. Add 500 μL of PBS and mix by inverting. Place the tube into the magnetic rack, leave the beads in the rack for 30 seconds, and remove the supernatant.
4. Repeat Step 3 two more times to wash the beads.

5. To the bead pellet in the tube, add 500 μL of 2% BSA in PBS. Mix the beads on a 360° rotator for 1 hour at room temperature. This step blocks unspecific binding. Keep the beads in this blocking solution until biotinylated phage solution is ready.
6. Place the tube with beads in 2% BSA from the previous step on the magnetic rack for 30 seconds. Remove the solution and add 200 μL of biotinylated phage with approximately 10^5–10^6 PFU/mL concentration of phage. Remove the tube from the rack and mix the beads by rotating on a 360° rotator for 30 minutes.
7. During the capture, titer biotinylated phage solution. As the phage concentration is 10^5–10^6 PFU/mL, it might be necessary to perform a 1 : 10 (for 10^5 PFU/mL) or 1 : 100 (for 10^6 PFU/mL) dilution before titering. Combine 10 μL of diluted solution with 190 μL of overnight *E. coli* ER2738 culture and plate it on an agar overlay (see Protocol 1). This measurement produces the titer designated as "before capture (BC)" or PFU^{BC}.
8. After phage capture on streptavidin beads, place the tube on the magnetic rack; capture the beads and titer 10 μL of the supernatant. This measurement produces the titer designated as "after capture" or PFU^{AC}.
9. The next day, count the plaques for each plate. Calculate the biotinylation % capture as:

   % capture = $[(PFU^{BC} - PFC^{AC})/PFU^{BC}] \times 100$. If both *LacZ*α(+) and WT phages are used, count the blue and white plaques separately and determine % capture for both. The capture of the WT should be <10%. Values for all PFU measurements must be in middle-to-high double digits or low triple digits. If two dilutions were plated, use one that yields 50–200 plaques and do not use PFU values obtained from plates that have fewer than 10 plaques or more than 300 plaques because the confidence of the % capture calculated from such observations is low. If both "blue" and "white" plaques are used for analysis, it might be necessary to extrapolate the blue PFU and white PFU from plates formed at two different dilutions of the phage inoculum.
10. You can measure the confidence of capture value by acquiring multiple replicates of PFU^{BC} and PFU^{AC}. For example, sampling three instances of 10 μL of the solution in Step 7 and three instances of 10 μL of supernatant in Step 8 and titering them separately yields three values of PFU^{BC} and PFU^{AC}, which can be used to calculate the average and standard deviation (stdev) of PFU, average capture and a stdev of capture propagated from the stdev of PFU values. The PFU values that deviate significantly from 50 to 200 plaques per plate may yield 20% or higher variance for titering; if PFU^{BC} and PFU^{AC} samples have low counts and high variability, the propagated variance of capture may reach 50% or more, and the confidence of the conclusions from such observations needs to be treated accordingly.

Protocol 4: Capture of biotinylated phage: evaluation by qPCR

Materials

1. Phage or phage library biotinylated by any of the methods described in this chapter.
2. Bead blocking solution: 2% (w/v) BSA in PBS. Sterile filter and store at 4 °C.
3. Phage blocking/dilution buffer 0.1% (w/v) BSA in PBS. Sterile filter and store at 4 °C.
4. Streptavidin Magnasphere™ paramagnetic particles (binding capacity: 2 pmol/μL) (Promega, catalog no. Z5481).
5. Magnetic rack (New England Biolabs, catalog no. S1509S).
6. Orbital shaker.
7. 360° rotator.
8. Zeba desalting columns (ThermoFisher Scientific, catalog no. 89882).
9. Bio-Rad CFX96 qPCR C1000 Thermal cycler (or similar).
10. Nuclease free water.
11. Autoclaved 1.5 mL microcentrifuge tube, 8-strip PCR tubes and flat PCR caps.

Procedure

1. As in previous procedure, this procedure uses a phage library, or a phage clone previously modified by any biotinylation reaction. The concentration of the biotinylated phage should be known at least approximately and the phage should be diluted to 10^7–10^9 PFU/mL for the capture procedure described below.
2. More than in Protocol 3 (capture + titering), it is very important in this protocol to remove the excess of biotinylation reagent from phage using dialysis or Zeba desalting column (see modification protocol). Any biotin-containing reagent will saturate biotin-binding sites on Streptavidin Magnasphere™ paramagnetic particles and prevent binding of biotinylated phage. See further discussion in "**Binding capacity and influence of contaminants.**"
3. Briefly vortex or mix by inversion a bottle of Promega streptavidin magnetic beads to create a uniform slurry. Use pipette to transfer 50 μL of this slurry to a 1.5 mL microcentrifuge tube.
4. Add 500 μL of PBS and invert the tube to mix the beads with PBS. Place the tube into the magnetic rack and leave the beads in the rack for

30 seconds and remove the supernatant using either a pipette or a vacuum aspirator.

5. Repeat Step 3 two more times to wash the beads.
6. To the bead-pellet in the tube, add 500 μL of 2% BSA in PBS. Mix the beads on a 360° rotator for 1 hour at room temperature. This step blocks the beads with 2% BSA. Keep the beads in this blocking solution until biotinylated phage solution is ready. Before adding the biotinylated phage solution to the beads, transfer 10 μL of biotinylated phage solution to a separate tube. This is your BC sample.
7. Place the tube with beads in 2% BSA from the previous step on the magnetic rack for 30 seconds. Remove the solution and add 200 μL of biotinylated phage with approximately 10^5–10^6 PFU/mL concentration of phage. Remove the tube from the rack and mix the beads by rotating on a 360° rotator for 30 minutes.
8. After phage capture on streptavidin beads, place the tube on the magnetic rack and wait 30 seconds. Transfer the supernatant to a clean tube. This is your after capture (AC) sample.
9. Determine the Cq value of the BC and AC samples in triplicates using qPCR, following Protocol 2. Keep samples on ice. Always leave one tube for negative control (no DNA, water alone).
10. The difference between Cq values of BC and AC samples ($\Delta Cq = {}^{BC}Cq - {}^{AC}Cq$) are used to determine the difference in DNA concentration caused by streptavidin capture of biotinylated phages. You can estimate the capture rate as $2^{(-\Delta Cq)}$ with $\Delta Cq = {}^{BC}Cq - {}^{AC}Cq$.
11. Base of "2" in the previous step assumes ideal PCR conditions. We recommend running a 10-fold dilution (5–8 dilutions) for one phage sample with $>10^{11}$ PFU/mL concentration. The result will be used as a calibration curve to convert Cq to DNA concentration and such conversion may improve the calculation of the capture rate.

Comparison of biotin capture by PFU versus qPCR: pro and con

Detection considerations: Reliable detection of phage by qPCR occurs at $>10^4$ particles per reaction (e.g. 2 μL of 10^7 PFU/mL solution used as template for qPCR). Dilution of phage below 10^7 PFU/mL is possible but it requires extra skill and caution to reliably detect differences by qPCR. Reliable detection of phage by titer occurs at ~10^2 plaques per well (e.g. 10 μL of 10^4 PFU/mL solution). Hence, dilution of phage to 10^4 PFU/mL yields

(Continued)

(Continued)

an optimal detection by PFU. Less is better for PFU. As samples for PFU assays are significantly more dilute, the assays are not sensitive to presence of residual biotinylating reagents.

Time: qPCR requires 20–30 minutes set up and 1 hour run time. PFU assay is developed overnight.

Multiplexing: up to 96 qPCR reactions can be set up relatively easily. Analysis of 1, 10, or 100 qPCR requires only minor increase in effort. Analysis of plates for PFU assay scales linearly. Reasonable recommended limit is 20 plates per day.

Binding capacity and influence of contaminants: Biotin-capture protocols in this chapter employ 50 μL of Streptavidin Magnasphere™ paramagnetic particles (binding capacity: 2 pmol/μL). This amount can capture 100 pmol of biotinylated molecules (10^{-10} mole or 6×10^{13} molecules). If only biotinylated phage particles are present in the capture solution, in principle 50 μL of beads can capture up to 10^{12} phage particles (in practice, this amount is always several orders lower and rarely exceeds 10^{9} particles). However, in addition to biotinylated phage, the solution *always* contains other biotin-containing entities (e.g. reagents used for modification of phage). These residual reagents will block biotin-binding sites and prevent binding of biotinylated phage. For illustration: a typical biotinylation reaction may employ 10^{12} PFU/mL of phage and 1 mM biotinylation reagent. After reaction, the solutions may be diluted as following:

A. Factor of 10^{3}, phage 10^{9} PFU/mL, biotinylation reagent 1 μM, 200 nmoles in 200 μL.
B. Factor of 10^{5}, phage 10^{7} PFU/mL, biotinylation reagent 10 nM, 2 nmoles in 200 μL.
C. Factor of 10^{7}, phage 10^{5} PFU/mL, biotinylation reagent 0.1 nM, 20 pmoles in 200 μL.

Note that in Dilution C, the amount of residual amount of biotinylation reagent is below the capacity of beads, but both Dilution A and Dilution B contain the residual amount of biotinylated reagents that exceeds the binding capacity (100 pmol) by factor of 20 in B and 2000 in A. If Dilution A or B is combined with streptavidin beads, biotinylated phage will not be captured. If reaction mixture cannot not be sufficiently diluted, biotinylation reagent must be thoroughly removed by Zeba desalting column or, in some cases, overnight dialysis. Zeba desalting column is used ubiquitously throughout this chapter but we recommend carefully reading the limitations of the desalting columns;

we have observed that Zeba desalting column can leave 10–30% of the biotinylating reagents in the flow-through and even multiple Zeba column desalting steps can leave behind enough of biotinylating reagent to inhibit the downstream capture. Processing of such samples will lead a false negative conclusion: "*no apparent biotinylation in the sample*".

Comparison to other techniques:

Other methods have been reported for measuring the yield of cPTM on clonal M13 phage or phage libraries. They can be classified as library-centric and clone-centric technologies. Noren and co-workers [18] and later Schultz and co-workers used Western blot techniques to detect incorporation of biotin into pIII protein [2]. Ng et al. employed SDS-PAGE to detect ligation of fluorescein-hydrazine to a library of aldehyde-peptides on pIII protein [19]. Lima et al. employed SDS-PAGE to detect enzymatic ligation of multi-domain proteins to pIII and pVIII proteins [20]. Limit of detection of direct SDS-PAGE is 1–10 billion M13 particles, whereas Western blot in principle can improve the limit of detection by 3–5 orders of magnitude. Both technologies employ densitometry and are best suitable for qualitative "yes/no" assessments. Our lab attempted Western blot techniques, and we found that quantification of yields relies on a nontrivial skill in running quantitative Western blot. Due to prolonged preparation of blots, the scalability of such approach is limited. We found SDS-PAGE suitable as an occasional final "confirmation" of the success of cPTM; but the runtime and limited scalability of SDS-PAGE make it unsuitable for optimization of cPTM.

Various forms of mass spectrometry (MS) can be used to perform clone-centric analyses of reactions on M13 phage. One must always remember the limitation of extrapolating the yield of reaction on one clone to the yield of modification of the entire library. The rate of cPTM depends on the sequence of peptide, and such rates can vary by factor of 100–1000 [21]. In several publications, we observed that the yields determined on clonal peptides are always higher than yields on peptide libraries [17,19,22–25]. Nevertheless, the near atom-scale resolution of MS technique makes it attractive for monitoring of chemical modification. The end-user has to strike a balance between conclusions drawn from modification of clone (precision) and library (relevance). Winter, Heinis, and co-workers employed electrospray ionization mass spectrometry (ESI-MS) of pIII protein to monitor alkylation of Cys residues [26]. Such modification required purification of pIII protein in bacteria and it served only as a proxy for

(Continued)

(Continued)

chemical modification on intact M13 phage. A single publication from Heinis and co-workers [27] and then a publication by Mayer and co-workers [28] used ESI-MS to characterize modification of peptides cleaved from pIII protein in clonal phage. Due to low amounts of cleaved peptides, such an approach has been difficult to generalize (Heinis, personal communication); ESI-MS analysis of peptides cleaved from pIII protein was unsuccessful in our hands. Sojitra et al. [29] and Lin et al. employed MALDI to monitor chemical glycosylation of pVIII proteins on intact M13 phage. Such modification, unfortunately, works only for a sufficiently large display copy number (>20 copies of chemical modification per virion).

4.3 N-terminal Oxidation and Functionalization to Incorporate Unnatural Fragments

Chemical post-translational modifications of phage libraries by the incorporation of pharmacophores through regioselective ligation to N-terminal aldehydes have been used in seven publications to date [21,23,24,30–33], whereas general modification of N-terminal aldehydes on proteins has been used by many groups and have been a topic of a recent review [34]. This modification requires phage-displayed peptide libraries bearing N-terminal serine or threonine residues. The method starts from the generation of an N-terminal aldehyde through the oxidation of N-terminal serine or threonine with sodium periodate to generate an N-terminal oxaloyl [19]. N-terminal Ser/Thr is not found in the native coat protein of M13 phage, and therefore, Ser/Thr-derived aldehyde serves as a unique site for introducing a broad range of chemical modifications [23,35]. Presence of aldehyde can be detected by modification with aminooxybiotin (AOB) and subsequent capture of biotinylated phage (Figure 4.6a). Sodium periodate is a strong oxidant, and hence, control of the oxidation time and concentration of $NaIO_4$ is important to minimize detrimental effect of $NaIO_4$ on phage (Figure 4.6b–c). For example, optimal concentration of periodate is 100 μM, whereas viability of phage decreases by 50% when 500 μM periodate is used. For the same reasons, oxidation reaction must be stopped by a quencher, such as glutathione.

The generated N-terminal oxaloyl is a versatile starting material for many reactions that modify aldehydes. Hydroxylamine pharmacophores (R-ONH_2) in the presence of anilinium acetate catalyst can modify phage but they can also be detrimental to phage viability if concentration of reagent is too high (Figure 4.6b–c). For example, 1 mM AOB modifies 80% of phage particles and has only modest impact on viability of phage. Increasing concentration to 5 mM AOB increases the yield of

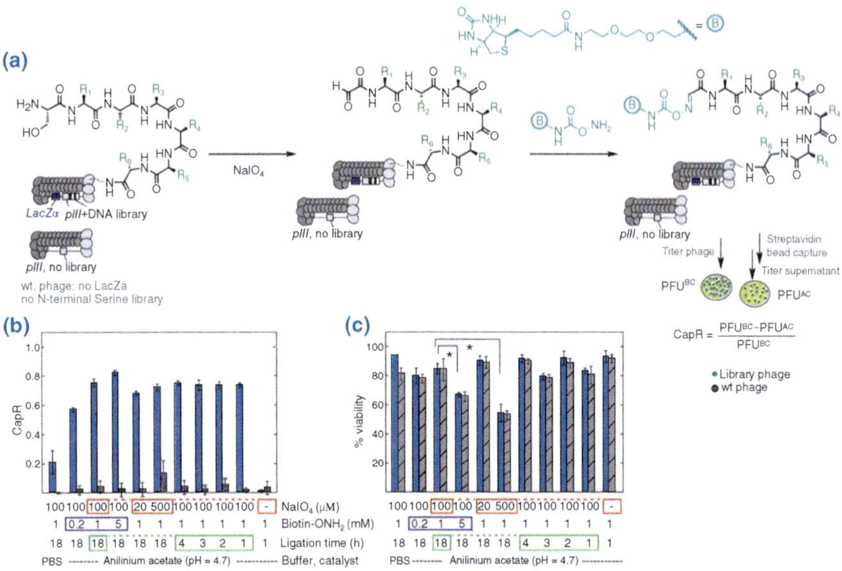

Figure 4.6 Modification of phage-displayed peptide library. (a) The number of available aldehyde groups for oxime-ligation was determined using aminooxy biotin (AOB) derivative that grafted biotins onto phage-displayed peptide libraries. WT phage that does not code for *LacZα* gene or display any peptides acted as a negative control to ensure the regioselectivity of the reaction. Results describe how capture rate (CapR) of both library (blue) and WT phage (gray) (b) and the infectivity of these phage particles (c) change in different reaction conditions. Note that unnecessarily high, 500 μM concentration of periodate yields non-specific modification of WT phage and a decrease in infectivity of phage. *Source*: (b, c) Adapted from ACS Chem. Biol., 2012, 7 (9), 1482–1487 Copyright © 2012 American Chemical Society. [19]

modification by a few percent but significantly decreases the viability of the phage (Figure 4.6b–c). Once modification of phage by AOB is characterized, the same bond-forming reaction, oxime ligation, can incorporate other R-ONH_2 moieties (Figure 4.7).

The efficiency of incorporation of hydroxylamine pharmacophores (R-ONH_2), for example, glycan with anomeric hydroxylamine group (Figure 4.7), is monitored by "pulse-chase" approach. First, glycan hydroxylamine is added to the aldehyde phage (pulse), and after 60 minutes, AOB is added to the same phage pool/sample to cap all unreacted aldehyde groups (chase) (Figure 4.7). Other reactions such as 2-amino benzamidoxime (ABAO) modification [32] and Wittig reactions to incorporate C—C bonds with pharmacophores bearing stabilized ylides [21,24] are not covered in this chapter, but they were developed and monitored analogously. The key monitoring technique in all cases was either biotin capture or biotin pulse-chase approach (Figure 4.7).

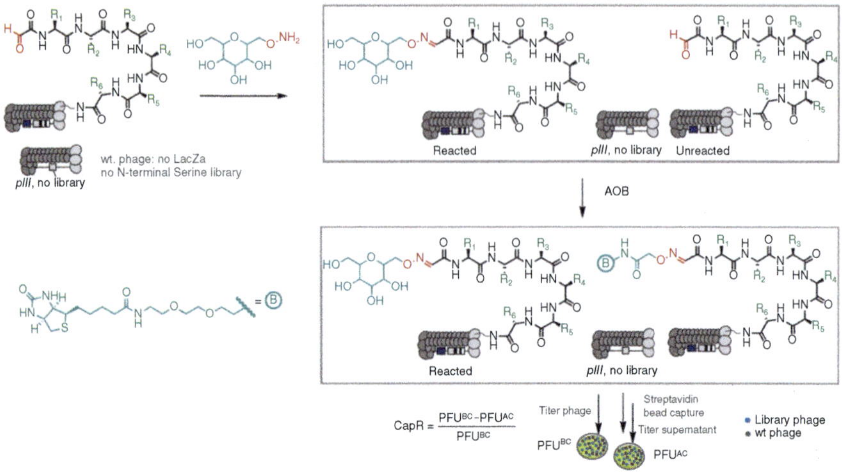

Figure 4.7 Functionalizing phage-displayed peptide libraries that display N-terminal oxaloyl group with hydroxylamine derivatives as monitored by "pulse-chase" biotinylation. Oxime-ligation with carbohydrate derivatives ("pulse") in the presence of WT phage as an internal control. Following the reaction, the unreacted oxaloyl groups can be monitored using aldehyde reactive biotin to determine the modification ratio.

Protocol 5: Modification of phage with N-terminal serine using oxidation and oxime ligation

Materials

1. Phage libraries with N-terminal Ser or Thr, WT phage as a background control.
2. Reagents needed for titering of phage (see Protocol 1).
3. Reagent needed for capture of biotinylated phage (see Protocol 3).
4. Blocking solution: 0.1% (w/v) BSA in PBS. Store at 4 °C.
5. Sodium Periodate (100× in Milli-Q water, MW 213.89): 1.28 mg/mL = 6 mM.
6. Glutathione (100× in Milli-Q water, MW 307.32) 15.4 mg/mL = 50 mM.
7. Hydroxylamine derivatives (e.g. mannose).
8. Biotin capture reagent AOB (aminooxy biotin), *N*-(aminooxyacetyl)-*N*′-(D-biotinoyl) hydrazine, trifluoroacetic acid salt (2 mM, 2× in 200 mM anilinium acetate buffer, pH 4.7). Store at −20 °C.
9. Zeba desalting columns (ThermoFisher Scientific, catalog no. 89882).
10. Phosphate-buffered saline 10× PBS (pH 7.4): 0.119 M phosphates, 1.37 M sodium chloride, and 0.027 M potassium chloride.

Procedure

1. Prepare a fresh 6 mM sodium periodate solution in MQ water. Solution may be stored at −20 °C but if result become suboptimal, we recommend making a fresh solution.
2. If phage libraries are stored in 20–50% glycerol, the glycerol must be removed by dialysis prior to modification because glycerol will react with $NaIO_4$. We recommend dialysis against PBS with 2–3 changes of buffer. Note that Zeba desalting column or similar columns cannot remove all glycerol.
3. Transfer 99 μL of a phage library at a concentration between 10^{12} and 10^{13} PFU/mL in PBS to 1.5 mL microcentrifuge tube. Add 1 μL of 6 mM sodium periodate solution, vortex the tube briefly and incubate it on ice for 5 minutes.
4. Add glutathione (3 μL of 50 mM in water) into the oxidized phage solution to quench the reaction. Vortex the tube briefly and incubate it at room temperature for 10 minutes.
5. The oxidation step is complete. You have a stock of oxidized phage library (~100 μL).
6. Transfer 5 μL of the oxidized phage solution into a 0.5-mL tube, for quantification of the fraction of aldehyde-terminated phages. Add AOB (5 μL, 2 mM) to the tube containing the phage solution (5 μL). Vortex the tube briefly and incubate it at room temperature for 1 hour.
7. Quench the reaction by adding 90 μL of blocking solution and immediately perform several serial dilutions by combining 10 μL of the solution + 990 μL of blocking solution to yield the 10^5 PFU/mL solution of biotinylated phage for biotin capture. For example, if the initial concentration of phage in Step 3 was 10^{12} PFU/mL, the concentration is 10^{11} PFU/mL in the quenched reaction, and it is 10^9, 10^7, and 10^5 PFU/mL in the first, second and third serial dilution. The third serial dilution, in this case, can be use directly as an input for biotin capture (see Protocol 3).
8. The rest of the library from Step 5 can be modified with any pharmacophore with a hydroxylamine functional group. Add $R\text{-}ONH_2$ (95 μL, 2 mM, 2×) into the tube containing the oxidized phage solution (95 μL). Vortex the tube briefly and incubate it at room temperature for 1 hour.
9. Quench the reaction by passing it through Zeba desalting column equilibrated with PBS pH 7.4. Perform 1–3 rounds of cleanup with Zeba desalting column, each time with freshly equilibrated column, to ensure proper removal of any traces of aniline, acetic acid, $R\text{-}ONH_2$, $NaIO_4$, and GSH.

(Continued)

(Continued)

The oxime ligation step is complete! You have a stock of R-modified phage library (~200 μL) in PBS.

10. Transfer 5 μL of the modified phage solution from Step 9, into a 0.5-mL tube. This sample will be used for quantification of the amount aldehyde-terminated phages that were not labelled with R-ONH_2. Add AOB (5 μL, 2 mM, 2×) into the tube containing the phage solution (5 μL). Vortex the tube briefly and incubate it at room temperature for 1 hour.
11. Quench the reaction by adding 90 μL of blocking solution and immediately perform several serial dilution by combining 10 μL of the solution + 990 μL of blocking solution to yield the 10^5 PFU/mL solution of biotinylated phage for biotin capture. See Step 7 for further details.
12. Two biotin capture procedures using solution from Step 7 (after oxidation but before any modification) and from Step 14 (after oxidation and modification by R-ONH_2) enable the determination of the modification percentage. For example, if solution from Step 7 shows 80% capture and solution from Step 11 shows 10% capture, the library is 80−10 = 70% modified with R-ONH_2.

4.4 Macrocyclization of Phage-displayed Peptides Bearing Cysteines

Cysteines are the most nucleophilic amino acid residues. The pKa of cysteine residue is 8–9 (depending on the environment); cysteines thus can react with electrophilic residues under neutral or mildly basic conditions. Cysteines on phage-displayed peptide libraries, thus, can undergo nucleophilic substitution reactions under biocompatible conditions in aqueous media. The procedure described in this chapter will require a library that bears two cysteines in the peptide sequence, for example, the NEB Ph.D.C7C library. This type of phage-displayed library is produced by bacteria in an oxidized (disulfide) form due to an oxidizing environment of bacterial periplasm where the assembly of phage takes place. Therefore, a reducing agent (e.g. TCEP) must be used to convert the disulfide bonds to generate the free sulfhydryls (thiols) (Figure 4.8). Following the disulfide reduction, the available free thiols can be quantified through nucleophilic substitution reaction with iodoacetamide analog of biotin, followed by biotin capture assay to monitor the percentage of modification (Figure 4.8).

This section describes one-step macrocyclization reaction of phage-displayed peptides that contain two cysteines to incorporate an unnatural cross-linker 1,3-dichloroacetone (DCA); or DCA-derived cross-linker that bears a pre-positioned

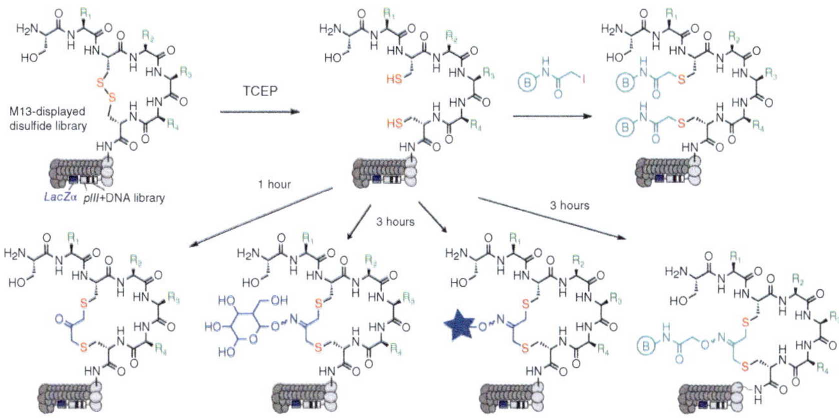

Figure 4.8 Disulfide reduction and modification of phage-displayed peptide libraries with dichloroacetone and dichlorooxime reagents. Pre-forming the dichlorooxime complex allowed the grafting of biotin, mannose, and other glycans, as well as other pharmacophores, including covalent warheads and biotin into phage-displayed peptide libraries. *Source*: Ref. [36]/Springer Nature.

unnatural pharmacophore such a glycan oxime (DCO-glycan). This cPTM is based on the DCA fragment reported by Dawson and co-workers [37]. Monitoring the yield of such reaction is possible using capture of biotinylated phage. Specifically, BIA or dichloroacetone-derived oxime (DCO) with biotin handle (DCO-biotin) reacts selectively with Cys in displayed peptides and the number of phage particles that are modified by biotin can be easily monitored by measuring the number of phage particles before and after their exposure to streptavidin agarose (Figure 4.9).

DCA ligation was published in libraries bearing small ring disulfides such as $SXCX_3C$ and $SXCX_4C$ displayed on M13 phage (X represents 19 any natural amino acids that are not Cys) [22], and it works in libraries bearing medium [36] and large rings (e.g. $CX_{12}C$). The same reaction can be used for installation of various pharmacophores through DCO-ligation, where the unnatural chemical fragments are pre-installed onto the DCA linchpin. To ensure the regioselectivity of these reactions, the library phage (containing the *LacZα* gene and appearing as blue-colored plaques) should be mixed with WT phage particles that do not display any peptides (do not express the *LacZα* gene and appear as white-colored plaques).

The cysteine alkylation reaction takes place at a mildly basic pH (pH 8.5) in the presence of a reducing agent such as TCEP, to convert disulfides to free thiols. A biotin derivative of DCA has been used to measure the rate of this reaction on a phage-displayed peptide library (Figure 4.9), whereas the blue-white phage assay confirms the regioselectivity of the reaction. The negligible capture ratio in the absence of TCEP confirms that DCA reacts with reduced cysteines and not

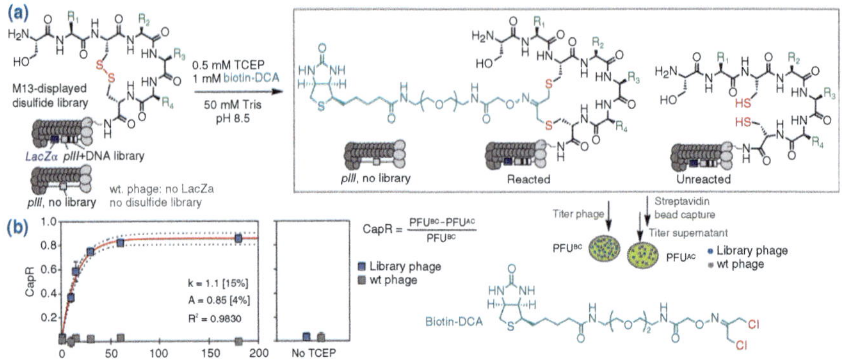

Figure 4.9 Kinetics of macrocyclization of phage-displayed peptide library. (a) The macrocyclization reaction efficiency was determined using a biotin-DCO complex that allowed to graft biotins onto phage-displayed peptide libraries. WT phage that does not code for *LacZα* gene or display any peptides acted as a negative control to ensure the regioselectivity of the reaction. (b) Kinetics of the macrocyclization reaction. The macrocyclization only takes place in the presence of TCEP as a reducing agent, confirming the selectivity toward cysteines rather than other nucleophilic residues. *Source*: Reproduced from [22]/with permission of Royal Society of Chemistry.

with other nucleophiles present in phage-displayed peptides such as N-terminal amines. Lack of reaction of DCA with WT phage confirms the excellent regioselectivity toward solvent-accessible disulfides on pIII-displayed peptides as opposed to buried disulfides inside of the pIII protein [22]. An 80% of the phage population retained viability following the 3-hour long cysteine alkylation reaction further confirming minimal crosslinking to proteins related to phage infectivity (Figure 4.9b). To summarize, modification with DCO occurs only when two conditions are satisfied: (1) disulfide displayed on the N-terminus of pIII protein is solvent-accessible [33,38]; and (2) free thiol is available through TCEP reduction.

The "pulse-chase" approach described in the previous protocol can be used to monitor the reaction between the phage-displayed peptides and DCA derivatives other than biotin (Figure 4.10). In this case, a commonly used biotinylating reagent, BIA, which alkylates cysteines, is used as the "chase" reagent. Following a DCA-based alkylation reaction on phage-displayed peptides, an aliquot of the library-WT phage mixture can be subjected to a reaction with BIA. Any unreacted cysteines will react with BIA through cysteine alkylation, grafting biotin moieties onto the phage-displayed peptides. These phage particles bearing biotins can then be captured with streptavidin magnetic beads to delineate the library that is modified with the desired DCA derivatives from libraries that bear unreacted cysteines. The WT (white) particles, biotin-DCO or BIA, act as the "negative control" for the reactivity and capture.

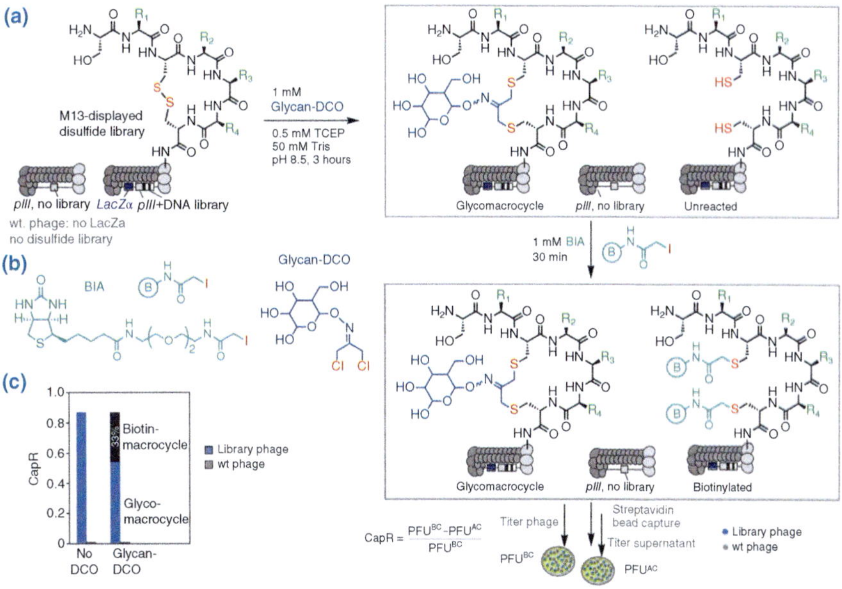

Figure 4.10 Macrocyclization of phage-displayed peptide libraries with DCO derivatives. (a) Macrocyclization with carbohydrate-DCO derivatives at room temperature for 3 hours, in the presence of WT phage as an internal control. Following the reaction, the unreacted cysteines can be monitored by using biotin-iodoacetamide (BIA) to determine the modification ratio. (b) Reagents used for "pulse-chase" monitoring described in (c). *Source*: Reproduced from [22]/with permission of Royal Society of Chemistry.

Protocol 6: Modification of phage with two Cys residues using DCO ligation

Materials

1. Phage libraries with two cysteine residues abbreviated as CXnC; WT phage as a background control.
2. Reagents needed for titering of phage (see Protocol 1).
3. Reagent needed for capture of biotinylated phage (see Protocol 3).
4. Tris(2-carboxyethyl)phosphine hydrochloride (TCEP-HCl) (prepare a fresh 100 mM solution in sterile water).
5. One of biotin chase reagents: (1) EZ-Link™ Iodoacetyl-PEG2-Biotin (ThermoFisher Scientific, catalog no. 21334). Prepare a solution of 100 mM in sterile water. The solution is abbreviated as BIA or (2) *O*-(2-oxo-2-(2-(5-((3a*S*,4*S*,6a*R*)-2-oxohexahydro-1*H*-thieno[3,4-*d*]imidazol-4-yl)pentanoyl) hydrazineyl)ethyl)hydroxylammonium. Prepare a solution of 100 mM in sterile water. This solution is abbreviated as biotin-DCO.
6. DCO-modified target reagent (e.g. DCA alone, DCO-biotin, DCO-mannose).

(Continued)

(Continued)

Procedure

1. Prepare a mixture of libraries and WT phage (CX_nC:WT) in 1:1 ratio to yield a final titer of 2×10^{11} pfu mL^{-1}. Adjust the solution's pH to 8.5 using Tris buffer (1 M in sterile water).
2. To 99 μL of phage solution in pH 8.5 Tris buffer, add TCEP (1 μL, 100 mM) and DCO-derivatives (1 μL, 100 mM in water). Incubate the reaction mixture for 3 hours at room temperature.
3. During this incubation period, equilibrate a Zeba desalting column with PBS solution (wash with PBS 3× with centrifugation at 1500×g for 3 minutes). Remove the last wash by centrifugation and decant the residual buffer in the collection tube.
4. Add the reaction mixture to a Zeba desalting column, purify and desalt by centrifugation (1500×g, 3 minutes). Collect the flowthrough into a clean tube. Repeat this Step 2 more times with freshly equilibrated Zeba desalting column. If you modified the phage with DCO-biotin at this stage, follow instructions to perform biotin capture (see Protocol 3).
5. Set up a biotinylation reaction with BIA. Take 10 μL of the reaction mixture and add 10 μL of Tris buffer (pH 8.5). Top the reaction up to 99 μL with 79 μL of water and add 1 μL of BIA. Incubate the reaction mixture for 30 minutes at room temperature.
6. Dilute the phage mixture until the approximate titer is 10^8 PFU/mL. That is, take 10 μL of the reaction mixture and mix it with 990 μL PBS. Repeat this process two more times to get a dilution factor of 10^6.
7. Mix 200 μL of the diluted mixture with 50 μL of washed and blocked magnetic streptavidin beads and perform biotin capture (see Protocol 3).

4.5 Macrocyclization and Installation of Warheads in Phage-displayed Libraries

This section focuses on forming macrocycles on phage-displayed peptide libraries by cross-linking cysteines with a reactive linchpin while providing a reactive handle for further functionalization. Similar to the DCA reaction described in the previous protocol, this procedure utilizes an orthogonal cysteine alkylating reagent with two ketone groups. The goal is to create a shelf-stable macrocyclic intermediate for the divergent synthesis of pharmacophore-bearing phage libraries [17]. The TCEP reduction exposes solvent-accessible free cysteines for alkylation with 1,5-dichloro-2,4-pentadione (DPD). The cysteine alkylation grafts a 1,3-diketone reactive chemical handle providing a substrate for a Knorr pyrazole reaction. The approach grafts 1,3-diketones on disulfide-bearing peptides displayed on phage. Reduced cysteine residues irreversibly react with 1,5-dichloro-2,4-pentadion linchpins to incorporate the shelf-stable, biorthogonal 1,3-diketone handles. These reactive handles can then be further functionalized with hydrazine-bearing pharmacophores via well-established Knorr pyrazole synthesis reactions to incorporate various pharmacophores and unnatural fragments onto these macrocyclic libraries (Figure 4.11). This method is a late-stage functionalization strategy in contrast to other techniques described in Section 4.5, where the functionalization co-occurs with macrocyclization. Separating functionalization and macrocyclization is beneficial because macrocyclization reactions are not trivial, and robust reactions incorporating reactive handles after the macrocyclization reactions (i.e. late-stage functionalization) are preferred.

We utilized various hydrazine derivatives to graft different functionalities onto the macrocycles at a mildly acidic pH (pH 4.5) in an ammonium acetate buffer. Most hydrazine derivatives exhibited high toxicity toward the phage particles, decreasing their infectivity by over 99.99% after only a few minutes of

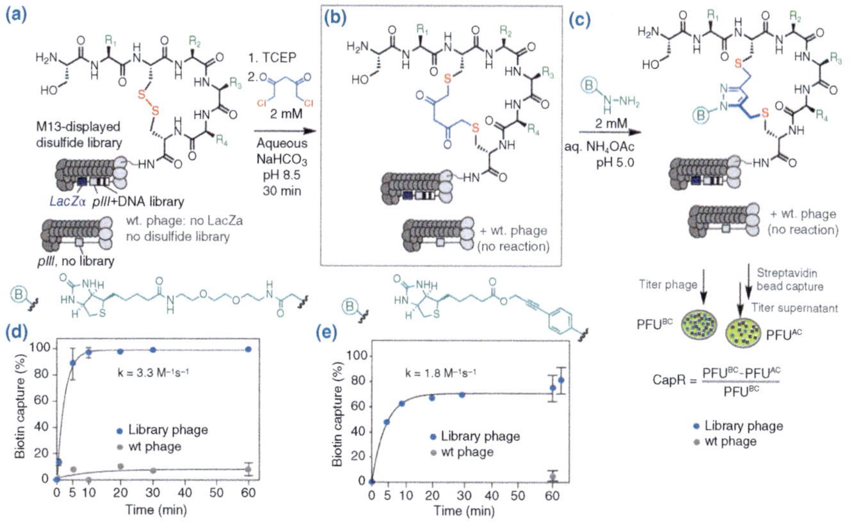

Figure 4.11 Macrocyclization of genetically encoded phage libraries. (a–b) M13-phage-displayed disulfide library is reduced by TCEP to yield reactive thiols and then reacted with DPD to form 1,3-diketone bearing macrocyclic library. (c) The 1,3-diketone linchpin bearing macrocycles were further functionalized with hydrazines e.g. biotin hydrazine. A biotin capture experiment was used to measure the conversion of the 1,3-diketone. (d) The rate of conversion of the M13 library bearing 1,3-diketone was calculated using capture of biotinylated phage clones, using alkyl biotin hydrazine reagent. (e) The rate of conversion of the M13 library bearing 1,3-diketone was calculated using capture of biotinylated phage clones, using phenyl biotin hydrazine reagent. *Source*: Adapted from [17]/American Chemical Society.

exposure. We were able to mitigate this toxicity by adding 5 mM EDTA to the reaction mixture to chelate bivalent metal cations that have been reported to mediate the toxicity of hydrazine derivatives [17]. Incorporating EDTA to the reaction mixture enabled the second step of the cPTM sequence with minimal toxicity effects toward the infectivity of phage.

The macrocyclization reaction through cysteine alkylation can be monitored through a biotin capture reaction with a hydrazine functionalized biotin. One can determine the kinetics of Knorr pyrazole reaction on phage using various biotin hydrazines. Such measurements show that aliphatic hydrazine reacts faster than aromatic and yielded higher overall conversion (Figure 4.11d,e). When grafting unnatural fragments other than biotin, a combination of direct biotinylation and "pulse-chase" biotinylation can confirm the success of both macrocyclization and Knorr pyrazole ligation of unnatural fragments. In "pulse-chase" approach the hydrazines act as the "pulse" followed by a "chase" with biotin hydrazine to quantify any unreacted diketone fragments (Figure 4.12). For example, modification of a disulfide library of $SXCX_3C$ (where X is any amino acid other than cysteine)

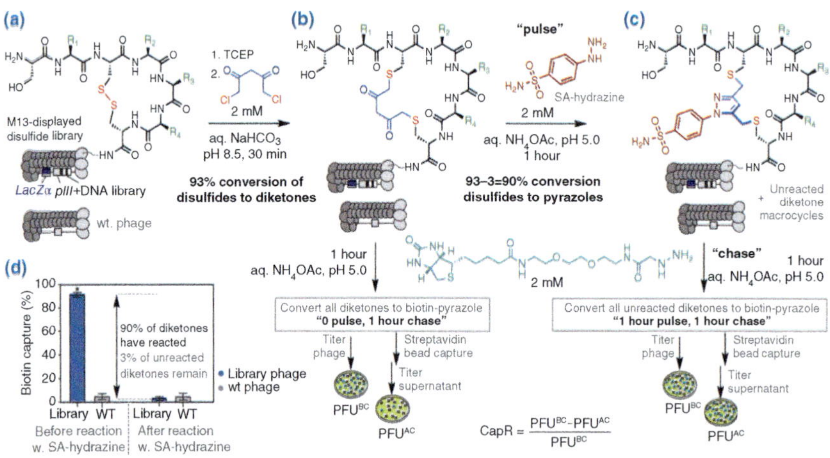

Figure 4.12 Pulse-chase monitoring of reactions in 1,3-diketone bearing phage-displayed macrocyclic libraries. (a) M13-phage-displayed disulfide library is reacted with TCEP and then DPD to form 1,3-diketone library. (b) Reaction of a portion of DKMP library with biotin hydrazine for 1 h detects that 93% of clones contain reactive 1,3-diketone groups. (c) Reaction of DKMP library with 4-hydrazine benzenesulfonamide (4-HBS) for 1 h ("pulse") followed by biotin hydrazine for 1 h ("chase") allows detection of residual 1,3-diketone groups that were not consumed by 4-HBS. (d) From 93% library with reactive 1,3-diketones, <3% were not consumed after 1 h reaction with 4-HBS and can be biotinylated; these results indicate that 90% of the library were modified by 4-HBS (M ± SD, $n = 3$, $^*p = 7 \times 10^{-5}$). Note that WT phage present in the same solution is not biotinylated in any of the experiments, indicating that only the disulfide library is modified with diketones and/or hydrazine derivatives. *Source*: Adapted from [17]/ American Chemical Society.

with 1,5-dichloro-2,4-pentadion and further exposing this library to biotin hydrazine indicated that over 90% of the peptides displayed on phage bear 1,3-diketone reactive handles (Figure 4.12).

Protocol 7: Modification of phage with two Cys residues to yield macrocyclic diketone-containing library

Materials

1. Phage libraries $MX_MCX_NCX_L$, where M is a constant amino acid such as alanine or serine, X is any amino acid, and M, N, L are integers; note that M or L can be 0 and we have not tested this protocol for $N < 3$.
2. WT phage as a background control.
3. Reagents needed for titering of phage (see Protocol 1).
4. Reagent needed for capture of biotinylated phage (see Protocol 3).
5. 1 M $NaHCO_3$ buffer (pH 8.5). Dissolve 84 g of $NaHCO_3$ in 1 L of sterile water and adjust the pH to 8.5. Check pH to be between 8 and 8.5 before each experiment. Filter using a 0.22 μm filter to sterilize.
6. 1 M NH_4OAc buffer (pH 4.5). Dissolve 77 g of NH_4OAc in 1 L of sterile water and adjust the pH to 4.5 using acetic acid. Filter using a 0.22 μm filter to sterilize.
7. Tris(2-carboxyethyl)phosphine hydrochloride (TCEP-HCl): Prepare fresh 200 mM TCEP solution by dissolving 5.7 mg of TCEP-HCl in 100 μL of water.
8. 200 mM solution of DPD (1,5-dichloro-2,4-pentadione) in DMF. Prepare by dissolving 3.4 mg of DPD in 100 μL of DMF. Solution is stable at −20 °C for 6 months.
9. Ethylenediaminetetraacetic acid (EDTA), 500 mM, pH 8.0 in sterile water.
10. Zeba desalting columns (ThermoFisher Scientific, catalog no. 89882).
11. Biotin chase reagent (biotin hydrazine): *N*-(14-hydrazineyl-13-oxo-6,9-dioxa-3,12-diazatetradecyl)-5-((3a*S*,4*S*,6a*R*)-2-oxohexahydro-1*H*-thieno [3,4-*d*]imidazol-4-yl)pentanamide, HCl salt. Prepare 200 mM solution by dissolving 9.6 mg of biotin hydrazine in 100 μL of water. Solution is stable at −20 °C for 6 months.

Procedure

1. Mix desired disulfide libraries and WT phage (1:1 ratio) to yield a final titer of 2×10^{11} pfu mL^{-1} or greater. In 1.5 mL microcentrifuge tube combine 88 μL of the phage solution with 1 μL of fresh 200 mM TCEP

(Continued)

(Continued)

solution and add 10 μL of 1 M sodium bicarbonate buffer, pH 8.5. Mix the solution with mild vortexing. Incubate the reaction mixture at room temperature for 30 minutes.

2. Add 1 μL of this 200 mM DPD solution to the reduced phage library. Incubate at room temperature for 30 minutes. During the reaction, the color of the reaction mixture changes to a dark yellow after ~15 minutes. After 30 minutes (do not exceed), load the reaction mixture on an equilibrated Zeba desalting column and elute by centrifugation at 2000×g. The macrocyclized phage library should have a light-yellow colored solution and the majority of colored compounds remain on Zeba desalting column.
3. To check the macrocyclization modify phage-diketone with biotin hydrazine, transfer 10 μL of the diketone library into a 1.5 mL microcentrifuge tube. Add 78 μL of autoclaved water, 1 μL of 500 mM EDTA, and 1 μL of the biotin hydrazine. (Note: EDTA must be added before any hydrazine derivative to avoid the toxicity of hydrazine to phage). Decrease the pH of the reaction by adding 10 μL of 1 M ammonium acetate, pH 5. Mix and incubate the reaction at room temperature for 1 hour.
4. After 1 hour, dilute the reaction mixture to 10^5–10^6 PFU/mL. Process the solution as described in Protocol 3 to determine the capture ratio.
5. The diketone-functionalized phage library can be modified by hydrazine derivatives on the same day. It can be also stored at 4 °C for a month or at −20 °C (as a glycerol stock) for many months without compromising the reactivity of 1,3-diketone.

Diketone-containing macrocyclic library displayed on phage is uniquely stable: 1,3-diketone functionality does not react with any functional groups present on phage and it is not fouled even after a prolonged storage; however, 1,3-diketone can be rapidly modified by any hydrazine derivative (Figure 4.12b,c). Monitoring of such modification is performed by standard "pulse-chase" approach using biotin hydrazine as a chase reagent. For example, a "pulse-chase" of diketone library modified by 4-hydrazine benzenesulfonamide (4-HBS, "pulse") followed by a "chase" with biotin hydrazine again indicated over 90% of the phage-displayed peptides were functionalized with the benzenesulfonamide pharmacophore (Figure 4.12). Both steps of the reaction exhibit regioselectivity toward solvent exposed peptides displayed on phage rather than any coat proteins, as demonstrated using WT phage as an internal negative control.

Protocol 8: Functionalization of diketone library with different hydrazine derivatives

Reagents

1. Hydrazine derivatives (200 mM), in DMF (e.g. 4-hydrazinebenzenesulfonamide).
2. Diketone-modified library (see Protocol 7).
3. Reagents needed for titering of phage (see Protocol 1).
4. Reagent needed for capture of biotinylated phage (see Protocol 3).
5. 1 M NH_4OAc buffer (pH 4.5). Dissolve 77 g of NH_4OAc in 1 L of sterile water and adjust the pH to 4.5 using acetic acid. Filter using a 0.22 μm filter to sterilize.
6. Ethylenediaminetetraacetic acid (EDTA), 500 mM, pH 8.0 in sterile water.
7. Zeba desalting columns (ThermoFisher Scientific, catalog no. 89882).
8. Biotin chase reagent (biotin hydrazine): *N*-(14-hydrazineyl-13-oxo-6,9-dioxa-3,12-diazatetradecyl)-5-((3a*S*,4*S*,6a*R*)-2-oxohexahydro-1*H*-thieno[3,4-*d*]i midazole-4-yl)pentanamide, HCl salt. Prepare 200 mM solution by dissolving 9.6 mg of biotin hydrazine in 100 μL of water. Solution is stable at −20 °C for 6 months.

Procedure

1. Modification with hydrazine derivatives follows the same Knorr pyrazole reaction in pH 5 ammonium acetate buffer as described in Protocol 7, Step 3.
2. In a 1.5 mL microcentrifuge tube, combine 88 μL of the diketone-modified library, 10 μL of 1 M NH_4OAc buffer (pH 4.5), 1 μL of EDTA (500 mM, pH 8.0), and 1 μL of the hydrazine solution (200 mM in DMF). Note that EDTA must always be added to phage before any hydrazine derivative.
3. Vortex the reaction mixture gently and incubate at room temperature for 1 hour. During this time, equilibrate a Zeba desalting column three times with PBS (pH 7.4).
4. After 1 hour, add the reaction mixture onto the Zeba desalting column and centrifuge (1500×g, 3 minutes). Collect the flow through in a sterile tube. Most small molecule compounds (excess hydrazine, EDTA) should remain on Zeba desalting column; however, note the desalting column discussion in section "**Binding capacity and influence of contaminants**."

(Continued)

(Continued)

5. Modified libraries can be used immediately in selection or stored for several days at 4 °C. For long-term storage, we recommend supplementing the solution with 50% glycerol and storing at −20 °C. Phage equipped with reactive warheads and unstable modifications should be used immediately.
6. To determine the yield of modification, modified library can be subjected to biotin hydrazine followed by Protocol 3 to determine the capture ratio. If all diketones are consumed, the library should not be modified by biotin hydrazine and should not be captured.
7. Sample 10 μL of the modified library from Step 4. Combine with 78 μL of autoclaved water, 1 μL of 500 mM EDTA, and 1 μL of the biotin hydrazine in a 1.5 mL microcentrifuge tube. (Note: EDTA must be added before biotin hydrazine to avoid the toxicity to phage.) Decrease the pH of the reaction by adding 10 μL of 1 M ammonium acetate, pH 5. Mix and incubate the reaction at room temperature for 1 hour.
8. After 1 hour, dilute the reaction mixture to 10^5–10^6 PFU/mL. Process the solution as described in Protocol 3 to determine the capture ratio.

4.6 Outlook

Genetically encoded libraries of macrocyclic peptides that contain chemical functionalities beyond the traditional 20-amino acid repertoire have firmly established their position as a critical tool in discovery of compounds that can address the undruggable targets. This chapter focuses on diverse examples of cPTM that introduce cyclization and/or introduce unnatural moieties in phage-displayed libraries. Forgo the obvious downstream utility (selection), this chapter pays special attention to the design of reaction monitoring that confirms the outcome of such transformations in phage libraries. Effective chemical transformation of million-to-billion of different peptides is not a trivial task, and we now start to understand the breadth of reactivity that different peptides can exhibit in such a library [21]. Based on the inherent complexity of chemical modification of a "*library of starting materials*," an argument can be made that reducing the number of chemical steps performed on phage-displayed peptides increases the practicality of cPTM. A one-step macrocyclization and installation of unnatural pharmacophore by DCO described in Section 4.4 is an example of such an efficient one-step approach. A different argument can be made for separation of macrocyclization, storage of macrocyclic library, and on-demand functionalization of such library with any pharmacophore using trivial and robust ligation chemistry, as described in Section 4.5. The practicality of divergent "late-stage" functionalization of cPTM libraries is apparent if one examines the availability of hydrazine vs. DCO-warheads: thousands of hydrazine derivatives are available commercially whereas DCO derivatives are not commercially available and must be prepared by the user.

Today, it is possible to produce genetically encoded-macrocyclic libraries with unnatural fragments without cPTM by direct translation *via* UAA mutagenesis [4,5,7], metabolic suppression [39,40], and flexizyme technology in mRNA-displayed libraries [12,41]. However, the unique advantage of cPTM and late-stage chemical modifications is the ability to introduce large functionalities such as fluorophores, metal chelators, pegylation, lipidation, reactive warheads, and simple [42] and complex glycans [29,30]: such groups might not be possible to introduce *via* translational machinery. Practitioners of UAA-containing libraries on phage, mRNA, and yeast frequently upgrade their libraries using cPTM; hence, a cPTM must always be viewed not as a competitive but rather an orthogonal manifold for diversification of genetically encoded libraries.

Genetically encoded libraries with covalent [43] or reversible covalent warheads [44,45] are one opportunity for cPTM to "shine" because incorporation of covalent warheads through translational machinery may be particularly problematic. The reactivity of such warheads toward nucleophiles present in the libraries themselves makes synthesis and characterization of libraries with covalent warhead an exciting challenge. Bogyo and co-workers employed a one-step DCO chemistry described in Section 4.4 to produce macrocyclic libraries with a built-in Michael acceptor that can covalently react with Cys residues in proteins [43]. In theory, however, if the cysteine reactivity of chloro-oxime and the Michael acceptor warhead are similar, both can engage with Cys residues in the library (Figure 4.13a), and little can be done to prevent this. In this example, we recognized an interesting advantage of the two-step divergent synthesis as the opportunity to separate nucleophilic reactivity using diketone reactive handles to modify cysteine residues for installation of reactive chemical fragments.

For example, Section 4.5 demonstrated that macrocyclization reaction can cap >90% of the solvent-accessible cysteines, and hence the late-stage incorporation of an electrophilic pharmacophore is not hampered by presence of unreactive Cys (Figure 4.13b). Second reaction proceeds in acidic environment (pH 4.5) where nucleophilicity of Cys and other proteogenic residues is shut off. Hence, during installation, the reactive electrophile is protected from any potential nucleophilic attacks. The installation of the warhead, thus, may be driven to completion and the resulting library can be stored in mildly acidic media without the risk of fouling of the reactive electrophile. And then, on demand, it can be turned on by placing such library into a neutral-basic buffer that contains a protein target (Derda Lab, unpublished observation). Such reaction engineering, multistep protection

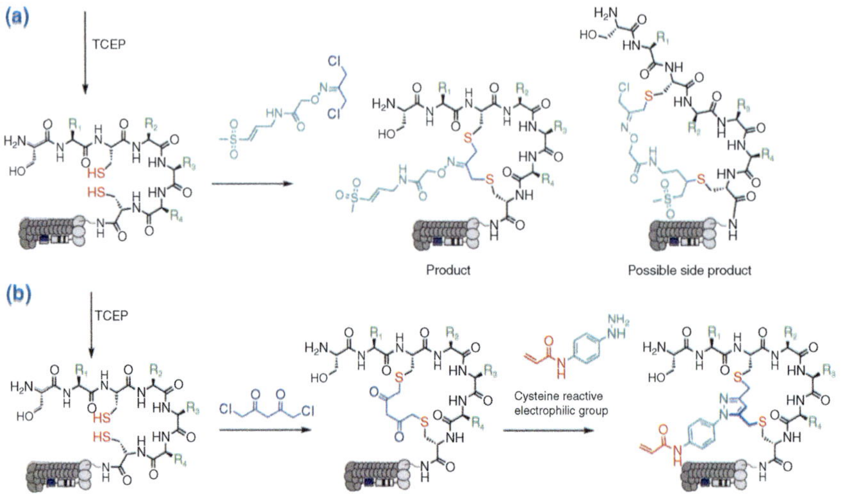

Figure 4.13 Grafting cysteine reactive warheads through cysteine macrocyclization reaction. (a) One-step modification using a cysteine macrocyclization reaction to graft cysteine reactive warheads/pharmacophores onto phage libraries used by Bogyo and co-workers. In such an approach, there is a possibility for the warhead to react with the cysteines due to its electrophilic properties. (b) The two-step reaction allows the incorporation of cysteine reactive groups onto phage-displayed peptide libraries. Separation of the macrocyclization and functionalization avoids the reactions between the warhead and the cysteines on phage-displayed peptide libraries.

and deprotection of reactive moieties, is a century-old and familiar concept in organic synthesis, and we have no doubt that it will be applied more and more in diversification of genetically encoded libraries.

References

1 Smith, G.P. Filamentous fusion phage: novel expression vectors that display cloned antigens on the virion surface. *Science*. 1985;228(4705):1315–1317.

2 Tian, F., Tsao, M.L., and Schultz, P.G. A phage display system with unnatural amino acids. *J Am Chem Soc*. 2004;126(49):15962–15963.

3 Liu, C.C., Mack, A.V., Tsao, M.L., et al. Protein evolution with an expanded genetic code. *Proc Natl Acad Sci USA*. 2008;105(46):17688–17693.

4 Oller-Salvia, B. and Chin, J.W. Efficient phage display with multiple distinct non-canonical amino acids using orthogonal ribosome-mediated genetic code expansion. *Angew Chem Int Ed Engl*. 2019;58(32):10844–10848.

5 Owens, A.E., Iannuzzelli, J.A., Gu, Y., and Fasan, R. MOrPH-PhD: an integrated phage display platform for the discovery of functional genetically encoded peptide macrocycles. *ACS Cent Sci*. 2020;6(3):368–381.

6 Strable, E., Prasuhn Jr., D.E., Udit, A.K., et al. Unnatural amino acid incorporation into virus-like particles. *Bioconjugate Chem.* 2008;19(4):866–875.
7 Wang, X.S., Chen, P.C., Hampton, J.T., et al. A genetically encoded, phage-displayed cyclic-peptide library. *Angew Chem Int Ed Engl.* 2019;58(44):15904–15909.
8 Szewczuk, Z., Rebholz, K.L., and Rich, D.H. Synthesis and biological activity of new conformationally restricted analogues of pepstatin. *Int J Pept Protein Res.* 1992;40(3–4):233–242.
9 White, C.J. and Yudin, A.K. Contemporary strategies for peptide macrocyclization. *Nat Chem.* 2011;3(7):509–524.
10 Cromm, P.M., Spiegel, J., and Grossmann, T.N. Hydrocarbon stapled peptides as modulators of biological function. *ACS Chem Biol.* 2015;10(6):1362–1375.
11 Fang, Z., Song, Y., Zhan, P., Zhang, Q., and Liu, X. Conformational restriction: an effective tactic in 'follow-on'-based drug discovery. *Future Med Chem.* 2014;6(8):885–901.
12 Passioura, T., Katoh, T., Goto, Y., and Suga, H. Selection-based discovery of druglike macrocyclic peptides. *Annu Rev Biochem.* 2014;83:727–752.
13 Sawyer, T.K., Partridge, A.W., Kaan, H.Y.K., et al. Macrocyclic alpha helical peptide therapeutic modality: a perspective of learnings and challenges. *Bioorg Med Chem.* 2018;26(10):2807–2815.
14 Morrison, C. Constrained peptides' time to shine? *Nat Rev Drug Discov.* 2018;17(8):531–533.
15 Murray, C.W. and Rees, D.C. The rise of fragment-based drug discovery. *Nat Chem.* 2009;1(3):187–192.
16 Derda, R. and Ng, S. Genetically encoded fragment-based discovery. *Curr Opin Chem Biol.* 2019;50:128–137.
17 Ekanayake, A.I., Sobze, L., Kelich, P., et al. Genetically encoded fragment-based discovery from phage-displayed macrocyclic libraries with genetically encoded unnatural pharmacophores. *J Am Chem Soc.* 2021;143(14):5497–5507.
18 Love, K.R., Swoboda, J.G., Noren, C.J., and Walker, S. Enabling glycosyltransferase evolution: a facile substrate-attachment strategy for phage-display enzyme evolution. *ChemBioChem.* 2006;7(5):753–756.
19 Ng, S., Jafari, M.R., Matochko, W.L., and Derda, R. Quantitative synthesis of genetically encoded glycopeptide libraries displayed on M13 phage. *ACS Chem Biol.* 2012;7(9):1482–1487.
20 Lima, G.M., Atrazhev, A., Sarkar, S., et al. DNA-encoded multivalent display of chemically modified protein tetramers on phage: synthesis and in vivo applications. *ACS Chem Biol.* 2022;17(11):3024–3035.
21 Yan, K., Triana, V., Kalmady, S.V., et al. Learning the structure-activity relationship (SAR) of the Wittig reaction from genetically-encoded substrates. *Chem Sci.* 2021;12(42):14301–14308.

22 Ng, S. and Derda, R. Phage-displayed macrocyclic glycopeptide libraries. *Org Biomol Chem*. 2016;14(24):5539–5545.

23 Tjhung, K.F., Kitov, P.I., Ng, S., et al. Silent encoding of chemical post-translational modifications in phage-displayed libraries. *J Am Chem Soc*. 2016;138(1):32–35.

24 Triana, V. and Derda, R. Tandem Wittig/Diels-Alder diversification of genetically encoded peptide libraries. *Org Biomol Chem*. 2017;15(37):7869–7877.

25 Kalhor-Monfared, S., Jafari, M.R., Patterson, J.T., et al. Rapid biocompatible macrocyclization of peptides with decafluoro-diphenylsulfone. *Chem Sci*. 2016;7(6):3785–3790.

26 Heinis, C., Rutherford, T., Freund, S., and Winter, G. Phage-encoded combinatorial chemical libraries based on bicyclic peptides. *Nat Chem Biol*. 2009;5(7):502–507.

27 Chen, S.Y., Touati, J., and Heinis, C. Tracking chemical reactions on the surface of filamentous phage using mass spectrometry. *Chem Commun*. 2014;50(40):5267–5269.

28 Oppewal, T.R., Jansen, I.D., Hekelaar, J., and Mayer, C. A strategy to select macrocyclic peptides featuring asymmetric molecular scaffolds as cyclization units by phage display. *J Am Chem Soc*. 2022;144(8):3644–3652.

29 Sojitra, M., Sarkar, S., Maghera, J., et al. Genetically encoded, multivalent liquid glycan array (LiGA). *bioRxiv*. 2020:10.1101/2020.03.24.997536.

30 Chou, Y., Kitova, E.N., Joe, M., et al. Genetically-encoded fragment-based discovery (GE-FBD) of glycopeptide ligands with differential selectivity for antibodies related to mycobacterial infections. *Org Biomol Chem*. 2018;16(2):223–227.

31 Ng, S., Lin, E., Kitov, P.I., et al. Genetically encoded fragment-based discovery of glycopeptide ligands for carbohydrate-binding proteins. *J Am Chem Soc*. 2015;137(16):5248–5251.

32 Kitov, P.I., Vinals, D.F., Ng, S., Tjhung, K.F., and Derda, R. Rapid, hydrolytically stable modification of aldehyde-terminated proteins and phage libraries. *J Am Chem Soc*. 2014;136(23):8149–8152.

33 Ng, S., Jafari, M.R., and Derda, R. Bacteriophages and viruses as a support for organic synthesis and combinatorial chemistry. *ACS Chem Biol*. 2012;7(1):123–138.

34 Spears, R.J. and Fascione, M.A. Site-selective incorporation and ligation of protein aldehydes. *Org Biomol Chem*. 2016;14(32):7622–7638.

35 Ng, S., Bennett, N.J., Schulze, J., Gao, N., Rademacher, C., and Derda, R. Genetically-encoded fragment-based discovery of glycopeptide ligands for DC-SIGN. *Bioorg Med Chem*. 2018;26(19):5368–5377.

36 Chen, S., Lovell, S., Lee, S., Fellner, M., Mace, P.D., and Bogyo, M. Identification of highly selective covalent inhibitors by phage display. *Nat Biotechnol*. 2021;39(4):490–498.

37 Assem, N., Ferreira, D.J., Wolan, D.W., and Dawson, P.E. Acetone-linked peptides: a convergent approach for peptide macrocyclization and labeling. *Angew Chem Int Ed Engl.* 2015;54(30):8665–8668.

38 Jafari, M.R., Deng, L., Kitov, P.I., et al. Discovery of light-responsive ligands through screening of a light-responsive genetically-encoded library. *ACS Chem Biol.* 2014;9:443–450.

39 Van Deventer, J.A., Yuet, K.P., Yoo, T.H., and Tirrell, D.A. Cell surface display yields evolvable, clickable antibody fragments. *ChemBioChem.* 2014;15(12):1777–1781.

40 Stieglitz, J.T., Kehoe, H.P., Lei, M., and Van Deventer, J.A. A robust and quantitative reporter system to evaluate noncanonical amino acid incorporation in yeast. *ACS Synth Biol.* 2018;7(9):2256–2269.

41 Huang, Y., Wiedmann, M.M., and Suga, H. RNA display methods for the discovery of bioactive macrocycles. *Chem Rev.* 2019;119(17):10360–10391.

42 Jongkees, S.A.K., Umemoto, S., and Suga, H. Linker-free incorporation of carbohydrates into in vitro displayed macrocyclic peptides. *Chem Sci.* 2017;8(2):1474–1481.

43 Chen, S., Lovell, S., Lee, S., Fellner, M., Mace, P.D., and Bogyo, M. Identification of highly selective covalent inhibitors by phage display. *Nat Biotechnol.* 2021;39:490–498.

44 McCarthy, K.A., Kelly, M.A., Li, K., et al. Phage display of dynamic covalent binding motifs enables facile development of targeted antibiotics. *J Am Chem Soc.* 2018;140(19):6137–6145.

45 Zheng, M.M., Chen, F.J., Li, K.C., Reja, R.M., Haeffner, F., and Gao, J.M. Lysine-targeted reversible covalent ligand discovery for proteins via phage display. *J Am Chem Soc.* 2022;144(34):15885–15893.

5

Computational Peptide Design for Diverse Structures and Functions

Stephen Rettie[1,2], David Juergens[2] and Gaurav Bhardwaj[2,3]

[1]*Molecular and Cellular Biology Program Biology Program, University of Washington, Seattle, Washington, USA*
[2]*Institute for Protein Design, University of Washington, Seattle, Washington, USA*
[3]*Department of Medicinal Chemistry, University of Washington, Seattle, Washington, USA*

Peptides perform a wide range of natural functions, from critical roles as hormones that control metabolism to acting as natural antibiotics and insecticides [1]. Over the last decade, peptides have regained considerable interest as a therapeutic modality, with around 70 approved peptide drugs and hundreds more currently under clinical trials [1–4]. This renewed interest is fueled by the recent success of peptide-based therapeutics like GLP-1 agonists, significant advances in approaches for the discovery and design of new peptide-based therapeutic candidates, and the unique potential of peptides to fill the gap between small molecule drugs and large biologics [1,2]. Peptide therapeutics can integrate the desirable drug-like properties of small molecules, such as stability and ease of delivery, with the high selectivity and potency of biologics. While it remains challenging to achieve robustly, the potential to develop membrane-traversing peptides that can disrupt protein–protein interactions (PPIs) also provides opportunities to target proteins and pathways currently considered difficult to drug, or even undruggable, with other therapeutic modalities [5,6]. The growing interest in peptide therapeutics has led to several recent advances in discovering and developing peptides with diverse functional properties. While considerable efforts have focused on improving the scale and speed of high-throughput library-based screening methods, there has also been significant progress in methods for accurate *in silico* design of peptides with diverse structures and functions.

The ability to computationally design peptides that fold into precisely crafted structures and perform desired functions complements the traditional methods

Peptide Science: Chemical Ligation, Lead Generation, and Therapeutic Advances, First Edition. Fa Liu.

for peptide therapeutics discovery. Multiple approaches for peptide therapeutic discovery start with natural peptides as starting points and optimize them to achieve the desired functional properties [1,2,7]. However, the chemical and structural diversity of natural peptides that are well-characterized and can be used as starting scaffolds in drug discovery efforts is limited. In principle, computational methods can rapidly design peptides with shapes, sizes, and functions not covered by the naturally occurring peptide scaffolds [8–10]. High-throughput screening of large peptide libraries is commonly used to discover peptide binders to therapeutic targets [2,11,12]. However, current library-based methods can only cover a small fraction of the chemical and structural diversity available to peptides. The state-of-the-art approaches, such as mRNA display, can screen libraries with up to 10^{12} members [11,13]; however, even the number of unique 10-mer peptides with canonical amino acids is ~10^{13}, and ~10^{15} if noncanonical amino acids like D-amino acids are also included. Computational methods can guide the chemical diversity of library-based screening methods by extensively sampling the chemical and structural diversity of putative binders *in silico* first and then guiding the experimental efforts on the structures and sequences that are more likely to have the desired functional properties. Computational peptide design is a rapidly growing field, with several new methods described over the last 5–10 years [8–10,14–18]. Additionally, the rise of deep learning (DL) methods has significantly accelerated progress in this area. While it is difficult to comprehensively describe all available methods here, we have tried to describe the key concepts and ideas underlying different approaches for designing peptides in this chapter.

5.1 The Various Types of Peptides

Before we cover different methods and approaches for designing peptides, it is important to discuss the diverse classes of peptides, as different computational methods are suitable for only particular types of peptides and not others (Figure 5.1). Moreover, different classes of peptides also demonstrate different benefits and limitations, which dictates their potential applications. “Linear peptides” are short stretches of amino acids and the most synthetically tractable (Figure 5.1a); however, they also tend to be very prone to proteolysis due to their typically flexible structure and exposed termini [19]. To address these issues, several “crosslinkers” have been developed that lock the structure of peptides into rigid “stapled peptides” that demonstrate improved protease stability and, in some cases, even better cellular permeability [20–26] (Figure 5.1b). “Cyclic peptides” also constrain the peptide structure and improve structural rigidity, protease stability, and permeability of peptides [27] (Figure 5.1c–f). Cyclization can be through backbone-to-backbone covalent bonds (e.g. N-to-C terminal peptide bond), backbone-to-sidechain bonds (e.g. thioether bonds between the Cysteine residue

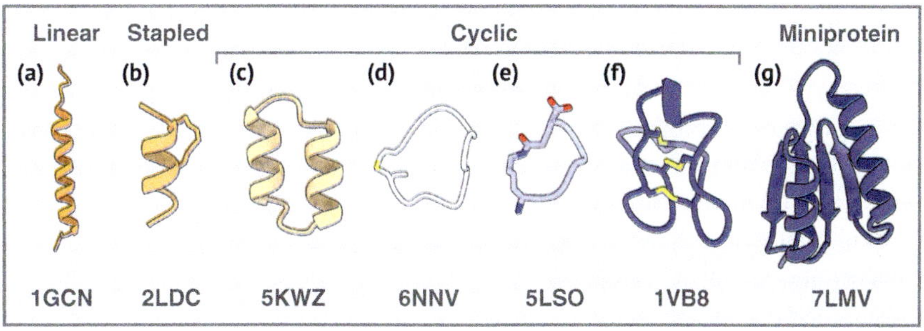

Figure 5.1 Selected examples of different types of peptides. Experimentally determined structures for select examples of different types of peptides. (a) Glucagon, a natural linear peptide hormone. (b) Estrogen receptor-binding stapled peptide SP1. (c) *De novo* designed head-to-tail cyclic peptide. (d) PD-L1 binding thioether cyclized sidechain-to-backbone cyclic peptide. (e) U2AF homology motif domain inhibitor, sidechain-to-sidechain cyclic peptide. (f) Natural, constrained cyclotide peptide from the plant root. (g) Computationally designed miniprotein inhibitor of integrin α5β6.

and N-terminal amino acids), or sidechain-to-sidechain bonds (e.g. lactam bridges). Cyclization is also common in the naturally occurring peptides. For example, plant cyclotides typically feature multiple disulfide bonds and N-to-C terminal cyclization (Figure 5.1f), resulting in very rigid and thermostable structures [28,29]. Small hyperstable proteins composed of 40–70 amino acids, commonly called "miniproteins" [30,31] (Figure 5.1g), are also a great alternative to antibodies and nanobodies. Typically, such miniproteins are linear peptides that fold into compact, ideal folds featuring combinations of α-helix and β-sheets. Such minibinders have been adapted to bind to diverse protein targets, and several new DL-based methods have been developed recently to design miniprotein binders with very high accuracy and success rates [32–35].

5.2 *In Silico* Peptide Design: From Physics-based Methods to DL

Computational design methods aim to design amino acid sequences that fold into a three-dimensional (3D) structures that meet some desired functional constraints [36]. The core idea underlying computational peptide and protein design is to understand and exploit the relationship between the sequence, structure, and function of proteins and peptides. For example, designing high-affinity peptide binders for a therapeutic target (a frequently desired function) requires identifying an amino acid sequence that folds into a structure with high shape and chemical complementarity to the target, presenting amino acid sidechains and backbone in a manner that makes energetically favorable contacts with the target at the binding interface [37,38]. Similarly, designing membrane-permeable and orally bioavailable peptides (another desired function) requires identifying amino acid sequences that fold into structures that do not make extensive contact with surrounding water molecules and can shield their polar groups while traversing nonpolar lipid membranes [39].

Computational methods for designing peptides can be broadly divided into two categories: classical "physics-based" methods and new "deep learning" methods. Physics-based methods approach the design problem by identifying amino acid sequences that minimize the energy of the peptide or peptide-protein complex in the desired structure. The energy or score functions employed in such calculations are derived from physically based equations or statistical potentials from analyzing available 3D structures of proteins and peptides. DL methods approach the same design challenge by training on the available data of experimentally determined structures to learn the patterns and relationships between the sequence and structure of proteins [40,41]. This encoded information can be leveraged post-training to accurately and rapidly predict and design the 3D structures of given sequences. These structure prediction networks have also been inverted to sample the sequence and structure simultaneously and to design new proteins and peptides for a desired function. While still in the early stages, DL methods for peptide design have shown considerable advantages over the classical methods in speed and accuracy [14,16,42,43]. However, given the large amount of data required for training such DL models, their applications are currently limited to only some classes of peptides. For example, most DL methods do not readily apply to design with noncanonical amino acids (NCAAs), as insufficient training data is available for peptides and proteins with NCAAs. Physics-based methods could complement DL methods for such cases and provide a hybrid way to address such challenges.

5.3 Physics-based Methods

While computational methods could be used to design peptides for various functions, a frequent design task is identifying peptides that bind a selected target of interest with high affinity. In such cases, one can use the physics-based computational peptide design methods to design amino acid sequences that fold into binding-competent conformations and have high shape and chemical complementarity to the intended target. Typically, computational design with physics-based methods requires searching for an amino acid sequence with the desired binding-competent structure as its lowest energy conformation and a big energy gap between this state and alternative ("off-target") conformations [9,36]. While designing target-binding peptides with such an approach, the objective also includes improving the calculated binding energy and shape complementarity of the interface between the designed peptide and the selected target protein [37,38]. Several strategies have been developed to design peptide binders using classical physics-based methods, including "motif-guided" and "*de novo* design" approaches. Motif-guided methods start with a fragment of other protein structures known to bind the target and try to improve the affinity of these "motifs" for the target. Alternatively, *de novo* design methods aim to design new peptide binders from scratch. The choice between these methods depends on the availability of sequence, structure, and experimental binding data for a given motif and the selected target of interest.

5.3.1 Motif-based Design with Physics-based Methods

For several therapeutic targets of interest, it is typically possible to identify a short peptide motif from natural interacting partners of the target or previously described binders, which can provide an ideal starting point for further improvement to its affinity or stability. Multiple databases of ideal helix-, β-sheet, or loop-based motifs for several targets are also available to mine for binding motifs [44–46]. Usually, the selected motifs are short stretches of amino acids that make several key contacts to the target and have been experimentally verified or are predicted with high confidence to bind the target. Two approaches can be followed to improve the binding affinity between the selected motif and the target protein. First, the motif can be stabilized and rigidified by adding crosslinkers. This can also be combined with sequence mutations and optimization to achieve superior target binding affinities. Second, the motif can be grafted on other rigid scaffolds, and the sequence of the other scaffold residues can be redesigned to make additional contacts with the target (Figure 5.2).

Typically, the linear motifs isolated from other native binding partners are flexible and do not readily fold into the binding-competent structures they adopt as part of the larger proteins from which they were extracted. Crosslinkers that bridge two or three different motif sidechains can rigidify the structure of the motifs (Figure 5.2). These stapled structures show improved binding affinity by reducing the entropic cost of going from a largely flexible motif in solution to the target-bound state [21,24]. The guidelines for crosslinkers that staple helices are extensively used and well established in terms of the helix positions they can staple (e.g. i to $i + 3$ position or i to $i + 7$ position) and which amino sidechains to use (e.g. cysteines, etc.) [21,24,25]. The stapling strategies for motifs with other secondary structures have not been explored as extensively. In such cases, computational methods can simultaneously sample all possible conformers for a crosslinker and scan the motif structure to find ideal geometric distances and angles for placing the selected crosslinker. Such computational methods can be generalized to staple motifs with diverse structures and incorporate any crosslinker, as demonstrated by Kritzer et al. [47].

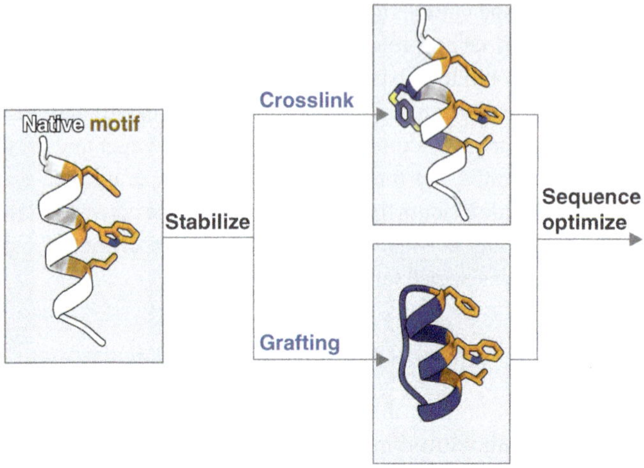

Figure 5.2 Motif-based design and optimization of peptides. Bioactive motifs from larger proteins can be extracted and stabilized in peptide form by (i) stapling them into rigid binding-competent conformations using crosslinkers, or (ii) grafting them on a pre-enumerated set of scaffold peptides that fold into a stable structure. Further, the amino acid sequence of the motif can be mutated to stabilize peptide structures or improve the function.

An alternative to stabilizing the motif by adding crosslinkers is to instead graft the motif on larger scaffold peptides that already fold into rigid hyperstable structures [48]. Many naturally constrained peptides, such as cyclotides, knottins, and conopeptides, have multiple disulfide bonds and/or backbone cyclization, making them rigid and resistant to heat and chemical denaturation [28,49]. Additionally, an extensive library of computationally designed hyperstable miniproteins and peptides that show high folding propensity, thermal stability, and resistance to chemical denaturation has also been described previously [8,9,50]. Together, these natural and designed hyperstable scaffolds cover diverse sizes and shapes, making them ideal for grafting motifs with broad structural and chemical diversity. For structure-guided grafting of motifs on a pre-existing set of scaffolds, computational methods rapidly scan for the stretches of amino acids on the scaffold peptides that can be aligned and replaced with the selected short motifs [48]. After grafting, one can also redesign the sequence of the scaffold around the motif to make any additional contacts with the target. Typically, the hyperstable scaffolds demonstrate high mutational tolerance and can still fold into the original shape after these changes. However, all the grafted designs are usually filtered *in silico* after grafting based on the computational metrics of interface quality (calculated binding affinity, shape complementarity, and interface surface area) and the propensity of the new grafted design to fold into the binding-competent structure. Such motif grafting strategies have been implemented in the Rosetta Macromolecular modeling suite to successfully design high-affinity binders for multiple therapeutic proteins [48].

5.3.2 *De Novo* Design with Physics-based Methods

While successful against some targets, the motif-based approaches are limited to targeting well-characterized proteins with known interaction partners and motifs. For many therapeutic targets, no starting motifs are available to initiate the motif-guided design calculations. Methods that can design peptide binders from scratch are required in such cases. While *de novo* design is technically more challenging, several strategies, including Rosetta kinematic closure and molecular dynamics (MD), have been described for designing peptide monomers and binders from scratch.

5.3.2.1 Peptide Design with Rosetta Generalized Kinematic Closure

Conceptually, the rotatable bonds and atoms in peptide backbones are analogous to the joints and links of the robotic systems. Given this similarity, the kinematics algorithms from the field of robotics have been adapted to enumerate the shapes of protein loop regions and cyclic peptides [51,52]. We previously described a generalized kinematic closure algorithm (GenKIC) to accurately design cyclic peptides composed of canonical and noncanonical amino acids [9]. We later extended this method to extensively enumerate the chemical and structural diversity available to small macrocyclic peptides and design membrane-traversing peptides with sizes beyond the traditional rule-of-five space that describes membrane-permeable molecules [8,39]. Typically, the GenKIC approach starts with a polyglycine chain with N rotatable bonds and uses random perturbations for "N-6" bonds, followed by calculating an analytical KIC-based solution for the six leftover bonds to yield a closed cyclic peptide backbone. The random perturbations of the backbone dihedral angles during each KIC attempt led to broad structural diversity in cyclic peptides generated across the design campaigns. After backbone generation, amino acid sequences are designed using physics-based methods to minimize the energy of each sampled backbone (Figure 5.3, top). Finally, the conformation energy landscapes of designed sequences are calculated to confirm if they are predicted to fold into their designed structures as their lowest-energy conformation and considerable energy gap between the designed conformation and alternative conformations. The GenKIC approach can be used to design target-binding peptides in multiple ways. First, the GenKIC approach can rapidly enumerate a vast library of cyclic peptides that can act as scaffolds for motif-grafting approaches described earlier. Indeed, Hosseinzadeh et al. [8] have used it to design thousands of macrocyclic peptides between 7 and 14 amino acids predicted to fold into their designed structures with high confidence. A similar approach has also been demonstrated to extensively enumerate diverse peptides capable of crossing cell membranes and gastrointestinal (GI) barrier for oral delivery [39]. Second, the KIC approach can also be extended to sample peptide sequences and structures directly in the target pocket to design cyclic peptide binders. In such cases, the KIC-based peptide backbone sampling starts from a known motif or an anchor residue (instead of a randomly initialized peptide chain) (Figure 5.3, bottom). During the backbone sampling step, the motif and anchor residues can be left unperturbed to ensure that the critical interactions from motifs and anchor residues stay fixed in the final designed interface. The sequence design for the sampled backbones is guided by the total energy of the complex and interface metrics like calculated binding affinity and number of interface contacts. This approach was used to design binders against histone deacetylase (HDACs) [37], New Delhi beta-lactamase (NDM-1) [38], and

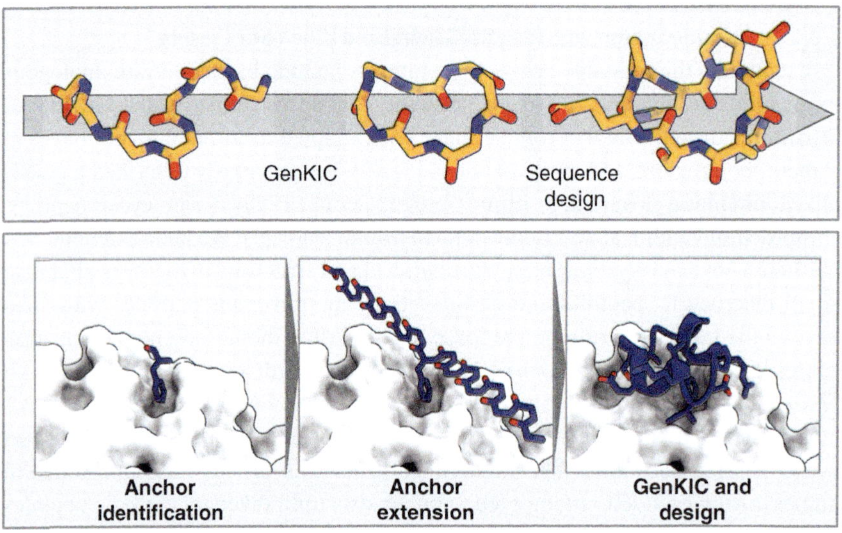

Figure 5.3 Design of peptide monomers and binders using Rosetta generalized kinematic closure (GenKIC) approach. Overall schematics for GenKIC-based design of peptide monomers (top) and binders (bottom). For designing monomeric peptides, the approach starts with a randomly initialized polyglycine chain, which is subsequently converted to a cyclic peptide backbone using random perturbations of the backbone dihedral angles and kinematic closure algorithm. Next, amino acid sequences are designed onto each sampled cyclic peptide backbone to drive folding into stable structures. This method can also be used to design cyclic peptides in the context of a binding pocket. Using one or more amino acids as an "anchor," cyclic peptides can be sampled using GenKIC and then sequence designed to improve the binding affinity and shape complementarity between the target and the designed cyclic peptide.

PD-L1 [53]. While the GenKIC approach has been used to design diverse monomeric cyclic peptides *de novo* [8,9,39], in all aforementioned cases of binder design with GenKIC, the anchor residues were picked from the previously described interactions. However, in principle, this approach can be initiated with computationally docked or predicted anchor residues, thus providing a path to *de novo* peptide binder design.

5.3.2.2 *De Novo* Peptide Binder Design with Physics-based Methods

An alternative to the GenKIC-based anchor extension approach is the *de novo* approach described by Baker and colleagues for designing miniprotein binders [33,34]. This approach starts by docking individual amino acids to the target surface to identify amino acid interactions with favorable interaction energies (Figure 5.4). The next step in the pipeline is to dock tens of thousands of pre-enumerated peptides that are predicted to fold into designed structures (termed "scaffolds") against the target to identify those which can accommodate the favorable amino acid interactions identified in the previous step (Figure 5.4). Next, the sequence of the docked scaffold peptide is redesigned to improve the binding affinity and shape complementarity. Designs are downselected based on the computational metrics of interface quality and the propensity of the designed sequence to fold into the desired binding-competent conformation. This approach has been used to successfully design very high-affinity binders against multiple diverse protein targets [33,34]. While these efforts have focused on using miniprotein scaffolds, the overall concept and design methods can be readily adapted for designing cyclic peptide binders

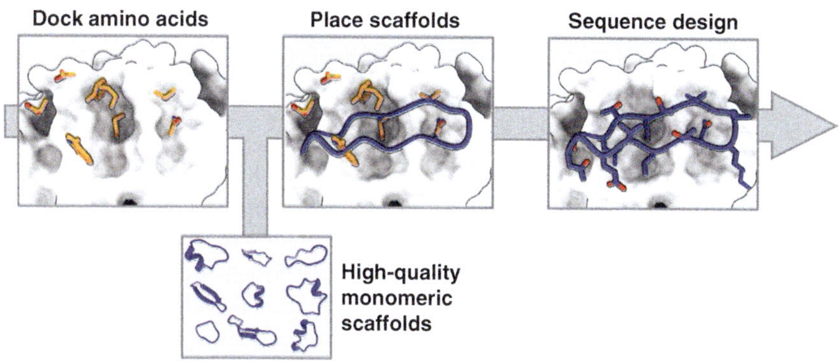

Figure 5.4 *De novo* peptide binder design using physics-based methods. Overall pipeline for *de novo* peptides design using a pre-enumerated library of cyclic peptides. First, amino acids are docked against the target to find high-affinity interactions. Next, the peptide scaffolds are placed to include these high-affinity interactions, followed by sequence design to improve the shape and complementarity between the designed peptides and the selected target.

de novo by using the pre-enumerated set of natural and computationally designed cyclic peptide scaffold libraries.

5.3.2.3 Molecular Dynamics-based Peptide Design

While the GenKIC and Rosetta-based design approaches are fast and have yielded successful peptide monomers and binders, these approaches have several limitations. First, they do not sufficiently account for the role of water molecules at the interface involved in bridging key interactions. Second, these approaches largely focus on designing peptides that fold into unique low-energy structures. As seen in some peptide binders identified from nature or large library-based screening approaches, peptide binders can be flexible and adopt multiple conformations [54]. Some of these limitations stem from how the designs are filtered in the Rosetta GenKIC approach [8,9,39] and how the energy functions are calculated and implemented [55,56]. MD simulations in explicit aqueous environments can address these challenges by appropriately accounting for the water molecules and calculating the structural ensembles adopted by a given peptide instead of focusing on a single low-energy structure. However, such MD simulations are time-intensive and cannot yet be applied to generate and select tens of thousands of design candidates. It is feasible, however, to apply extensive MD simulations for in silico validation of a few downselected candidates before the experimental validation. Approaches based on accelerated MD simulations have also been described for designing small macrocyclic peptides [10,57,58]. More recently, these simulations have also provided additional synthetic data sets for training new, faster DL models capable of predicting structure ensembles of macrocyclic peptides [16,59].

5.4 DL Methods for Peptide Design

Deep learning methods, like AlphaFold, RoseTTAFold, and ESMFold, have been transformative for protein structure prediction and design [40,41,60]. In addition to significantly improving the ability to predict protein structure from their sequences, these methods have also enabled design of new proteins for diverse functional applications [35,61–65]. Despite this considerable progress in structure prediction and design of large proteins, similar advances for predicting and designing peptides have been slow. This is largely due to the limited availability of training data for small peptides containing aforementioned crosslinks and nonnatural amino acids. However, several DL-based methods for peptide structure prediction and design have been described recently [15,16,39,42,43], and they typically fall into two broad categories: (i) structure-based methods and (ii) sequence-based large language models (LLMs). Structure-based models are trained on available experimentally determined X-ray, NMR, and cryo-EM biomolecular structures. These models learn to create functional peptides by generating peptide structures that have shapes and chemical composition required for the desired function. Contrastingly, sequence-based methods train on billions of amino acid sequences available in large databases and implicitly learn to predict functional sequences by recapitulating the characteristics associated with sequences evolved by nature. It should be noted that one is not restricted to using purely structure- or sequence-based models for a given design challenge. They can be easily combined at the polypeptide generation and/or downselecting step, and the nature of this combination may depend on the particular design challenge at hand.

Developing effective ways to represent polypeptides for a DL model is a critical component of training and using DL models for designing peptides. Peptides can be encoded and represented as one-dimensional (1D) arrays of their amino acid sequences, 1D strings that denote their chemical formula and bond connectivity, or 1D arrays of the bond angles and distances. The amino acid sequence in proteins can be treated analogously to words in a human language, and protein language models (PLMs) can be trained analogously to LLMs such as OpenAI GPT4. However, details beyond the amino acid sequence may sometimes be required while training DL models. In such cases, formats representing the chemical formula and connectivity between those different chemical elements, such as SMILES strings [66,67], may provide additional context. Typically, the amino acid sequence or the SMILES strings are converted to a one-hot encoded format representing these sequences as a matrix input. For structure-based methods, peptides can be encoded as their 3D coordinates or as an array of backbone and sidechain torsional angles. Alternatively, proteins and peptides can be represented as graphs with atoms as nodes and the connections between those atoms as edges, or as two-dimensional (2D) distance maps that compile the all-vs-all distances between atoms or residues in a protein or peptide. The choice of the representation style typically depends on the type of model being trained and the amount of information available.

5.4.1 Structure-based DL Methods for Peptide Design

Deep learning-based protein structure prediction methods, such as AlphaFold and RoseTTAFold, have shown considerable improvements in speed and accuracy over conventional physics-based methods [40,41]. These methods take a sequence of interest, its multiple sequence alignment (MSA), and available homologous structures as inputs and predict the 3D structure for that given sequence. Generally, the neural network can extract coevolutionary information from the MSA and use it to infer pairwise (and higher-order) spatial proximity ("contacts") between various amino acids, which it manifests in the predicted 3D structure. Recent state-of-the-art generative design methods, such as RFdiffusion, also rely on the neural architectures and (sometimes) pretrained weights of structure prediction networks as denoisers in the stepwise denoising process implemented in diffusion models [35]. While very powerful for larger proteins, it is not clear how readily these networks transfer to the structure prediction and design of peptides. However, much of the biophysical phenomena that dictates the relationship between protein sequence, structure, and function should carry over to the space of peptides – at least for those composed solely of canonical L-amino acids. Multiple studies have benchmarked AF2 for structure prediction of linear peptides and peptide-protein complexes and demonstrated promising but limited utility [68,69].

One example of the transferability of protein structure prediction training to the modeling and design of peptides is that of AfCycDesign [14]. Despite the original AlphaFold network not being trained on cyclic peptides, Rettie et al. [14] demonstrated that the relative positional encodings in AlphaFold can be modified to predict cyclic peptide structures. These modifications tell the network that the peptide is cyclic; i.e. the would-be last amino acid in a linearized version of the peptide is followed by the first amino acid. Benchmarking studies showed that this modified AlphaFold ("AfCycDesign") can predict cyclic peptide structures with high accuracy and confidence. Taking a step further, they adapted the network to redesign the sequence of a fixed peptide backbone and even design a new cyclic peptide *de novo* [14]. To design new peptides, the approach starts with a random sequence and predicts its structure. A "loss function" (usually network confidence, and/or agreement to a desired structure) computes the compatibility of the predicted sequence and structure (Figure 5.5). Backpropagation of the loss through the network then updates the sequence to improve the loss for the predicted structure. This procedure is performed iteratively, with the goal of converging to a sequence input that causes the network to confidently predict the structure (Figure 5.5). Typically, starting with different initial random sequences results in diverse sequences and structures. Studies with this approach have shown that it can be used to design diverse sequences and structures rapidly, and the X-ray structures of these designed peptides match very closely with their design models [14].

Multiple diffusion-based models have also been described recently for structure prediction and design of proteins [35,70] and peptides [15,71]. Typically, these models are trained by corrupting natural polypeptide structures to varying degrees and tasking the network with recovering the native structure. After training, such models can generate new structures by starting with randomly initialized atom coordinates and denoising them iteratively [35,72–75]. Similar methods have been developed to generate sequences [76] or jointly generate sequences and structures [77]. Denoising diffusion models, such as RFdiffusion, have shown remarkable success in the design of protein monomers, assemblies, and target-binding macrocycles [35,65].

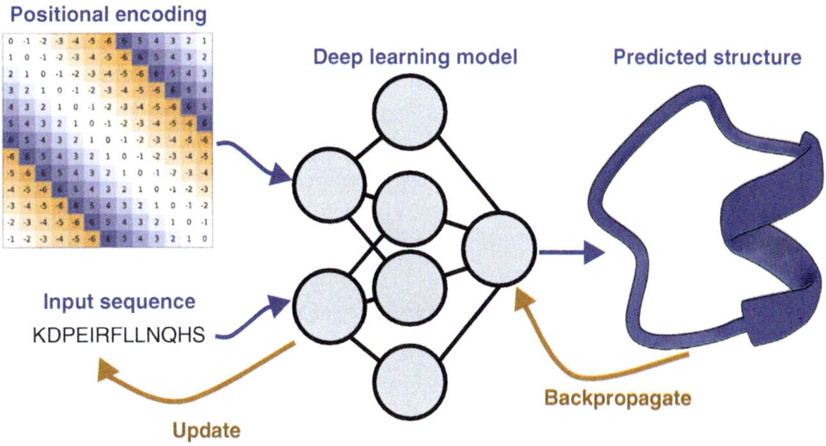

Figure 5.5 Cyclic peptide design using DL models. The relative positional encoding of amino acids in DL models can be modified with cyclic offsets to treat the terminal residues as neighbors, enabling structure prediction and design of cyclic peptides with DL methods. For designing new peptides, one can typically start with the prediction (forward pass) of a random initial sequence and then backpropagate through the network to update the sequence to improve the confidence metrics of prediction. The process is repeated until a sequence with sufficiently high prediction metrics is generated.

This approach has since enabled the design of nanomolar to picomolar affinity miniprotein binders against several therapeutic targets [35,65]. Although powerful, RFdiffusion is currently not trained to design small-sized peptides. However, other diffusion models, such as PepFlow, have been described recently, which enable peptide structure prediction and can estimate structural ensembles of input peptides [71]. We believe this to be a very exciting area of work that will continue to expand and could result in significant improvements in methods for designing peptides for diverse functions.

5.4.2 PLMs for Peptide Design

Protein language models have also been recently leveraged to design linear peptide binders for targets [17,18]. PLMs are conceptually analogous to LLMs that try to predict the next word in a sentence or fill in the missing words in a sentence. Although trained on purely sequence information without structure, it has been demonstrated that PLMs can estimate co-evolutionary information when given a single protein sequence [60]. This co-evolutionary information can be used as constraints to predict protein structures. Such PLMs enable rapid structure prediction and have been used to predict the structures of more than 600 million metagenomic sequences [60]. Beyond the structure prediction of proteins, information in such PLMs can also be leveraged to design new proteins and peptides for functional properties. Specifically, pipelines that iterate between the sequence design and the structure prediction steps can generate proteins that fold into designed structures and assemblies. These PLMs have also been extended for designing peptide binders to therapeutic targets by either identifying peptide binders from natural or predicted interaction partners or by designing peptide binders *de novo* [17,18]. Chatterjee and colleagues described an approach that uses language models to identify the per-residue probability of being a binder in natural PPI, followed by leveraging these per-residue probabilities to identify stretches of amino acids with a high probability of binding to the selected target [17]. More recently, language models such as ESM-2 were fine-tuned to generate *de novo* peptide binders by tethering the peptide to the target terminus, masking the peptide sequence, and using the PLMs to generate sequences that are predicted to bind and have low sequence entropy at each amino acid position [18]. Notably, since these methods can start from a target sequence alone, they can be used to design peptide binders to targets with unknown structures. However, a potential limitation of such methods is the difficulty in extending them to chemical and structural diversity in regions of low evolutionary information. For example, these methods do not work as well at predicting structures for sequences that do not have many homologous sequences (such as designed proteins and peptides), and it is difficult to adapt them to peptides and proteins with noncanonical amino acids as there is not enough information to train such models for that task. Overall, the current focus in this field is on training increasingly large models that can take multimodal inputs and incorporate structural information in addition to the sequence.

5.5 Optimizing for Membrane Permeability and Stability

Beyond the target binding specificity and selectivity, developing peptide-based therapeutics also requires optimizing for additional drug-like properties, such as serum stability, permeability, nonimmunogenicity, and metabolic stability. Protease and serum stability are generally better for peptides with noncanonical amino acids and rigid structures [2,3,19,25]. N-to-C terminal cyclization also improves stability by increasing resistance to exopeptidases and constraining the macrocyclic structure. The metabolic stability of peptides is more complex and remains underexplored. While some studies have recently explored the stability of peptides in intestinal and liver microsomes [78], the amount of publicly available data on peptide sequences/structures and their experimentally measured metabolic stability is currently too small to train general predictive computational models for metabolic stability or develop design principles for peptides with improved metabolic stability.

However, computational peptide design can improve the metabolic stability of peptides in two ways. First, it is typically possible to computationally generate multiple candidates with diverse shapes and sizes to bind to a selected target. These diverse hits could serve as a risk-mitigation strategy as these different hits may span different levels of metabolic stability. Second, in cases where experiments identify metabolic hotspots and liabilities, computational methods can be used to redesign the hits to remove those liabilities without affecting the target binding and other functions.

While most currently approved peptide-based therapeutics are injectables, there has been an increased focus on exploring the design principles for peptides with enhanced membrane permeability and oral bioavailability. The ability to cross membranes enables binder design against intracellular targets, which are currently considered undruggable with large-sized biologics. The ability to cross the GI barrier enables oral delivery of peptide therapeutics. Inspired by the natural membrane-traversing macrocycles like cyclosporine [79], there has been considerable work towards developing design principles for peptide membrane permeability and accurately designing membrane-traversing peptides with diverse shapes and sizes [39,80–85]. For achieving passive membrane permeability by design, the computational approaches have focused on two key structural features. First, to reduce the penalty of removing bound water molecules while traversing across the lipid membranes, computational methods focus on generating peptides without exposed and unsatisfied hydrogen bond donors. This can be achieved by designing peptides in shapes that have extensive inter-molecular hydrogen bonding and *N*-methylation of amino acid positions that are exposed and non-hydrogen bonded (Figure 5.6). The second approach has focused on designing chameleonic peptides that can switch conformations while traversing the membranes. Such chameleonic peptides are designed to have distinct low-energy conformations, with one state preferred in the lipid membranes and the other preferred in the aqueous environment [39,81,82,86] (Figure 5.6). The design principles of peptides for both active and passive transport have been extensively covered in other review papers [27,87–91].

5.6 Remaining Challenges

Despite considerable progress, several challenges remain in accurately designing peptides for diverse functions. While physics-based methods for peptide design are currently more generalizable and can span the rich chemical diversity of NCAAs, they are limited by the accuracy of their underlying energy functions. Methods like Rosetta GenKIC are fast and enable rapid exploration of a vast chemical and structural diversity to design diverse monomer peptide scaffolds and binders

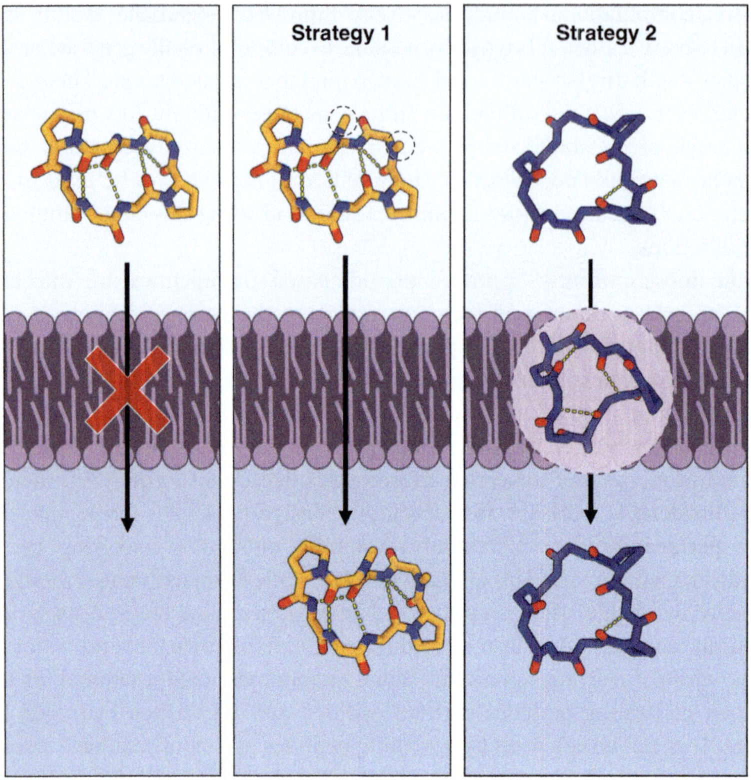

Figure 5.6 Computational design for membrane-traversing peptides. Two different strategies for designing passively permeable cyclic peptides include: (left) masking hydrogen-bond donors in a given peptide, either by the design of internal hydrogen bond networks or including *N*-alkyl amino acids such as proline or *N*-methylated amino acids; (right) designing chameleonic peptides with two distinct low-energy conformations. Typically, one conformation is designed to have fewer or no exposed and unsatisfied hydrogen bond donors than the other conformation. The peptide adopts the more polar conformation with more exposed hydrogen bond donors in the aqueous environment but can switch to the alternative conformation while crossing the membranes.

using motif-based approaches. However, the *de novo* design of peptide binders with these physics-based methods has had low success rates and still needs to be extensively validated experimentally. Moreover, such methods do not explicitly consider water molecules or typically do not design for multiple states accurately, thus limiting their accuracy against many therapeutically relevant targets. MD simulation-based approaches can include water molecules; however, they are slow and cannot be readily used currently to generate tens of thousands of design candidates against a selected target.

DL-based approaches have been transformative in designing medium- to large-size proteins, enabling the highly accurate design of miniprotein binders to arbitrary protein targets. However, the lack of large training data for peptide structures or target-bound peptide complexes, especially for peptides composed of noncanonical amino acids, has slowed the development of DL methods for designing peptide-based binders to therapeutic targets. Methods that adapt the information from protein structure prediction and tools for designing peptides have shown promise and considerable improvements in speed and accuracy over the physics-based methods; however, such methods are currently limited to designing with canonical amino acids. The rise of "all-atom" models in protein design that enable the inclusion of nonprotein and nonstandard atom types, such as RoseTTAFold All Atom and AlphaFold3 [72,92], is promising for expanding peptide design to include NCAAs. However, the designs from these methods will need to be validated extensively, and models may require further fine-tuning based on the experimental data. Toward that goal, we will need to implement design, build, test, and learn cycles that are considerably faster and scalable to a large number of design candidates. This could be achieved by integrating design methods with high-throughput display-based screening approaches; however, this remains experimentally challenging for peptides primarily composed of noncanonical amino acids.

5.7 Conclusion

Physics-based and DL methods have demonstrated significant progress in accurately designing structured peptides, peptide-based binders, and peptides with improved membrane permeability. These computational methods complement the experimental approaches for peptide screening and natural product discovery. While many challenges remain, the rapid and ongoing progress in DL methods will significantly improve the accuracy and speed of peptide therapeutic discovery and optimization. DL methods are also rapidly progressing toward robust *de novo* design of peptide binders to target proteins. We believe that future work in the field will focus on expanding DL methods to include building blocks beyond the standard amino acids and integrating such methods with high-throughput screening methods to collect more data on successes and failures and train improved models.

More work should also focus on training predictive models for properties beyond target binding, such as membrane permeability, immunogenicity, and metabolic stability. Initially, these could be trained and implemented as separate models; however, the goal should be to build a joint model that can simultaneously design and optimize for multiple drug-like properties. Computational biologists should also make an effort to make the new methods easily accessible to a diverse set of users, enabling the broad applicability of these new computational methods. While the early stages of peptide therapeutics discovery relied on natural product discovery, semisynthetic optimization, and random library screening, we believe that computational methods will play a significant role in developing the next generation of peptide-based therapeutics, and it is entirely plausible that we will have a computationally designed peptide therapeutics in the near future.

References

1 Muttenthaler, M., King, G.F., Adams, D.J., and Alewood, P.F. Trends in peptide drug discovery. *Nat Rev Drug Discov*. 2021;20:309–325. https://doi.org/10.1038/s41573-020-00135-8.

2 Vinogradov, A.A., Yin, Y., and Suga, H. Macrocyclic peptides as drug candidates: recent progress and remaining challenges. *J Am Chem Soc*. 2019;141(10):4167–4181.

3 Wang, L., Wang, N., Zhang, W., et al. Therapeutic peptides: current applications and future directions. *Signal Transduct Target Ther*. 2022;7(1):48.

4 Zhang, H. and Chen, S. Cyclic peptide drugs approved in the last two decades (2001–2021). *RSC Chem Biol*. 2022;3(1):18–31.

5 Villar, E.A., Beglov, D., Chennamadhavuni, S., et al. How proteins bind macrocycles. *Nat Chem Biol*. 2014;10(9):723–731.

6 Tsomaia, N. Peptide therapeutics: targeting the undruggable space. *Eur J Med Chem*. 2015;94:459–470.

7 Zorzi, A., Deyle, K., and Heinis, C. Cyclic peptide therapeutics: past, present and future. *Curr Opin Chem Biol*. 2017;38:24–29.

8 Hosseinzadeh, P., Bhardwaj, G., Mulligan, V.K., et al. Comprehensive computational design of ordered peptide macrocycles. *Science*. 2017;358(6369):1461–1466.

9 Bhardwaj, G., Mulligan, V.K., Bahl, C.D., et al. Accurate de novo design of hyperstable constrained peptides. *Nature*. 2016;538(7625):329–335.

10 Slough, D.P., McHugh, S.M., and Lin, Y.S. Understanding and designing head-to-tail cyclic peptides. *Biopolymers*. 2018;109(10):e23113.

11 Passioura, T. and Suga, H. A RaPID way to discover nonstandard macrocyclic peptide modulators of drug targets. *Chem Commun*. 2017;53(12):1931–1940.

12 Obexer, R., Walport, L.J., and Suga, H. Exploring sequence space: harnessing chemical and biological diversity towards new peptide leads. *Curr Opin Chem Biol*. 2017;38:52–61.

13 Li, X., Craven, T.W., and Levine, P.M. Cyclic peptide screening methods for preclinical drug discovery. *J Med Chem*. 2022;65(18):11913–11926.

14 Rettie, S. A. et al. Cyclic peptide structure prediction and design using AlphaFold2. *Nat. Commun.* 16, 1–15 (2025).

15 Xie, X., Valiente, P.A., Kim, J., and Kim, P.M. HelixDiff, a score-based diffusion model for generating all-atom α-helical structures. *ACS Cent Sci.* 2024;10(5):1001–1011.

16 Miao, J., Descoteaux, M.L., and Lin, Y.S. Structure prediction of cyclic peptides by molecular dynamics + machine learning. *Chem Sci.* 2021;12(44):14927–14936.

17 Brixi, G., Ye, T., Hong, L., et al. SaLT&PepPr is an interface-predicting language model for designing peptide-guided protein degraders. *Commun Biol.* 2023;6(1):1081.

18 Chen, T., Pertsemlidis, S., Watson, R., et al. PepMLM: target sequence-conditioned generation of peptide binders via masked language modeling. *ArXiv*. 2023. Available from: https://www.ncbi.nlm.nih.gov/pubmed/37873004.

19 Al Musaimi, O., Lombardi, L., Williams, D.R., and Albericio, F. Strategies for improving peptide stability and delivery. *Pharmaceuticals*. 2022;15(10):1283. https://doi.org/10.3390/ph15101283.

20 Walensky, L.D., Kung, A.L., Escher, I., et al. Activation of apoptosis in vivo by a hydrocarbon-stapled BH3 helix. *Science*. 2004;305(5689):1466–1470.

21 Walensky, L.D. and Bird, G.H. Hydrocarbon-stapled peptides: principles, practice, and progress. *J Med Chem*. 2014;57(15):6275–6288.

22 Bird, G.H., Mazzola, E., Opoku-Nsiah, K., et al. Biophysical determinants for cellular uptake of hydrocarbon-stapled peptide helices. *Nat Chem Biol.* 2016;12(10):845–852.

23 Blackwell, H.E. and Grubbs, R.H. Highly efficient synthesis of covalently cross-linked peptide helices by ring-closing metathesis. *Angew Chem Int Ed Engl.* 1998;37(23):3281–3284.

24 Bluntzer, M.T.J., O'Connell, J., Baker, T.S., Michel, J., and Hulme, A.N. Designing stapled peptides to inhibit protein-protein interactions: an analysis of successes in a rapidly changing field. *Pept Sci*. 2021;113(1):e24191. https://doi.org/10.1002/pep2.24191.

25 Li, Y., Wu, M., Fu, Y., et al. Therapeutic stapled peptides: efficacy and molecular targets. *Pharmacol Res*. 2024;203:107137.

26 Lau, Y.H., de Andrade, P., Wu, Y., and Spring, D.R. Peptide stapling techniques based on different macrocyclisation chemistries. *Chem Soc Rev*. 2015;44(1):91–102.

27 Nielsen, D.S., Shepherd, N.E., Xu, W., Lucke, A.J., Stoermer, M.J., and Fairlie, D.P. Orally absorbed cyclic peptides. *Chem Rev*. 2017;117(12):8094–8128.

28 de Veer, S.J., Kan, M.W., and Craik, D.J. Cyclotides: from structure to function. *Chem Rev*. 2019;119(24):12375–12421.

29 Troeira Henriques, S. and Craik, D.J. Cyclotide structure and function: the role of membrane binding and permeation. *Biochemistry*. 2017;56(5):669–682.

30 Baker, E.G., Bartlett, G.J., Porter Goff, K.L., and Woolfson, D.N. Miniprotein design: past, present, and prospects. *Acc Chem Res*. 2017;50(9):2085–2092.

31 Crook, Z.R., Nairn, N.W., and Olson, J.M. Miniproteins as a powerful modality in drug development. *Trends Biochem Sci*. 2020;45(4):332–346.

32 Chevalier, A., Silva, D.A., Rocklin, G.J., et al. Massively parallel de novo protein design for targeted therapeutics. *Nature*. 2017;550(7674):74–79.

33 Cao, L., Goreshnik, I., Coventry, B., et al. De novo design of picomolar SARS-CoV-2 miniprotein inhibitors. *Science*. 2020;370:426–431. https://doi.org/10.1126/science.abd9909.

34 Cao, L., Coventry, B., Goreshnik, I., et al. Design of protein-binding proteins from the target structure alone. *Nature*. 2022;605(7910):551–560.

35 Watson, J.L., Juergens, D., Bennett, N.R., et al. De novo design of protein structure and function with RFdiffusion. *Nature*. 2023;620(7976): 1089–1100.

36 Huang, P.S., Boyken, S.E., and Baker, D. The coming of age of de novo protein design. *Nature*. 2016;537(7620):320–327.

37 Hosseinzadeh, P., Watson, P.R., Craven, T.W., et al. Anchor extension: a structure-guided approach to design cyclic peptides targeting enzyme active sites. *Nat Commun*. 2021;12(1):3384.

38 Mulligan, V.K., Workman, S., Sun, T., et al. Computationally designed peptide macrocycle inhibitors of New Delhi metallo-β-lactamase 1. *Proc Natl Acad Sci USA*. 2021;118(12):e2012800118.

39 Bhardwaj, G., O'Connor, J., Rettie, S., et al. Accurate de novo design of membrane-traversing macrocycles. *Cell*. 2022;185(19)3520–3532.e26. https://doi.org/10.1016/j.cell.2022.07.019.

40 Jumper, J., Evans, R., Pritzel, A., et al. Highly accurate protein structure prediction with AlphaFold. *Nature*. 2021;596(7873):583–589.

41 Baek, M., DiMaio, F., Anishchenko, I., et al. Accurate prediction of protein structures and interactions using a three-track neural network. *Science*. 2021;373(6557):871–876.

42 Grambow, C.A., Weir, H., Diamant, N.L., et al. RINGER: rapid conformer generation for macrocycles with sequence-conditioned internal coordinate diffusion. *arXiv* [q-bio.BM]. 2023. Available from: http://arxiv.org/abs/2305.19800.

43 Grambow, C.A., Weir, H., Cunningham, C.N., Biancalani, T., and Chuang, K.V. CREMP: conformer-rotamer ensembles of macrocyclic peptides for machine learning. *arXiv* [q-bio.BM]. 2023. Available from: http://arxiv.org/abs/2305.08057.

44 Gavenonis, J., Sheneman, B.A., Siegert, T.R., Eshelman, M.R., and Kritzer, J.A. Comprehensive analysis of loops at protein-protein interfaces for macrocycle design. *Nat Chem Biol*. 2014;10(9):716–722.
45 Tsai, C.Y., Salawu, E.O., Li, H., et al. Helical structure motifs made searchable for functional peptide design. *Nat Commun*. 2022;13(1):102.
46 Watkins, A.M. and Arora, P.S. Anatomy of β-strands at protein-protein interfaces. *ACS Chem Biol*. 2014;9(8):1747–1754.
47 Pace, J.R., Lampkin, B.J., Abakah, C., et al. Stapled β-hairpins featuring 4-mercaptoproline. *J Am Chem Soc*. 2021;143(37):15039–15044.
48 Silva, D.A., Correia, B.E., and Procko, E. Motif-driven design of protein–protein interfaces. In: *Computational Design of Ligand Binding Proteins* (ed. B.L. Stoddard). New York, NY: Springer New York; 2016. pp. 285–304.
49 Crook, Z.R., Girard, E.J., Sevilla, G.P., et al. Ex silico engineering of cystine-dense peptides yielding a potent bispecific T cell engager. *Sci Transl Med*. 2022;14(645):eabn0402.
50 Rocklin, G.J., Chidyausiku, T.M., Goreshnik, I., et al. Global analysis of protein folding using massively parallel design, synthesis, and testing. *Science*. 2017;357(6347):168–175.
51 Coutsias, E.A., Seok, C., Jacobson, M.P., and Dill, K.A. A kinematic view of loop closure. *J Comput Chem*. 2004;25(4):510–528.
52 Mandell, D.J., Coutsias, E.A., and Kortemme, T. Sub-angstrom accuracy in protein loop reconstruction by robotics-inspired conformational sampling. *Nat Methods*. 2009;6(8):551–552.
53 Guardiola, S., Varese, M., Roig, X., Sánchez-Navarro, M., García, J., and Giralt, E. Target-templated de novo design of macrocyclic d-/l-peptides: discovery of drug-like inhibitors of PD-1. *Chem Sci*. 2021;12(14):5164–5170.
54 Malde, A.K., Hill, T.A., Iyer, A., and Fairlie, D.P. Crystal structures of protein-bound cyclic peptides. *Chem Rev*. 2019;119(17):9861–9914. https://doi.org/10.1021/acs.chemrev.8b00807.
55 Leman, J.K., Weitzner, B.D., Lewis, S.M., et al. Macromolecular modeling and design in Rosetta: recent methods and frameworks. *Nat Methods*. 2020;17(7):665–680.
56 Park, H., Bradley, P., Greisen P Jr, et al. Simultaneous optimization of biomolecular energy functions on features from small molecules and macromolecules. *J Chem Theory Comput*. 2016;12(12):6201–6212.
57 McHugh, S.M., Yu, H., Slough, D.P., and Lin, Y.S. Mapping the sequence-structure relationships of simple cyclic hexapeptides. *Phys Chem Chem Phys*. 2017;19(4):3315–3324.
58 Slough, D.P., Yu, H., McHugh, S.M., and Lin, Y.S. Toward accurately modeling N-methylated cyclic peptides. *Phys Chem Chem Phys*. 2017;19(7):5377–5388.

59 Hui, T., Descoteaux, M.L., Miao, J., and Lin, Y.S. Training neural network models using molecular dynamics simulation results to efficiently predict cyclic hexapeptide structural ensembles. *J Chem Theory Comput*. 2023; 19(14) 4757–4769. https://doi.org/10.1021/acs.jctc.3c00154.

60 Lin, Z., Akin, H., Rao, R., et al. Evolutionary-scale prediction of atomic-level protein structure with a language model. *Science*. 2023;379(6637):1123–1130.

61 Norn, C., Wicky, B.I.M., Juergens, D., et al. Protein sequence design by conformational landscape optimization. *Proc Natl Acad Sci USA*. 2021;118(11):e2017228118. https://doi.org/10.1073/pnas.2017228118.

62 Wang, J., Lisanza, S., Juergens, D., et al. Scaffolding protein functional sites using deep learning. *Science*. 2022;377(6604):387–394.

63 Wicky, B.I.M., Milles, L.F., Courbet, A., et al. Hallucinating symmetric protein assemblies. *Science*. 2022;378(6615):56–61.

64 Yeh, A.H.W., Norn, C., Kipnis, Y., et al. De novo design of luciferases using deep learning. *Nature*. 2023;614(7949):774–780.

65 Vázquez Torres, S., Leung, P.J.Y., Venkatesh, P., et al. De novo design of high-affinity binders of bioactive helical peptides. *Nature*. 2024;626(7998):435–442.

66 Weininger, D. SMILES, a chemical language and information system. 1. Introduction to methodology and encoding rules. *J Chem Inf Comput Sci*. 1988;28(1):31–36.

67 Weininger, D. SMILES. 3. DEPICT. Graphical depiction of chemical structures. *J Chem Inf Comput Sci*. 1990;30(3):237–243.

68 Chang, L. and Perez, A. Ranking peptide binders by affinity with AlphaFold. *Angew Chem Int Ed Engl*. 2023;62(7):e202213362.

69 Tsaban, T., Varga, J.K., Avraham, O., Ben-Aharon, Z., Khramushin, A., and Schueler-Furman, O. Harnessing protein folding neural networks for peptide-protein docking. *Nat Commun*. 2022;13(1):176.

70 Ingraham, J.B., Baranov, M., Costello, Z., et al. Illuminating protein space with a programmable generative model. *Nature*. 2023;623(7989):1070–1078.

71 Abdin, O. and Kim, P.M. PepFlow: direct conformational sampling from peptide energy landscapes through hypernetwork-conditioned diffusion. *bioRxiv*. 2023. https://doi.org/10.1101/2023.06.25.546443.

72 Krishna, R., Wang, J., Ahern, W., et al. Generalized biomolecular modeling and design with RoseTTAFold All-Atom. *Science*. 2024;384(6693):eadl2528.

73 Lin, Y. and AlQuraishi, M. Generating novel, designable, and diverse protein structures by equivariantly diffusing oriented residue clouds. *arXiv* [q-bio.BM]. 2023. Available from: http://arxiv.org/abs/2301.12485.

74 Lin, Y., Lee, M., Zhang, Z., and AlQuraishi, M. Out of many, one: designing and scaffolding proteins at the scale of the structural universe with genie 2. *arXiv* [q-bio.BM]. 2024. Available from: http://arxiv.org/abs/2405.15489.

75 Yim, J., Trippe, B.L., De Bortoli, V., et al. SE(3) diffusion model with application to protein backbone generation. *arXiv* [csLG]. 2023. Available from: https://arxiv.org/pdf/2302.02277.

76 Alamdari, S., Thakkar, N., van den Berg, R., et al. Protein generation with evolutionary diffusion: sequence is all you need. *bioRxiv*. 2023 [cited 2024 Jul 26]. p. 2023.09.11.556673. Available from: https://www.biorxiv.org/content/10.1101/2023.09.11.556673v1.

77 Yim, J., Campbell, A., Mathieu, E., et al. Improved motif-scaffolding with SE(3) flow matching. *Transact Mach Learn Res*. 2024 [cited 2024 Jul 26]. Available from: https://openreview.net/pdf?id=fa1ne8xDGn.

78 Guo, Y., Ji, S., Wang, W., et al. An integrated strategy for assessing the metabolic stability and biotransformation of macrocyclic peptides in drug discovery toward oral delivery. *Anal Chem*. 2022;94(4):2032–2041.

79 Limbach, M.N., Antevska, A., Oluwatoba, D.S., et al. Atomic view of aqueous cyclosporine a: unpacking a decades-old mystery. *J Am Chem Soc*. 2022;144(28):12602–12607.

80 Golosov, A.A., Flyer, A.N., Amin, J., et al. Design of thioether cyclic peptide scaffolds with passive permeability and oral exposure. *J Med Chem*. 2021;64(5):2622–2633.

81 Kamenik, A.S., Linker, S.M., and Riniker, S. Matching simulations and experiments of conformationally flexible cyclic peptides: steps toward a holistic view of passive membrane permeability. In: *Approaching the Next Inflection in Peptide Therapeutics: Attaining Cell Permeability and Oral Bioavailability* (*ACS Symposium Series*; vol. 1417). American Chemical Society; 2022. pp. 137–54.

82 Ono, S., Naylor, M.R., Townsend, C.E., et al. Cyclosporin A: conformational complexity and chameleonicity. *J Chem Inf Model*. 2021;61(11):5601–5613.

83 Naylor, M.R., Ly, A.M., Handford, M.J., et al. Lipophilic permeability efficiency reconciles the opposing roles of lipophilicity in membrane permeability and aqueous solubility. *J Med Chem*. 2018;61(24):11169–11182.

84 Kelly, C.N., Townsend, C.E., Jain, A.N., et al. Geometrically diverse lariat peptide scaffolds reveal an untapped chemical space of high membrane permeability. *J Am Chem Soc*. 2021;143(2):705–714.

85 Furukawa, A., Schwochert, J., Pye, C.R., et al. Drug-like properties in macrocycles above MW 1000: backbone rigidity versus side-chain lipophilicity. *Angew Chem Int Ed Engl*. 2020;59(48):21571–21577.

86 Witek, J., Keller, B.G., Blatter, M., Meissner, A., Wagner, T., and Riniker, S. Kinetic models of cyclosporin A in polar and apolar environments reveal multiple congruent conformational states. *J Chem Inf Model*. 2016;56(8):1547–1562.

87 Bockus, A.T. and Lokey, R.S. Bioactive and membrane-permeable cyclic peptide natural products. In: *Practical Medicinal Chemistry with Macrocycles*. Hoboken, NJ, USA: John Wiley & Sons, Inc.; 2017. pp. 101–132.

88 Räder, A.F.B., Weinmüller, M., Reichart, F., et al. Orally active peptides – is there a magic bullet? *Angew Chem Int Ed Engl*. 2018;57(44):14414–14438. https://doi.org/10.1002/anie.201807298.

89 Wang, C.K. and Craik, D.J. Cyclic peptide oral bioavailability: lessons from the past. *Biopolymers*. 2016;106(6):901–909.

90 Golosov, A.A., Flyer, A.N., and Monovich, L.G. Design and discovery of orally bioavailable macrocycles: toward orally bioavailable peptide therapeutics. In: *Approaching the Next Inflection in Peptide Therapeutics: Attaining Cell Permeability and Oral Bioavailability* (*ACS Symposium Series*; vol. 1417). American Chemical Society; 2022. pp. 199–222.

91 Peraro, L. and Kritzer, J. Getting in: emerging methods and design principles for cell-penetrant peptides. *Angew Chem Int Ed Engl*. 2018;130(37)12042–12057. https://doi.org/10.1002/ange.201801361.

92 Abramson, J., Adler, J., Dunger, J., et al. Accurate structure prediction of biomolecular interactions with AlphaFold 3. *Nature*. 2024;630(8016):493–500.

Section III

Peptide Drug Lead Generation and Optimization

6

Peptide Discovery: Lead Generation

Yvonne Angell[1], Wendy Hartsock[2], Hans Melo[3] and Timothy M. Reichart[4]

[1]*WuXi TIDES, San Francisco, CA, USA*
[2]*Aralez Bio, Berkeley, CA, USA*
[3]*Menten AI, San Francisco, CA, USA*
[4]*Department of Chemistry, Hampden-Sydney College, Hampden-Sydney, VA, USA*

6.1 Introduction

There are 80 FDA-approved peptide drugs on the market (Table 6.1), and the number of new approvals has been steadily increasing over the past two decades. In addition, there are more than 170 peptides in active clinical development, with many more in preclinical studies. The scientific discoveries and innovations that led to the recent increased success of peptides as therapeutics has been an exciting and remarkable journey.

For decades, the development of peptide therapeutics was considered a high and costly hurdle, and peptides were instead synthesized primarily for use as tool compounds for biological assay development or as the starting point for discovery and development of small molecule drugs. In fact, the field of peptidomimetics emerged from this and has been widely utilized in the pharmaceutical industry to discover and develop new drugs by applying chemical approaches to transform peptides into more druggable, small molecule-like compounds.

There are several key scientific achievements and technological advancements that contributed to the rise of peptide therapeutics on the market. Some of the most notable achievements in the peptide therapeutic field are the first isolation of insulin from animal pancreas in 1921 by Frederick Banting [1], which was further developed by Frederick and Charles Best [2], solution-phase chemical synthesis of oxytocin in 1953 [3] and arginine vasopressin in 1954 by Vincent du

Peptide Science: Chemical Ligation, Lead Generation, and Therapeutic Advances, First Edition. Fa Liu.

Table 6.1 FDA-approved peptide drugs.

Approval year	Country	Generic (brand) name	Company	Mode of action	Therapeutic use	Route of administration	Peptide structure	Mode of discovery
1975	USA	Calcitonin-salmon (Miacalcin)	Novartis	Calcitonin receptor agonist	Paget's disease and postmenopausal osteoporosis	Nasal spray	32 aa (Cys1-Cys7)	Natural hormone
1977	USA	DDAVP (1-deamino-8-D-arginine vasopressin), Minirin	Lilly	Selective agonist for V2 receptors expressed in the renal collecting duct (CD)	Hemophilia A and von Willebrand disease (vWD), bed wetting in children	Nasal spray	1-deamino-8-D-arginine vasopressin, 8 aa (Cys3-Cys8)	Natural hormone
1982	USA	Insulin (Humulin, Humalog, Lantus, Levemir	Novo Nordisk, Eli Lilly, Sanofi	Glucose homeostasis	Diabetes	SC	Two S-S connected linear chains	Natural hormone
1983	USA	Cyclosporin (Restasis)	Allergan	Calcineurin inhibitor	Dry eye disease	Eye drops	Cyclic peptide (head-to-tail), 11 AA, D-AA, N-Me AA, noncanonical AA	Isolated from fungus *Tolypocladium inflatum*
1985	USA	Leuprolide (Lupron), Leuprorelin (Eligard)	AbbVie, Astellas, Takeda	GnRH analog	Prostate and breast cancer	SC	Linear peptide, 9 AA, noncanonical AA, -CONHEt	Natural hormone
1988	USA	Octreotide (Sandostatin)	Novartis	Somatostatin analog	Cancer, acromegaly	SC	8 AA, cycl (S-S), D-AA, -CH2OH	Natural hormone
1996	USA	Glatiramer (Copaxone, Glatopa)	Teva, Sandoz	Induce the peripheral Th2 cells to reduce inflammation	Multiple sclerosis	SC	Random mix of Glu,Ala,Lys,Tyr -COOH polypeptides	Copolymerization, rational design
1997	USA	Geref (Sermorelin)	EMD Serono	Growth hormone analog	Idiopathic growth hormone deficiency (GHD) in children	SC	(GRF [1–44]-NH2)	Natural hormone

2000	Japan	Taltirelin (Ceredist)	Mitsubishi Tanabe	Thyrotropin-releasing hormone (TRH) analog (TRH-R superagonist)	Adult spinal muscular atrophy	Oral	Peptidomimetic (3 aa)	Natural hormone TRH analog
2000	EU	Tractocile (Atosiban)	Ferring	Inhibitor of oxytocin and vasopressin	Premature labor	IV	Cyclic peptide (S-S), 9 AA, -CONH2	Natural hormone
2000	USA	Vasostrict	Endo Pharma	Synthetic arginine vasopressin	Hypotonic shock	IV	Cyclic peptide (S-S), 9 AA, -CONH2	Natural hormone
2000	USA	Bivalirudin (Angiomax)	Medicines Company	Direct thrombin inhibitor	Unstable angina	IV	Linear peptide (20 aa)	Synthetic congener of the naturally occurring drug hirudin, found in the saliva of the medicinal leech *Hirudo medicinalis*
2000	USA	Cyclosporin (Sandimmune, Neoral)	Sandoz/Novartis	Calcineurin inhibitor	Immune disease (psoriasis, colitis ulcerosa), organ transplantation	Oral soft gelatin capsules, oral solution, injection for IV	Cyclic peptide (head-to-tail), 11 AA, D-AA, N-Me AA, noncanonical AA	Natural product, isolated from fungus *Tolypocladium inflatum*
2001	USA	Nesiritide (Natrecor)	Scios	Stimulates cGMP	Congestive heart failure	IV	Disulfide bridged cyclic peptide (32 aa)	Natural hormone recombinant form
2001	USA	Triptorelin (Trelstar)	Debiopharm	GnRH agonist	Hormone responsive cancer	IM	Linear peptide (10 aa)	Natural hormone analog
2002	USA	Teriparatide (Forteo)	Lilly	Activates osteoblasts and osteoclasts	Osteoporosis	SC	Linear, 34 AA, recombinant produced	Natural hormone N-terminus

(Continued)

Table 6.1 (Continued)

Approval year	Country	Generic (brand) name	Company	Mode of action	Therapeutic use	Route of administration	Peptide structure	Mode of discovery
2003	Germany and the Netherlands	Abarelix (Plenaxis)	Praecis	GnRH antagonist	Prostate cancer	IM	Linear peptide (10 aa), 7 UAA	Modified natural hormone
2003	USA	Enfuvirtide (Fuzeon)	Roche	HIV fusion inhibitor, activator of FPR1	HIV-1	SC	Linear peptide (36 aa)	Derived from natural protein sequence of HIV-1 gp41 CHR domain
2003	USA	Daptomycin (Cubicin)	Cubist	disrupts bacterial cell membrane function resulting in bacterial cell death	systemic and life-threatening infections caused by Gram-positive infections	IV	Lipopeptide, cyclic (10 aa), with 3 aa exocyclic, N-terminal C10 fatty acid	Discovered from the actinomycete *Streptomyces roseosporus*
2004	USA	Ziconotide (Prialt)	Elan	Selective N-type voltage-gated calcium channel blocker	Severe chronic pain	IT	ω-Conotoxin peptide (3 S-S bridges, 25 aa)	Derived from the cone snail *Conus magus*
2005	USA	Exenatide (Byetta, Bydureon)	Amylin/Eli Lilly	GLP1 agonist	Diabetes Type II	SC	Linear peptide, 39 AA, -CONH2	Recombinant exendin-4 from *Heloderma suspectum*
2005	USA	Pramlintide (Symlin)	Amylin/Astra Zeneca	Amylin analog	Diabetes Type I and II	SC	Single disulfide bridged cyclic peptide (37 aa)	Natural hormone
2007	USA	Lanreotide (Somatuline LA)	Ipsen	Somatostatin analog	Acromegaly	IM	Single disulfide bridged cyclic peptide (8 aa)	Natural hormone

2008	USA	Degarelix (Firmagon)	Ferring	GnRH antagonist	Prostate Cancer	SC	Linear peptide (10 aa)	Natural hormone
2008	USA	Romiplostim (Nplate) (AMG531)	Amgen	TPOR agonist	immune thrombocytopenia (ITP)		Peptibody composed of 4 TPO-R agonist peptides linked by glycine spacers to the C-terminus of the Fc carrier domain	Phage display
2009	USA	Ecallantide (Kalbitor) (DX88)	Dyax Corporation	Plasma kallikrein inhibitor	Hereditary angioedema	SC	Developed from a Kunitz domain through phage display to mimic antibodies inhibiting kallikrein	Phage display
2009	USA	Telavancin (Vibativ)	Theravance/ Astellas	Antibacterial agent	Skin infection	IV	Lipoglycopeptide	Semisynthetic derivative of vancomycin
2009	USA	Romidepsin (Istodax)	Celgene/Teva	HDAC inhibitor	Cutaneous T-cell lymphoma	Oral	Depsipeptide	Natural product derived from
2010	USA	Liraglutide (Victoza)	Novo Nordisk	GLP-1 receptor agonist	Diabetes Type II	SC	Linear peptide, 31 AA, COOH, acylated with palmitic acid	Natural hormone
2010	USA	Tesamorelin (Egrifta)	Thera Technologies	GH receptor agonist	Lipodystrophy in HIV	SC	Linear analog of human GHRH with a *trans*-3-hexenoic acid group (44 aa)	Natural hormone
2011	USA	Boceprevir (Victrelis)[a]	Schering/Merck	NS3/4A protease inhibitor	Hepatitis C Virus genotype 1	Oral	Peptidomimetic (3 aa)	Rational design based on an HCV protease substrate sequence

(Continued)

Table 6.1 (Continued)

Approval year	Country	Generic (brand) name	Company	Mode of action	Therapeutic use	Route of administration	Peptide structure	Mode of discovery
2011	USA	Telaprevir (Incivek)	Vertex & JNJ	NS3/4A protease inhibitor	Hepatitis C Virus genotype 1	Oral	Peptidomimetic (4 aa)	Computational and med chem design from X-ray crystal structural knowledge of the active site of NS3/NS4A protease in the active form
2011	USA	Icatibant (Firazyr)	Shire/Takeda	Bradykinin B2 receptor antagonist	Hereditary angioedema	SC	Peptidomimetic (10 aa)	Natural hormone
2012	USA	Sinapultide (Lucinactant)	Discovery	Pulmonary surfactant	RDS in premature infants	ITD	Linear peptide sinapultide (KL4 acetate, 21 aa), combined with dipalmitoyl phosphatidylcholine, palmitoyl oleoyl phosphatidyl glycerol, and palmitic acid	Synthetic mimetic of natural SP-B
2012	USA	Pasireotide (Signifor)	Novartis	Somatostatin analog	Cushing's disease	SC	Cyclic peptide (6 aa)	Natural hormone
2012	USA	Peginesatide (Omontys)[a]	Affymax/Takeda	EPOR agonist	Anemia in CKD with dialysis	SC	Dimeric (2 × 22 aa), single disulfide bridged, pegylated (40 kDa branched PEG) peptide	Phage display

2012	USA	Carfilzomib (Kyprolis)	Onyx/Amgen	Selective proteasome inhibitor	Refractory multiple myeloma	IV	Linear peptidomimetic with N-terminal epoxy ketone, noncanonical (4 aa)	Synthetic epoxomicin (found in Actinomyces) derivative
2012	USA	Linaclotide (Linzess, Constella)	Ironwood/ Forest/Allergan/ Almirall	guanylate cyclase-2C (GC-C) agonist	Irritable bowel syndrome with constipation	Oral	Cyclic peptide (3 S-S), 14 AA	*E. coli* heat stable enterotoxin analog
2012	USA	Afamelanotide (Scenesse)	Clinuvel	Melanocortin 1 receptor (MC1-R) agonist	Erythropoietic protoporphyria	SC	Linear peptide (13 aa)	Analog of α-MSH ([Nle4,D-Phe7]-α-MSH)
2012	EU	Lixisenatide (Lyxumia)	Sanofi	GLP-1 receptor agonist	Type-II diabetes	SC	Linear peptide (44 aa), des-38-proline-exendin-4-(1–39)-peptidylpenta-L-lysyl-L-lysinamide	GLP-1 analog
2012	USA	Teduglutide (Gattex)	NPS	GLP-2 receptor agonist	Adult short bowel syndrome	SC	Linear peptide (33 aa), peptidomimetic	GLP-2 analog
2014	USA	Dulaglutide (Trulicity)	Eli LIlly	GLP-1 receptor agonist	Type-II diabetes	SC	Linear peptide, 30 AA, fusion with IgG4-Fc	GLP-1 analog
2017	USA	Plecanatide (Trulance)	Synergy	Activation of guanylate cyclase-C	Gastrointestinal laxative	Oral	16-Amino acid peptide structurally identical to human uroguanylin (with exception of a single-amino acid substitution	Guanylate cyclase-C (GC-C) agonist

(Continued)

Table 6.1 (Continued)

Approval year	Country	Generic (brand) name	Company	Mode of action	Therapeutic use	Route of administration	Peptide structure	Mode of discovery
2017	USA	Etelcalcetide (Parsabiv)	KAI/Amgen/Ono	Activation of CaSR on parathyroid chief cells	Secondary hyperparathyroidism in adult patients with chronic kidney disease on hemodialysis	IV	7 aa linear all D-aa peptide	Natural hormone
2017	USA	Abaloparatide (Tymlos)	Ipsen/Radius Health	Selective activation of the parathyroid hormone 1 receptor	Osteoporosis	SC	PTH 1-34	Natural hormone
2017	USA	Semaglutide (Ozempic)	Novo Nordisk	GLP-1 receptor agonist	Treatment of type 2 diabetes mellitus	SC	31 aa lipidated peptide	Modified natural hormone
2017	USA	Macimorelin (Macrilen)	Aeterna Zentaris	Growth hormone secretagogue receptor (ghrelin receptor) agonist	For the diagnosis of adult growth hormone deficiency	Oral	3 AA (Aib-Trp-gTrp-CHO)	Natural hormone (ghrelin) mimetic
2017	USA	Angiotensin II (Giapreza)	La Jolla Pharm Co.	Acts on the CNS to increase ADH production, vasoconstrictor	Hypotension resulting from septic shock or other critical conditions	IV	7 aa linear peptide	Natural hormone
2019	USA	Semaglutide (Rybelsus)	Novo Nordisk	Glucagon-like Peptide-1 agonist	Treatment of type 2 diabetes mellitus	Oral	31 aa lipidated peptide	Modified natural hormone
2020	USA	Sogroya (somapacitan)	Novo Nordisk		GHD in adults	SC	GH[Leu10Cys]	
2020	USA	Octreotide (Mycapssa)	Amryt	Somatostatin analog	Acromegaly	Oral	8 AA, cycl (S-S), D-AA, -CH2OH	Modified natural hormone

2021	USA	Vosoritide (Voxzogo)	Biomarin	Natriuretic peptide receptor B agonist(NPRB)	Achondroplasia	SC	39 aa, cycl (S-S)	C-type natriuretic peptide natural hormone analog
2021	USA	Lonapegsomatropin-tcgd (Skytrofa)	Ascendis	Prodrug of Somatropin	Growth hormone deficiency	SC	rhGH-PEG	Modified natural hormone
2021	USA	Pegcetacoplan (Empaveli, Aspaveli)	Apellis, Swedish Orphan Biovitrum AB	C3 complement inhibitor	Paroxysmal nocturnal hemoglobinuria (PNH)	SC	Cyclic peptide (S-S), 13 AA, -CONH2, Acetyl-, dimer, PEGylated	Phage display
2022	USA	Tirzepatide (Mounjaro, Zepbound)	Eli Lilly	Dual GIP and GLP-1 receptor agonist	Type II diabetes	SC	Lipidated, linear peptide, -CONH2	Med chem
2022	USA	Lutetium 177Lu Vipivotide Tetraxetan (Pluvicto)	ABX, Endocyte, Novartis	PSMA targeting ligand	Antineoplastic activity against PSMA-expressing tumor cells	IV	177Lu-PSMA-617	Radioconjugate
2023	USA	Zilucoplan (ZILBRYSQ)	Ra Pharma, UCB	C5 complement inhibitor	Adult patients with acetylcholine receptor antibody positive (AChR-Ab+) generalized myasthenia gravis (gMG)	SC	15 AA macrocyclic peptide, (CH2CH2O)23-C16	mRNA display
2023	USA	Pegcetacoplan (Syfovre)	Apellis Pharmaceuticals, Swedish Orphan Biovitrum AB	C3 complement inhibitor	Geographic atrophy	Intravitreal injection	Cyclic peptide (S-S), 13 AA, -CONH2, Acetyl-, dimer, PEGylated	Phage display
2023	USA	Trofinetide (Daybue)	Neuren, Acadia	Analog of the neuropeptide (1–3) IGF-1	Rett Syndrome	Oral solution	Alpha-methyl proline tripeptide (G-P-E) analog	Analog of the neuropeptide (1–3) IGF-1

(Continued)

Table 6.1 (Continued)

Approval year	Country	Generic (brand) name	Company	Mode of action	Therapeutic use	Route of administration	Peptide structure	Mode of discovery
2023	USA	Rezafungin (Rezzayo)	Cidara and Melinta	New generation of broad-spectrum, long-acting echinocandin antifungal (inhibits the 1,3-β-D-glucan synthase complex in the fungal cell wall)	Candidemia and invasive candidiasis fungal infections	IV infusion	Cyclic hexapeptide analog of anidulafungin (Micafungin), C5 ornithine position is replaced with a choline aminal ether	Derived from fungi FR901379, and produced by *Coleophoma empetri*
2023	USA	Posluma (flotufolastat F 18)	Blue Earth Diagnostics	18F-labeled PSMA ligand and radiodiagnostic imaging agent	Detection and localization of prostate cancer in PET imaging	IV injection	Flotufolastat F 18 gallium DOTAGA complex bound to nonradioactive gallium	Radioconjugate
				Peptide drugs in the clinic				
Phase 3		MK-0616	Merck	PCSK9 inhibitor	Lowers low-density lipoprotein (LDL) cholesterol	Oral	Macrocyclic peptide	mRNA display
Phase 2		TCMCB07	Endevica Bio	MC4R antagonist	Cachexia	SC	Cyclized, synthetic peptidomimetic	Modified natural hormone
Phase 3		JNJ-2113	JnJ	IL-23R antagonist	Moderate-to-severe plaque psoriasis	Oral	Macrocyclic peptide	Phage display
Phase 1		LUNA18	Chugai	KRAS inhibitor	Oncology	Oral	Macrocyclic peptide	mRNA display

[a]Withdrawn.

Vigneaud [4], and the invention of solid-phase peptide synthesis (SPPS) by Robert B. Merrifield in 1963 using a solid support resin to enable peptide synthesis at high speed, yield, and purity [5]. Since then, there have been numerous inventions of novel coupling reagents [6] and orthogonal protecting group strategies [7,8] to accelerate peptide synthesis as well as advances in solid support resins and linkers for SPPS [9–12]. Furthermore, fragment synthesis approaches, particularly native chemical ligation [13] and expressed protein ligation (EPL) [14,15], which allow the selective ligation of unprotected polypeptides from synthetic and recombinant origin, have advanced the assembly of long peptide and protein sequences. Additional technological advancements enabling discovery and development of peptides as key therapeutic agents include the automation of peptide synthesis, improved methods for peptide purification and characterization, development of drug delivery technologies to increase the efficacy and bioavailability of peptides, and novel strategies for therapeutic peptide hit discovery, design, and hit-to-lead optimization.

This chapter will focus on the successful and emerging novel approaches to peptide therapeutic hit discovery and lead generation. These strategies fall into three general categories. The first category is inspiration from nature. Traditionally, peptides have been discovered as natural receptor agonists or antagonists. In these cases, the natural molecules (or fragments) themselves are used as the starting points, or "hits," for drug design and hit-to-lead optimization. A second category is screening of artificial peptide libraries, which can be comprised of either a random design for diversity and screening against a wide variety of targets, or predesigned libraries taking into account some structural knowledge of the target binding site or protein–protein interface. A third, more recent category is computational screening of virtual peptide libraries and *de novo* design of peptide therapeutics involving purely computational tools.

Since the completion of the human genome project in 2003 [16], a tremendous number of novel drug targets have been identified, leading to a high demand for more high-throughput hit discovery strategies. This demand has led to the development of many novel approaches to peptide hit discovery. In this chapter, we will discuss some of the traditional approaches to peptide therapeutic hit discovery and compare and contrast these strategies with the recent innovations in peptide hit discovery.

The majority of the marketed peptide therapeutics are based on natural receptor agonists or antagonists and will be discussed in Section 6.1.1. Some notable examples include insulin, oxytocin, vasopressin, leuprorelin, and somatostatin. In addition to those marketed peptides based on natural hormones, there are five marketed peptide therapeutics that were discovered using an artificial peptide library hit discovery approach, either phage display or mRNA display. In addition, there are many other novel artificial peptide hit discovery platform technologies, including plate-based or bead-based rational design or random library synthesis, split 'n pool libraries, DNA-encoded library (DEL) screening, functional selection screening, bacterial display and computational or *de novo* design. These novel platform approaches are all capable of generating large numbers of peptide hits, which coupled with advances in drug delivery technologies are predicted to result in an increasing number of approved peptide drugs over the next decade.

6.1.1 Inspiration from Nature

When seeking a molecule that has a specific biological function, it is logical to start one's search with naturally occurring molecules that have that function. In the case of peptides, those peptides can be fragments of proteins, whole peptides (e.g. hormones), or molecules derived from those pieces. At the same time, the peptide and/or analogs must be accessible by large-scale production. Therefore, the history of peptide drugs can be divided into different eras that track the development of enabling technologies, both for synthesis and for discovery. The first peptide drug, insulin, was isolated from whole animal tissue. The development of peptide chemical synthesis, first in solution and then in solid phase, allowed the development of synthetic peptide drugs, first from native sequences and then from synthetic analogs. The combination of the two, exemplified by the process chemistry found in the chemical synthesis of Fuzeon, dramatically expanded the accessibility of synthetic peptide drug discovery platforms in both academia and industry.

One way to organize the history of peptide drug development is by enabling technology. The earliest peptide drugs were taken directly from nature without modification. Insulin, commercially available to the public starting in 1923, was initially harvested from animal pancreas extracts (typically bovine and porcine), which required more than 10 pounds of animal pancreas to produce enough insulin for one patient for 1 year [17]. It is not surprising that the first FDA-approved product from recombinant DNA technology was recombinant insulin in 1982 [18] after the discovery by City of Hope and Genentech of a method to produce insulin using recombinant DNA technology in 1978 [19,20]. Although beyond the scope of this chapter, several excellent reviews on the history of insulin drug development, including substantial peptide medicinal chemistry to produce longer-lasting medicines, were published around the 100-year anniversary of insulin's discovery [21–23].

The first peptide drug obtained by chemical synthesis was oxytocin, a nine-amino acid long peptide initially investigated for its role in inducing contractions in labor. It was first synthesized in 1953 by Vincent du Vigneaud, whose achievement earned him the 1955 Nobel Prize in Chemistry; this was followed by its commercial introduction in 1962 in order to induce childbirth. Du Vigneaud synthesized a related peptide hormone, vasopressin (Figure 6.1), in 1954, which also became commercially available in 1962 [24].

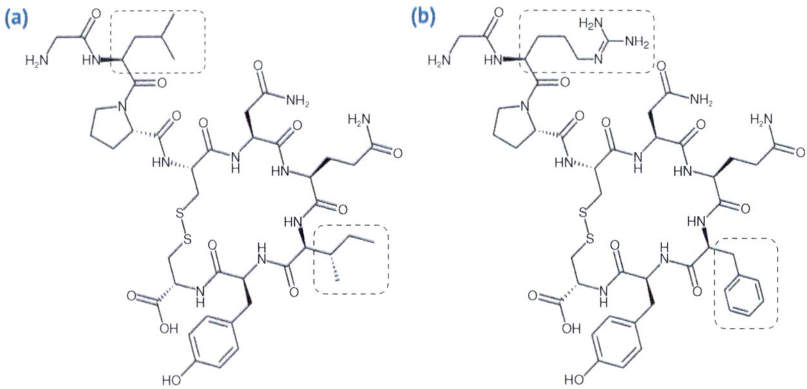

Figure 6.1 (a) Structure of oxytocin. (b) Structure of vasopressin.

Although the first synthetic peptide drugs were synthesized using traditional solution methods, the field of peptide chemistry as a whole took a gigantic leap forward with R. Bruce Merrifield's development of SPSS, first published in 1963 [5], and an achievement that earned him the 1984 Nobel Prize in Chemistry. Synthesizing a peptide on a solid support and washing away unreacted products instead of requiring separate workup procedures after each synthetic step drastically increased the speed, reliability, and efficiency of peptide synthesis. Additionally, the development of automated peptide synthesizers [25] further increased the speed, reliability, and throughput of this essential enabling technology [26,27].

Of course, not all peptide drugs are naturally occurring sequences. It did not take long after the commercial introduction of naturally occurring peptides for medicinal chemistry to develop synthetic analogs to introduce desired properties or to ameliorate undesired ones. In 1971, the structure of the native gonadotropin-releasing hormone (GnRH) was determined [28,29]. This hormone plays an essential role in the reproductive system, and recognition of this important discovery quickly resulted in Roger Guillemin and Andrew Schally winning shares of the Nobel Prize in Physiology or Medicine in 1977. Potential therapeutic applications were immediately recognized; however, the short half-life of the native hormone (2–4 minutes in serum) necessitated the development of longer-lasting analogs, such as leuprorelin, with a serum half-life of about 3 hours [30]. Two modest chemical differences – substitution of D-leucine for glycine and modification of the C terminus to an ethyl amide (Figure 6.2) – are only possible through chemical synthesis. These modest chemical changes result in a large biological and physiological effect through enhanced binding affinity for the anterior pituitary receptor and enhanced resistance to proteolytic degradation, resulting in a molecule

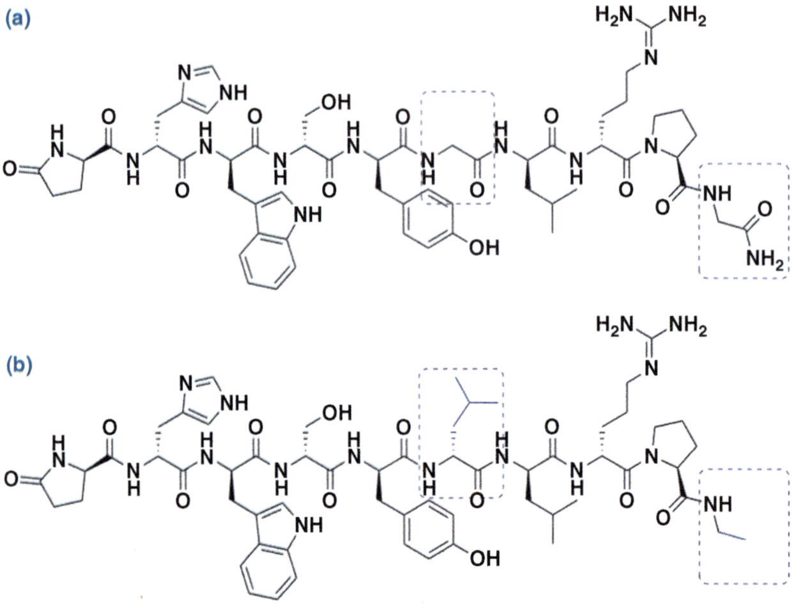

Figure 6.2 (a) Structure of native gonadotropin-releasing hormone (GnRH). (b) Structure of leuprorelin, with modifications from native sequence outlined.

80 times more potent than native GnRH. Although leuprorelin was initially approved by the FDA in 1984 to treat hormone-responsive cancers, it has subsequently been approved for a variety of other clinical applications, including the treatment of prostate cancer, endometriosis, and uterine fibroids, as well as being used during *in vitro* fertilization [31]. It is still under investigation for a variety of additional indications ranging from Alzheimer's disease to functional bowel disease [31,32].

The increasing pace of technological development used in drug discovery has made it difficult to neatly map more recent drugs to single specific technological advancements. However, several key technologies used in modern drug discovery include high-performance liquid chromatography (HPLC), mass spectrometry, and peptidomics/proteomics. HPLC, first made commercially available in 1967, was quickly adopted to separate, purify, and analyze synthetic peptides at high speed and high efficiency, and is more broadly applicable than other separation techniques [33]. HPLC was shortly coupled to electrospray ionization mass spectrometry (ESI-MS), to more rapidly identify a complex mixture of peptides in solution [34]. These highly sensitive tools, in combination, allow for the discovery of individual components of highly complex mixtures of molecules that are a rich source of peptide lead compounds. One example of this process was used in the

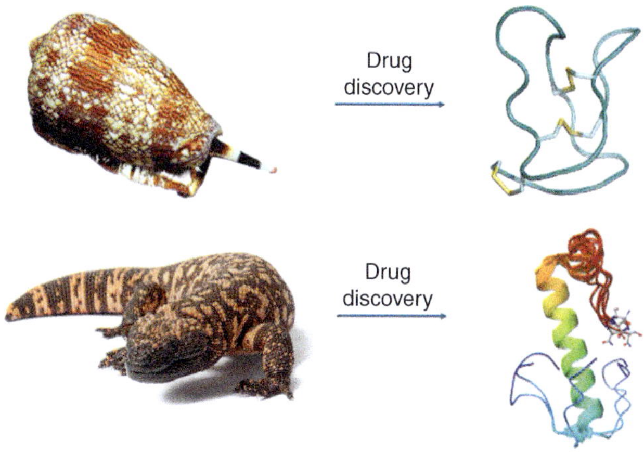

Figure 6.3 Peptide drug discovery from venomous organisms. (a) Isolation of peptide toxins from cone snails led to the development of ziconotide. (b) Isolation of peptide toxins from gila monsters led to the development of exenatide.

discovery of ziconotide, a drug approved in 2004 as a potent nonopioid analgesic to treat chronic severe pain. A 25-amino acid peptide with three disulfide bonds, ziconitide is an ω-conotoxin that was identified through the HPLC and MS-MS analysis of the venom of the cone snail Conus magus (Figure 6.3), which contains over 100 separate peptide toxins [35]. Ziconotide works as a venom by blocking Ca channels in the prey animal; its mode of action in humans is to selectively block presynaptic neuronal N-type Ca channels in the spinal cord [36].

One notable peptide drug dodging the tendency of peptides to not be suitable to oral administration is linaclotide, approved in 2012 by the FDA for the treatment of irritable bowel syndrome with constipation (IBS-C) and abdominal pain. Linaclotide is a synthetic 14-amino acid analog of the conserved regions of bacterial heat-stable enterotoxins that derive its heat and proteolytic stability from three intramolecular disulfide bonds [37]. This structure also provides the stability needed for oral administration. Linaclotide functions similarly to endogenous hormones, binding and activating guanylate cyclase-C in intestinal epithelial cells, which increases cGMP levels, which in turn induces fluid secretion and accelerates intestinal transit, resulting in reduced abdominal pain and constipation associated with IBS-C [38]. Although effective, further development of linaclotide analogs via peptide medicinal chemistry has already shown the possibility for significant extensions gastrointestinal stability with minimal change in potency [39].

A recent explosion in glucagon-like peptide 1 (GLP-1) analogs such as liraglutide and semaglutide deserves mention here, in part because of the effectiveness in patient weight loss [40]. Semaglutide is a fantastic example of the power of peptide drug discovery and of peptide medicinal chemistry. GLP-1 is a naturally occurring peptide hormone that acts to reduce blood sugar levels by enhancing insulin secretion and is therefore an appealing drug candidate. Its short serum half-life, however, has led to many discovery programs to increase the effectiveness primarily by increasing serum half-life. Semaglutide has several modifications from the natural GLP-1 to enhance half-life, including incorporating an unnatural Aib residue and conjugation at K26 with glutamic acid and a 17-carbon carboxylic acid, which allow semaglutide to bind to serum albumin and see drastically improved serum half-life allowing weekly injections [41].

As in the field of small molecule drug discovery, there remain vigorous research programs in isolating naturally occurring peptides with pharmacological effects from a staggering variety of organisms found around the world, ranging from Australian cone snails to Peruvian green tarantulas [42]. The various -omics revolutions, including genomics, proteomics, peptidomics, and metabolomics, have drastically accelerated the pace of this research, both guiding where to look for novel molecules as well as the identification of those molecules. We anticipate these efforts discovering new chemical modalities to serve as future peptide hit generation successes.

6.1.2 Artificial Libraries

The development of novel hit discovery technologies has advanced rapidly to keep pace with the high unmet medical need for molecules to address targets classified as "undruggable" by small molecules. The discovery of novel drug hits as starting points for targets where there is not a natural receptor agonist or antagonist as a starting point "hit" presents a significant challenge. Recent advances in peptide display methods (Figure 6.4), including phage display, ribosomal display, mRNA display, yeast cell display, one-bead one-compound, DNA-encoded, peptide nucleic acid encoded, and new peptide-encoded technology coupled with advances in parallel and automated peptide synthetic methods have spurred the emergence of numerous innovative peptide hit discovery platform approaches. These platforms rely on building diverse libraries and either a screening or selection technique to identify a phenotype of interest [43–49]. Selection-based platforms generally couple genotype, with phenotype to link function with DNA survival, wherein DNA coding for only the fittest variant is identified. Display-based libraries typically identify molecules with high target-binding affinities through multiple rounds of selection. Some of these approaches are described in more detail.

The five marketed peptide therapeutics discovered using artificial peptide library hit discovery strategies each took a novel path from discovery to the market (Figure 6.5). Romiplostim, ecallantide, peginesatide, and pegcetacoplan were optimized from hits discovered using phage display library screening. Furthermore,

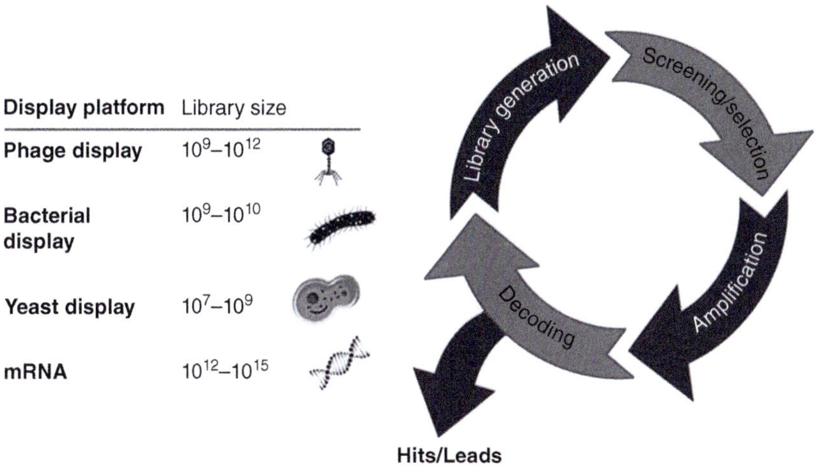

Figure 6.4 Peptide hit discovery display technologies. The size and diversity of the displayed peptide libraries varies depending on the platform. Regardless of the display strategy employed for hit discovery, each technique depends on rounds of library generation followed by screening, amplification and decoding. As hits are identified, they may be fed back into the system for further optimization to generate leads.

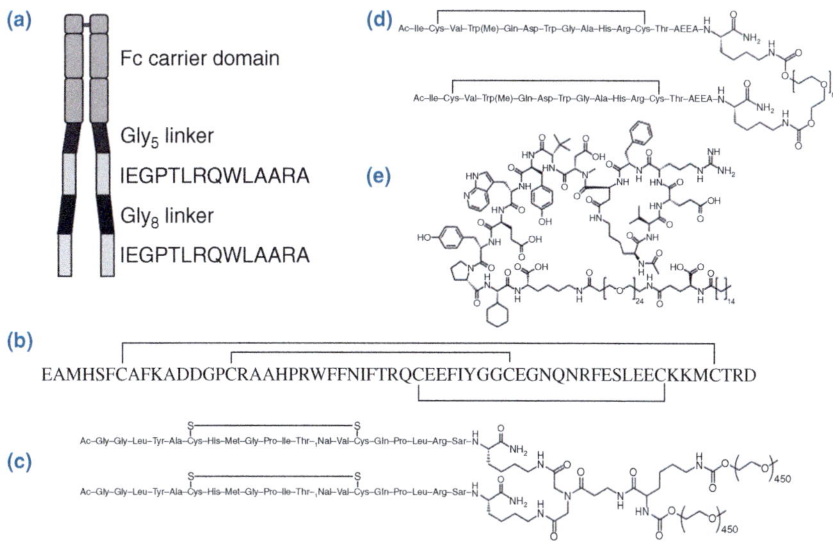

Figure 6.5 Chemical structure of approved peptide drugs discovered using display approaches. (a) Chemical structure of romiplostim, (b) chemical structure of ecallantide, (c) chemical structure of peginesatide, (d) chemical structure of pegcetacoplan, (e) chemical structure of zilucoplan. Fc, fragment crystallizable.

FDA-approved zilucoplan, and MK-0616 as well as LUNA 18, currently in phase 3, were developed from hits originally identified through mRNA display technology. There are yet more examples in the clinic, which have advanced from optimization of phage display hits, such as several bicyclic peptide-toxin conjugate molecules, peptides targeting inflammation, and many other novel molecules in various phases of development.

6.1.3 Phage Display

Phage display and advances on this technology have revolutionized peptide hit discovery by utilizing the potential of molecular diversity to unlock new possibilities in targeted drug discovery. Phage display is a powerful tool for discovery of novel peptides that bind to protein targets of interest and can be applied to a wide range of target classes [50]. In 1985, George Smith first pioneered phage display technology for the presentation of short peptides on the surface of bacteriophages [51], and the technique has been further developed by other researchers. In fact, George Smith and Greg Winter were awarded a share of the 2018 Nobel Prize in chemistry for their contributions to phage display [52]. The power of phage display is the ability to display multiple copies of a large number of peptides to enable rapid identification and selection of binders to a protein target of interest. After each selection round, mutations are introduced into the enriched combinatorial library, and the next generation of peptides is screened for improved affinity and/or activity.

Phage display has some limitations. In general, phage display is limited by the complexity of the sequence libraries that can be generated; for example, phage display libraries typically are limited by transfection efficiency to less than 10^9 independent members. In contrast, other screening methods such as chemical-based libraries or cell-based screening allow testing of hundreds or thousands of synthetic compounds, or tens of thousands of plated colonies. In fact, 10^9 variants in a phage display library is actually a small part of the available theoretical sequence space [53].

A series of publications from the Derda Lab in 2012–2016 described chemical modification of the N-terminal residue on phage-displayed peptides. They started from commodity linear phage-displayed libraries peptide libraries and employed chemical posttranslational modifications (cPTM) to produce cPTM-libraries with diverse unnatural functionalities such as monosaccharides [54–56], complex glycans [57], pro-fluorescent functionalities [58], covalently-reactive functionalities such as electron-deficient alkenes [59], and small molecule warheads [60]. These cPTM phage libraries from Derda Lab contributed fundamental knowledge to the growing field of Genetically Encoded Fragment-Based Discovery (GE-FBD) [61]. These publications benchmarked a universal biotin capture methodology for monitoring the reaction rate and regioselectivity of any cPMT in any phage displayed library and built the foundation for monitoring chemical reactivity in hundreds of thousands of peptides by next-generation sequencing [62,63]. These publications also built the foundation for the first example of DNA-encoding of cPTM in phage libraries [60].

In 2017, Derda Lab built on S_NAr chemistry popularized by the Pentelute Lab [64] to identify two commercially available perfluoro aromatics, perfluoropyridine (fPy) and decafluoro-diphenylsulfone (DFS), which can be used to produce billion-scale phage-displayed libraries modified by perfluoromatic linchpins [65]. In 2023, they not only demonstrated the use of DFS-macrocyclic libraries to discover macrocycles that bind to albumin but also uncovered one of the major liabilities of DFS-macrocycles [66]. DFS reacts with biological thiols even after their incorporation into the peptide. Problems can be circumvented by late-stage replacement of DFS by less reactive isosteres or by expanding S_NAr chemistry to other type of cross-linkers: for example, in 2020, Chuanliu Wu et al. [67] developed other classes of S_NAr-reactive linchpins for modification of phage-displayed libraries. Perfluoropyridine (fPy) has not been used in phage-displayed libraries since 2016, but it is due to be revisited after a 2023 publication by Ben Davis' lab identified fPy-modified Cys residues as precursors to chiral radicals en route to stereoretentive radical reactions on proteins [68]. Analogous radical chemistry, in principle, can be employed in billion-scale fPy-modified peptide libraries and the feasibility for creating such libraries has been highlighted in a 2016 report of Derda Lab [65].

In 2016–2023 the Derda Lab transferred these developments from linear to cyclic architectures and built a core for development of macrocyclic genetically encoded libraries with grafted unnatural pharmacophores. It employs DiChloroAcetone (DCA) chemistry described by the Dawson lab for modification of synthetic peptides in 2015; Ng and Derda employed biotin capture technology [54] to measure the rate of modification of phage-displayed libraries by DCA [69]; they demonstrated regioselectivity of such modification and observed that DCA is a one of the most benign reagents for the modification of the peptides displayed on phage. Unlike classical xylene-derived cross-linkers developed by Timmerman et al. [70] and re-purposed by Heinis et al., in traditional modification of phage-display libraries [71], DCA exhibited no toxic impact on infectivity of phage even at higher concentrations. Ng and Derda then upgraded DCA from a mere cross-linker to a pharmacophore-functionalized linchpin that can perform a simultaneous cyclization and grafting of a pharmacophore onto phage libraries. They attached a series of carbohydrates onto DCA linchpin and formed DiChloro-acetone-derived-Oxime (DCO) cross-linkers. Reacting phage libraries with DCO-glycan produced a series of macrocyclic libraries with grafted glycans. Although the 2016 publication did not use DCA or DCO-modified libraries for any discovery procedures, subsequent publications from the Heinis (Kale 2018) and Bogyo Labs [72] employed DCO and DCA linchpins to discovery potent macrocyclic, bicyclic, and covalent inhibitors, all building on DCA-DCO framework.

To date, the Derda Lab has disclosed two reports that describe bicyclization of phage-displayed libraries [73]. Both break away from complex C3v-symmetry imposed on bicyclic peptides by reports of Timmerman [70] and Heinis-Winter [71]. The Derda Lab pioneered the use of C2-symmetric linchpin that has two thiol-reactive groups and one group that reacts with N-terminus to form bicyclic libraries [74]. They demonstrated their utility by selecting inhibitors for growth factors [74] (Wong 2021) and proteases [73]. The main design principle of bicyclization by Derda Lab involves separation of reactivities in assembly of phage-displayed bicycles: N-terminal reactivity is driven in acidic conditions, whereas Cys-reactivity is driven in mildly basic conditions. In 2023, the Jianmin Gao lab perfected the concept of C2-linchpin-modified bicycles displayed on phage [75].

While genetic code expansion for incorporation of noncanonical amino acids as well as chemical modification efforts are helping to increase the diversity of phage display libraries beyond the 20 canonical amino acids [76], peptides from phage display are mostly limited to sequences containing all natural amino acids. Therefore, hit-to-lead optimization of peptide hits discovered through phage display screening platforms often presents a considerable challenge requiring significant work to improve peptide bioactivity, stability, and other more druggable attributes. All the marketed peptide drugs discovered through phage display technology required a significant lead optimization effort.

Nevertheless, phage display is a useful technique that has been widely applied to accelerate peptide hit discovery primarily for macrocycles and bicyclic peptides with particular success in oncology [53,77,78]. The approach developed by Christian Heinis at EPFL for bicyclic peptide phage display expression using fully synthetic short peptides constrained with a small molecule scaffold to form two loops that stabilize their structural geometry facilitates target binding with high affinity and selectivity, making bicycles attractive candidates for drug development [79]. Significant progress has been made using this approach. In fact, there are several bicyclic peptides and bicyclic peptide conjugates in various phases of development, including Bicycle toxin-conjugates and Bicycle tumor-targeted immune cell agonists™ (Bicycle TICA™ [80–83]. Phage display has also been successfully applied in discovery of disulfide-rich peptides [84–86].

6.1.4 mRNA Display

The unique power of mRNA display is that it combines *in vitro* display technology with the flexibility to incorporate a wide variety of nonnatural amino acids to produce novel drug-like peptides [53]. Unlike certain other display technologies, *in vitro* display does not require the use of a bacterial or yeast host, and it can produce libraries of 10–100 trillion members, allowing for the rapid discovery of highly potent candidate molecules. mRNA display was discovered in the lab of Jack W. Szostak [121–123] at Mass General Hospital and further developed by Jack Szostak and Hiroaki Suga [124,125]. The current technology allows for the generation of mRNA–peptide fusion libraries comprised of 10^{12}–10^{15} independent members. The technology is a robust system in which an mRNA becomes directly attached to the peptide or protein it encodes by a stable covalent linkage. The RNA acts as both a chimeric message and adapter, which when translated, the adapter end enters the ribosome and accepts the nascent peptide to generate a covalent mRNA–peptide fusion. Thus, the sequence information present in the peptide is encoded in the covalently attached mRNA. This allows the unreadable information in the protein portion to be read and recovered through the attached mRNA [121]. This is an innovative *in vitro* selection technology based on the physical linkage of a peptide to a nucleic acid tag (the mRNA that encoded it) that can be amplified by PCR and read by DNA sequencing. The peptide component may be linear or cyclized using a variety of chemistries. The throughput of this technology and library diversity has been greatly improved with the Flexizyme [126,127] and RaPID [128] advancements. More recently, researchers have expanded on this technology by demonstrating that the promiscuous Orthogonal aminoacyl-tRNA synthetases (ORSs), for example ORS *p*-CNF-RS, can incorporate noncanonical amino acids at amber codons in *in vitro* translation (IVT), including the novel substrate para-cyanopyridylalanine (*p*-CNpyrA), to enable a pyridine-thiazoline (pyr-thn) macrocyclization in mRNA display [129]. They used this approach to identify a potent macrocyclic binder that exhibits good selectivity for the deubiquitinase USP15 and close homologs over other ubiquitin-specific proteases (USPs), thus illustrating how promiscuous ORSs can both expand side-chain diversity and provide structural novelty in mRNA display libraries through a heterocycle forming macrocyclization [129].

mRNA-display is conceptually similar to phage display, which works by linking the peptide on the surface of the phage capsid with the amplifiable and readable DNA that encoded it contained within the phage DNA [130]. However, there are three main differences that are important for macrocycle discovery. The first is that all of the steps of mRNA-display are entirely *in vitro* and, therefore, library size is not limited by the need to transform bacteria. Transformation steps impose an upper limit on the diversity of phage display libraries of 10^9. By contrast, peptide mRNA-display libraries can be as large as 10^{13}, which enables complete coverage of macrocyclic peptide libraries (10^{12}) with approximately 10 copies of each variant. The oversampling afforded by the sequence capacity of mRNA display ensures that essentially every sequence has sufficient copies to be selected.

A second advantage of mRNA display is that it is monovalent, with only one copy of the peptide displayed on a given mRNA. The monovalent nature of peptide–mRNA fusions allows the displayed peptides to be enriched based solely on their target affinity. In contrast, phage particles display several copies of the same peptide, which produces an avidity effect that allows peptides with modest target affinities to enrich during the selection process [130]. Lastly, the synthesis of the peptide–mRNA fusion molecules can be done in a modified *in vitro* translation reaction that has been modified with a more drug-like set of unnatural amino acids [131–134]. This technology is being applied widely across peptide discovery teams for discovery of peptides for both extracellular as well as intracellular targets [135]. This technology has already demonstrated clinical success with Zilucoplan marketed as ZILBRYSQ.

6.2 Approved Drugs Discovered Through Phage Display

6.2.1 Romiplostim (NPlate)

The FDA approved romiplostim on August 22, 2008, for the treatment of chronic immune thrombocytopenia (ITP) in patients who have had an unfavorable response to corticosteroids, immunoglobulins, or splenectomy. Romiplostim is recommended in patients whose degree of thrombocytopenia increases the risk of bleeding. It is not recommended for normalizing the platelet count. Romiplostim is a thrombopoietin receptor agonist (TPO-RA) that has been used in adults with chronic immune thrombocytopenia (ITP) for more than 11 years [87]. Use of romiplostim was also approved in pediatric patients in the United States and the European Union, and use in adults has been extended to include those with newly diagnosed ITP. There are several other marketed TPO-RAs used to stimulate TPO receptors on hematopoietic stem cells to promote platelet production; however, romiplostim is the only TPO-RA that binds at the same site as endogenous TPO protein [87].

Romiplostim (molecular weight ~60 kDa) is a peptibody composed of four identical thrombopoietin peptides of 14 amino acids each that are chemically coupled by glycine spacers to the carboxy-terminus of the Fc carrier domain. The four TPO-R binding domains identified by screening mutagenesis phage display peptide libraries [88] bind with high affinity to the TPO-R (also known as the myeloproliferative leukemia protein, MPL, or CD110) and have no sequence homology with endogenous TPO [89–92]. Romiplostim binds to and activates the TPO-R on megakaryocyte precursors in bone marrow [90]. This sequence is unique and would have been nearly impossible to identify without phage display technology. It binds in the same manner as endogenous TPO and can displace TPO from its receptor [93,94]. Romiplostim activates many of the same pathways as TPO, leading to sustained improvement of platelet counts with continued treatment in patients with ITP [95].

Two families of small peptides that bind to the human thrombopoietin receptor and compete with the binding of the natural ligand thrombopoietin (TPO) were initially identified from recombinant peptide libraries. The sequences of these peptides were not found in the primary sequence of TPO. Screening libraries of variants of one of these families under affinity-selective conditions yielded a 14-amino acid peptide (Ile-Glu-Gly-Pro-Thr-Leu-Arg-Trp-Leu-Ala-Ala-Arg-Ala) with high affinity (dissociation constant approximately 2 nM) that stimulates the proliferation of a TPO-responsive Ba/F3 cell line with a median effective concentration (EC50) of 400 nM. Dimerization of this peptide by a carboxyl-terminal linkage to a lysine branch produced a compound with an EC50 of 100 picomolar, which was equipotent to the 332-amino acid natural cytokine in cell-based assays. The peptide dimer also stimulated the *in vitro* proliferation and maturation of megakaryocytes from human bone marrow cells and promoted an increase in platelet count when administered to normal mice [93] (Cwirla 1997). However, the final romiplostim drug structure is not dimerized at the C-terminus of the 14-aa sequences, but rather four copies of the sequence are linked through glycine spacers to the Fc carrier domain. Romiplostim, marketed as NPlate by Amgen, received FDA approval in 2008, and is the first FDA-approved drug resulting from a phage display hit discovery approach [96–99]. Romiplostim is produced by recombinant DNA technology in *Escherichia coli* (*E. coli*).

6.2.2 Ecallantide (Kalbitor)

The US Food and Drug Administration (FDA) approved ecallantide subcutaneous injection, marketed as Kalbitor by Dyax Corp in 2009 for the treatment of sudden and potentially fatal fluid build-up associated with hereditary angioedema (HAE) in patients aged 16 years and older [98]. Ecallantide, an inhibitor of the protein kallikrein, is a 60-mer polypeptide, which was based on the first Kunitz domain of human tissue factor pathway inhibitor and discovered through phage display. It mimics antibodies inhibiting kallikrein. It is used for the treatment of hereditary angioedema (HEA) and in the prevention of blood loss in cardiothoracic surgery [100]. Ecallantide is a 60 amino-acid recombinantly expressed large peptide, or small protein that binds selectively to plasma kallikrein and blocks its binding site, thereby inhibiting the generation of bradykinin from high-molecular mass kininogen [53,101,102]. Hereditary angioedema is a rare autosomal dominant disease characterized by a deficiency of C1 esterase inhibitor (C1-INH), a critical endogenous inhibitor of plasma kallikrein [103,104]. The disorder is associated with recurrent attacks of severe, painful swelling and is thought to be caused by unregulated activity of kallikrein and excessive bradykinin production [105,106]. Ecallantide is a potent ($Ki = 25$ pM), selective, and reversible inhibitor of plasma kallikrein that effectively inhibits the active plasma kallikrein generated during an HAE attack [107]. Ecallantide was discovered through phage display technology and is produced in the Pichia pastoris strain of yeast [51,101,108].

6.2.3 Peginesatide (Omontys)

Peginesatide (Omontys, Affymax), a peptide-based ESA [109,110], was approved in the United States in March 2012 for the treatment of anemia caused by chronic kidney disease in adults who are undergoing dialysis. Although peginesatide has no sequence homology with erythropoietin, [109] it stimulates the erythropoietin receptor, initiating a similar intracellular signaling cascade [111]. Peginesatide is pegylated, which enhances its biologic activity *in vivo*. Phase 2 data suggested that treatment with peginesatide (0.019–0.044 mg per kilogram of body weight, once a month) could maintain hemoglobin levels in patients with chronic kidney disease who were not receiving dialysis [112].

Peginesatide is a peptide drug designed to mimic the biological activity of Epo (erythropoietin). It binds to the human Epo-R (erythropoietin receptor) and activates the receptor to stimulate erythropoiesis in human red blood cell precursors [113]. In 2012, the FDA approved it to treat patients with renal failure–induced anemia. Peginesatide is a synthetic peptide that consists of three parts: two identical monomeric peptides with 21 amino acids, a novel linker derived from iminodiacetic acid and β-alanine, and a 40-kD branched polyethylene glycol (PEG) chain (Figure 6.5c). Conjugated monomeric peptides contribute pharmacologic activity, whereas PEGylation of the peptide prolongs the half-life and improves the pharmacokinetic behavior. The original hit peptide was identified by phage display from a library of disulfide-cyclized peptides [114] and subsequently further engineered to improve its pharmacological properties. In particular, it was dimerized, which afforded increased affinity for the erythropoietin receptor (EPOR) and correspondingly elevated potency in cell-based assays and in mice [115], and it was chemically conjugated to a 40-kDa branched PEG to prolong the *in vivo* half-life. The PEG moiety confers increased peptide solubility and stability and reduced renal clearance, thereby increasing plasma persistence [116,117].

6.2.4 Pegcetacoplan (Empaveli)

Pegcetacoplan (APL-2) is a member of the compstatin family of cyclic peptides, which selectively bind human and nonhuman primate C3 and C3b and prevent interaction with the C3 convertase through steric hindrance [118], The initial hit molecule was discovered in 1996 through a phage-displayed peptide library screen for C3b-binding [119]. Optimization to design derivatives with enhanced inhibitory activity (1MeW, Cp40) and the addition of polyethylene glycol moieties resulted in peptide drugs with longer half-life properties. Pegcetacoplan is a medication used to treat paroxysmal nocturnal hemoglobinuria (PNH) and geographic atrophy of the retina. It acts as a complement inhibitor, specifically targeting complement protein C3. Pegcetacoplan is the first FDA-approved drug to treat GA (geographic atrophy) secondary to AMD (age-related macular degeneration) [120].

6.3 Approved Drugs Discovered Through mRNA Display

6.3.1 Zilucoplan (ZILBRYSQ, Ra Pharma, UBC)

The clinical success of mRNA display as a viable platform technology for therapeutic peptide hit discovery has been demonstrated by the FDA approval of zilucoplan for use in the treatment of generalized myasthenia gravis (gMG) and paroxysmal nocturnal hemoglobinuria (PNH). This is the first drug on the market to come from mRNA display hit discovery. Zilucoplan, discovered by Ra Pharmaceuticals (now a wholly owned subsidiary of UCB), is a small (15 amino acid) macrocyclic peptide developed based on the ExtremeDiversity™ mRNA display platform [136]. It is lipidated with a palmitoyl (C16) lipid conjugated to the peptide through an ethylene glycol (CH2CH2O)23 spacer. Zilucoplan is a subcutaneous (SC), self-administered peptide inhibitor of complement component 5 (C5 inhibitor) for the treatment of adult patients with acetylcholine receptor antibody positive (AChR-Ab+) generalized myasthenia gravis (gMG). In 2019, the US FDA granted orphan drug designation to zilucoplan for the treatment of MG. Zilucoplan is a macrocyclic peptide that binds to the protein complement component 5 and inhibits its cleavage into C5a and C5b. Additionally, zilucoplan inhibits the binding of C5b with C6. Thus, zilucoplan appears to use a dual mechanism of action to prevent terminal complement pathway activation [137]. As a complement C5 inhibitor, zilucoplan is a targeted therapy that inhibits key components in the underlying pathophysiology of gMG, addressing the underlying mechanism of neuromuscular junction damage. Zilucoplan received its first approval, in Japan, in September 2023 for the treatment of gMG in adult patients who inadequately respond to steroids or other immunosuppressants and are positive for anti-acetylcholine receptor (AChR) antibodies [136]. Subsequently, zilucoplan was approved in the USA in October 2023 for the treatment of gMG in adult patients who are anti-AChR antibody positive [138], and in the EU in December 2023 as an add-on to standard therapy for the treatment of gMG in adult patients who are anti-AChR antibody positive [138]. With subcutaneous self-administration, zilucoplan has the potential to provide a more convenient treatment option for patients with gMG compared with intravenously administered C5 inhibitors [136]. The optimization of zilucoplan from initial hit to clinical candidate required significant optimization. From two mRNA display selections targeting full length complement C5, two active peptide hit series were identified, two double digit hits (RA100358, 12 nM IC50 in red blood cell (RBC) lysis assay and RA100379, 27.5 nM IC50) were selected in 2012. These two peptide series were combined to form a single hybrid lead molecule, which was then optimized to improve potency, stability and pharmacokinetic parameters. Through several key modifications, including *N*-methylation of Asn for protease stability, replacement of phenylglycine with cyclohexylglycine to avoid epimerization, and truncation of the C-terminal norvaline residue and replacement with lysine, followed by C-terminal lipidation on the lysine side chain for half-life extension the goal of once weekly dosing was achieved. Approximately 250 peptides were designed and synthesized during the hit-to-lead optimization resulting in the clinical candidate in 2014, RA101495 with potency 2.4 nM IC50 in the RBC lysis assay, plasma protein binding >99% in rat, monkey and human, and PK half-life ~ 180 hours in cyno dosed sc at 3 mg/kg.

6.3.2 Plate-based Array Synthetic Libraries

Despite improvements to classical display techniques such as the Flexizyme [126,127] and RaPID [128] technologies, there is still a need for more diverse libraries containing modifications that are commonly found in natural bioactive peptide molecules but cannot be incorporated into display libraries, methods that select for molecules with inherent properties such as proteolytic stability and cell permeability, and the ability to rapidly iterate and rationally mature a "hit" to a "lead" by exploring a predefined chemical space to further optimize therapeutic characteristics [139].

In 1991, Fodor et al. presented a new high-throughput hit discovery technology employing light-directed, spatially addressable parallel chemical synthesis [140–142]. They synthesized peptide libraries in the form of two-dimensional arrays, e.g. on a glass plate, each member of the combinatorial library being identified by its position on the array. By merging the techniques of solid-phase chemistry and photolithography, a spatially addressable array of peptides was displayed. The base plate is first functionalized with amino groups, which are protected with a photolabile leaving group. By masking regions of the plate and removing the protecting group from exposed regions with light, amino groups become available for coupling at selected, trackable sites on the plate. By a smart masking technique, excellent spatial resolution can be achieved on sites of only 50 m^2, thus making it possible to provide 40,000 discrete synthesis sites on a 1 cm × 1 cm square plate. The position of each peptide depends on the masking strategy. This method has been expanded to use in other fields, for example, for DNA sequencing and for diagnosing of genetic mutations ("DNA-on-the-chip"). The typical amounts of molecules of interest synthesized with this technology are in the range of mmol or nmol.

This technology has been further developed to enable rational peptide array library design combined with parallel screening and stepwise evolution to discover novel peptide binding hotspots on proteins. These hotspots can be methodically matured to create high-affinity, high-specificity binding peptides to a protein target in a consistent and digitally controlled process [139]. The method can be applied to synthesize both linear and cyclic peptides, as well as peptides composed of natural and nonnatural amino acids, thereby enabling screening in a greatly diverse chemical space. This digital, light-directed array technology [140,143,144] can be used to synthesize arrays containing peptides on an amine-functionalized slide and combines four characteristics that could facilitate its use as a novel binder discovery platform: (1) high feature density, resulting in 2.9 million unique molecules per array; (2) ability to accommodate broad chemical diversity, e.g. nonnatural amino acids; and (3) digitally controlled synthesis.

6.3.3 Split and Pool (Tea Bag) Approach

In 1985, Houghton published the so-called "tea-bag method" [145], which had been originally developed for the multiple peptide synthesis of approximately 150 different peptides in 50 mg amounts (with a peptide chain length of approximately 10 amino acids). For separation and isolation means of the polymeric support ("beads"), they are shrink-packed into a net-like, polypropylene bags. Each bag contains between 50 and 100 mg polymeric resin beads as supports, and the mean bead diameter has to exceed the mean pore diameter of the polymeric membrane of the bag. A simple handwritten and solvent-resistant label on the bag on the outside allows for coding and tracking. Washing steps, coupling, and cleavage of protecting groups are carried out in a common container (i.e. beaker or flask), with the coupling of different amino acids and the final cleavage of the resulting peptide carried out in a unique vessel for each bag, or a selection of polymeric bags containing the support resin beads. One characteristic of the tea-bag method is that the information about the spatial localization of each reaction step is ensured by labeling the bags. The flexibility, the relatively high amounts of resulting peptides, and the low equipment investment are advantageous. However, the intensity of manual lab work (washing, sorting, labeling, etc.) and the limitations regarding the number of manageable peptides lessen the scope of application of this special method.

Another unique approach, the "pin-synthesis" reported in 1984 by Geysen [146], operates with lower amounts of peptides and involves a parallel synthesis of molecular libraries of peptides, in which the peptides were created one amino acid at a time, beginning with an attachment to an array of amino-functionalized polyethylene rods (i.e. 8 × 12 matrix, each pin: 4 mm diameter, 40 mm length).

By knowing the specific reaction history of each pin, the resulting peptide on each pin is known. All reaction steps are performed by handling the array of pins either in separate wells for each pin or huge containers for the whole array. Because of the very small amount of peptide produced (~300 nmol) and the lack of such sensitive test assays at this time, there was no need to separate the peptide from the top of the pin. By working in this way, the positional information of each material was known at each reaction step. This methodology has the advantage of enabling both synthesis and screening on the same platform and thus has a large impact on immunochemical assays for peptide–antibody interactions. This technology has also been applied to other fields, such as a fully combinatorial synthesis of solid-state materials for functional inorganic materials, where a "single bead reactor" was developed using the Split&Pool (S&P) principle [147].

6.3.4 Split-and-pool DNA-encoded Libraries Approach (Wendy)

Encoded split-and-pool combinatorial libraries via genetically barcoded building blocks to generate DELs of 10^{12} peptides through alternating steps of peptide synthesis and DNA elongation were introduced by Brenner and Lerner in 1992 [148] (PNAS 1992). This technique has since been widely adopted by the pharmaceutical industry and adapted primarily for the generation of small molecule libraries. However, with advances in DEL technology such as DNA-templated [149,150] and DNA-reported techniques, there has been a resurgence in peptide libraries, particularly macrocyclic peptide libraries [151,152]. DELs can be generated on solid supports or in solution and used to find a single pharmacophore or multiple pharmacophores made up of DNA tagged fragments [153] There are a variety of encoding strategies and selection techniques that have been reviewed [49]. Finding compatible chemistry between peptide chain elongation and DNA polymerization has been challenging and some have opted for Peptide Nucleic Acid-encoded libraries that are compatible with SPPS [154,155]. An advantage of DEL over some of the display techniques is the ability to incorporate unlimited noncanonical amino acids and other moieties to access a broad and drug-like chemical space. However, as with most hit platforms, optimization to the lead is typically required [153].

6.3.5 Functional Selection for Peptide Hit Identification

Intracellular approaches to peptide hit identification in eukaryotic systems offer the advantage of engaging protein targets in their native states. In addition to relevant posttranslational modifications such as phosphorylation, secondary metabolite cofactors, and redox states, any protein binding partners that are required for activity, whether normal or aberrant, are also represented in the cellular milieu, further recapitulating physiologically relevant conditions.

The utilization of the reverse yeast two-hybrid system to identify inhibitors of protein–protein interactions (PPIs) in the ToRPPIDO system exploits functional selection to identify otherwise inaccessible mechanisms of PPI disruption. In ToRPPIDO, a negative selection reporter is activated upon complex formation of the targeted proteins, leading to reversible cell death. When an inhibitor of the targeted PPI binds and disrupts the complex, cell death is avoided, enabling cell survival. This system is not biased toward orthosteric sites often identifying binders of loops and other allosteric sites that can also act as intramolecular glues. While the system is also modality agnostic, peptide libraries can be directly expressed in the host cell and can even undergo downstream processing such as cyclization. Synthetic molecules can also be delivered to the yeast cells, allowing for incorporation of noncanonical residues. A variety of PPIs can be targeted using ToRPPIDO, including scaffolding proteins, PPIs within multisubunit complexes, broad specificity enzymes such as nucleases, phosphatases, and proteases in addition to transcription factors and splicing factors [156].

A similar yeast hybrid system was also developed as a first-in-class approach to identify protein degraders through E3 ligase-mediated ubiquitination. Yeast has the unique benefit of having a conserved eukaryotic ubiquitin-proteasome system and can be further engineered to encompass many of the 600 predicted human E3 ligase specificity factors. The ToRNEDO system allows for identification and functional validation of the target, E3 ligase, and hit compound from the primary screen. This is a result of the ability to multiplex the system through the expression of a barcoded E3 ligase library in the host cell, thus selecting for functional E3 ligase engagement simultaneously with hit compound identification. As with ToRPPIDO, hit libraries may be recombinantly expressed in the host cell or synthesized and delivered to the cells [157].

The SICLOPPS (Split-Intein Circular Ligation of Peptides and ProteinS) approach for intracellular screening of macrocyclic peptides takes advantage of two-hybrid selection and has been applied to both prokaryotic and eukaryotic systems for functional screening *in vivo* [158,159]. SICLOPPS generates 10^{7-9} head-to-tail cyclized peptide libraries ranging from tetrapeptides to proteins and has been employed to identify inhibitors of hypoxia-inducible factors (HIFs) 1 and 2 [160]. As the system predominantly incorporates canonical amino acids, additional synthetic medicinal chemistry is likely necessary to drive efficacy and address any pharmacokinetic liabilities.

6.3.6 Bacterial Display

The display of peptides on bacterial surfaces, often incorporated at surface protein termini or in exposed loops, allows for the generation of 10^9–10^{10} membered libraries that can be screened using quantitative techniques such as fluorescence-activated cell sorting (FACS) [161–163]. This approach has been deployed to identify a range of peptide and protein hits, including antibody epitopes, peptide substrates, and cell targeting peptides [162]. More recently, Surface Localized Antimicrobial displaY (SLAY) was developed by Davies and colleagues to identify functional macrocyclic peptide inhibitors of bacterial growth in human serum [164,165]. Other advancements, such as Stabilized Peptide Engineering with *E. coli* Display (SPEED), allow for the incorporation of noncanonical amino acids via click chemistry to constrain peptides on the bacterial surface and identify optimized link locations [166].

While phage and mRNA display have had clinical successes, none of the current peptide hit discovery technologies meet all of the needs to accelerate peptide drugs to the clinic. However, as the value of peptide therapeutics is fully realized, these technologies are rapidly advancing, and improvements to existing technologies or the emergence of newer, more advanced cutting-edge technologies promise to fulfill the hit discovery needs in the near future.

6.3.7 Computational Approaches

As discussed above, peptide therapeutics, owing to their size, high specificity, efficacy, and low toxicity, have emerged as promising candidates for drugging challenging targets beyond the reach of small molecules and biologics. However, the inherent challenges associated with peptide drug development such as poor stability and bioavailability have hindered their clinical translation [167], and there remains significant need for new methods to help identify and optimize potential peptide therapeutics.

While library screening methods such as phage display, mRNA display, or synthetic combinatorial libraries can interrogate a seemingly large number of peptides against a target of interest, their main limitation is the size of the libraries (often limited to only the 20 canonical amino acids), which represents only a small fraction of the potential testable sequence space. For example, a 14-residue peptide built from different amino acids (e.g. a mix of the 20 natural amino acids and 10 noncanonical) could have ~10^{20} possible sequences, which is far greater than the upper limit of mRNA display of ~10^{13} [168]. Because the number of peptides that can bind potently and selectively to a specific target is very limited, the chances of identifying a hit peptide using screening methods are very low [169]. In practice, scientists need to run multiple screening campaigns and several rounds of optimization and chemical modifications before a suitable peptide candidate is found, if successful at all.

Computational methods offer a powerful alternative to overcome these obstacles and expedite the design and optimization of peptide-based therapeutics. The main advantage of this approach is the much larger sequence space that can be interrogated compared to high-throughput screening methods, as well as the ability to model and design with noncanonical amino acids that may improve drug properties. In recent years, structure-based design methods have been developed that enable the design of peptides that not only bind to a drug target of interest but also offer other necessary drug properties [169,170].

6.3.8 *De Novo* Peptide Design

What does it mean to *design* a peptide?

In general, the peptide drug design problem can be described as finding a sequence of amino acids, among an astronomical number of possibilities, that will fold into a stable peptide structure capable of binding to a drug target of interest (usually a protein), as well as exhibiting other key desirable drug properties such as specificity, solubility, stability, and bioavailability. In particular, *de novo* computational design aims to generate novel peptide sequences completely from scratch without requiring a pre-existing peptide as a starting point.

6.3.8.1 Physics-based versus ML-based Approaches

Two distinct methodologies have emerged that offer different, yet complementary, strategies for generating potential peptide hits: physics-based modeling and machine learning. Physics-based approaches, grounded in fundamental principles of physics, chemistry, and biophysics, rely on computational algorithms and simulations to solve the peptide design problem [170]. In contrast, machine learning-based approaches generally rely on devising and training models on large datasets of experimentally characterized peptides or proteins to predict the function of existing peptides [171] or to design novel peptides [172,173]. As discussed later, both approaches offer advantages and disadvantages. For illustration purposes, we summarize below the general process for designing cyclic peptides using a state-of-the-art physics-based approach with specific references to the Rosetta Software Suite.

6.3.9 Physics-based Design: The Process of Designing Cyclic Peptides

For structure-based peptide design, the starting point is ideally a high-resolution structure of the drug target of interest. These are typically obtained through X-ray crystallography, nuclear magnetic resonance, or cryogenic electron microscopy methods. Computationally predicted structures may also be obtained through homology modeling or ML-based structure prediction such as RoseTTAFold [174] or Alphafold2 [175]. However, the use of low-quality or low-confidence predicted structures increases the risk of failure due to inaccurate structures, making high-resolution crystal structures the preferred choice.

Once the target structure has been identified and prepared, design can begin. Standard practice using the Rosetta Design Suite is to first identify one or a couple of functional groups, *anchors*, that are known to bind at the interface of interest. In the absence of known binding functional groups, peptide designers may use computational methods such as Rotamer Interaction Field or rely on their knowledge and intuition to propose suitable functional groups or amino acids [176]. These anchors are used to define starting points for creating the peptide binder. The anchors are then extended by adding amino acids to each end so as to "grow" the peptide backbone to a desired length. The N- and C-termini are then connected to each other via a chemical bond (i.e. amide bond, disulfide, thioether, etc.).

The next step is to sample conformations of the backbone against the drug target of interest. Rosetta includes a generalized method, GeneralizedKIC, which samples backbone conformations while allowing the anchor groups to remain relatively static at the binding interface while ensuring the closure of the cyclic peptide [177,178]. Once potential backbones have been identified, side-chain amino acids are added to stabilize the backbone and improve binding to the target. This is known as sequence design.

In physics-based approaches, the sequence design problem can be expressed as a complex combinatorial optimization problem, where given a fixed backbone, a list of designable positions, and a set of side-chain identities, the goal is to find a sequence of identities that minimizes an objective function. Many algorithms have been proposed to solve this problem, including exhaustive enumeration and deterministic algorithms [178]; however, as design problems increase in size, the solution space becomes intractable. Fortunately, heuristic methods such as Monte Carlo sampling provide a quick means to approximate a solution and provide a practical approach to solving large problems [179]. For example, the Rosetta software's Packer implements a heuristic method based on Monte Carlo random walks [180]. More recently, quantum computing algorithms have also been proposed for solving the design problem [181]. Once a sequence has been identified, further computational methods are then used to allow backbone flexibility and improve conformational selection.

Note that the design process alone does not maximize binding to the target, but rather optimizes a sequence to stabilize a given conformation. In order to drive the scoring function to optimize binding to the target, additional techniques that add more weight to the energy at the binding interface are implemented [182]. The resulting output is a virtual library of 10^4–10^5 peptide designs.

The final step is to screen through the virtual library to extract the best candidates to take forward for wet lab validation. Here, a number of filters and metrics (e.g. shape complementarity, ΔG of binding, hydrogen bond satisfaction, fold propensity, etc.) are applied according to the design requirements to select the most likely candidates to succeed in the wet lab.

6.4 A Successful Example

Scientists have used computational *de novo* design successfully to design cyclic peptide that binds to therapeutic targets. For example, Mulligan and colleagues applied a similar design protocol as described above to computationally design peptide macrocycles that inhibit New Delhi Metallo-β-Lactamase 1 (NDM-1), an antibiotic resistance factor that degrades penicillin and other related molecules [183]. Seven cyclic peptides were selected from the design pool and tested *in vitro* for binding and inhibiting NDM-1. Six of those peptides were in fact stronger inhibitors than the starting point control inhibitor. X-ray crystallography confirmed that the designs matched closely the predicted conformation with <1Å rmsd accuracy, and for two of the designs, even the binding mode was an almost perfect match (see Figure 6.6).

This study is significant for several reasons. First, it showcases the *de novo* design of cyclic peptides designed to bind a specific target; second, it demonstrated the incorporation of noncanonical amino acids using L- and D-amino acids to expand the building blocks available for design; and finally, researchers were able to predict the rank order of IC50 values from computational fold propensity metrics. The latter further validated their hypothesis that minimizing the entropic cost of binding by explicitly designing cyclic peptides to favor a binding-competent conformation would improve binding affinity. More generally, this work demonstrates a potentially generalizable protocol for cyclic peptide design that can be applied across a variety of targets, including many challenging drug targets of significant unmet medical need. In fact, other related studies have built on this work to design potent cyclic peptides with therapeutic potential such as histone deacetylase inhibitors [184] and PPI-disrupting peptides [185].

6.4.1 Current Challenges and Future Directions in Computational Peptide Design

While recent years have seen immense progress in computational methods for peptide design, many challenges remain [170]. For physics-based design, one challenge is modeling the Gibbs free energy. When designing cyclic peptide binders, researchers aim to maximize the Gibbs free energy between the bound state and all other states. However, enthalpy is quantum mechanical in nature and modeling it requires solving complex quantum mechanical functions, which is challenging and prohibitively expensive to do in practice for large molecules such as peptides. While classical approximations such as force fields exist and are generally used, their accuracy is low and thus it constitutes a challenge for design accuracy. Aside from computing enthalpies accurately, entropic challenges such

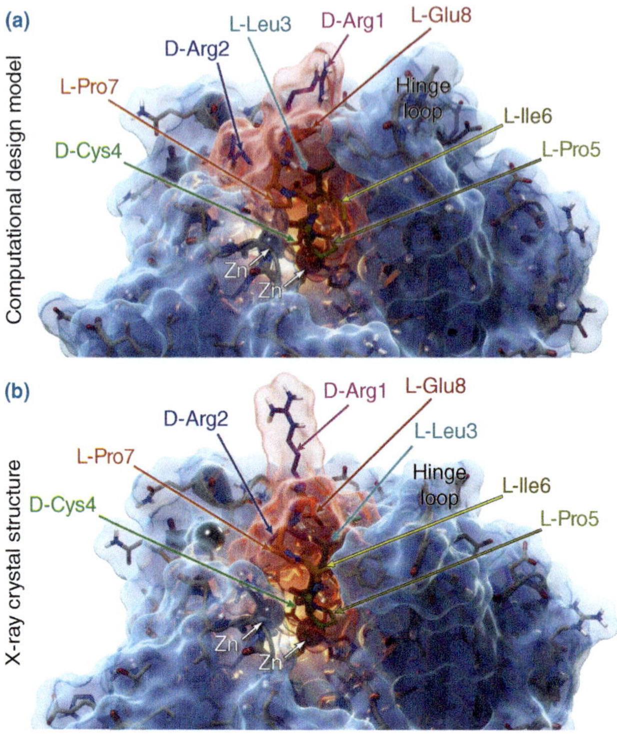

Figure 6.6 Comparison of computational design model (a) and X-ray crystal structure (b) of computationally designed peptide macrocycle NDM1i-1G (pink) bound to the New Delhi metallo-β-lactamase 1 (NDM1, blue). X-ray crystal structure shows chains B and D of PDB structure 6XBF; side chains for residues not resolved in the electron density (d-Arg2, Asp66, and Met67) have been generated with Rosetta. The NDM-1 active site is illuminated, with active-site zinc atoms labeled. The NDM-1 hinge loop, which presents a hydrophobic surface to which the NDM1i-1G peptide binds, is indicated. *Source*: Adapted from Mulligan et al. (2021).

as the hydrophobic effect remain. Rosetta and other software implement approximations and different solutions to get around these challenges, and efforts are ongoing to improve accuracy while maintaining computational costs feasible.

For ML-based design approaches, and deep learning in particular, one of the main challenges is the requirement to train on large datasets. In contrast to the large body of protein structures available, peptide structures are much more limited, and databases of peptides with characterized biochemical functions are also very scarce. Even when small datasets are available, the data is often noisy and biased, making it not useful for extracting generalizable ML models. Nevertheless, researchers have been able to use "small data" machine learning methods such as random forest or support vector machines to leverage small datasets to inform design [186–188]. Unfortunately, these methods do not generalize very well beyond the training set and are largely limited to specific cases. There is a significant need for developing methods that are capable of leveraging small datasets, but which can generalize or extend beyond specific use cases.

A specific challenge for designing peptides is the fact that, in contrast to proteins, peptides can be built from noncanonical amino acids that exhibit different properties. This means that, unfortunately, generative models trained on protein data do not generalize to peptides containing nonnatural amino acids. Again, the lack of training data containing peptides with noncanonical amino acids prohibits the development of large generalizable machine-learning models based on experimental data. This is in contrast to physics-based methods, which, although not trivial, are able to model and design with noncanonical amino acids.

On the other hand, while physics-based methods excel in capturing molecular-level details and mechanistic insights, machine learning techniques offer scalability, speed, and the ability to navigate high-dimensional sequence-structure spaces efficiently. Integrating these complementary approaches holds the promise of revolutionizing peptide drug discovery and design, enabling the development of next-generation therapeutics with unprecedented precision and efficacy. A potential approach is to use physics-based methods to inform or extend the data available to machine learning models. One potential advantage of this approach is the potentially lower cost and computational efficiency of running machine learning models rather than expensive computational simulations. For example, molecular dynamics simulations could be used to generate data, which is then fed to a deep learning model for training.

Another promising technology for advancing design and optimization of peptides is quantum computing. For example, where classical computers typically solve difficult combinatorial problems by iterating through many possibilities, quantum computers have the potential to represent all possible solutions to a problem posed by a superposition of quantum states [188,189]. Recently, researchers have explored different ways to leverage this technology in the context of peptide design, including representing the problem as a combinatorial optimization problem solvable on quantum annealers [181,190], or on gate-based quantum computers [191,192], as well as solving the multi-body docking problem on quantum hardware [193]. As quantum hardware continues to evolve, larger and more complex problems may be solvable on these hard devices and may eventually offer an advantage over classical hardware. As current technology matures and new methods become available in the coming years, many of these challenges will be addressed and improved methods for peptide design may accelerate drug discovery and development of next-generation peptide therapeutics that address unmet medical needs.

6.5 Conclusion

Long recognized as key mediators of important biological effects, peptides hold a prominent position in drug discovery. Recently, the steadily growing interest in peptides as therapeutics, coupled with the need for novel drug candidates to address "undruggable" targets, has led to the emergence of powerful peptide hit discovery platforms and technologies. The numerous advancements in peptide hit discovery will enable improved peptide drug starting points for hit-to-lead optimization and accelerate the path to lead candidate selection. Peptides promise to present exciting opportunities for therapeutic intervention.

The ability to generate and analyze large peptide libraries efficiently could accelerate the development of new therapeutics with improved properties, thereby enhancing the treatment options for many diseases where there is an unmet medical need. The specific successful examples provided by novel hit discovery platform technologies presented here highlight some of the emerging capability in the peptide field to refine and optimize peptide drugs, potentially leading to more effective and longer-lasting treatments.

References

1 Banting, F.G., Best, C.H., Collip, J.B. Campbell, W.R., and Fletcher, A.A. Pancreatic extracts in the treatment of diabetes mellitus. *Can Med Assoc J.* 1922;12(3):141–146.

2 Scott, D.A. and Best, C.H. The preparation of insulin. *Ind Eng Chem.* 1925;17:238–240.

3 du Vigneaud, V., Ressler, C., Swan, J.M., Roberts, C.W., and Katsoyannis, P.G. The synthesis of an octapeptide amide with the hormonal activity of oxytocin. *J Am Chem Soc.* 1953;75:4879–4880.

4 du Vigneaud, V., Gish, D.T., and Katsoyannis, P.G. A synthetic preparation possessing biological properties associated with arginine-vasopressin. *J Am Chem Soc.* 1954;76:4751–4752.

5 Merrifield, R.B. Solid phase peptide synthesis. I. Synthesis of a tetrapeptide. *J Am Chem Soc.* 1963;85(14):2149–2154.

6 Albericio, F. and Carpino, L.A. Coupling reagents and activation. *Methods Enzymol.* 1997;289:104–126.

7 Isidro-Llobet, A., Álvarez, M., and Albericio, F. Amino acid-protecting groups. *Chem Rev.* 2009;109(6):2455–2504.

8 Behrendt, R., White, P., and Offer, J. Advances in Fmoc solid-phase peptide synthesis. *J Pept Sci.* 2016;22(1):4–27.

9 Fields, G.B. and Noble, R.L. Solid phase peptide synthesis utilizing 9-fluorenylmethoxycarbonyl amino acids. *Chem Biol Drug Des.* 1990;35(3):161–214.
10 Chan, W.C. and White, P.D. (eds.) *Fmoc Solid Phase Peptide Synthesis: A Practical Approach (Practical Approach Series)*. 1st ed. Oxford University Press; 2000.
11 Kreutzer, A.G., Salveson, P.J., Yang, H., and Guaglianone, G. Standard practices for Fmoc-based solid-phase peptide synthesis in the Nowick laboratory (Version 1.7.2). 2020, pp. 1–17.
12 Benoiton, N.L. *Chemistry of Peptide Synthesis.* Taylor & Francis Group; 2006.
13 Dawson, P.E., Muir, T.W., Clark-Lewis, I., and Kent, S.B. Synthesis of proteins by native chemical ligation. *Science.* 1994;266(5186):776–779.
14 Hayes, H.C., Luk, L.Y.P., and Tsai, Y.H. Approaches for peptide and protein cyclisation. *Org Biomol Chem.* 2021;19:3983–4001.
15 Muir, T.W., Sondhi, D., and Cole, P.A. Expressed protein ligation: a general method for protein engineering. *Biochemistry.* 1998;95(12):6705–6710.
16 Human Genome Sequencing Consortium. Finishing the euchromatic sequence of the human genome. *Nature.* 2004;5:931–945.
17 Rosenfeld, L. Insulin: discovery and controversy. *Clin Chem.* 2002;48(12):2270–2288.
18 Human insulin receives FDA approval. *FDA Drug Bull.* 1982;12(3):18–19.
19 Johnson, I.S. Human insulin from recombinant DNA technology. *Science.* 1983;219(4585):632–637.
20 Chance, R.E. and Frank, B.H. Research, development, production, and safety of biosynthetic human insulin. *Diabetes Care.* 1993;16(Suppl 3):133–142.
21 Sugumar, V., Ang, K.P., Alshanon, A.F., et al. A comprehensive review of the evolution of insulin development and its delivery method. *Pharmaceutics.* 2022;14(7): 1406.
22 Jarosinski, M.A., Dhayalan, B., Chen, Y.S., Chatterjee, D., Varas, N., and Weiss, M.A. Structural principles of insulin formulation and analog design: a century of innovation. *Mol Metab.* 2021;52: 101325.
23 Kurtzhals, P., Nishimura, E., Haahr, H., et al. Commemorating insulin's centennial: engineering insulin pharmacology towards physiology. *Trends Pharmacol Sci.* 2021;42(8):620–639.
24 du Vigneaud, V. Trail of sulfur research: from insulin to oxytocin. *Science.* 1956;123(3205):967–974.
25 Merrifield, R.B. Automated synthesis of peptides. *Science.* 1965;150(3693):178–185.
26 Pedersen, S.L. and Jensen, K.J. Instruments for automated peptide synthesis. In: *Peptide Synthesis and Applications (Methods in Molecular Biology; vol. 1047)* (eds. K.J. Jensen, P.T. Shelton, and S.L. Pedersen). Humana Press; 2013; pp. 215–224.

27 Mäde, V., Els-Heindl, S., and Beck-Sickinger, A.G. Automated solid-phase peptide synthesis to obtain therapeutic peptides. *Beilstein J Org Chem*. 2014;10:1197–1212.

28 Matsuo, H., Baba, Y., Nair, R.M., Arimura, A., and Schally, A.V. Classic pages in obstetrics and gynecology. Structure of the porcine LH- and FSH-releasing hormone. I. The proposed amino acid sequence. *Biochem Biophys Res Commun*. 1971;43:1334–1339.

29 Burgus, R., Butcher, M., Amoss, M., et al. Primary structure of the ovine hypothalamic luteinizing hormone-releasing factor (LRF) (LH-hypothalamus-LRF-gas chromatography-mass spectrometry-decapeptide-Edman degradation). *Proc Natl Acad Sci U S A*. 1972;69:278–282.

30 Conn, P.M. and Crowley Jr, W.F. Gonadotropin-releasing hormone and its analogues. *N Engl J Med*. 1991;324(2):93–103.

31 Wilson, A.C., Vadakkadath Meethal, S., Bowen, R.L., and Atwood, C.S. Leuprolide acetate: a drug of diverse clinical applications. *Expert Opin Investig Drugs*. 2007;16(11):1851–1863.

32 Butler, T., Goldberg, J.D., Galvin, J.E., et al. Rationale, study design and implementation of the LUCINDA Trial: Leuprolide plus Cholinesterase Inhibition to reduce Neurologic Decline in Alzheimer's. *Contemp Clin Trials*. 2021;107: 106488.

33 Horvath, C.G., Preiss, B.A., and Lipsky, S.R. Fast liquid chromatography: an investigation of operating parameters and the separation of nucleotides on pellicular ion exchangers. *Anal Chem*. 1967;39(22):1422–1428.

34 Loo, A., Udseth, H.R., and Smith, R.D. Peptide and protein analysis by electrospray ionization-mass spectrometry and capillary electrophoresis-mass spectrometry. *Anal Biochem*. 1989;179(2):404–412.

35 Safavi-Hemami, H., Brogan, S.E., and Olivera, B.M. Pain therapeutics from cone snail venoms: from ziconotide to novel non-opioid pathways. *J Proteomics*. 2019;190:12–20.

36 McGivern, J.G. Ziconotide: a review of its pharmacology and use in the treatment of pain. *Neuropsychiatr Dis Treat*. 2007;3:69–85.

37 Lin, J.E., Valentino, M., Marszalowicz, G., et al. Bacterial heat-stable enterotoxins: translation of pathogenic peptides into novel targeted diagnostics and therapeutics. *Toxins (Basel)*. 2010;2(8):2028–2054.

38 Castro, J., Harrington, A.M., Hughes, P.A., Currie, M.G., Blackshaw, L.A., and Brierly, S.M. Linaclotide inhibits colonic nociceptors and relieves abdominal pain via guanylate cyclase-C and extracellular cyclic guanosine 3′,5′-monophosphate. *Gastroenterology*. 2013;145:1334–1346.

39 Emidio, N.B., Tran, H.N.T., Andersson, A., et al. Improving the gastrointestinal stability of linaclotide. *J Med Chem*. 2021;64:8384–8390.

40 Ghusn, W., De la Rosa, A., Sacoto, D., et al. Weight loss outcomes associated with semaglutide treatment for patients with overweight or obesity. *JAMA Netw Open*. 2022;5(9): e2231982.

41 Knudsen LB and Lau, J. The discovery and development of liraglutide and semaglutide. *Front Endocrinol*. 2019:10: 155.

42 Middleton, R.E., Warren, V.A., Kraus, R.L., et al. Two tarantula peptides inhibit activation of multiple sodium channels. *Biochemistry*. 2002;41:14734–14747.

43 Packer, M.S. and Liu, D.R. Methods for the directed evolution of proteins. *Nat Rev Genet*. 2015;16:379–394.

44 Bozovi, K. and Bratkovi, T. Evolving a peptide: library platforms and diversification strategies. *Int J Mol Sci*. 2020;21(1): 215.

45 Josephson, K., Ricardo, A., and Szostak, J.W. mRNA display: from basic principles to macrocycle drug discovery. *Drug Discov Today*. 2014;19(4):388–399.

46 Ullman, C.G., Frigotto, L., and Cooley, R.N. In vitro methods for peptide display and their applications. *Briefings in Functional Genomics*. 2011;10(3):125–134.

47 Gala, A., Comor, L., Horvatic, A., et al. Library-based display technologies: where do we stand? *Mol Biosyst*. 2016;12(8):2342–2358.

48 Pitt, A. and Nims, Z. Peptide display technologies. In: *Cyclic Peptide Design* (ed. G. Goetz), pp. 285–298. Springer; 2019.

49 Fan, Y., Feng, R., Zhang, X., et al. Encoding and display technologies for combinatorial libraries in drug discovery: the coming of age from biology to therapy. *Acta Pharm Sin B*. 2024;14(8):3362–3384.

50 Song, B.P.C., Ch'ng, A.C.W., and Lim, T.S. Review of phage display: a jack-of-all-trades and master of most biomolecule display. *Int J Biol Macromol*. 2024;256(2): 128455.

51 Smith, G.P. Filamentous fusion phage: novel expression vectors that display cloned antigens on the virion surface. *Science*. 1985;228(4705):1315–1317.

52 The Nobel Prize in Chemistry 2018. NobelPrize.org. *Nobel Prize Outreach AB 2024*. Sun. 18 Aug 2024. Available from: https://www.nobelprize.org/prizes/chemistry/2018/summary/.

53 Nixon, A.E., Sexton, D.J., and Ladner, R.C. Drugs derived from phage display: from candidate identification to clinical practice. *MAbs*. 2014;6(1):73–85.

54 Ng, S., Jafari, M.R., Matochko, W.L., and Derda, R. Quantitative synthesis of genetically encoded glycopeptide libraries displayed on M13 phage. *ACS Chem Biol*. 2012;7:1482–1487.

55 Ng, S., Lin, E., Kitov, P.I., et al. Genetically encoded fragment-based discovery of glycopeptide ligands for carbohydrate-binding proteins. *J Am Chem Soc*. 2015;137(16):5248–5251.

56 Vinals, D.F., Kitov, P.I., Tu, Z., et al. Selection of galectin-3 ligands derived from genetically encoded glycopeptide libraries. *Pept Sci*. 2019;111(1): e24097.

57 Chou, Y., Kitova, E.N., Joe, M., et al. Genetically-encoded fragment-based discovery (GE-FBD) of glycopeptide ligands with differential selectivity for antibodies related to mycobacterial infections. *Org Biomol Chem.* 2018;16:223–227.

58 Kitov, P.I., Vinals, D.F., Ng, S., Tjhung, K.F., and Derda, R. Rapid, hydrolytically stable modification of aldehyde-terminated proteins and phage libraries. *J Am Chem Soc.* 2014;136:8149–8152.

59 Triana, V. and Derda, R. Tandem Wittig/Diels-Alder diversification of genetically encoded peptide libraries. *Org Biomol Chem.* 2017;15:7869–7877.

60 Tjhung, K.F., Kitov, P.I., Ng, S., et al. Silent encoding of chemical post-translational modifications in phage-displayed libraries. *J Am Chem Soc.* 2016;138(1):32–35.

61 Derda, R. and Ng, S. Genetically encoded fragment-based discovery. *Curr Opin Chem Biol.* 2019;50:128–137.

62 Yan, K., Triana, V., Kalmady, S.V., et al. Learning the structure-activity relationship (SAR) of the Wittig reaction from genetically-encoded substrates. *Chem Sci.* 2021;12:14301–14308.

63 He, B., Tjhung, K.F., Bennett, N.J., et al. Compositional bias in naïve and chemically-modified phage-displayed libraries uncovered by paired-end deep sequencing. *Sci Rep.* 2018;8(1): 1214.

64 Spokoyny, A.M., Zou, Y., Ling, J.J., Yu, H., Lin, Y.S., and Pentelute, B.L. A perfluoroaryl-cysteine S(N)Ar chemistry approach to unprotected peptide stapling. *J Am Chem Soc.* 2013;135(16):5946–5949.

65 Kalhor-Monfared, S., Jafari, M.R., Patterson, J.T., et al. Rapid biocompatible macrocyclization of peptides with decafluoro-diphenylsulfone. *Chem Sci.* 2016;7(6):3785–3790.

66 Wong, J.Y.K., Ekanayake, A.I., Kharchenko, S., et al. Genetically encoded discovery of perfluoroaryl macrocycles that bind to albumin and exhibit extended circulation in vivo. *Nat Commun.* 2023;14(1): 5654. https://doi.org/10.1038/s41467-023-41427-y. Erratum in: *Nat Commun.* 2023;14(1):7587.

67 Zheng, X.J., Liu, W.D., Liu, Z.Y., Zhao, Y.B., and Wu, C.L. Biocompatible and rapid cyclization of peptides with 2,4-difluoro-6-hydroxy-1,3,5-benzenetricarbonitrile for the development of cyclic peptide libraries. *Bioconjugate Chem.* 2020;31:2085–2091.

68 Fu, X.P., Yuan, Y., Jha, A., et al. Stereoretentive post-translational protein editing. *ACS Cent Sci.* 2023;9(3):405–416.

69 Ng, S. and Derda, R. Phage-displayed macrocyclic glycopeptide libraries. *Org Biomol Chem.* 2016;14:5539–5545.

70 Timmerman, P., Beld, J., Puijk, W.C., and Meloen, R.H. Rapid and quantitative cyclization of multiple peptide loops onto synthetic scaffolds for structural mimicry of protein surfaces. *ChemBioChem.* 2005;6:821–824.

71 Heinis, C., Rutherford, T., Freund, S., and Winter, G. Phage-encoded combinatorial chemical libraries based on bicyclic peptides. *Nat Chem Biol.* 2009;5:502–507.
72 Chen, S., Lovell, S., Lee, S., Fellner, M., Mace, P.D., and Bogyo, M. Identification of highly selective covalent inhibitors by phage display. *Nat Biotechnol.* 2021;39(4):490–498.
73 Yan, K., Miskolzie, M., Bañales Mejia, F., et al. Late-stage reshaping of phage-displayed libraries to macrocyclic and bicyclic landscapes using multipurpose linchpin. *ChemRxiv*. 2024; https://doi.org/10.26434/chemrxiv-2024-08847-v2. This content is a preprint and has not been peer-reviewed.
74 Wong, J.Y., Mukherjee, R., Miao, J., et al. Genetically-encoded discovery of proteolytically stable bicyclic inhibitors for morphogen NODAL. *Chem Sci.* 2021;12(28):9694–9703.
75 Chen, F.J., Pinnette, N., Yang, F., and Gao, J.M. A cysteine-directed proximity-driven crosslinking method for native peptide bicyclization. *Angew Chem Int Ed*. 2023;62(31): e202306813.
76 Hampton, J.T. and Liu, W.R. Diversification of phage-displayed peptide libraries with noncanonical amino acid mutagenesis and chemical modification. *Chem Rev*. 2024;124:6051–6077.
77 Brown, K.C. Peptidic tumor targeting agents: the road from phage display peptide selections to clinical applications. *Curr Pharm Des.* 2010;16(9):1040–1054.
78 Umlauf, B.J., Mercedes, J.S., Chung, C.Y., and Brown, K.C. Identification of a novel lysosomal trafficking peptide using phage display biopanning coupled with endocytic selection pressure. *Bioconjug Chem*. 2014;25(10):1829–1837.
79 Heinis, C. and Winter, G. Encoded libraries of chemically modified peptides. *Curr Opin Chem Biol*. 2015;26:89–98.
80 Bendell, J.C., Wang, J.S.-Z., Bashir, B., et al. BT5528-100 phase I/II study of the safety, pharmacokinetics, and preliminary clinical activity of BT5528 in patients with advanced malignancies associated with EphA2 expression. *J Clin Oncol.* 2020;38: TPS3655.
81 Bennett, G., Brown, A., Mudd, G., et al. MMAE delivery using the bicycle toxin conjugate BT5528. *Mol Cancer Ther*. 2020;19(7):1385–1394.
82 Upadhyaya, P., Kristensson, J., Lahdenranta, J., et al. Discovery and optimization of a synthetic class of Nectin-4-targeted CD137 agonists for immuno-oncology. *J Med Chem*. 2022;65:9858–9872.
83 Banerji, U., Cook, N., Evans, T.R.J., et al. A cancer research UK phase 1/2a trial of BT1718 (a first in class bicycle drug conjugate) given intravenously in patients with advanced solid tumours. *J Clin Oncol*. 2018;36(15): TPS2610.
84 Chen, F.J., Pinnette, N., and Gao, J. Strategies for the construction of multicyclic phage display libraries. *ChemBioChem*. 2024;25(9): e202400072.

85 Wang, C.K. and Craik, D.J. Designing macrocyclic disulfide-rich peptides for biotechnological applications. *Nat Chem Biol*. 2018;14:417–427.
86 Chen, S. and Heinis, C. Phage selection of bicyclic peptides based on two disulfide bridges. In: *Protocol in Peptide Libraries*. Springer; 2014. pp. 119–137.
87 Gilreath, J., Lo, M., and Bubalo, J. Thrombopoietin receptor agonists (TPO-RAs): drug class considerations for pharmacists. *Drugs*. 2021;81:1285–1305.
88 Cwirla, S.E., Balasubramanian, P., Duffin, D.J., et al. Peptide agonist of the thrombopoietin receptor as potent as the natural cytokine. *Science*. 1997;276(5319):1696–1699.
89 Kuter, D.J. New thrombopoietic growth factors. *Clin Lymphoma Myeloma*. 2009;9(Suppl 3):S347–S356.
90 Wang, B., Nichol, J.L., and Sullivan, J.T. Pharmacodynamics and pharmacokinetics of AMG 531, a novel thrombopoietin receptor ligand. *Clin Pharmacol Ther*. 2004;76(6):628–638.
91 Kuter, D.J. The biology of thrombopoietin and thrombopoietin receptor agonists. *Int J Hematol*. 2013;98(1):10–23.
92 Bussel, J.B., Soff, G., Balduzzi, A., Cooper, N., Lawrence, T., and Semple, J.W. A review of romiplostim mechanism of action and clinical applicability. *Drug Des Devel Ther*. 2021;15:2243–2268.
93 Ghanima, W., Cooper, N., Rodeghiero, F., Godeau, B., and Bussel, J.B. Thrombopoietin receptor agonists: ten years later. *Haematologica*. 2019;104(6):1112–1123.
94 Broudy, V.C. and Lin, N.L. AMG531 stimulates megakaryopoiesis in vitro by binding to Mpl. *Cytokine*. 2004;25(2):52–60.
95 EMA. European Medicines Agency. Romiplostim summary of product characteristics. Available from: https://ec.europa.eu/ health/documents/community-register/2021/20210122150420/anx_150420_en.pdf. [cited 2021 Apr 15].
96 FDA approves romiplostim (Nplate) for the treatment of thrombocytopenia: What's new from OODP [Office of Oncology Drug Products]. Available from: www. fda.gov/cder/Offices/OODP/whats new/romiplostim.htm. [cited 2009 Mar 21].
97 Romiplostim, prescribing information. Thousands Oaks, CA: Amgen; August 2008. Available from: www.nplate.com. [cited 2009 Mar 21].
98 Waknine, Y. FDA Approvals: Nplate, Aloxi, Vidaza. Medscape. September 4, 2008. Available from: www.medscape com/viewarticle/580032. [cited 2009 Mar 21].
99 Hubulashvili, D. and Marzella, N. Romiplostim (Nplate), a treatment option for immune (idiopathic) thrombocytopenic purpura. *P T*. 2009;34(9):482–485.

100 Lehmann, A. Ecallantide (DX-88), a plasma kallikrein inhibitor for the treatment of hereditary angioedema and the prevention of blood loss in on-pump cardiothoracic surgery. *Expert Opin Biol Ther*. 2008;8(8):1187–1199.

101 Bradbury, A.R., Sidhu, S., Dübel, S., and McCafferty, J. Beyond natural antibodies: the power of in vitro display technologies. *Nat Biotechnol*. 2011;29:245–254.

102 Markland, W., Ley, A.C., Lee, S.W., and Ladner, R.C. Iterative optimization of high-affinity protease inhibitors using phage display. 1. Plasmin. *Biochemistry*. 1996;35:8045–8057.

103 Stolz, L.E. and Horn, P.T. Ecallantide: a plasma kallikrein inhibitor for the treatment of acute attacks of hereditary angioedema. *Drugs Today (Barc)*. 2010;46:547–555.

104 Joseph, K., Tuscano, T.B., and Kaplan, A.P. Studies of the mechanisms of bradykinin generation in hereditary angioedema plasma. *Ann Allergy Asthma Immunol*. 2008;101:279–286.

105 Nussberger, J., Cugno, M., Amstutz, C., Cicardi, M., Pellacani, A., and Agostoni, A. Plasma bradykinin in angiooedema. *Lancet*. 1998;351:1693–1697.

106 Kaufman, N., Page, J.D., Pixley, R.A., Schein, R., Schmaier, A.H., and Colman, R.W. Alpha 2-macroglobulin-kallikrein complexes detect contact system activation in hereditary angioedema and human sepsis. *Blood*. 1991;77:2660–2667.

107 Bernstein, J.A. and Qazi, M. Ecallantide: its pharmacology, pharmacokinetics, clinical efficacy and tolerability. *Expert Rev Clin Immunol*. 2010;6:29–39.

108 Guterman, S., Kent, R.B., Ladner, R., Ley, A., Markland, W., and Roberts, B.L. Direct evolution of novel binding proteins. *US Patent* 5223409A (Patent) 1991.

109 Woodburn, K.W., Holmes, C.P., Wilson, S.D., et al. Absorption, distribution, metabolism and excretion of peginesatide, a novel erythropoiesis-stimulating agent, in rats. *Xenobiotica*. 2012;42:660–670.

110 Livnah, O., Stura, E.A., Johnson, D.L., et al. Functional mimicry of a protein hormone by a peptide agonist: the EPO receptor complex at 2.8 Å. *Science*. 1996;273:464–471.

111 Fan, Q., Leuther, K.K., Holmes, C.P., et al. Preclinical evaluation of Hematide, a novel erythropoiesis stimulating agent, for the treatment of anemia. *Exp Hematol*. 2006;34:1303–1311.

112 Macdougall, I.C., Wiecek, A., Tucker, B., et al. Dose-finding study of peginesatide for anemia correction in chronic kidney disease patients. *Clin J Am Soc Nephrol*. 2011;6:2579–2586.

113 Schmid, H. Peginesatide for the treatment of renal disease-induced anemia. *Expert Opin Pharmacother*. 2013;14:937–948.

114 Wrighton, N.C., Farrell, F.X., Chang, R., et al. Small peptides as potent mimetics of the protein hormone erythropoietin. *Science*. 1996;273(5274):458–463.

115 Wrighton, N.C., Balasubramanian, P., Barbone, F.P., et al. Increased potency of an erythropoietin peptide mimetic through covalent dimerization. *Nat Biotechnol.* 1997;15:1261–1265.

116 Harris, J.M. and Chess, R.B. Effect of pegylation on pharmaceuticals. *Nat Rev.* 2003;2:214–221.

117 Knauf, M.J., Bell, D.P., Hirtzer, P., Luo, Z., Young, J.D., and Katre, N.V. Relationship of effective molecular size to systemic clearance in rats of recombinant interleukin-2 chemically modified with water-soluble polymers. *J Biol Chem.* 1998;263:15064–15070.

118 Mastellos, D.C., Yancopoulou, D., Kokkinos, P., et al. Compstatin: a C3-targeted complement inhibitor reaching its prime for bedside intervention. *Eur J Clin Invest.* 2015;45(4):423–440.

119 Sahu, A., Kay, B.K., and Lambris, J.D. Inhibition of human complement by a C3-binding peptide isolated from a phage-displayed random peptide library. *J Immunol.* 1996;157(2):884–891.

120 Gerber, G.F. and Brodsky, R.A. Pegcetacoplan for paroxysmal nocturnal hemoglobinuria. *Blood.* 2022;139(23):3361–3365.

121 Roberts, R.W. and Szostak, J.W. RNA-peptide fusions for the in vitro selection of peptides and proteins. *Proc Natl Acad Sci USA.* 1997;94(23):12297–12302.

122 Suga, H., Cowan, J.A., and Szostak, J.W. Unusual metal ion catalysis in an acyl-transferase ribozyme. *Biochemistry.* 1998;37:10118–10125.

123 Suga, H., Lohse, P.A., and Szostak, J.W. Structural and kinetic characterization of an acyl transferase ribozyme. *J Am Chem Soc.* 1998;120:1151–1156.

124 Lee, N., Bessho, Y., Wei, K., Szostak, J.W., and Suga, H. Ribozyme-catalyzed tRNA aminoacylation. *Nat Struct Biol.* 2000;7:28–33.

125 Goto, Y. and Suga, H. The RaPID platform for the discovery of pseudo-natural macrocyclic peptides. *Acc Chem Res.* 2021;54(18):3604–3617.

126 Ohuchi, M., Murakami, H., and Suga, H. The flexizyme system: a highly flexible tRNA aminoacylation tool for the translation apparatus. *Curr Opin Chem Biol.* 2007;11:537–542.

127 Goto, Y., Katoh, T., and Suga, H. Flexizymes for genetic code reprogramming. *Nat Protoc.* 2011;6:779–790.

128 Tanaka, Y., Hipolito, C., Maturana, A., et al. Structural basis for the drug extrusion mechanism by a MATE multidrug transporter. *Nature.* 2013;496:247–251.

129 Iskandar, S.E., Pelton, J.M., Wick, E.T., et al. Enabling genetic code expansion and peptide macrocyclization in mRNA display via a promiscuous orthogonal aminoacyl-tRNA synthetase. *J Am Chem Soc.* 2023;145(3):1512–1517.

130 Sidhu, S.S., Lowman, H.B., Cunningham, B.C., and Wells, J.A. Phage display for selection of novel binding peptides. *Methods Enzymol.* 2000;328:333–363.

131 Shimizu, Y., Inoue, A., Tomari, Y., et al. Cell-free translation reconstituted with purified components. *Nat Biotechnol.* 2001;19(8):751–755.

132 Forster, A.C., Weissbach, H., and Blacklow, S.C. A simplified reconstitution of mRNA-directed peptide synthesis: activity of the epsilon enhancer and an unnatural amino acid. *Anal Biochem.* 2001;297(1):60–70.

133 Josephson, K., Hartman, M.C., and Szostak, J.W. Ribosomal synthesis of unnatural peptides. *J Am Chem Soc.* 2005;127(33):11727–11735.

134 Hartman, M.C.T., Josephson, K., Lin, C.-W., and Szostak, J.W. An expanded set of amino acid analogs for the ribosomal translation of unnatural peptides. *PLoS One.* 2007;2(10): e972.

135 Faris, J.H., Adaligil, E., Popovych, N., et al. Membrane permeability in a large macrocyclic peptide driven by a saddle-shaped conformation. *J Am Chem Soc.* 2024;146(7):4582–4591.

136 Howard Jr., J.F., Bresch, S., Genge, A., et al. Safety and efficacy of zilucoplan in patients with generalised myasthenia gravis (RAISE): a randomised, double-blind, placebo-controlled, phase 3 study. *Lancet Neurol.* 2023;22(5):395–406.

137 Tang, G.Q., Tang, Y., Dhamnaskar, K., et al. Zilucoplan, a macrocyclic peptide inhibitor of human complement component 5, uses a dual mode of action to prevent terminal complement pathway activation. *Front Immunol.* 2023;14: 1213920.

138 UCB. UCB announces approval of RYSTIGGO[®] (rozanolixizumab) and ZILBRYSQ[®] (zilucoplan) for the treatment of adult patients with generalized myasthenia gravis in Japan [media release]. 2023. https://www.ucb.com/stories-media/Press-Releases/article/UCB-announces-approval-of-RYSTIGGOR-rozanolixizumab-and-ZILBRYSQR-zilucoplan-for-the-treatment-of-adult-patients-with-generalized-myasthenia-gravis-in-Japan. [cited 2023 Sep 25].

139 Lyamichev, V.I., Goodrich, L.E., Sullivan, E.H., et al. Stepwise evolution improves identification of diverse peptides binding to a protein target. *Nat Sci Rep.* 2017;7(12116):1–10.

140 Fodor, S.P.A., Read, J.L., Pirrung, M.C., Stryer, L., Lu, A.T., and Solas, D. Light-directed, spatially addressable parallel chemical synthesis. *Science.* 1991; 251: 767.

141 Winkler, J.L., Fodor, S.P.A., Buchko, C.J., Ross, D., Aldwin, L., and Modlin, D.N. Printing oligonucleotide arrays using deprotection agents solely in the vapor phase. WO 93/09668 (Patent) 1992.

142 Pirrung, M.C., Read, J.L., Fodor, S.P.A., and Stryer, L. Large scale photolithographic solid phase synthesis of polypeptides and receptor binding screening thereof. US5143854A (Patent) 1992.

143 Singh-Gasson, S., Green, R.D., Yue, Y., et al. Maskless fabrication of light-directed oligonucleotide microarrays using a digital micromirror array. *Nat Biotechnol.* 1999;17(10):974–978.

144 Pellois, J.P., Zhou, X., Srivannavit, O., Zhou, T., Gulari, E., and Gao, X. Individually addressable parallel peptide synthesis on microchips. *Nat Biotechnol.* 2002;20(9):922–926.

145 Houghten, R.A. General method for the rapid solid-phase synthesis of large numbers of peptides: specificity of antigen-antibody interaction at the level of individual amino acids. *Proc Natl Acad Sci U S A*. 1985;82(15):5131–5135.

146 Geysen, H.M., Meloen, R.H., and Barteling, S.J. Use of peptide synthesis to probe viral antigens for epitopes to a resolution of a single amino acid. *Proc Natl Acad Sci U S A*. 1984;81(13):3998–4002.

147 Klein, J., Zech, T., Newsam, J.M., and Schunk, S.A. Application of a novel Split&Pool-principle for the fully combinatorial synthesis of functional inorganic materials. *Appl Catal A: Gen*. 2003;254:121–131.

148 Brenner, S. and Lerner, R.A. Encoded combinatorial chemistry. *Proc Natl Acad Sci U S A*. 1992;89(12):5381–5383.

149 Gartner, Z.J. and Liu, D.R. The generality of DNA-templated synthesis as a basis for evolving non-natural small molecules. *J Am Chem Soc*. 2001;123(28):6961–6963.

150 Li, X. and Liu, D.R. DNA-templated organic synthesis: nature's strategy for controlling chemical reactivity applied to synthetic molecules. *Angew Chem Int Ed Engl*. 2004;43(37):4848–4870.

151 Plais, L. and Scheuermann, J. Macrocyclic DNA-encoded chemical libraries: a historical perspective. *RSC Chem Biol*. 2022;3:7–17.

152 Silvestri, A.P., Zhang, Q., Ping, Y., et al. DNA-encoded macrocyclic peptide libraries enable the discovery of a neutral MDM2-p53 inhibitor. *ACS Med Chem Lett*. 2023;14(6):820–826.

153 Gironda-Martínez, A., Donckele, E.J., Samain, F., and Neri, D. DNA-encoded chemical libraries: a comprehensive review with successful stories and future challenges. *ACS Pharmacol Transl Sci*. 2021;4(4):1265–1279.

154 Ottl, J., Leder, L., Schaefer, J.V., and Dumelin, C.E. Encoded library technologies as integrated lead finding platforms for drug discovery. *Molecules*. 2019;24(8): 1629.

155 Winssinger, N., Damoiseaux, R., Tully, D.C., Geierstanger, B.H., Burdick, K., and Harris, J.L. PNA-encoded protease substrate microarrays. *Chem Biol*. 2004;11(10):1351–1360.

156 Soloveychik, M. and Chahwann, C. Selective modulation of protein-protein interactions. US20200347462A1 (Patent) 2018.

157 Chahwann, C. and Soloveychik, M. Selective degradation of proteins. US20220298503A1(Patent) 2022.

158 Scott, C.P., Abel-Santos, E., Wall, M., Wahnon, D.C., and Benkovic, S.J. Production of cyclic peptides and proteins in vivo. *Proc Natl Acad Sci U S A*. 1999;96(24):13638–13643.

159 Kritzer, J.A., Hamamichi, S., McCaffery, J.M., et al. Rapid selection of cyclic peptides that reduce alpha-synuclein toxicity in yeast and animal models. *Nat Chem Biol*. 2009;5(9):655–663.

160 Miranda, E., Nordgren, I.K., Male, A.L., et al. A cyclic peptide inhibitor of HIF-1 heterodimerization that inhibits hypoxia signaling in cancer cells. *J Am Chem Soc*. 2013;135(28):10418–10425.

161 Bessette, P.H., Rice, J.J., and Daugherty, P.S. Rapid isolation of high-affinity protein binding peptides using bacterial display. *Protein Eng Des Sel*. 2004;17(10):731–739.

162 Daugherty, P.S. Protein engineering with bacterial display. *Curr Opin Struct Biol*. 2007;17(4):474–480.

163 Kenrick, S.A. and Daugherty, P.S. Bacterial display enables efficient and quantitative peptide affinity maturation. *Protein Eng Des Sel*. 2010;23(1):9–17.

164 Tucker, A.T., Leonard, S.P., DuBois, C.D., et al. Discovery of next-generation antimicrobials through bacterial self-screening of surface-displayed peptide libraries. *Cell*. 2018;172(3):618–628.

165 Randall, J.R., Groover, K.E., O'Donnell, A.C., Garza, J.M., Cole, T.J., and Davies, B.W. Adapting antibacterial display to identify serum-active macrocyclic peptide antibiotics. *PNAS Nexus*. 2023;2(8): pgad270.

166 Navaratna, T., Atangcho, L., Mahajan, M., et al. Directed evolution using stabilized bacterial peptide display. *J Am Chem Soc*. 2019;142(4):1882–1894.

167 Chen, C.H. and Lu, T.K. Development and challenges of antimicrobial peptides for therapeutic applications. *Antibiotics*. 2020;9(1): 24.

168 Obexer, R., Walport, L.J., and Suga, H. Exploring sequence space: harnessing chemical and biological diversity towards new peptide leads. *Curr Opin Chem Biol*. 2017;38:52–61.

169 Mulligan, V.K. and Hosseinzadeh, P. Computational design of peptide-based binders to therapeutic targets. In: *Approaching the Next Inflection in Peptide Therapeutics: Attaining Cell Permeability and Oral Bioavailability*. (American Chemical Society Symposium Series 1417): Red Hook, New York: Curran Associates Publishing; 2022. pp. 55–102.

170 Mulligan, V.K. Computational methods for peptide macrocycle drug design. In: *Peptide Therapeutics: Fundamentals of Design, Development, and Delivery*. Cham: Springer International Publishing; 2022. pp. 79–161.

171 Wan, F., Wong, F., Collins, J.J., and de la Fuente-Nunez, C. Machine learning for antimicrobial peptide identification and design. *Nat Rev Bioeng*. 2024;2:1–16.

172 Kosugi, T. and Ohue, M. Design of cyclic peptides targeting protein–protein interactions using AlphaFold. *Int J Mol Sci*. 2023;24(17): 13257.

173 Rettie, S.A., Campbell, K.V., Bera, A.K., et al. Cyclic peptide structure prediction and design using AlphaFold. *bioRxiv* [Preprint]. 2023 Feb 26: 2023.02.25.529956. https://doi.org/10.1101/2023.02.25.529956.

174 Baek, M., DiMaio, F., Anishchenko, I., et al. Accurate prediction of protein structures and interactions using a three-track neural network. *Science*. 2021;373(6557):871–876. https://doi.org/10.1126/science.abj8754.

175 Jumper, J., Evans, R., Pritzel, A., et al. Highly accurate protein structure prediction with AlphaFold. *Nature*. 2021;596(7873):583–589. https://doi.org/10.1038/s41586-021-03819-2.

176 Dou, J., Vorobieva, A.A., Sheffler, W., et al. De novo design of a fluorescence-activating β-barrel. *Nature*. 2018;561(7724):485–491. https://doi.org/10.1038/s41586-018-0509-0.

177 Bhardwaj, G., Mulligan, V.K., Bahl, C.D., et al. Accurate de novo design of hyperstable constrained peptides. *Nature*. 2016;538(7625):329–335.

178 Georgiev, I., Lilien, R.H., and Donald, B.R. The minimized dead-end elimination criterion and its application to protein redesign in a hybrid scoring and search algorithm for computing partition functions over molecular ensembles. *J Comput Chem*. 2008;29(10):1527–1542. https://doi.org/10.1002/jcc.20909.

179 Hastings, W.K. Monte Carlo sampling methods using Markov chains and their applications. *Biometrika*. 1970;57(1):97–109.

180 Leaver-Fay, A., Tyka, M., Lewis, S.M., et al. Rosetta3. In: *Methods in Enzymology*. Amsterdam, Netherlands: Elsevier; 2011. pp. 545–574.

181 Mulligan, V., Khipple, H., Melo, H., et al. Designing peptides on a quantum computer. *BioRxiv*. 2019;752485.

182 Cao, L., Coventry, B., Goreshnik, I., et al. Robust de novo design of protein binding proteins from target structural information alone. *BioRxiv*. 2021;2021–2009.

183 Mulligan, V.K., Workman, S., Sun, T., et al. Computationally designed peptide macrocycle inhibitors of New Delhi metallo-β-lactamase 1. *Proc Natl Acad Sci U S A*. 2021;118(12): e2012800118.

184 Hosseinzadeh, P., Watson, P.R., Craven, T.W., et al. Anchor extension: a structure-guided approach to design cyclic peptides targeting enzyme active sites. *Nat Commun*. 2021;12(1): 3384.

185 Schneider, J.A., Craven, T.W., Kasper, A.C., et al. Design of peptoid-peptide macrocycles to inhibit the β-catenin TCF interaction in prostate cancer. *Nat Commun*. 2018;9(1): 4396.

186 Manavalan, B., Shin, T.H., Kim, M.O., and Lee, G. AIPpred: sequence-based prediction of anti-inflammatory peptides using random forest. *Front Pharmacol*. 2018;9: 348997.

187 Sanders, W.S., Johnston, C.I., Bridges, S.M., Burgess, S.C., and Willeford, K.O. Prediction of cell-penetrating peptides by support vector machines. *PLoS Comput Biol*. 2011;7(7): e1002101.

188 Feynman, R.P. Simulating physics with computers. *Int J Theor Phys*. 1982;21(6-7):467–488.

189 Feynman, R.P. Quantum mechanical computers. *Found Phys*. 1986;16(6):507–531.

190 Ehmer, T., Karemore, G., and Melo, H. The quantum computing paradigm. In: *Computational Drug Discovery: Methods and Applications*, vol. 2. Wiley; VCH GmbH; 2024. pp. 627–678.

191 Galda, A., Mulligan, V.K., MacCormack, I., Crooks, G., and Melo, H. Peptide design with quantum approximate optimization algorithm. In: *APS March Meeting Abstracts*. 2022;B01–004.

192 Boulebnane, S., Lucas, X., Meyder, A., Adaszewski, S., and Montanaro, A. Peptide conformational sampling using the quantum approximate optimization algorithm. *npj Quantum Inf.* 2023;9(1): 70.

193 Pandey, M., Zaborniak, T., Melo, H., Galda, A., and Mulligan, V.K. Multibody molecular docking on a quantum annealer. arXiv preprint arXiv:2210.11401. 2022.

7

A Survey and Guide to the Development of Fatty Acid-derivatized Peptide Therapeutics

Michael T. Jacobsen[1], *Christian Poulsen*[2], *Katharina L. Kopp*[3], *Jørgen Olsen*[4] *and Nick Cox*[1]

[1]*Novo Nordisk, Chemical Biology, Lexington, Massachusetts, USA*
[2]*Novo Nordisk, Biophysics and Injectable Formulation, Måløv, Denmark*
[3]*Novo Nordisk, Functional Assays & Screening, Måløv, Denmark*
[4]*Novo Nordisk, Discovery ADME, Måløv, Denmark*

7.1 Introduction

The clinical impact of incretin-based drugs, e.g. semaglutide and tirzepatide, has fulfilled the promise of potent and long-acting peptide therapeutics. It highlights the utility of fatty acid acylation to upgrade peptides from drug molecules with high potency and selectivity but poor pharmacological properties into therapeutics that are safe, long-acting, and convenient to dose. Historically, peptide-based drug discovery via structure–activity relationship (SAR) campaigns is driven by cellular potency and binding assays, i.e. the most potent and high affinity compounds emerge as the leads. However, the most potent engineered compounds often do not result in the best drug molecules for a variety of reasons such as low solubility and aggregation problems (influenced by the inclusion of a fatty acid), chemical and metabolic instabilities (short *in vivo* half-lives), immunogenicity, lack of receptor selectivity, and manufacturability challenges. In this review, we first introduce the origins and basic concepts around fatty acid derivatization or acylation (protraction) of peptides to extend half-life. Second, we elaborate four case studies of natural peptides that were upgraded to fatty acid-derivatized, long-acting therapeutics: (1) GLP-1-based receptor agonists, (2) once-daily and once-weekly insulins, (3) amylin-based cagrilintide, and (4) peptide YY (PYY). In the four cases presented here, the SAR campaigns to advance these fatty acid-derivatized peptides required

Peptide Science: Chemical Ligation, Lead Generation, and Therapeutic Advances, First Edition. Fa Liu.

multiparametric optimization, balancing potency, affinity, selectivity, and stability after derivatization. Lastly, we provide practical instruction on optimizing peptide therapeutics, focusing on solubility, biophysical stability, chemical and metabolic stability, immunogenicity, general manufacturing considerations, and strategies for fatty acid derivatization. In addition to the compound's activity, these other parameters must be considered early on the path to achieving a successful clinical peptide candidate.

7.2 Basics of Fatty Acid Derivatization of Peptides

The concept of half-life extension of peptides by conjugation with a fatty acid, also known as derivatization, acylation, or lipidation (see extensive review [1]), was introduced in the mid-1990s [2]. Lipidation prolongs half-life primarily by reversible binding to human serum albumin (HSA). HSA is a 66 kDa protein [3], which is the most abundant protein in extracellular fluid, with serum concentrations of around 50 g/L [4,5]). Its half-life in circulation is commonly reported to be up to three weeks [6]. This long half-life is dictated by both its large size, which reduces renal elimination, and its recycling through binding to the neonatal Fc receptor (FcRn) [7]. HSA interacts with FcRn in a pH-dependent manner, where it binds and is recycled through the endosome at ~pH 5–6 and then released back into the plasma at pH 7.4 [8]. Additionally, HSA, being a multidomain protein, is the primary carrier or transporter of lipids and several other important solutes in the body. It possesses seven different fatty acid binding sites [9–11], with two principal sites, termed Sudlow's sites I and II [1,12], separate from its FcRn-binding site, in domain III [6]. Hence, HSA is an optimal carrier of lipidated species due to this reversible binding, its large size to avoid renal elimination, and long half-life, facilitated by recycling through interaction with FcRn.

The precise molecular control, consistent half-life extension, lack of immunogenicity, and overall safety profile of fatty acid derivatization have led to its emergence as the preferred method for half-life extension of peptides. Nevertheless, there are alternatives that should be noted [13], such as PEGylation with high molecular weight polyethylene glycol [14–16] or carbohydrate polymers [17,18]. Other high molecular weight half-life extension strategies, i.e. by reducing renal elimination, include fusions with long amino acid repeats: XTENylation [19,20] and PASylation [21,22]. Alternatively, one may even directly fuse the peptide or protein to an Fc domain or albumin itself [23–25]. Lastly, peptide fusion with nonlipid modalities that also bind to albumin, including cyclic peptides and proteins, is attracting increasing interest [26–29]. Nonetheless, this review focuses on the technical applications and benefits, along with hurdles, of developing fatty acid-derivatized peptides as therapeutics.

7.3 Evolution of GLP-1-based Incretins into Stable, Long-acting Therapeutics

Native GLP-1 is a 30-residue peptide secreted by L-cells in the intestines following eating, and it is a key hormone involved in the incretin effect: insulin release in response to glucose [30–33]. However, underivatized native GLP-1 (7–36 amide or 7–37 acid) (Figure 7.1) has a half-life of only a few minutes *in vivo* [34], primarily due to a high renal clearance and proteolysis by dipeptidyl peptidase IV (DPP-4) [35,36], which generates the main metabolite GLP-1 (9–36 amide) that possesses several orders of magnitude lower potency at GLP-1R [37]. Inspired by work on exendin-4 [38], which possesses a different N-terminal sequence (Gly 8 instead of Ala 8) and consequent less susceptibility to DPP-4 proteolysis, substitutions at position 8 in GLP-1 were tested in a more comprehensive manner. Here, substitutions with Gly, Aib, Ser, and Thr all resulted in significant *in vitro* and *in vivo* stabilization [39]. Over time, Aib [40] has been selected as the optimal

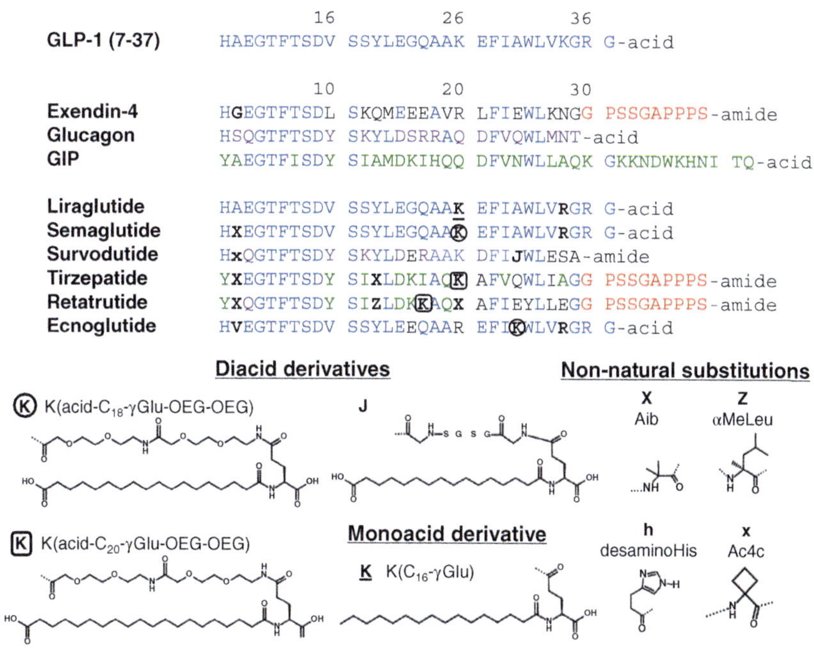

Figure 7.1 Potent, long-acting, and stable GLP-1-based therapeutics with multiple receptor agonism, including derivatizations and nonnatural substitutions. Amino acids are colored as follows: blue, originating with GLP-1; red, C-terminal extension (Cex) domain from exendin-4; purple, derived from glucagon; green, derived from GIP; black and bold, special amino acids or fatty acid-derivatized Lys, as shown on the bottom panel.

residue at position 8 (position 2 in derivatized, multi-agonists) due to its imparting DPP-4 resistance, while maintaining high GLP-1R affinity, as seen in several modern GLP-1 receptor agonists, e.g. semaglutide, tirzepatide, and retatrutide. Nevertheless, despite this metabolic remedy, the initial GLP-1 analogs were still rapidly eliminated through renal clearance. Thus, efforts to extend half-live via fatty acid derivatization and albumin binding were pursued [41].

One of the first reports to derivatize GLP-1 with a fatty acid was published in 1998 [42,43], where a C8 fatty acid was attached to a GLP-1 (7–37-acid) peptide in order to investigate a relationship between bioavailability and hydrophobic association. This peptide includes a desaminoHis at the first N-terminal position to reduce DPP-4 degradation [44], an Arg substitution at Lys 26, and C8 acylation at Lys 34 ε-amine. In this important work, they showed that acylation led to solvent- and temperature-dependent association phenomena, with three different states: (1) a stable α-helical form, (2) a self-association driven by β-sheet structure, and (3) a large micelle-like β-sheet-rich aggregate. In these cases, the bioavailability and PK profiles were shown to be highly influenced by whichever association state the compound originated. Nevertheless, significant challenges with the high degree of aggregation in this acylated compound were encountered. Other early studies on advancing GLP-1 and glucagon peptides also noted similar stability challenges [45,46].

Before further elaborating on the development of fatty acid-derivatized GLP-1s, a brief discussion around the development of the biophysically challenging glucagon molecule is warranted. Glucagon is one of the most effective treatments for hypoglycemia in diabetic patients [47] because of its ability to rapidly elevate glucose levels [48,49]. An important first step in the development of glucagon-based therapeutics was its successful synthesis [50]. However, glucagon possesses notoriously poor biophysical properties, with low solubility at physiological pH [51], and chemical/biophysical instability once dissolved [52,53]. In several important studies, Dimarchi et al. advanced multiple glucagon analogs using different stabilizing strategies, while retaining satisfactory glucagon receptor potency and *in vivo* activity. These reports are good case studies on multiparametric SAR optimization, with a high priority on improving the biophysical behavior [54].

In 2010 [55], the same group reported a new glucagon variant with a single substitution: Asn to Asp 28. This change decreased the calculated pI from 6.7 to 5.4. Consequently, it had increased solubility at pH 7.4, and showed much higher stability, up to several months after incubation at pH 7.4 – in contrast to its poor stability in typical formulation conditions of dilute acid. In this same report, they showed that the fusion of glucagon with full, and partial, C-terminal extensions, based on the so-called Cex domains originally from exendin-4 [38], significantly improved solubility (from 0.4 to 14.7 mg/mL) and enhanced stability for up to five months at 37 °C. Appending a Cex domain to incretin-based peptides has consistently improved their metabolic and biophysical stability [56]. Importantly, both Asp 28 substitution and Cex fusion minimally impacted glucagon receptor activity, while enhancing the biophysical stability. However, after more extensive analysis of the Asp 28 analog at elevated temperatures and extended periods of time (for several months), chemical instability and higher-order aggregates were identified, necessitating a follow-up study [57]. In this next round of optimization, the first beneficial mutation reported was the introduction of another anionic residue: mutation of Thr 29 to Glu, which doubled the solubility at pH 7.4 without impacting receptor activity. Next, Ala and Aib scans were conducted to identify positions that could benefit from additional conformational stability. In particular, substitution of Ser 16 with a more stable Aib [40,58] abolished the major chemical instability (hydrolysis of the Asp-Ser peptide bond through an aspartimide intermediate [59,60]) and provided greater overall aqueous stability. Comparison of the original Asp 28 analog with the new triple variant, "Glucagon Lead," Asp 28 + Glu 29 + Aib 16 showed an additional 40% greater stability after six months of incubation in pH 8.6 buffer. The introduction of stabilizing Aib residues significantly reduced the formation of high molecular weight aggregates in this study. Furthermore, an additional variant, termed IUB76, containing a total of seven changes relative to native glucagon, showed even better solution stability out to one year. Besides advancing glucagon therapies, lessons from these papers on molecular design, and how to critically evaluate the stability of peptide therapeutics, impacted the development of highly potent and stable dual and triple agonists [61,62].

Returning to the development of fatty acid-derivatized GLP-1s, liraglutide, the first GLP-1 drug appropriate for once-daily subcutaneous (SC) administration [63,64], the critical design steps were to identify suitable fatty acid derivatization sites that could extend half-life, while retaining good potency and biophysical stability. Here, an initial Ala scan [65] suggested potential derivatization sites, which were later supported by structural data on GLP-1 bound to its receptor [66,67]. In two studies on the installation of fatty acid protractors into GLP-1 peptides, Knudsen et al. [68,69] reported:

1. In several cases, attachment of fatty monoacids up to C16 in length did not significantly impact receptor potency.
2. Multiple variants containing C16 fatty acids showed half-lives >12 hours in pigs, compared to ~1 hour in nonderivatized control, with the length of the fatty acid scaling with the protraction effect.
3. Double derivatization, i.e. coupling two fatty acids at different positions, generally led to negative effects on potency.
4. Longer chain fatty acids (longer than C16) resulted in negative effects on potency.
5. Generally, different linkers and spacers between the peptide and fatty acid, and different acylation positions along the peptide backbone were tolerated, with some caveats.

Liraglutide, also known as NN2211 in several publications [63,64], retains an N-terminal His-Ala sequence [70], with an expectation that the fatty acid's binding to albumin helps to shield the N-terminal part of the compound from DPP-4 proteolysis. In addition to delaying renal elimination, the C16 fatty acid also imparts a delayed SC absorption [71,72]. Importantly, these later studies revealed that the protraction mechanism of the fatty monoacid modification in liraglutide is driven by its strongly associated heptamers, with additional contributions from albumin binding and membrane association [73].

The promising *in vivo* results with liraglutide, through its C16 fatty monoacid derivatization of GLP-1, motivated efforts to extend the dosing regimen from once-daily to once-weekly [41,74]. Here, there was a renewed focus on albumin binding by further inspecting the linkers, spacers, and fatty acid derivatizations. In an influential paper [74], Lau et al. detailed the discovery of fatty diacid modification with an optimal linker to extend half-life suitable for once-weekly SC dosing; lessons from this paper have laid the foundation for general half-life extension of peptides [1].

A high priority in this campaign was to establish a method to rank the "protraction effect," i.e. to evaluate albumin binding and protraction while avoiding animal PK studies for all analogs. Here, they introduced the "BR value" as an indirect measure; in this setup, the binding IC50 in the presence of 2% HSA is compared to the binding IC50 in the absence of HSA. Here, compounds with "tight" albumin binding that interferes with receptor engagement would be expected to have large BR values. Additionally, the receptor potencies for all new analogs were evaluated with respect to liraglutide and native GLP-1 (7–37). It should be noted here that the assumption of albumin-bound species being completely inactive is not entirely valid, as elaborated below in several different case studies. Nevertheless, evaluation of activity in the presence and absence of HSA is essential in evaluating the behavior of fatty acid-derivatized compounds. In this report, two important lessons were learned: (1) identification of the beneficial effects of diacid derivatization and (2) appreciation of the value of extensive linker screening.

They reported that installing short, discrete PEGs (termed OEG for an oligoethyleneglycol unit, also referred to as PEG2; Ado, 8-amino-3,6-dioxaoctanoic acid; or AEEA, 2-(2-(2-aminoethoxy)ethoxy)acetyl) between the peptide backbone and fatty acid had large effects on the receptor binding in the presence and absence of albumin. For example, an analog with just a C18(diacid) side chain had 15× worse affinity in the absence of albumin and a BR value of 104, whereas a C18(diacid)-γGlu-OEG-OEG side chain had only 1.5× worse affinity in absence of albumin but a large BR value of 527. In other words, in the former analog, the fatty acid had a major effect on the GLP-1R binding, which was even further affected by albumin binding. In the latter compound (with a longer spacer and better albumin binder), the GLP-1R binding was minimally affected in the absence of albumin, but then it was significantly impacted by albumin binding. Several rounds of engineering were conducted to find the right balance between receptor binding/potency and albumin binding. Similar to earlier works [68,69], the degree of albumin binding and protraction effect positively correlated with the length of the fatty acid. They settled on a set of top analogs for *in vivo* PK studies in rats and mini-pigs, leading to selection of the semaglutide compound. It contains C18(diacid)-γGlu-OEG-OEG acylation and an Aib 8 substitution, as well as an Arg 34 substitution. It possesses a much longer half-life of ~9 hours in rats compared to ~3 hours with liraglutide.

Ultimately, a half-life of 168 hours in humans, suitable for once-weekly SC dosing, was reported with semaglutide [75,76]. Semaglutide has shown proven efficacy in type 2 diabetes and provided more than 15% weight loss in obese individuals after one year [77], as well as demonstrating cardiovascular benefits [78].

Following the success of the derivatized GLP-1 agonist semaglutide and other combined agonists, Eli Lilly pioneered the tirzepatide compound, also referred to as LY3298176 [79–81], which is currently one of the most promising clinical dual incretin peptides. Dual receptor activity at GLP-1 and GIP can lead to superior glycemic and weight control [82], as demonstrated in consistent dose-dependent decreases in body weight [79]. Tirzepatide [83] contains two nonnatural Aib residues at positions 2 and 13, as well as acylation with a C20(diacid)-γGlu-OEG-OEG, suitable for once-weekly dosing in humans. Significant progress has been made toward dual receptor activity at GLP-1 and glucagon [84,85]. A representative analog is survodutide [86], which has a C18(diacid)-γGlu-GGSGSG acylation, multiple glucagon point substitutions, and uses a 1-aminocyclobutanecarboxylic acid (Ac4c) building block to install DPP-4 resistance at the second position.

In the engineering of retatrutide, also referred to as LY3437943 [87,88], the beneficial regulation of glucose metabolism and weight control was enhanced by applying triple agonism at GLP-1, GIP, and glucagon. The combination of triple agonism can cause even greater weight loss than mono or dual agonism [89–91]. In the case of retatrutide, three nonnatural residues were incorporated: Aib 2, Aib 20, and α-methyl-leucine at position 13 to optimize glucagon and GIP activity. Acylation with C20(diacid)-γGlu-OEG-OEG was again used for once-weekly dosing, however, at position 17. It is important to note that acylation with a C20-diacid (as seen in other incretins and insulins) favors once-weekly dosing, while C16 acylation, as used in SAR441255 [91], was suitable for only once-daily dosing in humans. One additional relevant GLP-1 analog is ecnoglutide [92], which has a receptor-biasing Val [93] at the second position and a C18(diacid)-γGlu-OEG-OEG acylation at position 24. Ecnoglutide is a cAMP-biased analog, which has a once-weekly half-life in humans and potently induces GLP-1R activity, with reduced receptor internalization – potentially beneficial to therapeutic efficacy [94].

Lastly, an important observation during the development of optimized triagonists (GLP-1, GIP, and glucagon) was the impact of fatty acid on receptor potency and selectivity, not just half-life. In other words, the fatty acid "protractor" itself can serve as part of the pharmacophore, not just influencing PK properties. For example, it was found that swapping a C-terminal C16-monoacid to a semaglutide-like C18-diacid resulted in differential loss of potency at all three receptors: ~3× at GLP-1R, ~7× at GIPR, and ~33× at the glucagon receptor [62]; similar results were seen during the development of dual GLP-1 and glucagon receptor agonists [95].

Cumulatively, these results on protracted GLP-1 and other incretins highlight several key points:

- Derivatization with fatty *monoacids* linked to the peptide backbone via γGlu (or other anionic groups) can extend half-life, but this protraction mechanism is primarily driven by self-association and membrane-binding, with less contribution from interaction with albumin.
- The length of the fatty acid derivative (either monoacid or diacid) typically scales with the half-life extension.
- Optimal peptide protraction is obtained when using a fatty diacid linked to the peptide backbone via γGlu (or other anionic groups) and an ideal spacer, usually composed of short discrete PEGs.
- The solubility benefit of the terminal carboxylate group when installing diacids, compared to monoacids, should not be dismissed.
- Fatty diacid lipidation provides several benefits from its albumin binding and consequent half-life extension, but the fatty acid position and its effect on potency (with and without albumin) need to be carefully screened.
- Additional attention should be paid to other sequence-specific metabolic liabilities (e.g. DPP-4 cleavage), whose enzymatic degradation might not always be sufficiently shielded by the fatty acid.

7.4 Diverse Mechanisms of Fatty Acid Derivatization to Extend Insulin Half-life

A discussion of fatty acid derivatization of insulins is essential, being that the first reported example of a fatty acid-derivatized peptide drug with extended half-life was in human insulin [2]. Several recent reviews [96–100] have discussed the molecular engineering of insulins including their design, *in vitro* and *in vivo* properties, and production. Hence, this section will focus on fatty acid derivatization to provide long-acting basal insulins. The reader is directed to these above reviews for extensive discussion around fast-acting, oral, hepatic-targeted, and glucose-sensitive insulins [101]. The development of insulins with long half-lives is also a representative case that the "most potent" compound is often not the "best" compound.

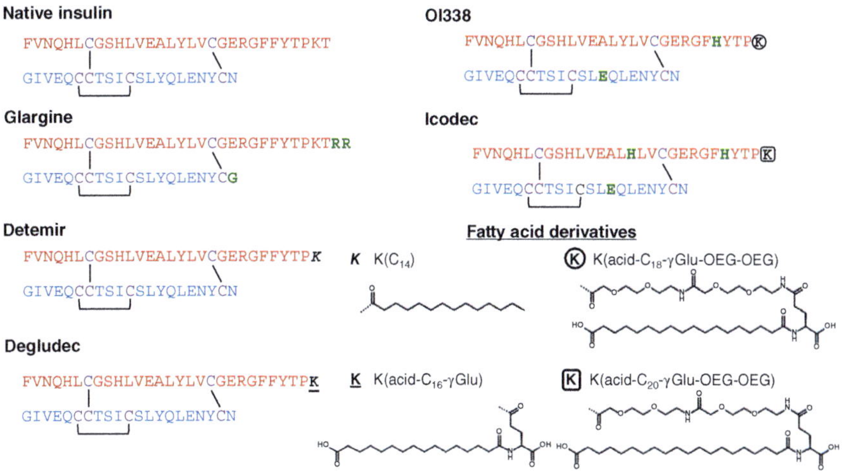

Figure 7.2 Engineered, long-acting forms of insulin. Amino acids are colored as follows: red, insulin B-chain; blue, insulin A-chain; purple, Cys residues involved in disulfides; green, residues different from native insulin sequence; black, Lys that are modified with fatty acids, where the derivatization at the particular ϵ-amine is indicated.

Native two-chain insulin (see Figure 7.2), like most nonfunctionalized peptides, has a short half-life of 5–15 minutes when injected intravenously [102]. Early efforts to "protract" the action of insulin were focused on delaying the release from the SC space following injection. In insulin glargine, the first basal insulin analog, two additional Arg residues are appended to the C-terminus of the B-chain, which increases the pI from 5.4 to ~6.8 [103–105]. Insulin glargine is dissolved at an acidic pH 4.0 in the vial and then "precipitates" at the injection site near pH 7.4. It then slowly releases into circulation as the compound redissolves over the next ~24 hours [106]. Additionally, an Asn to Gly replacement at the C-terminus of the A-chain was introduced to enhance stability [103]. The *in vivo* activity of insulin glargine was further enhanced by optimizing the formulation from 100 to 300 units/mL, which resulted in a longer half-life (up to 19 hours) and more evenly distributed activity over time [107].

The next efforts to time-extend the action of insulin were based on fatty acid modification, utilizing albumin binding as a protraction mechanism. The first clinically approved fatty acid-derivatized peptide is insulin detemir, which possesses a C14 fatty (mono)acid derivatization at position 29 of the B-chain [2,108] (the C-terminal Thr is removed). This acylated peptide showed a half-life of ~4–6 hours in humans following SC injection, and ~3 hours following IV injection [109]. This discrepancy in half-life was attributed to "flip-flop kinetics," where the rate of absorption is slower than the rate of elimination [110]. Insulin detemir was found to bind albumin and bundle into a dihexameric state, which delayed the SC absorption [109,111]. The half-life of this first fatty acid-modified insulin was consistent with once-to-twice daily dosing [112].

To further extend the half-life of insulin analogs, the focus shifted to stabilizing the soluble hexamer assemblies using fatty acid derivatization. In nature, insulin assembles as a hexamer in the pancreatic β-cells, coordinated by Zn(II) [113–115]; hexamers then dissociate into active monomers upon dilution into the blood. This natural mechanism of controlling insulin activity through quaternary structure inspired the design of improved insulin therapeutics, including both fast-acting and basal variants [97]. This concept of controlling higher-order (multi-hexamer) assembly originated from an early study with cholic acid acylated derivatives [116]. In an important study [117], Jonassen et al. prepared a series of insulins with different fatty diacid-γGlu modifications at B-chain position 29, with lengths ranging from 12 to 18 carbons and including various anionic substitutions (L-γGlu, D-γGlu, or L-αGlu). These analogs were then systematically characterized by CD spectroscopy, SEC, and MALS in different formulations and physiological conditions to evaluate higher-order association states. Using SEC, they showed that the various analogs (containing Glu linkers, thus omitting insulin detemir) associated as dihexamers in formulation buffers with phenol, cresol, and zinc acetate. Elimination of phenol, as would occur upon injection, led to the formation of soluble, higher-order assemblies (>5 MDa) in the case of C16-L-γGlu (insulin degludec) and C18-L-γGlu fatty acid derivatives. Monomers of active insulin then dissociate from the multi-hexamer via dissociation of Zn. The ability of C16-L-γGlu fatty acid side chains to thread between various hexamers (so-called "pearls on a string") was confirmed in crystal structures of insulin degludec [118]. The half-life of insulin degludec in human is ~25 hours following SC injection [119].

As described thus far, fatty acid derivatization in insulin detemir and degludec extended half-life primarily via altered self-association, not albumin binding. The benefits of fatty diacid technology to achieve significantly longer half-lives (up to once-weekly dosing) in insulin-based analogs will now be described. Two important advances were needed to achieve this breakthrough: (1) increasing the albumin binding affinity, as seen in fatty acid-derivatized incretins [74,79], and (2) decreasing the insulin receptor affinity, to slow elimination of the analog via receptor-mediated clearance [120,121]. Critically, it was previously shown that insulins of different receptor affinities had similar glucodynamic properties, but altered PK profiles, and lower affinity analogs showed longer half-lives [122].

Significant efforts to engineer an orally available insulin were recently reported [123,124]. To begin these studies, a library of insulins was incubated in proteolytic enzyme assays (chymotrypsin, pepsin, and carboxypeptidase A). Two stabilizing substitutions were selected: A-chain Tyr 14 Glu and B-chain Phe 25 His. This double-substituted insulin showed ~4-fold lower receptor binding, but 10-fold higher stability in rat duodenum compared to native insulin. Introduction of these same mutations also improved the solubility at pH 5.5–6.5, as found in the small intestine [125]. After derivatization with C18-diacid-γGlu-OEG-OEG at B-chain position 29, this analog (OI338) was shown to have ~50-fold lower receptor binding compared to native insulin (~500-fold when tested in the presence of 1.5% HSA). Nevertheless, this analog had comparable *in vivo* potency to human insulin (using hyperinsulinemic euglycemic clamp studies) in rat (45%), pig (99%), and dog (72%). The importance of the backbone substitutions (A-chain Tyr 14 Glu and B-chain Phe 25 His) to lower receptor binding and extend half-life was validated in a PK study in dogs, where the native backbone control (OI216) showed a half-life of ~4 hours, versus 17 hours for OI338. Ultimately, OI388 showed a mean half-life of ~55 hours in humans and ~3% bioavailability after oral dosing in a sodium caprate-based GIPET tablet [126]. Oral dosed OI338 [127] also successfully completed a Phase 2 study that showed glycemic control comparable to SC-dosed insulin glargine in patients with type 2 diabetes [128].

The molecular engineering to achieve once-weekly insulin (insulin icodec) [129] expanded on the lessons from the development of long-acting, orally bioavailable OI338. Again, half-life extension was driven by both fatty acid derivatization and reduced receptor affinity. First, the C18-fatty diacid in OI338 was swapped for a C20-fatty diacid, which was already previously shown to double the half-life from 6 to 12 hours in rats [124]. Second, a third substitution (B-chain Tyr 16 His) [130] was introduced to further reduce insulin receptor affinity ~4–5×. This additional mutation, in the context of the previous mutations (A-chain Tyr 14 Glu and B-chain Phe 25 His) and C20-fatty diacid derivatization extended half-life in rats from 12 to 26 hours, and in dogs from 33 to 60 hours. Importantly, this new ultra-long-acting insulin possesses a half-life of up to 196 hours in humans [131,132] suitable for once-weekly dosing, while retaining the necessary glucodynamic (glucose-lowering) activity.

Some of the general lessons from these studies on fatty acid derivatization of insulins include:

- Both fatty mono- and diacid derivatization alters the biophysical association of insulin by altering the monomer–hexamer–multihexamer equilibria, which then "protracts" the action of SC-injected insulins.
- Similar positive correlations between fatty acid chain length, degree of albumin binding, and *in vivo* half-lives were seen with insulins as were observed with incretins.
- Importantly, judicious point substitutions were made in the insulin backbone to slow receptor-mediated clearance, stabilize against protease resistance, and improve solubility. Although these mutations negatively affect receptor binding, their incorporation results in significantly extended half-lives and importantly retained similar *in vivo* efficacies.

7.5 Development of a Stable, Fatty Acid-derivatized Amylin Analog

The development of cagrilintide (see Figure 7.3), a long-acting amylin analog, is instructive concerning the multifactorial challenge of converting a potent but fragile (unstable and short-lived) peptide into a therapeutically relevant molecule [133]. Indeed, it possesses pronounced challenges with solubility, biophysical stability, and chemical stability that had to be overcome to attain a long-acting fatty acid-derivatized clinical candidate.

Amylin is co-secreted with insulin by the β-cells in the pancreas in response to eating [134,135]. It can cause beneficial effects on glucose regulation, energy expenditure, and satiety [136,137]. There is recent interest in combining its biological activities with GLP-1 and similar incretins, as they may possess potentially complementary, synergistic mechanisms (e.g. targeting different regions of the brain) [138–140]. However, amylin is a biophysically challenging molecule to develop due to its capacity to form amyloid-like aggregates, which are seen in the pancreases of patients with type 2 diabetes [141–143] and even possibly connected to the amyloid-β fibrils seen in Alzheimer's disease [144,145]. Amylin also has a short half-life following SC administration, with mean residence times of ~25 minutes [146].

The first amylin molecule approved for diabetes was Pramlintide, which contains three Pro substitutions originating from rat amylin at positions 25, 28, and

		10	20	30	37
Amylin (human)		KCNTATCATQ	RLANFLVHSS	NNFGAILSST	NVGSNTY-amide
Amylin (rat)		KCNTATCATQ	RLANFLVRSS	NNLGPVLPPT	NVGSNTY-amide
Pramlintide		KCNTATCATQ	RLANFLVHSS	NNFGPILPPT	NVGSNTY-amide
Calcitonin (salmon, sCT)		CSNLSTCVLG	KLSQELHKLQ	TY PRT	NTGSGTP-amide
Calcitonin (human, hCT)		CGNLSTCMLG	TYTQDFNKPH	TF PQT	AIGVGAP-amide
Davalintide		KCNTATCVLG	RLSQELHRLQ	TY PRT	NTGSNTY-amide
KBP-088	Ac-	CSNLSTCMLG	RLSQELHRLQ	TF PKT	DVGANAP-amide
KBP-336	Ac-	CASLSTC**X**LG	RLSAELHKⓀA	TY PKT	DVGANAP-amide
Cagrilintide (AM833)		***K***CNTATCATQ	RLA**E**FL**R**HSS	NNFG**P**IL**PP**T	NVGSNT**P**-amide

X Aib

Ⓚ K(acid-C_{18}-γGlu-OEG-OEG)

K K(acid-C_{20}-γGlu)
(derivatization is at the α-amine, not ε-amine)

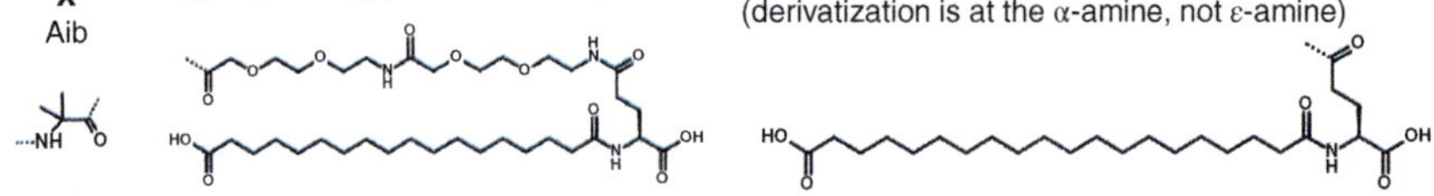

Figure 7.3 Optimization of stable, long-acting cagrilintide/amylin analogs. Amino acids are colored as follows: blue, originating from the human amylin sequence; red, Cys residues involved in intramolecular disulfide; black and bold, key substitutions and Lys that are modified with fatty acids, where the derivatization is indicated.

29 [147] to improve solubility and reduce formation of aggregates [148], while retaining similar activity as amylin. Pramlintide showed small (single-digit kg), but significant weight loss in insulin-treated diabetic [149] and obese [150,151] patients. Unfortunately, the improved solubility and stability did not significantly extend half-life, which was found to be in the range of 30–50 minutes, necessitating multiple injections per day [152].

The next major step in advancing the activity and half-life of amylin-based agonists, besides protein fusion [153,154], was to capitalize on a slower receptor off-rate. Here, research groups relied on amylin's homology and receptor interplay with salmon calcitonin (sCT). sCT has a slow off-rate from the CT family of receptors, CTRs [155]. In the case of davalintide (a 32 aa peptide), positional changes from sCT were introduced into the middle portion of human amylin (37 aa peptide), as well as deletion of five residues. Davalintide enhanced the inhibition of food intake and showed prolonged effects on gastric emptying and glucose levels (for up to 8 hours), both more effective than amylin controls in rat models; however, the *in vivo* half-life was not significantly extended [156,157]. These beneficial effects were explained by the slower off-rate for the amylin receptors (AMYRs). Several subsequent generations of amylin-calcitonin hybrids, termed DACRAs (dual amylin and calcitonin receptor agonists) have been pursued in an effort to install nonselectivity at both CTRs and AMYRs (composed of both CTR and a receptor activity-modifying protein RAMP1–3), leading to higher *in vivo* efficacy [158,159]. One representative DACRA is KBP-088, which is also N-terminally acetylated to improve metabolic stability [160,161]. However, a weak association of sCT with cancer and limited *in vivo* benefit have engendered caution when designing amylin analogs that have prolonged binding and duration of action at CTRs [162,163].

These studies to balance CTR and AMYR potency set the stage for the development of cagrilintide [133]. A sophisticated screening plan was devised to identify optimally potent and balanced analogs against the relevant receptors [155,164]. First, two different human receptors were used, hCTR and $hAMY_3R$; $hAMY_3R$ was selected over other isoforms due to a previous study showing relevance for RAMP3 in the area postrema of the brain where amylin mediates its anorexic effect [165,166]. Second, because the *in vivo* evaluation of activity was food intake in rats, the rat homologs of these same receptors were also tested: rCTR and $rAMY_3R$. Third, as previously done for characterizing fatty acid-derivatized compounds, the *in vitro* potency and binding experiments were tested in both the presence and absence of HSA.

During the early SAR work to identify long-acting amylin analogs, solubility challenges and injection site reactions with signs of necrosis were observed [133] when the compounds were formulated at pH 4, as was done with pramlintide [167]. Here, it was proposed that although the fatty acid-derivatized amylin analog could bind albumin at the injection site to prop up the solubility, there might not be enough local albumin at this site to saturate large doses of injected compound. Critically, fibrils of human amylin-based compounds have been shown to be cytotoxic [168]. This early result motivated design efforts to improve the solubility and reduce fibril formation in the formulation and SC space (pH 4–7.4). This compound possessed an isoelectric point of 7.4 and showed low solubility at near-neutral pH values. Two assays were then used to overcome this challenge. In the first assay, peptides were methodically screened in a solubility assay at different pH values (from 3 to 8) using buffered solutions. In the second assay, the fibril formation propensity of analogs was evaluated in a thioflavin T assay, similar to that used with insulin [169], where the lag time and % recovery were determined after mechanical stress for up to 45 hours.

Over 800 amylin-based analogs were screened in the campaign to develop cagrilintide [133]. Four important sequence changes were realized:

- Fatty acid derivatization at the N-terminus with C20-diacid-γGlu, to extend half-life.
- An additional salt bridge (i to $i + 3$) between positions Glu 14 and Arg 17, which improved solubility and eliminated a potential deamidation by product at Asn 14.
- Three Pro substitutions at positions 25, 28, and 29 to improve solubility and biophysical stability, as in both the rat amylin and pramlintide sequences. Substitution with a single Pro at position Ile 26 was inadequate to reduce fibril formation of the fatty acid-derivatized peptide, in contrast to a nonderivatized peptide [170]).
- A C-terminal Pro amide, as in the hCT and sCT sequences, to balance CTR potency.

Cagrilintide (also referred to as AM833) completed a phase 2 trial in 2020 and was shown to have a half-life in humans of 159–195 hours [171], sufficient for once-weekly dosing. In combination studies with semaglutide (CagriSema), mean body weight decreases of 15.6% were obtained after 32 weeks, compared to 5.1% for semaglutide alone and 8.1% for cagrilintide alone [172]. A similar fatty acid-derivatized DACRA compound (KBP-336) also recently showed beneficial impacts on weight loss and glucose control [173].

The challenge of achieving long-acting and potent amylin analogs revealed additional complexities not seen during the development of GLP-1 analogs. In this case, several important lessons can be noted:

- Physical and chemical stability challenges should be confronted early in the SAR campaign of fatty acid-derivatized peptides. Issues such as injection site reactions and chemical instabilities (Asn deamidation) can cripple a project, so early re-designs around these problems may be required.

- Traditional SAR optimizations seen in advancing nonlipidated peptides, such as helical stabilization through salt bridges, introduction of Pro substitutions to reduce aggregation, and elimination of chemical instabilities, remain vital to the development of fatty acid-derivatized peptides.
- Two conservative design principles should be mentioned:
  - Consider the downstream manufacturing requirements for the successful clinical candidate while conducting the SAR campaign. Here, multiple side-chain positions tolerated fatty acid derivatization with minimal impact on potency, but direct acylation at the N-terminus also resulted in a similar potency. Additionally, acylation at the N-terminus is easier to deploy when using chemical synthesis as it does not require orthogonal protecting groups or nonnatural amino acid substitutions.
  - As possible, minimize the number of mutations from the native human sequence, to reduce the risk of immunogenicity (see Immunogenicity section for more details).

7.6 Stabilization and Fatty Acid Derivatization of Peptide YY (PYY)

PYY (Peptide YY, Figure 7.4) is the last case study to be discussed before providing prescriptive advice on optimizing fatty acid-derivatized peptide therapeutics. Here, PYY is highlighted for several reasons:

- It possesses a short *in vivo* half-life due to renal elimination, similar to most peptides.
- It has multiple metabolic instabilities, driven by enzymatic processes.
- There are multiple defined chemical instabilities, which have been thoroughly studied and addressed.
- Its therapeutic window can be optimized through tuning of its receptor selectivity.
- Key studies provide educational insights into optimizing fatty acid derivatization for enhancing the *in vivo* PD and PK of peptides as an entire class.

PYY is an important satiety hormone [174], which is co-released with GLP-1 by L-cells in the intestines in response to eating [175,176]. Like GLP-1, PYY is rapidly cleaved at its N-terminus by DPP-4 [177] to generate PYY_{3-36}. PYY_{3-36} possesses higher selectivity to its receptor Y2R compared to the full-length hormone [178], has an anorexic effect [174,179], and has attracted great interest as a treatment for obesity and diabetes [180,181]. Additional N-terminal trimming by aminopeptidase P has also been observed [178,182], creating species such as PYY_{2-36}. Nevertheless, the primary metabolite seen in plasma [182,183] is PYY_{3-34}, which

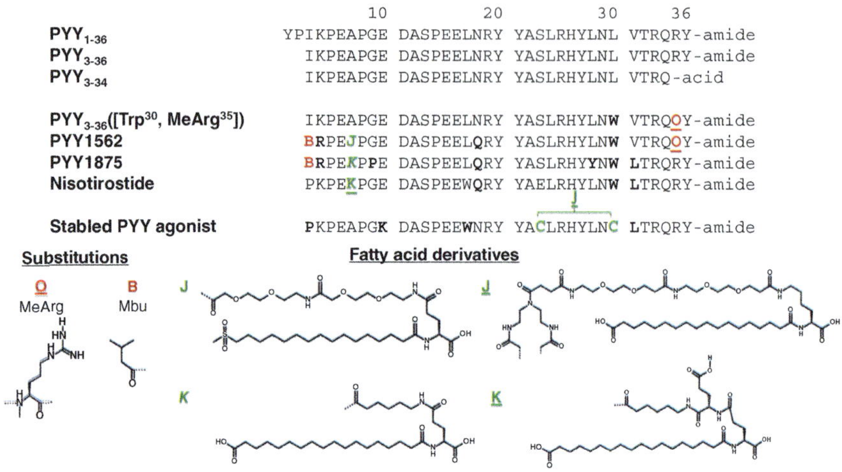

Figure 7.4 Optimization of representative PYY_{3-36} peptides. Amino acids are colored as follows: red, nonnatural substitutions; black and bold, important natural amino acid substitutions; green, fatty acid derivatizations at particular ϵ-amines as indicated, or Cys thiols in the stabled PYY agonist sequence.

is thought to be generated via DPP-4 cleavage at the N-terminus and sequential, two-step enzymatic processing at the C-terminus of the peptide. During the course of developing more stable, long-acting PYY analogs as potential therapeutics, it was important to address this *in vivo* instability at the C-terminus, as this region of the molecule (residues 25–36) is critical to its receptor interaction and subsequent activity [184,185]. Furthermore, to improve the overall therapeutic efficacy of the molecule, it is important to maximize the selectivity toward Y_2R, versus other members of the neuropeptide receptor family (Y_1R, Y_4R, and Y_5R) due to their counteracting effects on appetite and related biological outcomes [186–188]. Lastly, another specific challenge that had to be considered is the pI of the compound. In studies with structurally similar NPY analogs, the N-terminal part of the molecule was removed, leading to short and potent agonists; however, this trimming also generated high pI compounds that caused local site reactions when injected [189].

In an important work [190] to identify highly potent and selective analogs of PYY_{3-36}, an exhaustive SAR campaign was conducted to simultaneously optimize the receptor affinity and potency, selectivity, and metabolic stability. Initially, multiple substitutions were systematically investigated in the context of receptor activity and proteolytic stability. These variants, focused on the C-terminus of the molecule, included:

- Alternative C-termini (including methylamide and alcohol)
- Traditional amino acid scans at every position

- *N*-methyl amino acid scans
- β-homo amino acid scans (to retain the helical structure [191])
- Peptoid-like [192,193] variants
- Point substitutions with either slightly shorter or longer side chains (e.g. Arg to dinorArg, norArg or homoArg)
- Other nonproteogenic substitutions such as D-amino acids and cyclized side chains or peptide backbone via lactams.

Addition of a single methyl unit at the C-terminus of the peptide (change from a C-terminal amide to a methylamide) led to a >2500 loss in affinity for Y_2R, emphasizing the very tight interaction between the C-terminus and the receptor. Several of the changes abolished receptor activity, such as peptoid-like bonds at the amide nitrogen, across multiple residues. Other analogs demonstrated the ability to increase promiscuity at multiple receptors of the NPY family, such as Gln 34 MeGln (*N*-methylGln), which increased potency at Y_1R by 100×, while retaining a similar potency at Y_2R, compared to backbone control. Following a canonical amino acid scan, it was found that Leu 30 to Trp 30 retained strong potency at Y_2R, with worse potency at Y_1R and Y_5R. Additionally, it was recognized that the Arg 35 side chain was essential for retaining high affinity and selectivity, while a subtle backbone substitution of this Arg with *N*-methylArg enhanced proteolytic stability after incubation in pig plasma, with half-lives ranging from 200 minutes in PYY_{3-36} reference peptide, 620 minutes in the MeArg 35 single substitute, and up to 980 minutes in the Trp 30, MeArg 35 double substitute.

Building on the progress of the Trp 30, MeArg 35 double substitution of PYY_{3-36}, Østergaard et al. [194] methodically investigated the optimal fatty acid derivatization (position and composition) to extend half-life while maximizing potency and retaining selectivity. In the first series of peptides, a "fatty acid scan" of C18-(diacid)-γGlu-OEG-OEG on ϵ-amine of Lys was conducted across the sequence, and the potency at multiple receptors was evaluated. Binding to Y_2R in the presence and absence of albumin was also examined. Derivatization in the C-terminal part of the molecule was deleterious to receptor activity, albeit moderate Y_2R potency was maintained by derivatization in position 30. This result was supported by previous data on the close packing interactions between the peptide C-terminus and the receptor [190,195,196]. In contrast, derivatization at several N-terminal sites, such as positions 7 and 10, had minimal impact on Y_2R potency, both within 3× of control. However, the *in vivo* half-lives in mini-pigs were longer in cases of C-terminal modification (up to ~75 hours) compared to N-terminal modification (~10–20 hours). This result was explained following metabolic stability studies, which showed enzymatic proteolysis at the C-terminus in the N-terminal-protracted variants; in contrast, the C-terminal-protracted variants were shielded against this same proteolysis. Thus, a very careful balancing act of potency versus metabolic stability needs to be performed when installing fatty acids into peptides.

After completing the fatty acid scan across the PYY_{3-36} sequence, the composition of the fatty acid itself was then examined at a single position. Here, position 30 was selected to test various fatty acids, as derivatization at this position resulted in the longest *in vivo* half-life (~75 hours), while retaining receptor potency within 10× of nonprotracted control. Importantly, the length of the half-life directly scaled with the length of the fatty diacid protractor: C14 (4 hours), C16 (28 hours), C18 (76 hours), and C20 (98 hours) – all variants with modification at position 30 via γGlu-OEG-OEG linker. Additional control studies showed that the γGlu linker was vital to obtaining a long half-life; its omission reduced half-lives roughly 5×, depending on the exact fatty diacid length. Furthermore, the length of the OEG spacer (2–6 units) had minimal impact on the half-life in this context, while abutting the fatty acid γGlu immediately onto the peptide backbone (no OEG spacer) extended half-life even further to 97 hours – explained by additional peptide backbone shielding through albumin binding.

The critical results obtained with these derivatized PYY_{3-36} peptides were then further expanded using albumin binding assays, in order to disentangle the protraction effects of direct albumin binding (and reduced renal filtration) versus solely albumin shielding against proteolytic degradation. Here, a fluorescent label was installed at position 4, which was then used to evaluate albumin binding with fluorescence polarization. The compounds with short half-life (N-terminal modification at positions 7 and 10) showed similar albumin binding as those with long half-life (C-terminal modification at positions 30 and 33) – revealing that the differences in half-life were due to metabolic instabilities rather than differences in albumin binding. These studies also recapitulated the "protraction effect" of longer fatty acids, where the observed albumin binding affinity scaled with the length of the fatty acid.

The lessons from the earlier 2018 paper [190] were then applied to compounds that emerged from this fatty acid derivatization campaign. Hence, the protease-resisting Arg 35 MeArg substitution was re-introduced into a PYY_{3-36} analog with N-terminal acylation at position 4. Indeed, this combined construct possessed a very long half-life of 83 hours, compared to the Arg 35 control of 13 hours. This result confirmed that the much shorter half-lives of N-terminal-derivatized variants were due to proteolytic instability at the C-terminus of the molecule.

An important study [197] was next reported to obtain a highly potent, selective, and stable PYY_{3-36} analog (PYY1562). In this study, a rational approach was used to identify and overcome remaining chemical and biophysical liabilities in the lead compound, ultimately generating a protracted compound with commercially viable properties. To start, several PYY_{3-36} analogs with different fatty acid modifications and amino acid substitutions were prepared, and these variants were meticulously investigated for chemical degradation byproducts, including asparagine deamidation and high molecular weight peptides/proteins (HMWPs), followed by examination of self-association and any undesired aggregation.

The PYY_{3-36} sequence contains two native Asn residues at positions 18 and 29, which could undergo deamidation [198–200]. After testing accelerated chemical degradation at 37 °C for up to 50 days, multiple deamidation byproducts were observed at both positions, including conversions from L-Asn to L-Asp and L-isoAsp. Position Asn 18 could be substituted with a Gln [201], which was not detrimental to receptor activity. However, substitution of Asn 29 with Gln significantly impacted Y_2R potency. Nevertheless, retaining Asn 29 ultimately led to a relatively low sample loss of ~0.64% per week at 37 °C. Lastly, although not explicitly overcome, isomerization at Asp 11 (from L-Asp to L-isoAsp and D-isoAsp) was also observed to a very minor extent (0.18% per week at 37 °C). These same "accelerated chemical degradation" conditions also identified an HMWP, which was a covalent dimer through a benzoquinone that originated in the excipient formulation from oxidation of the antimicrobial phenol. This problem was solved by eliminating the two primary amines in the PYY molecule. First, at the N-terminus, Ile 3 was replaced with a desamino version, Mbu (3-methylbutanoyl). Second, Lys 4 was substituted with Arg. These changes reduced the accumulation of HMWP dimer from ~4% to less than 1% after six weeks at 37 °C (in the presence of phenol additive), and importantly they did not impact the secondary structure or receptor potency.

The two C-terminal substitutions at positions 30 (Leu to Trp) and 35 (Arg to MeArg) that were previously reported [190] to improve receptor selectivity and metabolic stability also enhanced the biophysical properties of the molecule, by reducing the solution turbidity as well as the size range of various self-associates. Finally, fatty acid derivatization was reinvestigated in the peptide backbone. Modification in the N-terminal part of the molecule led to rather minimal impacts on receptor potency (2–4×, depending on the position) and did not affect the helical structure, as measured by CD. These derivatized compounds did not show any signs of fibril formation or aggregation on storage, although a small degree of turbidity was seen in one compound. Importantly, these results contrast significantly with those seen in an earlier study [202], which analyzed the biophysical behavior of PYY_{3-36} derivatized with a C16-(monoacid)-γGlu at three different Lys side chains. In this extensive report, gels and multiple different nanostructures were seen depending on the position of the acylation. Although the monoacid derivatization could provide enhanced thermal stability, this report emphasizes the complex biophysical challenges that can be endured with a nonideal (monoacid) derivatization.

Novo Nordisk recently reported [203] another PYY analog (PYY1845) that shows noteworthy blood glucose lowering benefits, in addition to superior body weight loss in combination with semaglutide. In this work, a deep mutational scan was performed to identify fatty acid lipidated analogs with even better activity than the native (nonderivatized) PYY_{3-36}. These analogs were derivatized at position Lys 7 with a C18(diacid)-γGlu-Ahx side chain to generate ultra long-acting analogs. Additionally, Eli Lilly has recently disclosed [204] a PYY analog (nisotirostide) with a C18(diacid)-γGlu-αGlu-Ahx side chain, also derivatized at position Lys 7. This analog contains multiple canonical amino acid substitutions along the backbone, as well as an N-terminal Pro, instead of the Mbu (3-methylbutanoyl) group described above. Beneficial effects on food intake and body weight were reported in the patent.

In addition to the side chain fatty acid derivatizations reported above to advance a stable PYY_{3-36} analog, a noteworthy "stabling" (staple stabilizing) approach was also reported. In this method, a peptide staple (cross-link via the side chains of two introduced Cys) is introduced to reduce proteolysis and improve potency by conformational restriction [205–208]. The staple also includes a fatty acid branch to bind albumin and extend half-life. For this approach to work, careful introduction and engineering of the staple are required; in this report [209], a staple scan (introduction of paired Cys residues with spacing of i to $i + 7$) across the sequence was first performed, guided by structural analysis [210,211], especially avoiding the C-terminal part of the molecule. These variants were then tested in the Y_2R potency assay, resulting in select stapled constructs that showed negligible potency differences from the native sequence, as well as retaining high selectivity against the other NPY receptors. In the end, they identified a "stabled" analog with a symmetric staple between positions 23 and 30, containing a branched semaglutide-like side chain, C18(diacid)-γGlu-OEG-OEG. This analog showed similar *in vitro* potency to the native hormone, as well as an albumin binding shift in its potency. It also had a significantly extended half-life of 14 hours (from 1 hour in nonstapled control) in rats and beneficial synergism with a stapled GLP-1 analog in reducing food intake and body weight in mice.

7.7 Guidance for Advancing Fatty Acid-derivatized Peptides

Following the case studies provided above – GLPs, insulins, amylins, and PYYs – prescriptive information is given to characterize and optimize the "other," nonpotency peptide parameters. Emphasis is placed on detailing the basic experiments and parameters that need to be considered in parallel with traditional SAR campaigns. These sections include:

- 7.7.1: Peptide Solubility
- 7.7.2: Biophysical Stability
- 7.7.3: Immunogenicity

- 7.7.4: Metabolic and Chemical Stability
- 7.7.5: General Manufacturing Considerations
- 7.7.6: Specific Principles for Fatty Acid Derivatization

7.7.1 Routine Methods for Evaluating Peptide Solubility

On top of conventional biological properties such as cellular potency, receptor binding, and activity in the presence of albumin, several biophysical properties should be included in the SAR campaign. Importantly, endogenous peptides have not been subject to evolutionary pressure to obtain the requisite biophysical properties needed for a peptide-based drug. As reviewed elsewhere [212–214], the biophysical properties, solubility in particular, can strongly influence the *in vivo* activity, solution stability, and manufacturability. Optimization of the solubility profile of the therapeutic peptide at formulation and *in vivo* conditions is crucial for obtaining the intended pharmacological and stability profile of the final drug.

Unlike larger proteins or antibodies, solubility of peptides can be dramatically altered by substitution with only a few amino acids. Historically, low solubility through formulation optimization or molecular design of the endogenous peptide was a strategy to prolong the *in vivo* half-life. For instance, increased half-life of insulin was obtained in the 1940s and 1950s by bringing the insulin to a solid state at formulation conditions as exemplified with Zn and protamine insulin suspensions [215]. Later, extension of insulin half-life based on low solubility *in vivo* was obtained through molecular design of insulin glargine where two additional Arg residues in the C-terminus of the B-chain shifts the isoelectric point of insulin glargine from pH 5.4 to ~6.8, thereby causing the insulin to precipitate at physiological pH [216] (described above in the insulin section).

However, modern techniques – especially fatty acid acylation – have now been introduced to prolong half-life while keeping the peptide in the dissolved state and avoiding the disadvantages of suspensions (variability in efficacy, poor resuspendability, lump formation in formulation, adsorption to glass surfaces, etc., as described for insulin [215,217,218]). Consequently, high peptide solubility at various conditions (reflecting manufacturing conditions, formulation, and *in vivo*) is an important parameter to include in early SAR work. Solubility of peptides in aqueous media is strongly influenced by the charge properties of the molecule, mainly driven by the number of charged residues (Asp, Glu, Arg, Lys, His), the nature of the two termini, and their corresponding pKa values [219,220]. Based on these values, the theoretical net charge at different pH values can be calculated, as exemplified in [221]. Online tools (e.g. PepCalc.com) can be used to calculate the net charge versus pH relationship based on the natural amino acid composition, with the net charge as a first proxy of the solubility profile (high solubility at high numeric net charge).

Table 7.1 Example of a pH gradient based on corresponding acid/base pairs.

Buffer	20 mM citric acid/Na-citrate pKa 3.1 – 4.8 – 6.4				40 mM MES/Na-MES pKa 6.2		30 mM TRIS-HCl/TRIS pKa 8.1	
pH	2.0	3.0	3.5	4.0	5.5	6.5	7.5	8.5

Table 7.2 Example of a pH gradient based on McIlvaine buffer system.

Na_2HPO_4 (mM)	4	41	64	77	112	139	182	195
Citric acid (mM)	98	79	68	61	44	31	9	3
pH	2.2	3.0	3.6	4.0	5.4	6.4	7.4	8.0

This approach does not adjust for any modifications of the natural amino acids, attachment of nonpeptide groups (such as fatty acids and PEG spacers), secondary/tertiary structure, and self-association, which could potentially influence solubility. However, more sophisticated tools for predicting intrinsic solubility at neutral pH and room temperature of peptides containing nonnatural amino acids were recently published [222,223]. Given the limitations of *in silico* procedures, the solubility profile should be established experimentally at relevant conditions for manufacturing and formulation (including variable pH, salt, buffer, and presence of organic solvents). A representative experimental setup for examining the pH–solubility relationship can be achieved by preparing a pH gradient, by mixing corresponding acid/base pairs of buffering agents at different pKa values to cover the relevant pH range as exemplified in Table 7.1.

Alternatively, a pH gradient can be established using the McIlvaine buffer system (citrate/phosphate mixture) [224] (see Table 7.2).

These buffer systems can be combined with tuning of other parameters, such as different salt concentration, presence of organic solvents, and temperature, to expand the knowledge on solubility properties beyond variations in pH [221]. For instance, the maximum solubility at a subset of conditions reflecting the final drug product composition can be useful to examine in case very high-concentration peptide formulations are needed. Aliquots of the dissolved peptide are then mixed with these buffers, and the solubility profile is established based on quantification of in-solution peptide (for instance, using UV detection at 214 or 280 nm) or detection of precipitated material (for instance, using OD measurements or light scattering techniques).

7.7.2 Concepts and Methods for Biophysical Characterization

Depending on the peptide sequence, covalent modification (e.g. fatty acid derivatization), and other external parameters (such as pH, ionic strength, excipients, temperature, mechanical stress, shear forces, etc.), many peptides possess the ability to (self-)assemble, where two or more peptide molecules combine, driven by intermolecular attractive forces. This phenomenon suffers from imprecise nomenclature, where terms such as aggregates, oligomers, self-assemblies, gel formation, self-associates, etc., are applied in an unsystematic manner. An attempt to harmonize the nomenclature was published in 2011 where the phrase "aggregates" was suggested as the general term and defined as "any self-associated species, with monomer defined as the smallest naturally occurring and/or functional subunit," with the option of further classification based on size, reversibility, secondary/tertiary structure, covalent modifications, and morphology [225]. However, from a drug development perspective, it makes sense to apply another terminology to distinguish between formation of undesirable versus desirable (or acceptable) multimeric species, which will be referred to as "aggregates" and "self-associates," respectively, in the present review.

7.7.2.1 Aggregates: Undesired Multimeric Species

Formation of aggregates in peptide and protein pharmaceuticals has been reviewed in numerous papers, including detailed descriptions of the phenomena and underlying mechanisms [213,214,226–231]. A common understanding is that the initial point of aggregation often involves misfolding of a minor fraction of the peptide molecule population, induced by external parameters (exposure to hydrophobic interfaces, temperature, ionic strength, etc.), followed by formation of intermediate metastable oligomers that grow upon monomer addition to the final mature aggregates. Formation of the final aggregates is typically irreversible. One subtype of peptide aggregates termed amyloid fibrils (irreversible, highly ordered, cross-β-sheet-enriched fibers) can be followed by detection of changes in fluorescence properties of the fluorescent dye thioflavin T upon specific binding to the fibrillar species as originally described by Naiki et al. [232] and Le Vine [233], reviewed by Malmos [234], and later applied to a number of different peptides including insulin [169,235–237], amylin [133,238,239], and glucagon [240,241]. It should be noted that formation of aggregates is a phenomenon occurring both *in vivo* such as amyloid fibril formation in neurodegenerative diseases and diabetes [242–244] and *in vitro* leading to particle or fibril formation at drug formulation conditions [226,228–230]. Aggregate formation in protein and peptide-based pharmaceuticals is to be reduced as much as possible due to the potential risk of aggregates eliciting an immune response in humans [245–250].

7.7.2.2 Self-associates: Better-defined, Acceptable Multimeric Species

Formation of better-defined, acceptable self-associates often involves the peptide pool assembling into higher-order structures of monodisperse, well-defined sizes that are stable over time and reversible upon dilution. The self-association patterns and stoichiometry are by nature concentration-dependent but are also affected by the properties of the solvent (pH, ionic strength, etc.). Insulin is a classic example: it is monomeric at sub-μM concentrations, dimeric at μM concentration, and hexamer at higher concentrations in the presence of zinc [251]. As a result of the reduced conformational flexibility following hexamer formation, the insulin molecule becomes less prone to aggregation (fibril formation) and chemical degradation [251]. Following SC injection, hexameric insulin dissociates to dimeric and monomeric species due to dilution and diffusion of hexamer-stabilizing ligands (zinc and phenol) [252]. Based on differences in the *in vivo* absorption of the monomer–dimer–hexamer species, molecular engineering toward increased monomerization has led to development of insulin analogs with faster PK properties than human insulin [252–254].

Fatty acid derivatization of the peptide backbone can induce self-association, as exemplified by GLP-1 [42,255,256] and exendin-4/glucagon co-agonists with self-associate size scaling with the length of the acyl chain [257,258], but also for insulin [118,259] as well as PYY [202], where large effects on self-associate stoichiometry follow relatively small changes in primary peptide structure [197]. In most cases, the equilibria between different self-association states are believed to be reached within a short period of time (in seconds); however, reports on liraglutide at specific conditions have revealed that days are required to reach equilibrium [72,260].

Due to the potential effects of aggregates and self-associates on biological properties (for instance, immunogenicity and altered PK), these parameters are important to include in early SAR work. Techniques that rely on size determination are often applied when examining these phenomena. Formation of aggregates is normally explored at accelerated aggregation-inducing conditions to allow for several design cycles within a reasonable time span. The exact aggregation-inducing conditions to use often differ depending on the peptide, but could include exposure to elevated temperature, mechanical stress/shear force, high salt concentration, and incubation at a pH near the peptide's pI [261]. The extent of aggregate formation (following specified periods of time) can be used for ranking and then optimizing peptide analogs, as exemplified for PYY [197], amylin [133], and GLP-1/glucagon/GIP triagonist [262]. The analytical techniques for measuring aggregate formation have been reviewed elsewhere [230,263] but often consist of techniques covering a wide size range (see Table 7.3):

- Monomers/dimers/oligomers (size exclusion chromatography)

- Soluble aggregates (field flow fractionation, analytical ultracentrifugation, light scattering DLS/SLS, and small-angle X-ray scattering)
- Aggregates detectable by light microscopy/micro-flow-imaging, even up to visual inspection for larger species.

Table 7.3 Examples of analytical techniques for investigation of aggregates and self-associates.

Technique (typical size range)	Main applications	Comments
SEC (10^{-9} to 10^{-7} m)	• Size distribution and quantification • Reversibility (dilution in elution buffer)	• Potential bias from elution buffer, on-column dilution/interaction • Can be combined with light scattering detection (MALS, multi-angle light scattering)
DLS/SLS (10^{-9} to 10^{-6} m)	• DLS/dynamic light scattering: size distribution • SLS/static light scattering: average molar mass	• Sensitive to large particles • Removal of large particles/dust to be considered
SAXS (10^{-10} to 10^{-7} m)	• Size, average molar mass, shape	• Structural information in solution • Increasingly applied over the last decade
AF4, field-flow fractionation (10^{-9} to 10^{-5} m)	• Size distribution	• No stationary phase but potential risk of membrane interaction • Broad dynamic concentration range
AUC (10^{-9} to 10^{-6} m)	• Molar mass distribution	• Not high-throughput • Well-established gold standard technique
Microscopy (light: 10^{-6} to 10^{-4} m) (electron: 10^{-9} to 10^{-5} m)	• Visualization of larger aggregates (e.g. insulin fibrils by TEM)	• Small sample size
Micro-flow Imaging (MFI) (>10^{-6} m)	• Size, number, and shape/morphology	• Requires large sample volumes
Visual inspection (>10^{-5} to 10^{-4} m)	• Presence of large aggregates, turbidity/opalescence	• Nondestructive whole-sample assessment • Can be operator-dependent (subjective)

Peptide self-association is *per se* not generally needed to be minimized/eliminated, as exemplified by the benign endogenous zinc-stabilized hexameric storage form of insulin in pancreatic beta cells [264]. Consequently, studies that demonstrate reversibility of self-associates upon dilution (see for instance [265,266] for GLP-2 and insulin analogs, respectively) can provide early predictions of adequate absorption from the SC tissue prior to conducting *in vivo* studies. Additionally, studies demonstrating that the self-association pattern does not change over time or upon exposure to various external stress factors should be conducted (see for instance [197] for acylated PYY). Instruments used for exploration of self-association overlap to a large extent with those used for analyzing aggregates and typically rely on light/X-ray scattering (DLS/SLS and SAXS) or sedimentation properties (analytical ultracentrifugation). Typical analytical techniques used for investigating self-associates and aggregates are listed in Table 7.3.

Due to their relatively small size, peptides often display less pronounced folding behavior compared to larger proteins; thus, these structural properties are often not first-line properties [226]. However, secondary structural features such as formation of α-helix, β-sheets, and random coil are present. To mention a few examples of helix formation, α-helix structure has been published for GLP-1 [267] and acylated counterparts [72,255], as well as for exendin-4 [268] and related GLP-1/glucagon dual agonist [257]. An increased α-helix content following acylation has been reported for vasoactive intestinal peptide [269], whereas virtually no effect was observed for PYY [202]. Amylin has been shown to be predominantly unfolded but with increasing α-helix structure in the presence of increasing concentrations of SDS (biological membrane mimic) [270], suggesting that amylin adopts increased α-helical structure upon binding to the transmembrane receptor. Similarly, propensity to form α-helix has been published for sCT [271] and prolactin-releasing peptide [272]. The aforementioned secondary structure properties can be examined using circular dichroism (CD) [273] optionally combined with other techniques such as Fourier transform infrared (FTIR) spectroscopy, NMR, and Raman spectroscopy as reported elsewhere [268,274–278].

7.7.3 Assessment of Peptide Immunogenicity

Peptide and protein drugs have the potential to induce immune responses, which may result in induction of anti-drug antibodies (ADAs) that can impact patient safety as well as drug efficacy [279]. This has, in several cases, led to discontinuation of a therapeutic peptide or protein in clinical trials [280–282]. Unwanted immunogenicity can be triggered by the API itself (e.g. its amino acid sequence, structure, PTMs), the drug product (e.g. impurities, aggregates, or host cell proteins), or clinical parameters such as the route of administration—which are complicated by the patient's genetic background and immune status. ADA responses are common [283] and are often driven by the presence of T cell epitopes in the primary amino acid sequence of the API and the subsequent induction of a CD4+ T cell response. A T cell epitope is an antigenic peptide within the amino acid sequence of a protein that is recognized by a T cell receptor (TCR). Molecular engineering is important to enhance therapeutic peptide/protein potency and overall activity, as well as biophysical and chemical properties. However, immunogenicity and creation of neoepitopes that can be recognized by the immune system as "foreign" may thwart these ventures, resulting in antibodies to both the engineered and endogenous proteins (if there is an endogenous counterpart) [284]. An illustrative case is the development of recombinant factor VIIa (rFVIIa) protein [285]. NovoSeven (wild-type rFVIIa) has been an established treatment for bleeding in hemophilic patients, with a proven safety profile [286,287]. However, clinical development of an engineered version of rFVIIa, vatreptacog alfa, which possesses >100-fold faster enzymatic activity, and has only three amino acid substitutions (Met 298 Gln, Glu 296 Val, and Val 158 Asp) [288], had to be discontinued due to the emergence of ADAs [289]. Another engineered analog (Bayer's BAY 86-6150) includes six mutations (Pro 10 Gln, Lys 32 Glu, Ala 34 Glu, Arg 36 Glu, Thr 106 Asn, and Val 253 Asn) designed to increase platelet binding and extend half-life. Despite a favorable *in silico* immunogenicity assessment (NetMHCIIpan v2.1 [290]), this analog was also withdrawn due to identification of a neutralizing antibody that cross-reacted with the native protein [291].

An important aspect in the optimization of peptide/protein drugs is the de-immunization of potential T cell epitopes. Though no standardized regulatory guidance currently exists, consideration and documentation of nonclinical assessment of immunogenicity potential is encouraged by both EMA and FDA [284,292]. Furthermore, early intervention to assess and decrease any potential unwanted T cell response is critical to a successful, streamlined SAR campaign. Several tools to assess immunogenicity potential *in silico* and *in vitro* are currently available [293]. Three tools (see Table 7.4 and Figure 7.5) are described that can be used to identify and de-immunize T cell epitopes, including their uses and limitations: (1) *in silico* algorithms, (2) MAPPs assays, and (3) T cell activation assays.

***In silico* algorithms** predict T cell epitopes in the linear amino acid sequence based on their binding potential to MHC (HLA in humans) class II protein [290,294] and in more recent versions the potential of an epitope to be presented on MHC class II on the surface of antigen-presenting cells (APCs) [295]. These tools are particularly suitable for early engineering rounds as large numbers of sequences can be screened in a relatively short time, and the output provides information on T cell neoepitope burden and core sequence. The core sequence is a nine-amino-acid-long sequence in the MHC class II binding peptide (usually 13–25 aa long) that binds to the major groove of the MHC class II molecules. Positions 1, 4, 6,

Table 7.4 Comparison of three standard tools to assess immunogenicity potential based on T cell epitopes.

	Discovery stage/throughput (number of compounds)	Input	What is assessed	Application	Limitations
In silico prediction tools	Throughout all stages/high (1000s)	Amino acid sequence	Linear peptide binding potential to MHC class II	*In silico* de-immunization, screening of a large number of sequences and ranking with respect to immunogenic potential, supporting proteomics data by identification of core peptides, assigning neoepitopes vs self	Overpredictive No consideration of T cell recognition, ignorant of peptide structure, or modifications
In vitro MAPPs assay	Parent molecules identified/low (usually around 20)	Candidate molecule	Processing and presentation of drug-derived peptide on MHC class II	Narrowing *in silico* selection, mapping of clusters of T cell epitopes, identification of presented peptide	Low throughput, only presentation of antigen, no assessment of T cell response, costly, time-consuming
In vitro T cell assays	Ranking of frontrunners/low (usually around 20)	Candidate molecule/overlapping peptides spanning full sequence	Activation of T cells by drug-derived peptides (can be performed on full-length API or set of overlapping peptides covering API sequences)	To evaluate T cell response, rank peptide according to their immunogenic potential	No information on amino acid sequence if performed on full-length API, costly, time-consuming

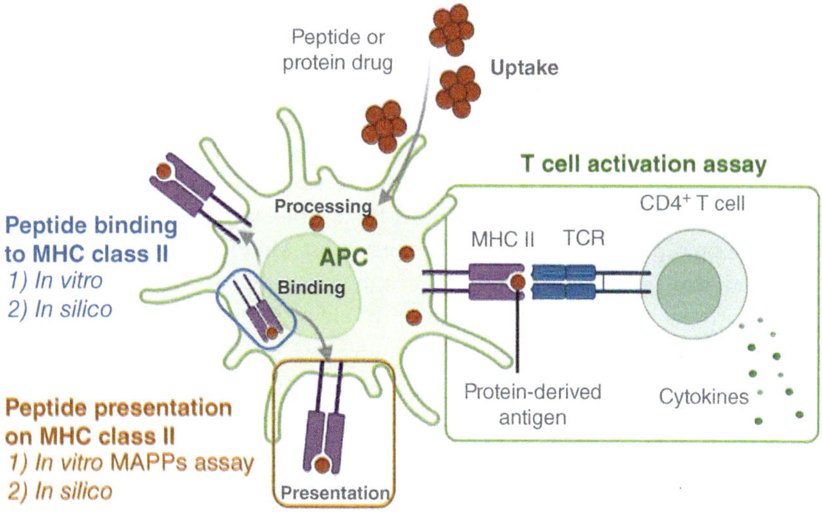

Figure 7.5 Assessment tools for T cell epitope-derived immunogenicity. When a peptide or protein drug is taken up by professional antigen-presenting cells (APCs), it is processed in the lyso/endosome where antigenic peptides are then loaded onto MHC class II molecules. MHC:peptide complexes are transported to the cell surface where they may be recognized by $CD4^+$ T cells with corresponding TCRs. The classical and best-characterized tools investigate immunogenicity potential via assessment of peptide binding (*in silico* and *in vitro*), peptide presentation on MHC (via MAPPs assay), and $CD4^+$ T cell response to presented peptide:MHC complexes, which in turn can elicit a B cell response, leading to ADAs. Figure created with BioRender.com.

and 9 of the core sequence contribute the most to the binding of the peptide to the open-ended peptide binding groove [296].

Prediction of T cell epitopes enables ranking of peptide variants according to their immunogenic potential, and several platforms for T cell epitope prediction are readily available [295–297]; hence, *in silico* tools are often used as the first step in immunogenicity risk assessment during early drug discovery [293].

Importantly, despite the fact that some versions of these algorithms today integrate MHC presentation of a sequence with epitope binding affinities, they remain overpredictive as they do not account for antigen processing, T cell receptor recognition of peptide: MHC complexes, T cell tolerance, or, importantly, the effect of fatty acid derivatization. Thus, the use of *in vitro* tools is important to identifying which *in silico* predicted epitopes are most relevant for decreasing the immunogenic potential of a peptide.

MHC-associated peptide proteomics (MAPPs) are routinely used to identify epitopes presented on MHC class II molecules on the surface of professional

APCs *in vitro* and thus are considered suitable to identify "relevant" T cell epitopes [298]. In these studies, monocytes are isolated from the peripheral blood of healthy donors and differentiated into dendritic cells *in vitro* [298]. Immature dendritic cells are then pulsed with peptide/protein and subsequently matured with LPS, resulting in the processing and presentation of any potential T cell epitopes. The cells are lysed, and peptide: MHC complexes are captured by immunoprecipitation with an MHC class II-specific antibody coupled to beads (magnetic or sepharose) [298,299]. Of note, most MAPPs assays are presently performed using pan HLA-DR-specific antibodies due to the availability of well-performing clones, lower expression levels of DP and DQ receptors, and sparse data on the relevance of HLA-DP and DQ in the induction of immunogenicity. However, recent studies have successfully integrated anti-HLA-DP and DQ clones in MAPPs assay [300]. When peptides eluted from the captured complexes are subsequently analyzed by mass spectrometry, the resulting peptide masses allow sequence identification via database search. Eventually, identified sequences can be mapped to the amino acid sequence of the API of interest, often revealing peptide clusters in specific regions of the protein [301]. The location of these clusters, combined with sequence information and knowledge about core binding peptides from *in silico* algorithms, provides valuable insight into which epitopes and amino acids to mutate, in order to reduce immunogenicity potential during further engineering rounds. Successful de-immunization can then be evaluated by performing side-by-side comparison of the parent molecule with its de-immunized variants in the MAPPs assay. Of note, these results are dependent on the individual's HLA type (the expression of an individual set of MHC class II alleles). Thus, it is crucial to perform the assay on a large enough set of donors, usually 15–30, to survey adequate HLA diversity, making MAPPs assays a time-consuming, costly, and laborious low-throughput assay.

While *in silico* tools and MAPPs assay can provide sequence information on potentially immunogenic epitopes, **T cell activation assays** can evaluate the lead compound's potential to induce activation of $CD4^+$ T cells, which is considered the main driver of unwanted immunogenicity. Like MAPPs assays, these assays are performed on primary cells from healthy donors, hence requiring investigation of a sufficiently large number of individuals (typically 30–50) and a significant amount of time and resources. As in the MAPPs assays, the number of molecules that can be assessed is limited by the setup and thus the assay is usually performed in later stages of drug discovery, when frontrunner molecules have been identified. Providing a setting that is arguably closest to "*in vivo*," T cell assays represent a system where more than one effector cell is present and exists in different formats, each exhibiting distinct advantages and disadvantages. Most commonly, T cell activation assays are performed either in (CD8-depleted) peripheral blood mononuclear cells (PBMCs, consisting of immature DCs, monocytes, [CD4+] T cells, and B cells) [293] or in DC:T cell restimulation assays [302], where sensitivity is increased by a restimulation step. In these assays, T cell activation is measured by either proliferation or cytokine secretion, in response to the peptide or protein therapeutic. The magnitude of the response in individual donors and the proportion of respondents allows ranking of candidate molecules with respect to their immunogenic potential.

These three tools provide the means to assess immunogenic potential, based on T cell epitopes in the primary amino acid sequence of the protein therapeutic, as well as novel epitopes introduced by peptide impurities related to synthesis methods and represent commonly used, well-characterized assays. The impact of product or process-related impurities (also termed innate immune response modulating impurities) on innate immune activation can be assessed *in vitro* in PBMC activation assays but are out of scope for this review [303]. Assessment of B cell epitopes is currently not common practice, even though direct recognition of a therapeutic by B cells is necessary for their activation and production of ADAs. This is largely due to the absence of feasible *in vitro* assays [294] as well as the superior performance of *in silico* tools for predicting T cell epitopes compared to B cell epitopes [304].

7.7.3.1 Impact of Fatty Acid Derivatization on Immunogenicity

Lastly, it must be highlighted that fatty acid derivatization can have a beneficial effect on reducing immunogenicity of engineered peptides and proteins. The type of fatty acid, and its position in the sequence, should be carefully considered during the design cycles, as elaborated throughout this review. In an early study [305] to measure the effect of fatty acid derivatization on peptide immunogenicity, exenatide (nonacylated) and liraglutide (C16 fatty acid-acylated) peptides were dosed in the Liraglutide Effect and Action in Diabetes (LEAD) Trials. Here, liraglutide showed significantly lower levels of antibody responses (~8 vs. ~61%) compared to exenatide. Furthermore, the presence of these low levels of anti-liraglutide antibodies did not impact its glycemic efficacy or safety (however, in this study, it cannot be unambiguously excluded that this was due to amino acid differences rather than derivatization). Importantly, in a systematic study on this property, Schulz et al. [306] reported that acylation with fatty acids minimized binding of peptides to class II HLAs and had a significant effect on reducing T cell activation. In this study, three model peptides, Influenza HA, exendin-4, and a 25mer peptide segment of vatreptacog alfa, were modified with different fatty acid derivatives, including 2, 12, and 18 carbon lengths. The observed binding affinities to class II HLAs were determined, showing that acylation, particularly with C18 fatty acid chains, generally reduces binding, although the effect is position- and HLA type-dependent. They next evaluated the T cell response to these peptides, again showing a highly beneficial effect with C18 derivatization; the degree of the deimmunizing effect depended on the exact position of the fatty acid in the sequence.

7.7.4 Evaluating Metabolic and Chemical Stability

Sufficient chemical [201] and metabolic stability of peptides and proteins is foundational to the success of any clinical candidate. Specifically, assessing the metabolic stability of peptides is a key component of a drug discovery program, as poor stability of a peptide analog can significantly impact the elimination half-life and bioavailability. Primary metabolic reactions involve cleavage of the peptide bond by exo- or endoproteases, although other reactions such as disulfide bond reduction, deamidation, and oxidation reactions are also observed [201]. Certain metabolic reactions may be predicted from the amino acid sequence, including degradation by DPP-4 (described above with GLP-1 and PYY) and cleavage of basic amino acids at the C-terminus by carboxypeptidases U/B [182]. Nevertheless, because of the wide variety of proteolytic enzymes in the body [307], a nonbiased approach for addressing proteolytic sites and metabolic stability is necessary. In this section, we will describe analytical methods, *in vitro* assays, and *in vivo* analyses for evaluating metabolic stability.

Liquid chromatography coupled to mass spectrometry (LC-MS) is the preferred analytical method for quantifying the parent peptide and identifying metabolites. An important, but not trivial, step is to develop a good sample preparation method for isolating the parent peptide and metabolites from the biological sample. Various methods such as protein precipitation and solid-phase extraction are available. Protein precipitation can be performed by adding reagents such as strong acids (e.g. trichloroacetic acid, perchloric acid), metal ions (e.g. zinc sulfate), salts (e.g. ammonium sulfate), or organic solvents to the biological sample [308]; for example, addition of 3–6 volumes of organic solvents (acetonitrile, methanol, or ethanol) to one part of plasma. Acid or base can be added to the organic solvent, which may improve recovery or remove background. The recovery, sensitivity, and reduction in background noise are all important parameters to consider.

The selection of UPLC columns and mobile phases is also important for the analytical workflow. A recommended starting point includes conventional reversed-phase columns based on C4, C8, or C18 stationary phases, with mobile phases composed of water, acetonitrile (or methanol), and formic acid (0.1–1%). Other volatile additives in the mobile phase such as ammonium acetate, ammonium formate, or ammonia can also be used. The quantification and identification of peptides/metabolites are best performed using a high-resolution mass spectrometer, such as a Q-TOF or instruments based on the Orbitrap technology. This ensures identification of metabolites from full scan and MS/MS spectra with a high mass accuracy, which is essential for confidently identifying drug metabolites.

Proteolytic degradation products may have different physical/chemical properties than the parent peptide, so sample preparation methods, chromatographic conditions, and MS settings should not be too selective to the parent compound. Samples from PK experiments are highly relevant for identifying *in vivo* metabolites. If doses are low and thus the resultant plasma concentrations are low, there may be additional requirements needed for the analytical methods, including the optimization of sample preparation, chromatographic conditions, and mass spectrometric parameters. A general rule of thumb for lower limit of quantification of the parent peptide is around 0.5–5 nM, based on a method consisting of a plasma protein precipitation followed by addition of organic solvent and LC/MS methods using conventional reversed phase columns.

The use of predictive *in vitro* metabolic stability assays is rather limited. Therefore, methods for evaluating the metabolic stability of peptides are needed, including evaluation in plasma, serum, whole blood, and with recombinant peptidases/proteases known to be involved in the clearance of the given peptide. While some peptides modified with fatty (mono) acids may undergo oxidation, which in theory could be linked to cytochrome P450 enzymes [309], assays based on drug metabolism in the liver (e.g. liver microsomes and hepatocytes) are more commonly associated with small molecules than with peptides and proteins.

Additionally, the stability of the peptide in plasma serves as a good starting point in a drug discovery project. The assay can be conducted using plasma from different species, including human and animals, especially those used in PK studies such as mouse, rat, dog, and mini-pig. Typically, time course experiments are conducted, and first-order half-lives are reported. For high-throughput screening campaigns, a simplified assay may be used, in which plasma stability is reported after one or two time points, such as 120 or 240 minutes. In our laboratories, GLP-1 (7–37) is often selected as a positive control in plasma stability experiments, where the half-life is reported, and the formation of the major metabolite GLP (9–37), via DPP-4 cleavage [35,36], is confirmed.

Key parameters in a plasma stability assay include substrate concentration, type of plasma, and incubation time. The substrate concentration is usually between 1 and 25 μM. As pH may become increasingly more alkaline during plasma incubations [310], a buffer is often added to stabilize pH during the experiment. Plasma is typically stabilized with either EDTA or heparin, and it is important to consider that some proteases, such as metalloproteases, may be inhibited by EDTA. For instance, initial *in vitro* stability experiments with PYY_{3-36} in EDTA-stabilized plasma failed to predict the PYY_{3-34} metabolite, which was the major one observed in mini-pigs and rhesus monkeys [182].

As mentioned above, the list of predictive *in vitro* systems is limited for peptides and proteins, but for the exploration of specific enzymatic pathways, peptide analogs can be incubated with defined recombinant enzymes. This approach is applicable when a clearance pathway has been identified, and it thus provides an efficient way to screen a high number of peptides. For native GLP-1 (7–37) or (7–36) amide, DPP-IV and neprilysin have shown to be important *in vivo* proteolytic enzymes [35,36,70], but, in general, there are many proteases that can be explored, so considerations regarding *in vivo* relevance are required.

Data from *in vitro* stability experiments are especially valuable when *in vivo* relevance has been established, as certain *in vitro* metabolic pathways may not always correlate with *in vivo* metabolic stability. Importantly, peptides that are acylated with fatty acids are often stable in a plasma stability assay due to a limited proteolytic activity in plasma and high plasma protein binding, which protects peptides against degradation. This same acylated peptide may still undergo *in vivo* enzymatic degradation due to other proteases located in other compartments, such as endothelial cells of different organs.

7.7.4.1 Notes on Relevant Chemical Instabilities

The multitude of chemical instabilities seen with peptide therapeutics is outside the scope of this review; hence reference is made to several, extensive reviews on this subject [201,226,311–313]. However, chemical instability is often related to a subset of potential labile amino acids including Asn (deamidation), Asp (isomerization), Met/Trp/Tyr (oxidation), and Cys (disulfide scrambling); mitigation is often achieved by substituting the labile amino acid with a structurally/functionally similar, but stable amino acid.

In the four cases of fatty acid acylated peptides described in this chapter (GLP-1 incretins, insulins, amylins, and PYYs), Asn deamidation was the most prevalent instability. In this reaction, an L-Asn residue can deamidate and isomerize, proceeding through a cyclic succinimide intermediate, into several different species [314,315], including L/D-Asp and L/D-isoAsp, which are usually chromatographically challenging to resolve. The degree of this byproduct is highly sequence (and structure-dependent), with the neighboring, C-terminal ($N+1$) residue being very influential: the Asn-Gly dipeptide is particularly bad [199,200]. The degree, and specific degradation byproduct, is also pH-, temperature-, and solvent-dependent [314,316]. Similar caution around Asp residues should also be exercised [317]. For an additional example, an analysis of exenatide [318] showed pH-dependent (pH 7.5–8.5) deamidation byproducts. Exenatide contains two potential deamidation sites: Asn 28 and Gln 12, where Asn 28 was determined to be the primary one [318]; the majority byproduct was an L-Asn to L-isoAsp, while a L-Asn to D-isoAsp byproduct was also observed. And, both deamidation byproducts at position Asn 28 show decreased activity at the receptor [319]. However, importantly, the substitution of Asn 28 with Gln results in an analog that possesses indistinguishable GLP-1R agonist activity and similar PK/PD profiles in rodents [319]. As demonstrated here (and showcased in PYY [197]), replacement of the troublesome Asn with a Gln was sufficient to resolve the problem. Alternatively, one may try to substitute the neighboring ($N+1$) residue to reduce the formation of the cyclic succinimide intermediate.

The same concept of accelerated aggregation-inducing conditions used for evaluating biophysical stability is also applied to chemical stability, termed forced degradation strategies [320]. These studies are used to "forecast" the stability of various peptide formulations. Representative conditions include hydrolytic (low and high pH), oxidative (with hydrogen peroxide or open environment), photolytic (under illuminating source), and thermal (high temperature). Analysis of these byproducts can be done using the same analytical techniques described above when considering metabolic stability.

7.7.5 General Manufacturing Considerations Related to Peptide Optimization

The concept of early multiparametric optimization in the development of fatty acid-derivatized peptides therapeutics has been described previously. Evers and Wagner et al. [221,262,274] detailed an "iterative optimization procedure" for identifying potent and balanced triple agonists that had robust physicochemical properties for once-daily dosing. Their design cycle included: (1) *in vitro* activity and selectivity, (2) physical stability, (3) formulation requirements, (4) chemical stability, and (5) PK/PD. In this review, we have shifted the focus slightly by emphasizing immunogenicity and manufacturability as critical early considerations in the peptide development lifecycle.

Manufacturability (scalability) considerations will be important to addressing the growing demand for peptide therapeutics, since the current leading compounds, semaglutide and tirzepatide, are produced by different means: semaglutide through semi-recombinant means and tirzepatide through total chemical (solid-phase) synthesis [321]. While manufacturability can be seen as a problem for the process chemist, careful forethought by the medicinal chemist can have an outsized impact, as subtle details in molecular structure can be the determining factor in the downstream selection of a production path and hence the ultimate API scalability.

It is outside the scope of this chapter to describe the complexity of selecting and optimizing a scale-up or production process. While more efficient [322,323], more sustainable [324–326], and more flexible processes (such as liquid-phase peptide synthesis [327–329], solid-phase synthesis with green solvents, and recombinant/enzymatic posttranslational modification to install noncanonical amino acids or fatty acids) exist, it is generally true that at present, production paths tend to fall into two categories:

1. Fully synthetic processes using conventional solid-phase synthesis. This includes fully linear syntheses as well as more convergent approaches that employ the condensation of smaller peptide fragments [330].
2. Semi- or fully recombinant processes wherein the majority of chemical bonds being formed and hence overall molecular weight of the API are generated using a recombinant expression technology, such as yeast or *Escherichia coli*, followed by acylation or modification of the N- or C-termini [331].

There may be many exceptions and while the degree of scalability of either approach is heavily dependent on the API in question, it is generally the case that fully synthetic processes require less development time and allow more precise control of multiple factors (like the regiochemistry of fatty acid derivatization and opportunities to incorporate noncoding amino acids); however, this comes at the expense of scalability. Solid-phase synthesis is much less atom-efficient than semi-recombinant production paths in most cases, largely due to the need for protective groups during the synthetic process, the heterogeneous nature of each reactive step often requiring stoichiometric excess of amino acid building blocks and coupling reagents, the nonconvergent nature of SPPS, and the requirement of very large solvent volumes during resin-washing steps to remove reaction byproducts and excess reagent. Conversely, while semi-recombinant processes can overcome some of these scalability and atom economy limitations, there are certain transformations that conventional expression technologies still cannot achieve at scale. Because of these factors, it is prudent to weigh scalability demands of the final drug product with the potential benefits that may come from the incorporation of molecular features that are only readily accessible by a fully synthetic process.

Additionally, an understanding of patient need is critical to assessing the relative importance of scalability against the required efficacy, safety, PK, and dosing route. A multitude of factors can determine the required scale of API production; while not exhaustive, some of the most common factors are the size of the diagnosable patient population, the predicted therapeutically relevant dose level, and the acceptable dosing frequency. Together with an estimated API capacity need, elements of molecular design will then influence the overall production path. Here we provide some factors that a peptide medicinal chemist should consider when attempting to optimize a fatty acid acylated peptide for manufacturability: the C-terminus, noncoding amino acids, and nucleophilic amino acids/reactive groups.

C-terminus. It is not always the case that choice of C-terminal acid versus amide will impact function, but there are numerous examples of bioactive peptides where this subtle difference can completely abolish potency [133,190]. Additionally, because C-terminal acids are protic and form a carboxylate at neutral pH, choice of the C-terminus can impact pI and overall net charge, influencing the peptide's solubility profile, which can, in turn, impact formulation conditions (as elaborated above). If it is necessary to select a peptide with a C-terminal amide as the lead candidate for development, this will impact the production path and scalability, likely forcing the adoption of a fully synthetic process [330], as semi-recombinant processes generally yield only C-terminal acid peptides that must then be modified. At present, the standard process for this transformation relies on enzymatic peptidolysis of a C-terminal glycine-extended precursor peptide using the PAM enzyme, which adds complexity and detracts from scalability [332]. Ultimately, if it is possible to achieve the required potency, safety, PK, and other aims with a C-terminal acid, this will generally allow for more flexibility and scalability in process development.

Noncoding Amino Acids. The use of noncoding or "non-natural" amino acids in the peptide backbone can impart significant benefits to potency as well as physical, chemical, and metabolic stability. However, judicious use of these amino acids is recommended if scalability is a priority. While there is significant ongoing work to develop techniques for the recombinant incorporation of noncoding amino acids (such as repurposing of stop codons, nonnatural base pairs, or other tools in synthetic biology), the widespread, scalable implementation of these technologies has yet to be realized. Additionally, if noncoding amino acids are deployed in a lead candidate, their position in the peptide sequence can have a significant impact on production path selection and hence scalability. Noncoding amino acids positioned more proximal to the N-terminus will increase the potential for a feasible semi-recombinant approach because the noncoding amino acid as well as the remaining N-terminal fragment containing a reactive leaving group can be generated through a fully synthetic process and then condensed with a C-terminal fragment originating from a recombinant production path. While it is true that this condensation approach could be used to introduce a noncoding amino acid at any position in the sequence, there are obvious diminishing returns in atom economy, efficiency, and scalability, contributed by the recombinant portion of the production process, the more proximal the noncoding amino acid is to the C-terminus.

Nucleophilic amino acids and other reactive functional groups. While there are several theoretically feasible synthetic techniques for incorporating fatty acid derivatization onto a putative peptide drug candidate, some of the most widely used and scalable approaches fall into three categories: acylation of the ε-amino group of a Lys or the α-amino group of the N-terminus to form an amide bond (recommended if possible), alkylation of a Cys to form a thioether (possible, but requires use of precursor peptides with a reduced Cys, which can add complexity in production, purification, and handling), and reductive alkylation of the peptide N-terminus to form a secondary alkyl amine (possible but requires use of a reducing agent and chemistry that is generally less efficient than amine acylation, and requires the fatty acid be positioned at the N-terminus). What these examples have in common is that they can potentially enable a highly regioselective peptide derivatization without the need for incorporation of a noncoding amino acid (such as an azide/alkyne Click substrate [333]), thereby increasing the likelihood of achieving a more recombinant-friendly production path. In each case, if scalability is a high priority, it is prudent to consider the ramifications of the frontrunner amino acid sequence, as this will influence the feasibility of achieving high regioselectivity in the derivatization step. If, for instance, the lead peptide requires the use of multiple Lys for reasons of bioactivity, and if the only allowable locations for a fatty acid are mid-sequence rather than at the N-terminus, it creates an inherent regioselectivity challenge where the fatty acid derivatization step could be incorporated on the wrong Lys position. Without describing an exhaustive list of scenarios, we conclude this section with a general recommendation that these considerations are taken into account during lead optimization if practical, and acknowledge that incorporation of multiple nucleophilic functional groups into the peptide backbone will have downstream implications on the available techniques for fatty acid derivatization, the overall production path, and the resulting achievable production scale.

7.7.6 Specific Advice for Peptide Fatty Acid Derivatization and Characterization

Several benefits of fatty acid derivatization of peptides have been elaborated in this review. Critically, multiple factors should be considered to identify the best-derivatized compounds in an SAR campaign:

1. Generally, when looking to extend half-life of peptides, derivatization with fatty diacids is the better choice over fatty monoacids, due to longer half-life extension and better solubility of the diacid format. Installation of fatty monoacids may introduce undesired membrane association, as well as solubility and biophysical challenges. However, if half-life extension through a depot effect after SC injection is desired, then a fatty monoacid may be the right choice.
2. For once-weekly SC dosing, C18 and 20-diacids are currently the preferred choice. These should be prepended immediately by an anionic residue (preferably L-γGlu) to maximize albumin binding [194]. Multiple L-γGlu can also be installed next to the protractor to improve solubility or biophysical properties, without negatively impacting albumin binding [95,334,335].

3. Protractor scanning, i.e. installing the fatty acid derivative at different positions within the peptide (or at the termini), should be performed to optimize the protraction effect, and this effect should be balanced against any impact on peptide activity.
4. The spacer between the fatty acid-L-γGlu and the peptide backbone should be optimized during the protractor scanning campaign. Good starting spacers are 1–4 OEG units, based on previous clinical peptide candidates. Introduction of a shorter spacer may result in better "shielding" of the sequence by albumin and consequent better protection against proteases, although this is likely to be sequence-dependent. Longer spacers can be used to simultaneously maximize receptor activity at the same time as retaining high affinity to albumin, although introduction of a spacer that is too long may remove some of the benefits of fatty acid derivatization (e.g. reduced immunogenicity and improved protease resistance).
5. Alternative anionic modifications at the distal end of the fatty acid (not just a carboxylic acid) may also be tried, such as sulfo- or tetrazole groups. However, derivatives with distal amino and hydroxyl groups showed shorter half-lives during the studies to optimize liraglutide half-life [69].
6. Fatty acid derivatization serves many roles in extending half-life and improving PK properties, but its effects on the actual pharmacodynamics of the compound should not be overlooked, particularly receptor selectivity [62,194].

As noted in the checklist above, the effect of fatty acid derivatization on peptide binding and activity at its receptor needs to be evaluated *in vitro*. The experimental design of these studies is vitally important to maximizing their value [74,133]. Firstly, lipidated peptides can stick nonspecifically to plastic surfaces, which will distort the actual concentrations in these experiments. Traditionally, one adds a small percentage of FBS (fetal bovine serum, which contains albumin) or BSA (bovine serum albumin) to the culture media. However, the fatty acid part of the molecule will bind to BSA, interfering with receptor binding, which may again distort the "real" concentration in solution. In early studies on evaluating the activity of protracted peptides, very dilute concentrations of albumin were added as a compromise to addressing this conundrum. Ultimately, it was determined that ovalbumin can be added to the buffer to minimize nonspecific sticking; it does not bind to the fatty acid part of the molecule. Thus, when evaluating either receptor binding or potency, two different conditions should be tested: first, the peptide should be tested in ovalbumin-containing media to get an assessment of the "pure" interaction between ligand and receptor; second, this same experiment should be repeated in the presence of HSA (rather than BSA) to better mimic the actual ligand-receptor conditions *in vivo*. Furthermore, when performing more precise examination of the albumin binding effect on derivatized peptides, care should be taken to select the right albumin product. Fatty acid-free albumin may be necessary to evaluate the most precise measure of the albumin binding effect because most commercial albumin sources already contain bound lipids. Additional insight into the albumin binding effect on analog activity *in vivo* can also be gleaned from testing different concentrations of HSA, e.g. 0%, 1.5%, and 4.5% [124].

7.8 Concluding Remarks

The impact of fatty acid acylation technology has been transformative to advancing peptide therapeutics. This safe molecular technology has enabled the robust advancement of multiple peptides with previously fragile backbones and short half-lives to become routinely administered therapeutics for reliable once-daily and once-weekly dosing in humans. In this chapter, we highlighted four success stories—GLP-1-based incretins, insulins, amylins, and PYYs—that showcased the challenges that were overcome, and lessons that were learned to achieving therapeutically suitable peptides. Importantly, a multiparametric perspective is needed to advance peptide therapeutics, which takes in the solubility, biophysical stability, immunogenicity, chemical and metabolic stability, and manufacturing considerations early in the SAR campaign.

Acknowledgments

We acknowledge fellow Novo Nordisk employees Thomas Kjeldsen, Thomas Kruse, and Søren Østergaard for their helpful comments and support during the writing of this chapter.

References

1 Kurtzhals, P., Ostergaard, S., Nishimura, E., and Kjeldsen T. Derivatization with fatty acids in peptide and protein drug discovery. *Nat Rev Drug Discov.* 2023;22(1):59–80.
2 Kurtzhals, P., Havelund, S., Jonassen, I., et al. Albumin binding of insulins acylated with fatty acids: characterization of the ligand-protein interaction and correlation between binding affinity and timing of the insulin effect in vivo. *Biochem J.* 1995;312(Pt 3):725–731.
3 Peters Jr., T. Serum albumin. *Adv Protein Chem.* 1985;37:161–245.

4 Ellmerer, M., Schaupp, L., Brunner, G.A., et al. Measurement of interstitial albumin in human skeletal muscle and adipose tissue by open-flow microperfusion. *Am J Physiol Endocrinol Metab*. 2000;278(2):E352–E356.

5 Ganong, W.F. *Review of Medical Physiology*. 11 ed. Los Altos, CA: Lange Medical Publications; 1983.

6 Nilsen, J., Trabjerg, E., Grevys, A., et al. An intact C-terminal end of albumin is required for its long half-life in humans. *Commun Biol*. 2020;3(1):181.

7 Chaudhury, C., Mehnaz, S., Robinson, J.M., et al. The major histocompatibility complex-related Fc receptor for IgG (FcRn) binds albumin and prolongs its lifespan. *J Exp Med*. 2003;197(3):315–322.

8 Sand, K.M., Bern, M., Nilsen, J., Noordzij, H.T., Sandlie, I., and Andersen, J.T. Unraveling the Interaction between FcRn and albumin: opportunities for design of albumin-based therapeutics. *Front Immunol*. 2014;5:682.

9 Curry S. Lessons from the crystallographic analysis of small molecule binding to human serum albumin. *Drug Metab Pharmacokinet*. 2009;24(4):342–357.

10 Hu, H., Ransdell, A.S., Qu, H., et al. Probing the binding mechanism of acylated peptides to human serum albumin. *ACS Chem Biol*. 2023;18(5): 1158–1167.

11 Curry, S., Mandelkow, H., Brick, P., and Franks N. Crystal structure of human serum albumin complexed with fatty acid reveals an asymmetric distribution of binding sites. *Nat Struct Biol*. 1998;5(9):827–835.

12 Sudlow, G., Birkett, D.J., and Wade, D.N. The characterization of two specific drug binding sites on human serum albumin. *Mol Pharmacol*. 1975;11(6):824–832.

13 Bech, E.M., Pedersen, S.L., and Jensen, K.J. Chemical strategies for half-life extension of biopharmaceuticals: lipidation and its alternatives. *ACS Med Chem Lett*. 2018;9(7):577–580.

14 Bailon, P. and Won, C.Y. PEG-modified biopharmaceuticals. *Expert Opin Drug Deliv*. 2009;6(1):1–16.

15 Jevsevar, S., Kunstelj, M., and Porekar, V.G. PEGylation of therapeutic proteins. *Biotechnol J*. 2010;5(1):113–128.

16 Gaberc-Porekar, V., Zore, I., Podobnik, B., and Menart V. Obstacles and pitfalls in the PEGylation of therapeutic proteins. *Curr Opin Drug Discov Devel*. 2008;11(2):242–250.

17 Gupta, S.K. and Shukla P. Glycosylation control technologies for recombinant therapeutic proteins. *Appl Microbiol Biotechnol*. 2018;102(24):10457–10468.

18 Sola, R.J. and Griebenow K. Glycosylation of therapeutic proteins: an effective strategy to optimize efficacy. *BioDrugs*. 2010;24(1):9–21.

19 Alters, S.E., McLaughlin, B., Spink, B., et al. GLP2-2G-XTEN: a pharmaceutical protein with improved serum half-life and efficacy in a rat Crohn's disease model. *PLoS One*. 2012;7(11):e50630.

20 Podust, V.N., Sim, B.C., Kothari, D., et al. Extension of in vivo half-life of biologically active peptides via chemical conjugation to XTEN protein polymer. *Protein Eng Des Sel.* 2013;26(11):743–753.

21 Ahmadpour, S. and Hosseinimehr, S.J. PASylation as a powerful technology for improving the pharmacokinetic properties of biopharmaceuticals. *Curr Drug Deliv.* 2018;15(3):331–341.

22 Gebauer, M. and Skerra A. Prospects of PASylation(R) for the design of protein and peptide therapeutics with extended half-life and enhanced action. *Bioorg Med Chem.* 2018;26(10):2882–2887.

23 Tagmose, T.M., Pedersen, K.M., Pridal, L., et al. Molecular engineering of efficacious mono-valent ultra-long acting two-chain insulin-Fc conjugates. *J Med Chem.* 2022;65(3):2633–2645.

24 Levin, D., Golding, B., Strome, S.E., and Sauna, Z.E. Fc fusion as a platform technology: potential for modulating immunogenicity. *Trends Biotechnol.* 2015;33(1):27–34.

25 Pilati, D. and Howard, K.A. Albumin-based drug designs for pharmacokinetic modulation. *Expert Opin Drug Metab Toxicol.* 2020;16(9):783–795.

26 Bao, W., Holt, L.J., Prince, R.D., et al. Novel fusion of GLP-1 with a domain antibody to serum albumin prolongs protection against myocardial ischemia/reperfusion injury in the rat. *Cardiovasc Diabetol.* 2013;12:148.

27 Jonsson, A., Dogan, J., Herne, N., Abrahmsen, L., and Nygren, P.A. Engineering of a femtomolar affinity binding protein to human serum albumin. *Protein Eng Des Sel.* 2008;21(8):515–527.

28 Zorzi, A., Linciano, S., and Angelini A. Non-covalent albumin-binding ligands for extending the circulating half-life of small biotherapeutics. *Medchemcomm.* 2019;10(7):1068–1081.

29 Adams, R., Joyce, C., Kuravskiy, M., et al. Serum albumin binding knob domains engineered within a V(H) framework III bispecific antibody format and as chimeric peptides. *Front Immunol.* 2023;14:1170357.

30 Nauck, M.A., Heimesaat, M.M., Orskov, C., Holst, J.J., Ebert, R., and Creutzfeldt, W. Preserved incretin activity of glucagon-like peptide 1 [7–36 amide] but not of synthetic human gastric inhibitory polypeptide in patients with type-2 diabetes mellitus. *J Clin Invest.* 1993;91(1):301–307.

31 Nauck, M.A. and Meier, J.J. The incretin effect in healthy individuals and those with type 2 diabetes: physiology, pathophysiology, and response to therapeutic interventions. *Lancet Diabetes Endocrinol.* 2016;4(6):525–536.

32 Nathan, D.M., Schreiber, E., Fogel, H., Mojsov, S., and Habener, J.F. Insulinotropic action of glucagonlike peptide-I-(7–37) in diabetic and nondiabetic subjects. *Diabetes Care.* 1992;15(2):270–276.

33 Kreymann, B., Williams, G., Ghatei, M.A., and Bloom, S.R. Glucagon-like peptide-1 7–36: a physiological incretin in man. *Lancet.* 1987;2(8571):1300–1304.

34 Deacon, C.F., Nauck, M.A., Toft-Nielsen, M., Pridal, L., Willms, B., and Holst, J.J. Both subcutaneously and intravenously administered glucagon-like peptide I are rapidly degraded from the NH_2-terminus in type II diabetic patients and in healthy subjects. *Diabetes*. 1995;44(9):1126–1131.

35 Mentlein, R., Gallwitz, B., and Schmidt, W.E. Dipeptidyl-peptidase IV hydrolyses gastric inhibitory polypeptide, glucagon-like peptide-1(7–36)amide, peptide histidine methionine and is responsible for their degradation in human serum. *Eur J Biochem*. 1993;214(3):829–835.

36 Deacon, C.F., Johnsen, A.H., and Holst, J.J. Degradation of glucagon-like peptide-1 by human plasma in vitro yields an N-terminally truncated peptide that is a major endogenous metabolite in vivo. *J Clin Endocrinol Metab*. 1995;80(3):952–957.

37 Kuc, R.E., Maguire, J.J., Siew, K., et al. Characterization of [^{125}I]GLP-1(9–36), a novel radiolabeled analog of the major metabolite of glucagon-like peptide 1 to a receptor distinct from GLP1-R and function of the peptide in murine aorta. *Life Sci*. 2014;102(2):134–138.

38 Eng, J., Kleinman, W.A., Singh, L., Singh, G., and Raufman, J.P. Isolation and characterization of exendin-4, an exendin-3 analogue, from Heloderma suspectum venom. Further evidence for an exendin receptor on dispersed acini from guinea pig pancreas. *J Biol Chem*. 1992;267(11):7402–7405.

39 Deacon, C.F., Knudsen, L.B., Madsen, K., Wiberg, F.C., Jacobsen, O., and Holst, J.J. Dipeptidyl peptidase IV resistant analogues of glucagon-like peptide-1 which have extended metabolic stability and improved biological activity. *Diabetologia*. 1998;41(3):271–278.

40 Toniolo, C. and Benedetti E. Structures of polypeptides from α-amino acids disubstituted at the α-carbon. *Macromolecules*. 1991;24(14):4004–4009.

41 Knudsen, L.B. and Lau J. The discovery and development of liraglutide and semaglutide. *Front Endocrinol (Lausanne)*. 2019;10:155.

42 Clodfelter, D.K., Pekar, A.H., Rebhun, D.M., et al. Effects of non-covalent self-association on the subcutaneous absorption of a therapeutic peptide. *Pharm Res*. 1998;15(2):254–262.

43 Chou, J.Z., Place, G.D., Waters, D.G., Kirkwood, J.A., and Bowsher, R.R. A radioimmunoassay for LY315902 an analog of glucagon-like insulinotropic peptide, and its application in the study of canine pharmacokinetics. *J Pharm Sci*. 1997;86(7):768–773.

44 Gallwitz, B., Ropeter, T., Morys-Wortmann, C., Mentlein, R., Siegel, E.G., and Schmidt, W.E. GLP-1-analogues resistant to degradation by dipeptidyl-peptidase IV in vitro. *Regul Pept*. 2000;86(1–3):103–111.

45 Gratzer, W.B. and Beaven, G.H. Relation between conformation and association state. A study of the association equilibrium of glucagon. *J Biol Chem*. 1969;244(24):6675–6679.

46 Senderoff, R.I., Kontor, K.M., Kreilgaard, L., et al. Consideration of conformational transitions and racemization during process development of recombinant glucagon-like peptide-1. *J Pharm Sci.* 1998;87(2):183–189.

47 Kedia N. Treatment of severe diabetic hypoglycemia with glucagon: an underutilized therapeutic approach. *Diabetes Metab Syndr Obes.* 2011;4: 337–346.

48 Pohl, S.L., Birnbaumer, L., and Rodbell M. Glucagon-sensitive adenyl cylase in plasma membrane of hepatic parenchymal cells. *Science.* 1969;164(3879):566–567.

49 Taborsky Jr., G.J., The physiology of glucagon. *J Diabetes Sci Technol.* 2010;4(6):1338–1344.

50 Mojsov, S. and Merrifield, R.B. Solid-phase synthesis of crystalline glucagon. *Biochemistry.* 1981;20(10):2950–2956.

51 Gratzer, W.B., Beaven, G.H., Rattle, H.W., and Bradbury, E.M. A conformational study of glucagon. *Eur J Biochem.* 1968;3(3):276–283.

52 Beaven, G.H., Gratzer, W.B., and Davies, H.G. Formation and structure of gels and fibrils from glucagon. *Eur J Biochem.* 1969;11(1):37–42.

53 Gratzer, W.B., Bailey, E., and Beaven, G.H. Conformational states of glucagon. *Biochem Biophys Res Commun.* 1967;28(6):914–919.

54 Chabenne, J.R., Mroz, P.A., Mayer, J.P., and DiMarchi, R.D. Structural refinement of glucagon for therapeutic use. *J Med Chem.* 2020;63(7): 3447–3460.

55 Chabenne, J.R., DiMarchi, M.A., Gelfanov, V.M., and DiMarchi, R.D. Optimization of the native glucagon sequence for medicinal purposes. *J Diabetes Sci Technol.* 2010;4(6):1322–1331.

56 Simonsen, L., Holst, J.J., Madsen, K., and Deacon, C.F. The C-terminal extension of exendin-4 provides additional metabolic stability when added to GLP-1, while there is minimal effect of truncating exendin-4 in anaesthetized pigs. *Regul Pept.* 2013;181:17–21.

57 Chabenne, J., Chabenne, M.D., Zhao, Y., et al. A glucagon analog chemically stabilized for immediate treatment of life-threatening hypoglycemia. *Mol Metab.* 2014;3(3):293–300.

58 Kumita, J.R., Weston, C.J., Choo-Smith, L.P., Woolley, G.A., and Smart, O.S. Prevention of peptide fibril formation in an aqueous environment by mutation of a single residue to Aib. *Biochemistry.* 2003;42(15):4492–4498.

59 Clarke S. Propensity for spontaneous succinimide formation from aspartyl and asparaginyl residues in cellular proteins. *Int J Pept Protein Res.* 1987;30(6):808–821.

60 Geiger, T. and Clarke S. Deamidation, isomerization, and racemization at asparaginyl and aspartyl residues in peptides. succinimide-linked reactions that contribute to protein degradation. *J Biol Chem.* 1987;262(2):785–794.

61 Knerr, P.J., Finan, B., Gelfanov, V., Perez-Tilve, D., Tschop, M.H., and DiMarchi, R.D. Optimization of peptide-based polyagonists for treatment of diabetes and obesity. *Bioorg Med Chem*. 2018;26(10):2873–2881.

62 Knerr, P.J., Mowery, S.A., Douros, J.D., et al. Next generation GLP-1/GIP/glucagon triple agonists normalize body weight in obese mice. *Mol Metab*. 2022;63:101533.

63 Agerso, H., Jensen, L.B., Elbrond, B., Rolan, P., and Zdravkovic M. The pharmacokinetics, pharmacodynamics, safety and tolerability of NN2211, a new long-acting GLP-1 derivative, in healthy men. *Diabetologia*. 2002;45(2):195–202.

64 Elbrond, B., Jakobsen, G., Larsen, S., et al. Pharmacokinetics, pharmacodynamics, safety, and tolerability of a single-dose of NN2211, a long-acting glucagon-like peptide 1 derivative, in healthy male subjects. *Diabetes Care*. 2002;25(8):1398–1404.

65 Adelhorst, K., Hedegaard, B.B., Knudsen, L.B., and Kirk O. Structure-activity studies of glucagon-like peptide-1. *J Biol Chem*. 1994;269(9):6275–6278.

66 Zhang, Y., Sun, B., Feng, D., et al. Cryo-EM structure of the activated GLP-1 receptor in complex with a G protein. *Nature*. 2017;546(7657):248–253.

67 Oddo, A., Mortensen, S., Thøgersen, H., et al. α-Helix or β-Turn? An investigation into N-terminally constrained analogues of glucagon-like peptide 1 (GLP-1) and exendin-4. *Biochemistry*. 2018;57(28):4148–4154.

68 Knudsen, L.B., Nielsen, P.F., Huusfeldt, P.O., et al. Potent derivatives of glucagon-like peptide-1 with pharmacokinetic properties suitable for once daily administration. *J Med Chem*. 2000;43(9):1664–1669.

69 Madsen, K., Knudsen, L.B., Agersoe, H., et al. Structure–activity and protraction relationship of long-acting glucagon-like peptide-1 derivatives: importance of fatty acid length, polarity, and bulkiness. *J Med Chem*. 2007;50(24):6126–6132.

70 Malm-Erjefält, M., Bjørnsdottir, I., Vanggaard, J., et al. Metabolism and excretion of the once-daily human glucagon-like peptide-1 analog liraglutide in healthy male subjects and its in vitro degradation by dipeptidyl peptidase IV and neutral endopeptidase. *Drug Metab Dispos*. 2010;38(11):1944–1953.

71 Steensgaard, D., Thomsen, J.K., Olsen, H.B., and Knudsen, L.B. The molecular basis for the delayed absorption of the once-daily human GLP-1 analogue, Liraglutide. *Diabetes*. 2008;57:A164.

72 Wang, Y., Lomakin, A., Kanai, S., Alex, R., and Benedek, G.B. Transformation of oligomers of lipidated peptide induced by change in pH. *Mol Pharm*. 2015;12(2):411–419.

73 Gao, X., Maziere, A., Beard, R., Klumperman, J., and Hannoush, R.N. Fatty acylation enhances the cellular internalization and cytosolic distribution of a cystine-knot peptide. *iScience*. 2021;24(11):103220.

74 Lau, J., Bloch, P., Schaffer, L., et al. Discovery of the once-weekly glucagon-like peptide-1 (GLP-1) analogue semaglutide. *J Med Chem*. 2015;58(18):7370–7380.

75 Jensen, L., Helleberg, H., Roffel, A., et al. Absorption, metabolism and excretion of the GLP-1 analogue semaglutide in humans and nonclinical species. *Eur J Pharm Sci*. 2017;104:31–41.

76 Shi, A., Xie, P., Nielsen, L.L., Skjoth, T.V., He, X., and Haugaard, S.P. Pharmacokinetics, safety and tolerability of once-weekly subcutaneous semaglutide in healthy Chinese subjects: a double-blind, phase 1, randomized controlled trial. *Adv Ther*. 2021;38(1):550–561.

77 Garvey, W.T., Batterham, R.L., Bhatta, M., et al. Two-year effects of semaglutide in adults with overweight or obesity: the STEP 5 trial. *Nat Med*. 2022;28(10):2083–2091.

78 Kosiborod, M.N., Petrie, M.C., Borlaug, B.A., et al. Semaglutide in patients with obesity-related heart failure and type 2 diabetes. *N Engl J Med*. 2024;390(15):1394–1407.

79 Coskun, T., Sloop, K.W., Loghin, C., et al. LY3298176, a novel dual GIP and GLP-1 receptor agonist for the treatment of type 2 diabetes mellitus: from discovery to clinical proof of concept. *Mol Metab*. 2018;18:3–14.

80 Frias, J.P., Nauck, M.A., Van, J., et al. Efficacy and safety of LY3298176, a novel dual GIP and GLP-1 receptor agonist, in patients with type 2 diabetes: a randomised, placebo-controlled and active comparator-controlled phase 2 trial. *Lancet*. 2018;392(10160):2180–2193.

81 Frías, J.P. Tirzepatide: a glucose-dependent insulinotropic polypeptide (GIP) and glucagon-like peptide-1 (GLP-1) dual agonist in development for the treatment of type 2 diabetes. *Expert Rev Endocrinol Metab*. 2020;15(6):379–394.

82 Finan, B., Ma, T., Ottaway, N., et al. Unimolecular dual incretins maximize metabolic benefits in rodents, monkeys, and humans. *Sci Transl Med*. 2013;5(209):209ra151.

83 Wang L. Designing a dual GLP-1R/GIPR agonist from tirzepatide: comparing residues between tirzepatide, GLP-1, and GIP. *Drug Des Dev Ther*. 2022;16:1547–1559.

84 Zhihong, Y., Chen, W., Qianqian, Z., et al. Emerging roles of oxyntomodulin-based glucagon-like peptide-1/glucagon co-agonist analogs in diabetes and obesity. *Peptides*. 2023;162:170955.

85 Hope, D.C.D., Vincent, M.L., and Tan, T.M.M. Striking the balance: GLP-1/glucagon co-agonism as a treatment strategy for obesity. *Front Endocrinol (Lausanne)*. 2021;12:735019.

86 Zimmermann, T., Thomas, L., Baader-Pagler, T., et al. BI 456906: Discovery and preclinical pharmacology of a novel GCGR/GLP-1R dual agonist with robust anti-obesity efficacy. *Mol Metab*. 2022;66:101633.

87 Coskun, T., Urva, S., Roell, W.C., et al. LY3437943, a novel triple glucagon, GIP, and GLP-1 receptor agonist for glycemic control and weight loss: from discovery to clinical proof of concept. *Cell Metab*. 2022;34(9):1234–1247.e9.

88 Jastreboff, A.M., Kaplan, L.M., Frias, J.P., et al. Triple-hormone-receptor agonist retatrutide for obesity – a phase 2 trial. *N Engl J Med*. 2023;389(6):514–526.

89 Finan, B., Yang, B., Ottaway, N., et al. A rationally designed monomeric peptide triagonist corrects obesity and diabetes in rodents. *Nat Med*. 2015;21(1):27–36.

90 Finan, B. and Douros, J.D. GLP-1/GIP/glucagon receptor triagonism gets its try in humans. *Cell Metab*. 2022;34(1):3–4.

91 Bossart, M., Wagner, M., Elvert, R., et al. Effects on weight loss and glycemic control with SAR441255, a potent unimolecular peptide GLP-1/GIP/GCG receptor triagonist. *Cell Metab*. 2022;34(1):59–74.e10.

92 Guo, W., Xu, Z., Zou, H., et al. Discovery of ecnoglutide – a novel, long-acting, cAMP-biased glucagon-like peptide-1 (GLP-1) analog. *Mol Metab*. 2023;75:101762.

93 van der Velden, W.J.C., Smit, F.X., Christiansen, C.B., et al. GLP-1 Val8: a biased GLP-1R agonist with altered binding kinetics and impaired release of pancreatic hormones in rats. *ACS Pharmacol Transl Sci*. 2021;4(1):296–313.

94 Jones B. The therapeutic potential of GLP-1 receptor biased agonism. *Br J Pharmacol*. 2022;179(4):492–510.

95 Ward, B.P., Ottaway, N.L., Perez-Tilve, D., et al. Peptide lipidation stabilizes structure to enhance biological function. *Mol Metab*. 2013;2(4):468–479.

96 Karas, J.A., Wade, J.D., and Hossain, M.A. The chemical synthesis of insulin: an enduring challenge. *Chem Rev*. 2021;121(8):4531–4560.

97 Kurtzhals, P., Nishimura, E., Haahr, H., et al. Commemorating insulin's centennial: engineering insulin pharmacology towards physiology. *Trends Pharmacol Sci*. 2021;42(8):620–639.

98 Kjeldsen, T., Andersen, A.S., Hubalek, F., et al. Molecular engineering of insulin for recombinant expression in yeast. *Trends Biotechnol*. 2023;42(4):464–478.

99 Zaykov, A.N., Mayer, J.P., and DiMarchi, R.D. Pursuit of a perfect insulin. *Nat Rev Drug Discov*. 2016;15(6):425–439.

100 Mathieu, C., Martens, P.-J., and Vangoitsenhoven R. One hundred years of insulin therapy. *Nat Rev Endocrinol*. 2021;17(12):715–725.

101 Hoeg-Jensen T. Review: Glucose-sensitive insulin. *Mol Metab*. 2021;46:101107.

102 Potocka, E., Baughman, R.A., and Derendorf H. Population pharmacokinetic model of human insulin following different routes of administration. *J Clin Pharmacol*. 2011;51(7):1015–1024.

103 Markussen, J., Diers, I., Engesgaard, A., et al. Soluble, prolonged-acting insulin derivatives. II. Degree of protraction and crystallizability of insulins substituted in positions A17, B8, B13, B27 and B30. *Protein Eng*. 1987;1(3):215–223.

104 Markussen, J., Diers, I., Hougaard, P., et al. Soluble, prolonged-acting insulin derivatives. III. Degree of protraction, crystallizability and chemical stability of insulins substituted in positions A21, B13, B23, B27 and B30. *Protein Eng.* 1988;2(2):157–166.

105 Markussen, J., Hougaard, P., Ribel, U., Sorensen, A.R., and Sorensen E. Soluble, prolonged-acting insulin derivatives. I. Degree of protraction and crystallizability of insulins substituted in the termini of the B-chain. *Protein Eng.* 1987;1(3):205–213.

106 Lepore, M., Pampanelli, S., Fanelli, C., et al. Pharmacokinetics and pharmacodynamics of subcutaneous injection of long-acting human insulin analog glargine, NPH insulin, and ultralente human insulin and continuous subcutaneous infusion of insulin lispro. *Diabetes.* 2000;49(12):2142–2148.

107 Becker, R.H., Dahmen, R., Bergmann, K., Lehmann, A., Jax, T., and Heise T. New insulin glargine 300 units · mL^{-1} provides a more even activity profile and prolonged glycemic control at steady state compared with insulin glargine 100 units · mL^{-1}. *Diabetes Care.* 2015;38(4):637–643.

108 Markussen, J., Havelund, S., Kurtzhals, P., et al. Soluble, fatty acid acylated insulins bind to albumin and show protracted action in pigs. *Diabetologia.* 1996;39(3):281–288.

109 Heinemann, L., Sinha, K., Weyer, C., Loftager, M., Hirschberger, S., and Heise T. Time-action profile of the soluble, fatty acid acylated, long-acting insulin analogue NN304. *Diabet Med.* 1999;16(4):332–338.

110 Yáñez, J.A., Remsberg, C.M., Sayre, C.L., Forrest, M.L., and Davies, N.M. Flip-flop pharmacokinetics–delivering a reversal of disposition: challenges and opportunities during drug development. *Ther Deliv.* 2011; 2(5):643–672.

111 Havelund, S., Plum, A., Ribel, U., et al. The mechanism of protraction of insulin detemir, a long-acting, acylated analog of human insulin. *Pharm Res.* 2004;21(8):1498–1504.

112 Heise, T. and Pieber, T.R. Towards peakless, reproducible and long-acting insulins. An assessment of the basal analogues based on isoglycaemic clamp studies. *Diabetes Obes Metab.* 2007;9(5):648–659.

113 Adams, M.J., Blundell, T.L., Dodson, E.J., et al. Structure of rhombohedral 2 zinc insulin crystals. *Nature.* 1969;224(5218):491–495.

114 Blundell, T.L., Cutfield, J.F., Cutfield, S.M., et al. Three-dimensional atomic structure of insulin and its relationship to activity. *Diabetes.* 1972;21(Suppl 2): 492–505.

115 Emdin, S.O., Dodson, G.G., Cutfield, J.M., and Cutfield, S.M. Role of zinc in insulin biosynthesis. Some possible zinc-insulin interactions in the pancreatic B-cell. *Diabetologia.* 1980;19(3):174–182.

116 Jonassen, I., Havelund, S., Ribel, U., et al. Biochemical and physiological properties of a novel series of long-acting insulin analogs obtained by acylation with cholic acid derivatives. *Pharm Res*. 2006;23(1):49–55.

117 Jonassen, I., Havelund, S., Hoeg-Jensen, T., Steensgaard, D.B., Wahlund, P.O., and Ribel U. Design of the novel protraction mechanism of insulin degludec, an ultra-long-acting basal insulin. *Pharm Res*. 2012;29(8):2104–2114.

118 Steensgaard, D.B., Schluckebier, G., Strauss, H.M., et al. Ligand-controlled assembly of hexamers, dihexamers, and linear multihexamer structures by the engineered acylated insulin degludec. *Biochemistry*. 2013;52(2):295–309.

119 Haahr, H. and Heise T. A review of the pharmacological properties of insulin degludec and their clinical relevance. *Clin Pharmacokinet*. 2014;53(9):787–800.

120 Sonne O. Receptor-mediated endocytosis and degradation of insulin. *Physiol Rev*. 1988;68(4):1129–1196.

121 Menting, J.G., Whittaker, J., Margetts, M.B., et al. How insulin engages its primary binding site on the insulin receptor. *Nature*. 2013;493(7431):241–245.

122 Ribel, U., Hougaard, P., Drejer, K., and Sorensen, A.R. Equivalent in vivo biological activity of insulin analogues and human insulin despite different in vitro potencies. *Diabetes*. 1990;39(9):1033–1039.

123 Hubalek, F., Refsgaard, H.H.F., Gram-Nielsen, S., et al. Molecular engineering of safe and efficacious oral basal insulin. *Nat Commun*. 2020;11(1):3746.

124 Kjeldsen, T.B., Hubalek, F., Tagmose, T.M., et al. Engineering of orally available, ultralong-acting insulin analogues: discovery of OI338 and OI320. *J Med Chem*. 2021;64(1):616–628.

125 Kararli, T.T. Comparison of the gastrointestinal anatomy, physiology, and biochemistry of humans and commonly used laboratory animals. *Biopharm Drug Dispos*. 1995;16(5):351–380.

126 Halberg, I.B., Lyby, K., Wassermann, K., Heise, T., Plum-Mörschel, L., and Zijlstra E. The effect of food intake on the pharmacokinetics of oral basal insulin: a randomised crossover trial in healthy male subjects. *Clin Pharmacokinet*. 2019;58(11):1497–1504.

127 Leonard, T.W., Lynch, J., McKenna, M.J., and Brayden, D.J. Promoting absorption of drugs in humans using medium-chain fatty acid-based solid dosage forms: GIPET. *Expert Opin Drug Deliv*. 2006;3(5):685–692.

128 Halberg, I.B., Lyby, K., Wassermann, K., Heise, T., Zijlstra, E., and Plum-Mörschel L. Efficacy and safety of oral basal insulin versus subcutaneous insulin glargine in type 2 diabetes: a randomised, double-blind, phase 2 trial. *Lancet Diabetes Endocrinol*. 2019;7(3):179–188.

129 Kjeldsen, T.B., Hubalek, F., Hjorringgaard, C.U., et al. Molecular engineering of insulin icodec, the first acylated insulin analog for once-weekly administration in humans. *J Med Chem*. 2021;64(13):8942–8950.

130 Ludvigsen, S., Roy, M., Thogersen, H., and Kaarsholm, N.C. High-resolution structure of an engineered biologically potent insulin monomer, B16 Tyr → His, as determined by nuclear magnetic resonance spectroscopy. *Biochemistry*. 1994;33(26):7998–8006.

131 Hövelmann, U., Brøndsted, L., Kristensen, N.R., et al. 237-OR: insulin icodec: an insulin analog suited for once-weekly dosing in type 2 diabetes. *Diabetes*. 2020;69(Supplement_1):237-OR.

132 Nishimura, E., Pridal, L., Glendorf, T., et al. Molecular and pharmacological characterization of insulin icodec: a new basal insulin analog designed for once-weekly dosing. *BMJ Open Diabetes Res Care*. 2021;9(1):e002301.

133 Kruse, T., Hansen, J.L., Dahl, K., et al. Development of cagrilintide, a long-acting amylin analogue. *J Med Chem*. 2021;64(15):11183–11194.

134 Young A. *Amylin: Physiology and Pharmacology*. Elsevier Academic Press; 2005.

135 Ludvik, B., Kautzky-Willer, A., Prager, R., Thomaseth, K., and Pacini G. Amylin: history and overview. *Diabet Med*. 1997;14(Suppl 2):S9–S13.

136 Hay, D.L., Chen, S., Lutz, T.A., Parkes, D.G., and Roth, J.D. Amylin: pharmacology, physiology, and clinical potential. *Pharmacol Rev*. 2015;67(3): 564–600.

137 Lutz, T.A. The role of amylin in the control of energy homeostasis. *Am J Physiol Regul Integr Comp Physiol*. 2010;298(6):R1475–R1484.

138 Liberini, C.G., Koch-Laskowski, K., Shaulson, E., et al. Combined Amylin/GLP-1 pharmacotherapy to promote and sustain long-lasting weight loss. *Sci Rep*. 2019;9(1):8447.

139 Roth, J.D., Erickson, M.R., Chen, S., and Parkes, D.G. GLP-1R and amylin agonism in metabolic disease: complementary mechanisms and future opportunities. *Br J Pharmacol*. 2012;166(1):121–136.

140 D'Ascanio, A.M., Mullally, J.A., and Frishman, W.H. Cagrilintide: a long-acting amylin analog for the treatment of obesity. *Cardiol Rev*. 2024;32(1):83–90.

141 Akter, R., Cao, P., Noor, H., et al. Islet amyloid polypeptide: structure, function, and pathophysiology. *J Diabetes Res*. 2016; 2016:2798269.

142 Cooper, G.J., Willis, A.C., Clark, A., Turner, R.C., Sim, R.B., and Reid, K.B. Purification and characterization of a peptide from amyloid-rich pancreases of type 2 diabetic patients. *Proc Natl Acad Sci USA*. 1987;84(23):8628–8632.

143 Westermark, P., Wernstedt, C., O'Brien, T.D., Hayden, D.W., and Johnson, K.H. Islet amyloid in type 2 human diabetes mellitus and adult diabetic cats contains a novel putative polypeptide hormone. *Am J Pathol*. 1987;127(3):414–417.

144 Morris, J.K. A tale of two amyloids. *Sci Transl Med*. 2017;9(406):eaao6119.

145 Wijesekara, N., Ahrens, R., Sabale, M., et al. Amyloid-β and islet amyloid pathologies link Alzheimer's disease and type 2 diabetes in a transgenic model. *FASEB J.* 2017;31(12):5409–5418.
146 Clodi, M., Thomaseth, K., Pacini, G., et al. Distribution and kinetics of amylin in humans. *Am J Physiol.* 1998;274(5):E903–E908.
147 Young, A.A., Vine, W., Gedulin, B.R., et al. Preclinical pharmacology of pramlintide in the rat: comparisons with human and rat amylin. *Drug Dev Res.* 1996;37(4):231–248.
148 Thompson, R.G., Pearson, L., Schoenfeld, S.L., Kolterman, O.G., and The Pramlintide in Type 2 Diabetes Group. Pramlintide, a synthetic analog of human amylin, improves the metabolic profile of patients with type 2 diabetes using insulin. *Diabetes Care.* 1998;21(6):987–993.
149 Pullman, J., Darsow, T., and Frias, J.P. Pramlintide in the management of insulin-using patients with type 2 and type 1 diabetes. *Vasc Health Risk Manag.* 2006;2(3):203–212.
150 Aronne, L., Fujioka, K., Aroda, V., et al. Progressive reduction in body weight after treatment with the amylin analog pramlintide in obese subjects: a phase 2, randomized, placebo-controlled, dose-escalation study. *J Clin Endocrinol Metab.* 2007;92(8):2977–2983.
151 Smith, S.R., Blundell, J.E., Burns, C., et al. Pramlintide treatment reduces 24-h caloric intake and meal sizes and improves control of eating in obese subjects: a 6-wk translational research study. *Am J Physiol Endocrinol Metab.* 2007;293(2):E620–E627.
152 Kolterman, O.G., Schwartz, S., Corder, C., et al. Effect of 14 days' subcutaneous administration of the human amylin analogue, pramlintide (AC137), on an intravenous insulin challenge and response to a standard liquid meal in patients with IDDM. *Diabetologia.* 1996;39(4):492–499.
153 Sisnande, T., Guerreiro, L.H., Braga, R.R., et al. Monoconjugation of human amylin with methylpolyethyleneglycol. *PLoS One.* 2015;10(10):e0138803.
154 Guerreiro, L.H., Guterres, M.F., Melo-Ferreira, B., et al. Preparation and characterization of PEGylated amylin. *AAPS PharmSciTech.* 2013;14(3):1083–1097.
155 Lutz, T.A., Tschudy, S., Rushing, P.A., and Scharrer E. Amylin receptors mediate the anorectic action of salmon calcitonin (sCT). *Peptides.* 2000;21(2):233–238.
156 Mack, C.M., Smith, P.A., Athanacio, J.R., et al. Glucoregulatory effects and prolonged duration of action of davalintide: a novel amylinomimetic peptide. *Diabetes Obes Metab.* 2011;13(12):1105–1113.
157 Mack, C.M., Soares, C.J., Wilson, J.K., et al. Davalintide (AC2307), a novel amylin-mimetic peptide: enhanced pharmacological properties over native amylin to reduce food intake and body weight. *Int J Obes (Lond).* 2010;34(2):385–395.

158 Andreassen, K.V., Larsen, A.T., Sonne, N., Mohamed, K.E., Karsdal, M.A., and Henriksen K. KBP-066A, a long-acting dual amylin and calcitonin receptor agonist, induces weight loss and improves glycemic control in obese and diabetic rats. *Mol Metab*. 2021;53:101282.

159 Larsen, A.T., Sonne, N., Andreassen, K.V., Karsdal, M.A., and Henriksen K. The calcitonin receptor plays a major role in glucose regulation as a function of dual amylin and calcitonin receptor agonist therapy. *J Pharmacol Exp Ther*. 2020;374(1):74–83.

160 Gydesen, S., Andreassen, K.V., Hjuler, S.T., Christensen, J.M., Karsdal, M.A., and Henriksen K. KBP-088, a novel DACRA with prolonged receptor activation, is superior to davalintide in terms of efficacy on body weight. *Am J Physiol Endocrinol Metab*. 2016;310(10):E821–E827.

161 Larsen, A.T., Sonne, N., Andreassen, K.V., Gehring, K., Karsdal, M.A., and Henriksen K. The dual amylin and calcitonin receptor agonist KBP-088 induces weight loss and improves insulin sensitivity superior to chronic amylin therapy. *J Pharmacol Exp Ther*. 2019;370(1):35–43.

162 Wells, G., Chernoff, J., Gilligan, J.P., and Krause, D.S. Does salmon calcitonin cause cancer? A review and meta-analysis. *Osteoporos Int*. 2016;27(1):13–19.

163 Fletcher, M.M., Keov, P., Truong, T.T., et al. AM833 is a novel agonist of calcitonin family G protein-coupled receptors: pharmacological comparison with six selective and nonselective agonists. *J Pharmacol Exp Ther*. 2021;377(3): 417–440.

164 Chelikani, P.K., Haver, A.C., and Reidelberger, R.D. Effects of intermittent intraperitoneal infusion of salmon calcitonin on food intake and adiposity in obese rats. *Am J Physiol Regul Integr Comp Physiol*. 2007;293(5):R1798–R1808.

165 Barth, S.W., Riediger, T., Lutz, T.A., and Rechkemmer G. Peripheral amylin activates circumventricular organs expressing calcitonin receptor a/b subtypes and receptor-activity modifying proteins in the rat. *Brain Research*. 2004;997(1):97–102.

166 Lutz, T.A., Mollet, A., Rushing, P.A., Riediger, T., and Scharrer E. The anorectic effect of a chronic peripheral infusion of amylin is abolished in area postrema/nucleus of the solitary tract (AP/NTS) lesioned rats. *Int J Obes Relat Metab Disord*. 2001;25(7):1005–1011.

167 Kenley, R.A., Tracht, S., Stepanenko, A., Townsend, M., and L'Italien J. Kinetics of pramlintide degradation in aqueous solution as a function of temperature and pH. *AAPS PharmSciTech*. 2000;1(2):E7.

168 Raleigh, D., Zhang, X., Hastoy, B., and Clark A. The β-cell assassin: IAPP cytotoxicity. *J Mol Endocrinol*. 2017;59(3):R121–R140.

169 Schlein M. Insulin formulation characterization-the Thioflavin T assays. *AAPS J*. 2017;19(2):397–408.

170 Abedini, A., Meng, F., and Raleigh, D.P. A single-point mutation converts the highly amyloidogenic human islet amyloid polypeptide into a potent fibrillization inhibitor. *J Am Chem Soc.* 2007;129(37):11300–11301.

171 Enebo, L.B., Berthelsen, K.K., Kankam, M., et al. Safety, tolerability, pharmacokinetics, and pharmacodynamics of concomitant administration of multiple doses of cagrilintide with semaglutide 2.4 mg for weight management: a randomised, controlled, phase 1b trial. *Lancet.* 2021;397(10286):1736–1748.

172 Frias, J.P., Deenadayalan, S., Erichsen, L., et al. Efficacy and safety of co-administered once-weekly cagrilintide 2.4 mg with once-weekly semaglutide 2.4 mg in type 2 diabetes: a multicentre, randomised, double-blind, active-controlled, phase 2 trial. *Lancet.* 2023;402(10403):720–730.

173 Larsen, A.T., Mohamed, K.E., Sonne, N., et al. Does receptor balance matter? – Comparing the efficacies of the dual amylin and calcitonin receptor agonists cagrilintide and KBP-336 on metabolic parameters in preclinical models. *Biomed Pharmacother.* 2022;156:113842.

174 Batterham, R.L., Cowley, M.A., Small, C.J., et al. Gut hormone PYY(3–36) physiologically inhibits food intake. *Nature.* 2002;418(6898):650–654.

175 Adrian, T.E., Ferri, G.L., Bacarese-Hamilton, A.J., Fuessl, H.S., Polak, J.M., and Bloom, S.R. Human distribution and release of a putative new gut hormone, peptide YY. *Gastroenterology.* 1985;89(5):1070–1077.

176 Habib, A.M., Richards, P., Rogers, G.J., Reimann, F., and Gribble, F.M. Co-localisation and secretion of glucagon-like peptide 1 and peptide YY from primary cultured human L cells. *Diabetologia.* 2013;56(6):1413–1416.

177 Eberlein, G.A., Eysselein, V.E., Schaeffer, M., et al. A new molecular form of PYY: structural characterization of human PYY(3–36) and PYY(1–36). *Peptides.* 1989;10(4):797–803.

178 Medeiros, M.D. and Turner, A.J. Processing and metabolism of peptide-YY: pivotal roles of dipeptidylpeptidase-IV, aminopeptidase-P, and endopeptidase-24.11. *Endocrinology.* 1994;134(5):2088–2094.

179 Neary, N.M., Small, C.J., Druce, M.R., et al. Peptide YY3–36 and glucagon-like peptide-17–36 inhibit food intake additively. *Endocrinology.* 2005;146(12):5120–5127.

180 Lafferty, R.A., Flatt, P.R., and Irwin N. Emerging therapeutic potential for peptide YY for obesity-diabetes. *Peptides.* 2018;100:269–274.

181 Lafferty, R.A., Flatt, P.R., and Irwin N. Established and emerging roles peptide YY (PYY) and exploitation in obesity-diabetes. *Curr Opin Endocrinol Diabetes Obes.* 2021;28(2):253–261.

182 Olsen, J., Kofoed, J., Østergaard, S., Wulff, B.S., Nielsen, F.S., and Jorgensen R. Metabolism of peptide YY 3–36 in Göttingen mini-pig and rhesus monkey. *Peptides.* 2016;78:59–67.

183 Toräng, S., Veedfald, S., Rosenkilde, M.M., Hartmann, B., and Holst, J.J. The anorexic hormone Peptide YY3–36 is rapidly metabolized to inactive Peptide YY3–34 in vivo. *Physiol Rep.* 2015;3(7):e12455.

184 Beck-Sickinger, A.G. and Jung G. Structure-activity relationships of neuropeptide Y analogues with respect to Y1 and Y2 receptors. *Biopolymers.* 1995;37(2):123–142.

185 Beck-Sickinger, A.G., Wieland, H.A., Wittneben, H., Willim, K.D., Rudolf, K., and Jung, G. Complete L-alanine scan of neuropeptide Y reveals ligands binding to Y1 and Y2 receptors with distinguished conformations. *Eur J Biochem.* 1994;225(3):947–958.

186 Cabrele, C., Langer, M., Bader, R., et al. The first selective agonist for the neuropeptide YY5 receptor increases food intake in rats. *J Biol Chem.* 2000;275(46):36043–36048.

187 Lecklin, A., Lundell, I., Salmela, S., Männistö, P.T., Beck-Sickinger, A.G., and Larhammar D. Agonists for neuropeptide Y receptors Y1 and Y5 stimulate different phases of feeding in guinea pigs. *Br J Pharmacol.* 2003;139(8):1433–1440.

188 Nguyen, A.D., Mitchell, N.F., Lin, S., et al. Y1 and Y5 receptors are both required for the regulation of food intake and energy homeostasis in mice. *PLoS One.* 2012;7(6):e40191.

189 Cross, L.J., Beck-Sickinger, A.G., Bienert, M., et al. Structure activity studies of mast cell activation and hypotension induced by neuropeptide Y (NPY), centrally truncated and C-terminal NPY analogues. *Br J Pharmacol.* 1996;117(2):325–332.

190 Ostergaard, S., Kofoed, J., Paulsson, J.F., Madsen, K.G., Jorgensen, R., and Wulff, B.S. Design of Y_2 receptor selective and proteolytically stable PYY_{3-36} analogues. *J Med Chem.* 2018;61(23):10519–10530.

191 Hayen, A., Schmitt, M.A., Ngassa, F.N., Thomasson, K.A., and Gellman, S.H. Two helical conformations from a single foldamer backbone: "split personality" in short alpha/beta-peptides. *Angew Chem Int Ed Engl.* 2004;43(4):505–510.

192 Zuckermann, R.N., Kerr, J.M., Kent, S.B.H., and Moos, W.H. Efficient method for the preparation of peptoids [oligo(N-substituted glycines)] by submonomer solid-phase synthesis. *J Am Chem Soc.* 1992;114(26):10646–10647.

193 Østergaard, S. and Holm, A. Peptomers: a versatile approach for the preparation of diverse combinatorial peptidomimetic bead libraries. *Mol Divers.* 1997;3(1):17–27.

194 Østergaard, S., Paulsson, J.F., Kofoed, J., et al. The effect of fatty diacid acylation of human PYY_{3-36} on Y_2 receptor potency and half-life in minipigs. *Sci Rep.* 2021;11(1):21179.

195 Burkert, K., Zellmann, T., Meier, R., et al. A deep hydrophobic binding cavity is the main interaction for different Y_2R antagonists. *ChemMedChem.* 2017;12(1):75–85.

196 Beck-Sickinger, A.G., Gaida, W., Schnorrenberg, G., Lang, R., and Jung G. Neuropeptide Y: identification of the binding site. *Int J Pept Protein Res.* 1990;36(6):522–530.

197 Poulsen, C., Pedersen, M.Ø., Wahlund, P.-O., et al. Rational development of stable PYY_{3-36} peptide Y_2 receptor agonists. *Pharm Res.* 2021;38(8):1369–1385.

198 Irudayanathan, F.J., Zarzar, J., Lin, J., and Izadi S. Deciphering deamidation and isomerization in therapeutic proteins: effect of neighboring residue. *MAbs.* 2022;14(1):2143006.

199 Robinson, N.E. and Robinson, A.B. Deamidation of human proteins. *Proc Natl Acad Sci USA.* 2001;98(22):12409–12413.

200 Robinson, N.E., Robinson, Z.W., Robinson, B.R., et al. Structure-dependent nonenzymatic deamidation of glutaminyl and asparaginyl pentapeptides. *J Pept Res.* 2004;63(5):426–436.

201 Cleland, J.L., Powell, M.F., and Shire, S.J. The development of stable protein formulations: a close look at protein aggregation, deamidation, and oxidation. *Crit Rev Ther Drug Carrier Syst.* 1993;10(4):307–377.

202 Hutchinson, J.A., Burholt, S., Hamley, I.W., et al. The effect of lipidation on the self-assembly of the gut-derived peptide hormone PYY_{3-36}. *Bioconjug Chem.* 2018;29(7):2296–2308.

203 Ostergaard, S., Jessen, C., Paulsson, J.F., et al. Deep mutational analysis of PYY_{3-36} leading to potent ultra-long acting PYY analogues with superior Y_2 receptor selectivity. 2025. https://doi.org/10.1126/scitranslmed.adq6392.

204 Briere, D.A., Lopes, D.C., and Muppidi, A. inventor; Eli Lilly, and Company, assignee. Protein tyrosine-tyrosine analogs and methods of using the same. 2021.

205 Yang, P.Y., Zou, H., Chao, E., et al. Engineering a long-acting, potent GLP-1 analog for microstructure-based transdermal delivery. *Proc Natl Acad Sci USA.* 2016;113(15):4140–4145.

206 Chang, Y.S., Graves, B., Guerlavais, V., et al. Stapled α-helical peptide drug development: a potent dual inhibitor of MDM2 and MDMX for p53-dependent cancer therapy. *Proc Natl Acad Sci USA.* 2013;110(36):E3445–E3454.

207 Walensky, L.D. and Bird, G.H. Hydrocarbon-stapled peptides: principles, practice, and progress. *J Med Chem.* 2014;57(15):6275–6288.

208 Walensky, L.D., Kung, A.L., Escher, I., et al. Activation of apoptosis in vivo by a hydrocarbon-stapled BH3 helix. *Science.* 2004;305(5689):1466–1470.

209 Lear, S., Pflimlin, E., Zhou, Z., et al. Engineering of a potent, long-acting NPY2R agonist for combination with a GLP-1R agonist as a multi-hormonal treatment for obesity. *J Med Chem.* 2020;63(17):9660–9671.

210 Kaiser, A., Müller, P., Zellmann, T., et al. Unwinding of the C-terminal residues of neuropeptide Y is critical for Y_2 receptor binding and activation. *Angew Chem Int Ed Engl.* 2015;54(25):7446–7449.

211 Nygaard, R., Nielbo, S., Schwartz, T.W., and Poulsen, F.M. The PP-fold solution structure of human polypeptide YY and human PYY_{3-36} as determined by NMR. *Biochemistry*. 2006;45(27):8350–8357.

212 Klepach, A., Tran, H., Ahmad Mohammed, F., and ElSayed, M.E.H. Characterization and impact of peptide physicochemical properties on oral and subcutaneous delivery. *Adv Drug Deliv Rev*. 2022;186:114322.

213 Zapadka, K.L., Becher, F.J., Gomes Dos Santos, A.L., and Jackson, S.E. Factors affecting the physical stability (aggregation) of peptide therapeutics. *Interface Focus*. 2017;7(6):20170030.

214 Furman, J., Chiu, M., and Hunter, M. Early engineering approaches to improve peptide developability and manufacturability. *AAPS J*. 2015;17(1): 111–120.

215 Brange, J., Skelbaek-Pedersen, B., Langkjær, L., et al. Galenics of insulin. *The Physico-chemical and Pharmaceutical Aspects of Insulin and Insulin Preparations*. 1 ed. Berlin, Heidelberg, New York: Springer-Verlag; 1987.

216 Kuerzel, G.U., Shukla, U., Scholtz, H.E., et al. Biotransformation of insulin glargine after subcutaneous injection in healthy subjects. *Curr Med Res Opin*. 2003;19(1):34–40.

217 Heise, T., Nosek, L., Rønn, B.B., et al. Lower within-subject variability of insulin detemir in comparison to NPH insulin and insulin glargine in people with type 1 diabetes. *Diabetes*. 2004;53(6):1614–1620.

218 Shnek, D.R., Hostettler, D.L., Bell, M.A., Olinger, J.M., and Frank, B.H. Physical stress testing of insulin suspensions and solutions. *J Pharm Sci*. 1998;87(11):1459–1465.

219 Grimsley, G.R., Scholtz, J.M., and Pace, C.N. A summary of the measured pK values of the ionizable groups in folded proteins. *Protein Sci*. 2009;18(1):247–251.

220 Thurlkill, R.L., Grimsley, G.R., Scholtz, J.M., and Pace, C.N. pK values of the ionizable groups of proteins. *Protein Sci*. 2006;15(5):1214–1218.

221 Evers, A., Pfeiffer-Marek, S., Bossart, M., et al. Peptide optimization at the drug discovery-development interface: tailoring of physicochemical properties toward specific formulation requirements. *J Pharm Sci*. 2019;108(4):1404–1414.

222 Oeller, M., Kang, R.J.D., Bolt, H.L., et al. Sequence-based prediction of the intrinsic solubility of peptides containing non-natural amino acids. *Nat Commun*. 2023;14(1):7475.

223 Frolov, A.I., Chankeshwara, S.V., Abdulkarim, Z., and Ghiandoni, G.M. pIChemiSt – free tool for the calculation of isoelectric points of modified peptides. *J Chem Inf Model*. 2023;63(1):187–196.

224 McIlvaine, T.C. A buffer solution for colorimetric comparison. *J Biol Chem*. 1921;49(1):183–186.

225 Narhi, L.O., Schmit, J., Bechtold-Peters, K., and Sharma, D. Classification of protein aggregates. *J Pharm Sci.* 2012;101(2):493–498.

226 Wang, W. Instability, stabilization, and formulation of liquid protein pharmaceuticals. *Int J Pharm.* 1999;185(2):129–188.

227 Wang, W. Protein aggregation and its inhibition in biopharmaceutics. *Int J Pharm.* 2005;289(1–2):1–30.

228 Roberts, C.J. Protein aggregation and its impact on product quality. *Curr Opin Biotechnol.* 2014;30(0):211–217.

229 Lapidus, L.J. Understanding protein aggregation from the view of monomer dynamics. *Mol Biosyst.* 2013;9(1):29–35.

230 Mahler, H.C., Friess, W., Grauschopf, U., and Kiese, S. Protein aggregation: pathways, induction factors and analysis. *J Pharm Sci.* 2009;98(9):2909–2934.

231 Wang, W., Nema, S., and Teagarden, D. Protein aggregation-pathways and influencing factors. *Int J Pharm.* 2010;390(2):89–99.

232 Naiki, H., Higuchi, K., Hosokawa, M., and Takeda, T. Fluorometric determination of amyloid fibrils in vitro using the fluorescent dye, thioflavine T. *Anal Biochem.* 1989;177(2):244–249.

233 LeVine, H. Thioflavine-t interaction with amyloid beta-sheet structures. *Amyloid.* 1995;2(1):1–6.

234 Gade Malmos, K., Blancas-Mejia, L.M., Weber, B., et al. ThT 101: a primer on the use of thioflavin T to investigate amyloid formation. *Amyloid.* 2017;24(1):1–16.

235 Groenning, M. Binding mode of Thioflavin T and other molecular probes in the context of amyloid fibrils-current status. *J Chem Biol.* 2009;3:1–18.

236 Groenning, M., Frokjaer, S., and Vestergaard, B. Formation mechanism of insulin fibrils and structural aspects of the insulin fibrillation process. *Curr Protein Pept Sci.* 2009;10(5):509–528(20).

237 Groenning, M., Norrman, M., Flink, J.M., et al. Binding mode of Thioflavin T in insulin amyloid fibrils. *J Struct Biol.* 2007;159(3):483–497.

238 Green, J., Goldsbury, C., Mini, T., et al. Full-length rat amylin forms fibrils following substitution of single residues from human amylin. *J Mol Biol.* 2003;326(4):1147–1156.

239 Milton, N.G.N. and Harris, J.R. Fibril formation and toxicity of the non-amyloidogenic rat amylin peptide. *Micron.* 2013;44(0):246–253.

240 Pedersen, J.S., Dikov, D., and Otzen, D.E. N- and C-terminal hydrophobic patches are involved in fibrillation of glucagon. *Biochemistry.* 2006;45(48):14503–14512.

241 Pedersen, J.S., Dikov, D., Flink, J.L., Hjuler, H.A., Christiansen, G., and Otzen, D.E. The changing face of glucagon fibrillation: structural polymorphism and conformational imprinting. *J Mol Biol.* 2006;355(3):501–523.

242 Alrouji, M., Al-Kuraishy, H.M., Al-Gareeb, A.I., et al. The potential role of human islet amyloid polypeptide in type 2 diabetes mellitus and Alzheimer's diseases. *Diabetol. Metab. Syndr.* 2023;15(1):101.

243 Westermark, G.T., Fändrich, M., Lundmark, K., and Westermark, P. Noncerebral amyloidoses: aspects on seeding, cross-seeding, and transmission. *Cold Spring Harb Perspect Med.* 2017;8(1):a024323.

244 Westermark, P., Andersson, A., and Westermark, G.T. Islet amyloid polypeptide, islet amyloid, and diabetes mellitus. *Physiol Rev.* 2011;91(3):795–826.

245 Wang, W., Singh, S.K., Li, N., Toler, M.R., King, K.R., and Nema, S. Immunogenicity of protein aggregates - Concerns and realities. *Int J Pharm.* 2012;431(1–2):1–11.

246 Mire-Sluis, A., Cherney, B., Madsen, R., Polozova, A., Rosenberg, A., and Smith, H. Analysis and immunogenic potential of aggregates and particles. *BioProcess Int.* 2011;9(10):38–43.

247 Rosenberg, A. Effects of protein aggregates: an immunologic perspective. *AAPS J.* 2006;8(3):E501–E507.

248 Carpenter, J.F., Randolph, T.W., Jiskoot, W., et al. Overlooking subvisible particles in therapeutic protein products: gaps that may compromise product quality. *J Pharm Sci.* 2009;98(4):1201–1205.

249 Kotarek, J., Stuart, C., De Paoli, S.H., et al. Subvisible particle content, formulation, and dose of an erythropoietin peptide mimetic product are associated with severe adverse postmarketing events. *J Pharm Sci.* 2016;105(3):1023–1027.

250 Barnard, J.G., Babcock, K., and Carpenter, J.F. Characterization and quantitation of aggregates and particles in interferon-beta products: potential links between product quality attributes and immunogenicity. *J Pharm Sci.* 2013;102(3):915–928.

251 Brange, J. and Langkjær, L. Insulin structure and stability. In: *Stability and Characterization of Protein and Peptide Drugs: Case Histories* (eds. Y.J. Wang and R. Pearlman). New York: Plenum Press; 1993. pp. 315–350.

252 Brange, J., Owens, D.R., Kang, S., and Vølund, A. Monomeric insulins and their experimental and clinical implications. *Diabetes Care.* 1990;13(9):923–954.

253 Brange, J., Ribel, U., Hansen, J.F., et al. Monomeric insulins obtained by protein engineering and their medical implications. *Nature.* 1988;333(6174): 679–682.

254 DeFelippis, M.R., Chance, R.E., and Frank, B.H. Insulin self-association and the relationship to pharmacokinetics and pharmacodynamics. *Crit Rev Ther Drug Carr Syst.* 2001;18(2):201–264.

255 Li, Q., Tangry, V., Allen, D.P., Seibert, K.D., Qian, K.K., and Wagner, N.J. Surface-mediated spontaneous emulsification of the acylated peptide, semaglutide. *Proc Natl Acad Sci USA.* 2024;121(5):e2305770121.

256 Frederiksen Tine, M., Sønderby, P., Ryberg Line, A., et al. Oligomerization of a glucagon-like peptide 1 analog: bridging experiment and simulations. *Biophys. J.* 2015;109(6):1202–1213.

257 Wolff, M., Schüler, A., Gast, K., et al. Self-assembly of exendin-4-derived dual peptide agonists is mediated by acylation and correlated to the length of conjugated fatty acyl chains. *Mol Pharm.* 2020;17(3):965–978.

258 van Witteloostuijn, S.B., Mannerstedt, K., Wismann, P., et al. Neoglycolipids for prolonging the effects of peptides: self-assembling glucagon-like peptide 1 analogues with albumin binding properties and potent in vivo efficacy. *Mol Pharm.* 2017;14(1):193–205.

259 Jensen, M.H., Wahlund, P.O., Toft, K.N., et al. Small angle x-ray scattering-based elucidation of the self-association mechanism of human insulin analogue LysB29(N^{ε}ω-carboxyheptadecanoyl) des(B30). *Biochemistry.* 2012;52(2):282–294.

260 Bothe, J.R., Andrews, A., Smith, K.J., Joyce, L.A., Krishnamachari, Y., and Kashi, S. Peptide oligomerization memory effects and their impact on the physical stability of the GLP-1 agonist liraglutide. *Mol Pharm.* 2019;16(5):2153–2161.

261 Hawe, A., Wiggenhorn, M., van de Weert, M., Garbe, J.H.O., Mahler, H.C., and Jiskoot, W. Forced degradation of therapeutic proteins. *J Pharm Sci.* 2012;101(3):895–913.

262 Evers, A., Pfeiffer-Marek, S., Bossart, M., et al. Multiparameter peptide optimization toward stable triple agonists for the treatment of diabetes and obesity. *Adv Therap.* 2020;3(9):2000052.

263 Tatford, O.C., Gomme, P.T., and Bertolini, J. Analytical techniques for the evaluation of liquid protein therapeutics. *Biotechnol Appl Biochem.* 2004; 40(Pt 1):67–81.

264 Dunn, M.F. Zinc-ligand interactions modulate assembly and stability of the insulin hexamer - a review. *BioMetals.* 2005;18(4):295–303.

265 Philo, J.S., Sydor, W., and Arakawa, T. The glucagon-like peptide 2 analog teduglutide reversibly associates to form pentamers. *J Pharm Sci.* 2020;109(1):775–784.

266 Gast, K., Schüler, A., Wolff, M., et al. Rapid-acting and human insulins: hexamer dissociation kinetics upon dilution of the pharmaceutical formulation. *Pharmaceutical Research.* 2017;34:2270–2286.

267 Přáda Brichtová, E., Krupová, M., Bouř, P., Lindo, V., Gomes Dos Santos, A., and Jackson, S.E. Glucagon-like peptide 1 aggregates into low-molecular-weight oligomers off-pathway to fibrillation. *Biophys J.* 2023;122(12):2475–2488.

268 Wolff, M., Gast, K., Evers, A., et al. A Conserved hydrophobic moiety and helix-helix interactions drive the self-assembly of the incretin analog exendin-4. *Biomolecules.* 2021;11(9):1305.

269 Longo, E., Santis, E.D., Hussain, R., et al. The effect of palmitoylation on the conformation and physical stability of a model peptide hormone. *Int J Pharm.* 2014; 472(1ΓÇô2):156–164.

270 Patil, S.M., Xu, S., Sheftic, S.R., and Alexandrescu, A.T. Dynamic alpha-helix structure of micelle-bound human amylin. *J Biol Chem.* 2009;284(18):11982–11991.

271 Merle, M., Lefevre, G., and Milhaud, G. Predicted secondary structure of calcitonin in relation to the biological activity. *Biochem Biophys Res Commun.* 1979;87(2):455–460.

272 DeLuca, S.H., Rathmann, D., Beck-Sickinger, A.G., and Meiler, J. The activity of prolactin releasing peptide correlates with its helicity. *Biopolymers.* 2013;99(5):314–325.

273 Kelly, S.M., Jess, T.J., and Price, N.C. How to study proteins by circular dichroism. *Biochim Biophys Acta.* 2005;1751(2):119–139.

274 Evers, A., Bossart, M., Pfeiffer-Marek, S., et al. Dual glucagon-like peptide 1 (GLP-1)/glucagon receptor agonists specifically optimized for multidose formulations. *J Med Chem.* 2018;61(13):5580–5593.

275 Khatun, S., Singh, A., Pawar, N., and Gupta, A.N. Aggregation of amylin: spectroscopic investigation. *Int J Biol Macromol.* 2019;133:1242–1248.

276 Iconomidou, V.A., Leontis, A., Hoenger, A., and Hamodrakas, S.J. Identification of a novel 'aggregation-prone'/'amyloidogenic determinant' peptide in the sequence of the highly amyloidogenic human calcitonin. *FEBS Lett.* 2013;587(6):569–574.

277 Jaikaran, E.T.A.S., Higham, C.E., Serpell, L.C., et al. Identification of a novel human islet amyloid polypeptide beta-sheet domain and factors influencing fibrillogenesis. *J Mol Biol.* 2001;308(3):515–525.

278 Zhou, C., Qi, W., Lewis, E.N., and Carpenter, J.F. Characterization of sizes of aggregates of insulin analogs and the conformations of the constituent protein molecules: a concomitant dynamic light scattering and Raman spectroscopy study. *J Pharm Sci.* 2016;105(2):551–558.

279 Karle, A.C., Wrobel, M.B., Koepke, S., et al. Anti-brolucizumab immune response as one prerequisite for rare retinal vasculitis/retinal vascular occlusion adverse events. *Sci Transl Med.* 2023;15(681):eabq5241.

280 Moreland, L.W., McCabe, D.P., Caldwell, J.R., et al. Phase I/II trial of recombinant methionyl human tumor necrosis factor binding protein PEGylated dimer in patients with active refractory rheumatoid arthritis. *J Rheumatol.* 2000;27(3):601–609.

281 Ridker, P.M., Tardif, J.C., Amarenco, P., et al. Lipid-reduction variability and antidrug-antibody formation with bococizumab. *N Engl J Med.* 2017;376(16):1517–1526.

282 Pfizer discontinues global development of bococizumab, its investigational PCSK9 inhibitor. Available from: https://www.pfizer.com/news/press-release/press-release-detail/pfizer_discontinues_global_development_of_bococizumab_its_investigational_pcsk9_inhibitor. Accessed on May 24, 2024.

283 Wang, Y.M., Wang, J., Hon, Y.Y., Zhou, L., Fang, L., and Ahn, H.Y. Evaluating and reporting the immunogenicity impacts for biological products—a clinical pharmacology perspective. *AAPS J.* 2016;18(2):395–403.

284 Administration TFaD. Immunogenicity testing of therapeutic protein products – developing and validating assays for anti-drug antibody detection. Available from: https://www.fda.gov/regulatory-information/search-fda-guidance-documents/immunogenicity-testing-therapeutic-protein-products-developing-and-validating-assays-anti-drug. Accessed on May 24, 2024.

285 Lamberth, K., Reedtz-Runge, S.L., Simon, J., et al. Post hoc assessment of the immunogenicity of bioengineered factor VIIa demonstrates the use of preclinical tools. *Sci Transl Med.* 2017;9(372):eaag1286.

286 Abshire, T. and Kenet, G. Recombinant factor VIIa: review of efficacy, dosing regimens and safety in patients with congenital and acquired factor VIII or IX inhibitors. *J Thromb Haemost.* 2004;2(6):899–909.

287 Kavakli, K., Makris, M., Zulfikar, B., et al. Home treatment of haemarthroses using a single dose regimen of recombinant activated factor VII in patients with haemophilia and inhibitors. A multi-centre, randomised, double-blind, cross-over trial. *Thromb Haemost.* 2006;95(4):600–605.

288 Persson, E., Kjalke, M., and Olsen, O.H. Rational design of coagulation factor VIIa variants with substantially increased intrinsic activity. *Proc Natl Acad Sci USA.* 2001;98(24):13583–13588.

289 Mahlangu, J.N., Weldingh, K.N., Lentz, S.R., et al. Changes in the amino acid sequence of the recombinant human factor VIIa analog, vatreptacog alfa, are associated with clinical immunogenicity. *J Thromb Haemost.* 2015;13(11):1989–1998.

290 Nielsen, M., Justesen, S., Lund, O., Lundegaard, C., and Buus, S. NetMHCIIpan-2.0 - improved pan-specific HLA-DR predictions using a novel concurrent alignment and weight optimization training procedure. *Immunome Res.* 2010;6:9.

291 Mahlangu, J., Paz, P., Hardtke, M., Aswad, F., and Schroeder, J. TRUST trial: BAY 86-6150 use in haemophilia with inhibitors and assessment for immunogenicity. *Haemophilia.* 2016;22(6):873–879.

292 European Medicines Agency. Guideline on immunogenicity assessment of therapeutics proteins. Available from: https://www.ema.europa.eu/en/documents/scientific-guideline/guideline-immunogenicity-assessment-therapeutic-proteins-revision-1_en.pdf. Accessed on May 24, 2024.

293 Jawa, V., Cousens, L.P., Awwad, M., Wakshull, E., Kropshofer, H., and De Groot, A.S. T-cell dependent immunogenicity of protein therapeutics: preclinical assessment and mitigation. *Clin Immunol.* 2013;149(3):534–555.

294 Ducret, A., Ackaert, C., Bessa, J., et al. Assay format diversity in pre-clinical immunogenicity risk assessment: toward a possible harmonization of antigenicity assays. *MAbs.* 2022;14(1):1993522.

295 Reynisson, B., Alvarez, B., Paul, S., Peters, B., and Nielsen, M. NetMHCpan-4.1 and NetMHCIIpan-4.0: improved predictions of MHC antigen presentation by concurrent motif deconvolution and integration of MS MHC eluted ligand data. *Nucleic Acids Res.* 2020;48(W1):W449–W454.

296 Wang, P., Sidney, J., Dow, C., Mothé B, Sette, A., and Peters, B. A systematic assessment of MHC class II peptide binding predictions and evaluation of a consensus approach. *PLoS Comput Biol.* 2008;4(4):e1000048.

297 Vita, R., Mahajan, S., Overton, J.A., et al. The immune epitope database (IEDB): 2018 update. *Nucleic Acids Res.* 2019;47(D1):D339–D343.

298 Jankowski, W., Kidchob, C., Bunce, C., Cloake, E., Resende, R., and Sauna, Z.E. The MHC associated peptide proteomics assay is a useful tool for the non-clinical assessment of immunogenicity. *Front Immunol.* 2023;14: 1271120.

299 Lee, M.V., Saad, O.M., Wong, S., et al. Development of a semi-automated MHC-associated peptide proteomics (MAPPs) method using streptavidin bead-based immunoaffinity capture and nano LC-MS/MS to support immunogenicity risk assessment in drug development. *Front Immunol.* 2023;14:1295285.

300 Hartman, K., Steiner, G., Siegel, M., et al. Expanding the MAPPs assay to accommodate MHC-II pan receptors for improved predictability of potential T cell epitopes. *Biology (Basel).* 2023;12(9):1265.

301 Karle, A.C. Applying MAPPs assays to assess drug immunogenicity. *Front Immunol.* 2020;11:698.

302 Siegel, M., Steiner, G., Franssen, L.C., et al. Validation of a dendritic cell and CD4+ T cell restimulation assay contributing to the immunogenicity risk evaluation of biotherapeutics. *Pharmaceutics.* 2022;14(12):2672.

303 Holley, C.K., Cedrone, E., Donohue, D., et al. An in vitro assessment of immunostimulatory responses to ten model innate immune response modulating impurities (IIRMIs) and peptide drug product, teriparatide. *Molecules.* 2021;26(24):7461.

304 Dalkas, G.A. and Rooman, M. SEPIa, a knowledge-driven algorithm for predicting conformational B-cell epitopes from the amino acid sequence. *BMC Bioinform.* 2017;18(1):95.

305 Buse, J.B., Garber, A., Rosenstock, J., et al. Liraglutide treatment is associated with a low frequency and magnitude of antibody formation with no apparent impact on glycemic response or increased frequency of adverse events:

results from the liraglutide effect and action in diabetes (LEAD) trials. *J Clin Endocrinol Metab*. 2011;96(6):1695–1702.

306 Schultz, H.S., Ostergaard, S., Sidney, J., Lamberth, K., and Sette, A. The effect of acylation with fatty acids and other modifications on HLA class II:peptide binding and T cell stimulation for three model peptides. *PLoS One*. 2018;13(5):e0197407.

307 Werle, M. and Bernkop-Schnürch, A. Strategies to improve plasma half life time of peptide and protein drugs. *Amino Acids*. 2006;30(4):351–367.

308 Polson, C., Sarkar, P., Incledon, B., Raguvaran, V., and Grant, R. Optimization of protein precipitation based upon effectiveness of protein removal and ionization effect in liquid chromatography-tandem mass spectrometry. *J Chromatogr B Analyt Technol Biomed Life Sci*. 2003;785(2):263–275.

309 Esposito, S., Krick, A., Pasquier, O., et al. Fatty acid acylated peptide therapeutics: discovery of omega-n oxidation of the lipid chain as a novel metabolic pathway in preclinical species. *J Pharm Biomed Anal*. 2023;227:115256.

310 Fura, A., Harper, T.W., Zhang, H., Fung, L., and Shyu, W.C. Shift in pH of biological fluids during storage and processing: effect on bioanalysis. *J Pharm Biomed Anal*. 2003;32(3):513–522.

311 Di, L. Strategic approaches to optimizing peptide ADME properties. *AAPS J*. 2015;17(1):134–143.

312 Powell, M.F., Stewart, T., Jr Otvos, L., et al. Peptide stability in drug development. II. Effect of single amino acid substitution and glycosylation on peptide reactivity in human serum. *Pharmaceutical Research*. 1993;10(9):1268–1273.

313 Manning, M.C., Chou, D.K., Murphy, B.M., Payne, R.W., and Katayama, D.S. Stability of protein pharmaceuticals: an update. *Pharm Res*. 2010;27(4):544–575.

314 Wakankar, A.A. and Borchardt, R.T. Formulation considerations for proteins susceptible to asparagine deamidation and aspartate isomerization. *J Pharm Sci*. 2006;95(11):2321–2336.

315 Kato, K., Nakayoshi, T., Kurimoto, E., and Oda, A. Mechanisms of deamidation of asparagine residues and effects of main-chain conformation on activation energy. *Int J Mol Sci*. 2020;21(19):7035.

316 Patel, K. and Borchardt, R.T. Chemical pathways of peptide degradation. III. Effect of primary sequence on the pathways of deamidation of asparaginyl residues in hexapeptides. *Pharm Res*. 1990;7(8):787–793.

317 Oliyai, C. and Borchardt, R.T. Chemical pathways of peptide degradation. IV. Pathways, kinetics, and mechanism of degradation of an aspartyl residue in a model hexapeptide. *Pharm Res*. 1993;10(1):95–102.

318 Benet, A., Halseth, T., Kang, J., et al. The effects of pH and excipients on exenatide stability in solution. *Pharmaceutics*. 2021;13(8):1263.

319 Schneider, E.L., Hearn, B.R., Pfaff, S.J., et al. A hydrogel-microsphere drug delivery system that supports once-monthly administration of a GLP-1 receptor agonist. *ACS Chem Biol.* 2017;12(8):2107–2116.

320 Patel, S., Vyas, V.K., and Mehta, P.J. A review on forced degradation strategies to establish the stability of therapeutic peptide formulations. *Int J Pept Res Ther.* 2023;29(2):22.

321 Kruse, T. and Ostergaard, S. Redefining peptide therapeutics with semaglutide. *Nat Chem.* 2024;16(2):296.

322 Mijalis, A.J., Thomas III, D.A., Simon, M.D., et al. A fully automated flow-based approach for accelerated peptide synthesis. *Nat Chem Biol.* 2017;13(5): 464–466.

323 Hartrampf, N., Saebi, A., Poskus, M., Gates, Z.P., et al. Synthesis of proteins by automated flow chemistry. *Science.* 2020;368(6494):980–987.

324 Martin, V., Egelund, P.H.G., Johansson, H., Thordal Le Quement, S., Wojcik, F., and Sejer Pedersen, D. Greening the synthesis of peptide therapeutics: an industrial perspective. *RSC Adv.* 2020;10(69):42457–42492.

325 Jad, Y.E., Acosta, G.A., Khattab, S.N., et al. 2-Methyltetrahydrofuran and cyclopentyl methyl ether for green solid-phase peptide synthesis. *Amino Acids.* 2016;48(2):419–426.

326 Kumar, A., Sharma, A., de la Torre, B.G., and Albericio, F. Scope and limitations of gamma-valerolactone (GVL) as a green solvent to be used with base for Fmoc removal in solid phase peptide synthesis. *Molecules.* 2019;24(21):4004.

327 Kitada, S., Fujita, S., Okada, Y., Kim, S., and Chiba, K. Hydrophobic tag-assisted liquid-phase synthesis of a growth hormone-inhibiting peptide somatostatin. *Bioorg Med Chem Lett.* 2011;21(15):4476–4479.

328 Okada, Y., Suzuki, H., Nakae, T., et al. Tag-assisted liquid-phase peptide synthesis using hydrophobic benzyl alcohols as supports. *J Org Chem.* 2013;78(2):320–327.

329 Liu, X., Zhang, N., Gu, X., et al. Total synthesis of semaglutide based on a soluble hydrophobic-support-assisted liquid-phase synthetic method. *ACS Comb Sci.* 2020;22(12):821–825.

330 Frederick, M.O., Boyse, R.A., Braden, T.M., et al. Kilogram-scale GMP manufacture of tirzepatide using a hybrid SPPS/LPPS approach with continuous manufacturing. *Org Process Res Dev.* 2021;25(7):1628–1636.

331 Staby, A., Steensgaard, D.B., Haselmann, K.F., et al. Influence of production process and scale on quality of polypeptide drugs: a case study on GLP-1 analogs. *Pharm Res.* 2020;37(7):120.

332 Kaufmann, P., Bergmann, A., and Melander, O. Novel insights into peptide amidation and amidating activity in the human circulation. *Sci Rep.* 2021;11(1):15791.

333 Meldal, M. and Tornøe, C.W. Cu-catalyzed azide–alkyne cycloaddition. *Chem Rev.* 2008;108(8):2952–3015.

334 Ramirez-Andersen, H.S., Behrens, C., Buchardt, J., et al. Long-acting human growth hormone analogue by noncovalent albumin binding. *Bioconjug Chem.* 2018;29(9):3129–3143.

335 Yang, B., Gelfanov, V.M., El, K., et al. Discovery of a potent GIPR peptide antagonist that is effective in rodent and human systems. *Mol Metab.* 2022;66:101638.

Section IV

Peptide Drug Case Studies

8

Insulin in Its Second Century: History, Evolution, and Aspirations

John P. Mayer[1], Kishore Thalluri[2], Ethan Mickelson[3], Michael H.B. Stowell[1] and Richard D. DiMarchi[2]

[1]*Department of Molecular, Cellular and Developmental Biology, University of Colorado, Boulder, CO, USA*
[2]*Department of Chemistry, Indiana University, Bloomington, IN, USA*
[3]*Department of Bioengineering, University of Washington, Seattle, WA, USA*

8.1 Introduction

The year 2021 marked a century since the discovery of insulin as the principal hormone in physiological glucose control and the realization of its miraculous ability to treat diabetes. The history of this landmark discovery, including the roles and relationships of the principal scientists in Toronto, was detailed in a recent review [1].

The demonstration that purified pancreatic extracts could rapidly reverse diabetic symptoms in canine and human prompted an interdisciplinary effort to characterize the insulin molecule, elucidate its mechanism of action, and establish reliable, large-scale manufacture to support its therapeutic use. These objectives promoted breakthroughs in macromolecule analysis and synthesis that included protein sequencing, X-ray crystallography, immunoassays (RIA), peptide synthesis, and rDNA protein expression. Insulin was often the first molecule employed to validate these ground-breaking methods, several of which have been recognized with Nobel prizes. In this respect, insulin has been a catalyst for a wide range of transformative research and has exerted enormous influence on a century of biomedical science [2].

While the discovery of insulin propelled fundamental advances in metabolism, biochemistry, molecular biology, and chemistry, the methodologies have enabled optimization of insulin as a medicine benefiting millions and millions of patients

Peptide Science: Chemical Ligation, Lead Generation, and Therapeutic Advances, First Edition. Fa Liu.

with diabetes. The first section of this review summarizes efforts to optimize insulin's time action to more closely mimic endogenous secretion through the introduction of prandial and basal analogs. The following section focuses on studies of insulin's interaction with the insulin receptor (IR) initially by means of X-ray crystallography and more recently by cryo-electron microscopy (cryo-EM). Next, we discuss advances in the chemical synthesis of insulin and related peptides, as these have enriched our understanding of the relationship between insulin's structure and biological action. We close with a discussion of evolving technologies that aim to deliver precise metabolic control through continuous glucose monitoring (CGM) and glucose-responsive insulin (GRI) analogs.

8.2 The Evolution of Insulin Therapy

Throughout its 100-year history, insulin has been the subject of continuous efforts to improve its production, purity, stability, pharmacology, safety, and convenience of administration. Persistent emphasis has been focused on the optimization of insulin's time action, prompted by the need to restore a more natural pharmacokinetic (PK) profile to improve diabetic outcomes and facilitate patient adherence to treatment guidelines. In a healthy state, plasma insulin is maintained at constant basal concentrations, punctuated by higher bolus levels in response to food intake. Therapeutically, insulin constitutes the most proven method of restoring glucose control in patients with elevated HbA1c values [3]. Clinical studies have established a relationship between elevated HbA1c values and long-term microvascular conditions, including nephropathy, proliferative eye disease, and neuropathology. Notably, even modest, persistent increases of 1% in HbA1c values in T2D patients have been correlated with significant increases in cardiovascular events, kidney dysfunction, and mortality [4]. Conversely, tighter glucose control has been associated with reduced incidence of these adverse outcomes in T2D patients [5,6]. Similar trends have also been observed in T1D patients treated with intensive glucose-lowering therapy, as reported in the landmark Diabetes Control and Complications Trial (DCCT) [7,8].

Prior to the introduction of rDNA-derived human insulin, animal-sourced insulins, specifically bovine and porcine, were the sole source of this life-saving hormone. Since then, two classes of insulin analogs, the long-acting basal and short-acting prandial, have evolved to play increasingly important roles in managing glucose levels either as stand-alone therapy or in combination (Figure 8.1). The most comprehensive approach employs integrated use of a short- and long-acting insulin in a "basal-bolus" regimen. Over the past four decades, extensive effort has been devoted to the discovery and development of prandial and basal analogs and their incorporation into optimized treatment algorithms. While this has provided

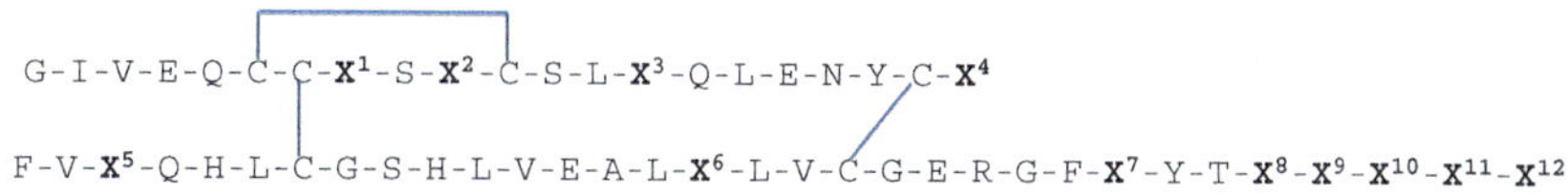

	X^1	X^2	X^3	X^4	X^5	X^6	X^7	X^8	X^9	X^{10}	X^{11}	X^{12}
Human	T	S	Y	N	N	Y	F	P	K	T		
Porcine										A		
Bovine	A	V								A		
Lispro								K	P			
Aspart								D		-		
Glargine				G							R	R
Glulisine					K				E	-		
Detemir									**1**	-		
Degludec									**2**	-		
Icodec			E			H	H		**3**	-		

1 is K-ε(C_{14}acid), **2** is K-ε(γE-C_{16}diacid), **3** is K-ε(2x-OEG-γE-C_{20}diacid)

Figure 8.1 Sequences of human, animal, and engineered insulin analogs used in diabetes treatment.

undeniable benefits to insulin-dependent patients in terms of glycemic control, intensive use of insulin has also been associated with the risk of hypoglycemia, weight gain, and progressive retinopathy [9,10]. Additionally, long-term studies [4] have reported a higher incidence of strokes, myocardial infarction, and other adverse CV events in T2D patients treated with intensive glucose management.

Ideally, insulin therapy should achieve optimal glucose control while minimizing the inherent risk and consequences of hypoglycemia [9]. Certain patients, however, require more customized approaches reflecting their individual health status and associated comorbidities. The addition of complementary drugs to insulin regimens, such as GLP-1 agonists, in the form of ideglira or lixilan, has been effective in minimizing insulin use and promoting weight loss [11]. Separately, the advances in glucose monitoring have proven instrumental in improving glucose control. The introduction of CGM has led to improved glycemic control with reduced variability in patients using multiple daily injections and insulin infusion pumps.

8.2.1 Prandial Insulins

The prandial, or "meal-time" insulins derive their name from their ability to control postprandial glucose (PPG) excursion, a challenging aspect of diabetes management as it represents the time of day when glucose levels fluctuate the most [12]. Insulin levels rise dramatically in response to food intake, peaking

45–60 minutes post meal, returning to baseline levels within four to six hours, depending upon the amount of food consumed. The prandial insulins were designed to mimic endogenous bolus insulin secretion, suppressing the PPG elevation and returning to baseline as glucose levels diminish as a function of insulin action.

8.2.1.1 First Generation

Insulin lispro (Humalog®, Eli Lilly, launched in 1996) was the first-in-class prandial insulin chemically optimized to provide faster onset and a shorter duration of action relative to native insulin [13,14]. Humalog, along with two subsequently introduced analogs, insulin aspart (Novolog®, Novo Nordisk, launched in 2000), and glulisine (Apidra®, Sanofi, launched 2006), clinically validated the prandial insulin concept of providing enhanced medicinal benefits and convenience in dosing closer to mealtime, either as a stand-alone therapy or as part of a basal/bolus regimen. Nevertheless, the first-generation analogs do not precisely mimic the rapid first-phase onset of physiological insulin action observed in the absence of diabetes. This has been attributed to delayed tissue absorption, a characteristic limitation of subcutaneous administration [15].

8.2.1.2 Second Generation

Faster acting insulin aspart (FIASP) was launched by Novo Nordisk [16,17] and involved the reformulation of insulin aspart (Novolog®) with niacin and arginine added to accelerate absorption. FIASP has proven superior to the conventional Novolog formulation in head-to-head comparative clinical studies. In a Phase 3 trial studying mealtime administration of the two formulations, FIASP demonstrated superiority in HbA1c and PPG control [18]. A pooled analysis of PK and pharmacodynamic data (PD) from six trials comparing FIASP and insulin aspart in T1D patients confirmed faster appearance, higher initial concentration, and an accelerated elimination for FIASP [19]. In a Phase 3 trial of poorly controlled T2D patients, FIASP was judged non-inferior to insulin aspart, which changed from initial HbA1c, but demonstrated lower PPG at the one-hour time point [20].

Eli Lilly's ultra-rapid lispro (URLi), launched in 2020 as Lyumjev®, is a novel formulation of lispro containing citrate to promote tissue permeability and treprostinil to facilitate vascular dilation. This combination results in a faster onset and shorter duration of action when compared to the conventional formulation of insulin lispro. A Phase 3 trial was conducted in T1D patients [21], randomized into three cohorts: mealtime URLi, post-meal URLi, and mealtime lispro. Mealtime URLi demonstrated superiority relative to conventional lispro in managing PPG as well as lowering the incidence of hypoglycemia in the post-meal period. In a euglycemic clamp study in T2D patients, URLi was pharmacokinetically detected prior to lispro, with a higher initial concentration and a shorter total exposure period.

The pharmacodynamic assessment also indicated a faster onset and a more pronounced insulin effect with a shorter duration of action [22].

Biochaperone lispro (BCLIS) is a novel prandial formulation of insulin lispro developed by Adocia, which contains the proprietary BC222 oligomer and citrate, added to enhance hexamer disassociation and monomer absorption from the site of injection. The BCLIS formulation has shown improved PPG in T1D patients when compared to conventional insulin lispro. When BCLIS was self-administered immediately prior to a mixed-meal test, the novel formulation produced greater PPG reduction relative to lispro. The ultra-rapid onset of BCLIS promotes dosing closer to mealtime, an increasingly preferable option from the patient's perspective [23]. A euglycemic clamp study conducted in T1D patients compared pump-administered BCLIS, FIASP, and insulin aspart. The glucose-infusion rate (GRI) one-hour post dosing as a primary endpoint was higher for BCLIS as compared to insulin aspart, but roughly equivalent to FIASP. The PK results aligned with the PD findings [24]. A unique characteristic of the Adocia polymer technology is that it allows co-formulation of insulins with otherwise incompatible aqueous solubility. For example, Biochaperone BC147 permits co-formulation of glargine at neutral pH with insulin lispro [25].

While all the second-generation prandial insulins involve reformulation of first-generation analogs (lispro and aspart), a report from the Chou group disclosed a novel insulin structure based upon the cone snail insulin. This analog, termed "Mini-Ins," is notable for a minimized structure devoid of the B-chain 22–30 portion noted for self-association and receptor signaling, and yet of equivalent potency to native insulin, as a monomeric insulin analog [26].

8.2.2 Basal Insulins

Basal insulin therapy is a well-established treatment paradigm for T1D and advanced T2D patients and is also extensively utilized in combination with prandial insulins as part of basal-bolus regimens [27]. An ideal basal insulin should have the following properties: a prolonged duration of action, a flat time-action profile, low day-to-day glycemic variability, and the potential for flexible dosing. While most recently diagnosed T2D patients are stabilized by oral antidiabetic drugs, clinical studies support the initiation of basal insulin therapy in these patients in order to restore glucose control and delay disease progression [4]. These beneficial effects are attributed to the restoration of β-cell function [28]. For example, the ADA and EASD both recommend initiation of basal insulin therapy in recently diagnosed T2D patients in instances when metformin treatment fails to control HbA1c [29].

Basal insulin therapy is intended to mimic a healthy physiological state in which the pancreas secretes low, yet consistent levels of endogenous insulin into portal circulation. This route facilitates rapid insulinization of the liver, effectively lowering glucose release by suppressing gluconeogenesis and glycogenolysis between meals and during the night. This direct, immediate effect on the liver is difficult to reproduce through subcutaneous insulin administration, which inherently results in peripheral *over*-exposure and hepatic *under*-exposure.

8.2.2.1 First Generation

The first insulins structurally modified for extended time action include insulin glargine (launched as Lantus® by Aventis in 2000) and insulin detemir (launched as Levemir® by Novo Nordisk in 2005). Insulin glargine is modified by the addition of two arginine residues at positions B31 and B32, which reduce solubility at physiological pH, and substitution of glycine for asparagine at A21 to maintain stability at acidic pH. Upon subcutaneous injection, insulin glargine precipitates, forming a depot from which it is then slowly released into the peripheral circulation with a duration of action approaching 24 hours [30,31] for the original U100 formulation. Insulin detemir (launched as Levemir® by Novo Nordisk in 2005) relies on an entirely different principle for protraction (one subsequently broadly adopted for other insulin and incretin analogs). Acylation of LysB29 with myristic acid enhances hexamer association and facilitates reversible binding to serum albumin, further prolonging time action with a flat PK profile [32] and a duration of action of 12–24 hours. [31,33]. Clinical comparison between detemir and glargine revealed comparable profiles, with similar glucose-lowering effects during the initial 12-hour time period [30,34] but with diminishing effects for detemir in the subsequent 12–24-hour period. Another study in T1D patients indicated that while insulin detemir and glargine in conjunction with meal-time insulin maintained similar glycemic control and rates of hypoglycemia, it also noted that most patients in the detemir cohort finished the study requiring twice-daily dosing [35]. In other studies, however, detemir consistently demonstrated lower weight gain and less within-subject variability relative to glargine.

8.2.2.2 Second Generation

The second generation of basal insulins represents incremental, yet clinically meaningful advancements and benefits to patients, including a longer duration of action and a peakless PK profile [36]. Insulin glargine U300 (iGlar300) launched as Toujeo® by Sanofi in 2015, was the first ultra-long-acting insulin introduced. The longer time-action of iGlar300 is attributed to the higher concentration of glargine (relative to the iGlar100 original formulation) within the subcutaneous depot, resulting in a slower and less variable dissolution rate. The new formulation demonstrated a superior performance relative to the original iGlar100 formulation in several clinical parameters. iGlar300 was shown to maintain glucose control for five hours longer than iGlar100 or ~30 hours [37]. The six-month EDITION-1 study of T2D patients using iGlar300 demonstrated equivalent glucose control based on HbA1c values, but with a lower incidence of nocturnal hypoglycemia [38]. A similar trend was observed in the longer EDITION-2 study when the two formulations were used in combination with oral anti-hyperglycemic agents [39]. A meta-analysis of Phase 3 studies with a T1D population also confirmed lower rates of hypoglycemia of iGlar300 relative to the iGlar100 formulation [40].

Insulin degludec (iDeg), launched in 2015 by Novo Nordisk as Tresiba®, the second ultra-long-acting insulin, is a desB30 insulin modified with γ-glutamic acid-linked C16 diacid, which facilitates non-covalent binding to serum albumin. Upon injection, the zinc–phenol-stabilized di-hexameric structure forms oligo-hexameric chains, which in turn disassociate into dimers and monomers with the gradual loss of zinc ions [41]. This unique time-extension mechanism facilitates a flat, peakless PK profile while maintaining glucose control of up to 42 hours following eight days of daily dosing [42]. While iGlar300 and iDeg achieve comparable clinical performance [43], recent head-to-head clinical studies have revealed some differences. For example, euglycemic clamp studies comparing 0.4 U/kg doses of iGlar300 and iDeg administered to T1D patients noted less pharmacodynamic and PK fluctuation for iGlar300. This difference, however, became statistically insignificant when the two insulins were dosed at 0.6 U/kg [44]. Another clamp study in T1D patients with the two insulins dosed at 0.4 U/kg, iDeg demonstrated more stable day-to-day and within-day glucose control than iGlarU300 [45]. The CONCLUDE trial designed to assess risk of hypoglycemia in T2D patients observed no difference in the primary endpoint of overall symptomatic hypoglycemia, but noted lesser rates within the secondary endpoints of nocturnal and severe hypoglycemia in the case of iDeg [46].

Surveys have shown that reducing both the frequency of insulin injections as well as the flexibility in their timing is an important factor in patient satisfaction and compliance. Therefore, the availability of a basal insulin administered less frequently than in the typical daily regimen may increase acceptance and adherence to insulin therapy. In the last few years, two once-weekly insulin candidates have advanced to late-stage clinical development [47]. The most advanced of these is insulin icodec from Novo Nordisk. The GluA13, HisB16, and HisB25 substitutions were introduced to enhance enzymatic stability while maintaining solubility and a much-reduced agonist potency [48]. Insulin icodec was found to have a mean half-life at a steady state of 155 hours [49]. Icodec has been thoroughly investigated during six comprehensive Phase 3 ONWARDS clinical trial programs [50]. ONWARDS Trials 1, 3, and 5 included T2D insulin-naive patients experiencing inadequate glucose control using oral agents or GLP-1 agonists. These trials compared initiation with icodec versus initiation with glargine (ONWARDS 1) [51] or iDeg (ONWARDS 3) [52]. ONWARDS 5 [53] compared the icodec's titration regimen with either iDeg or glargine (U100 or U300). All three trials unequivocally demonstrated once-weekly icodec as superior to once-daily glargine and iDeg based on HbA1c and time spent in an optimal glycemic range. ONWARDS 2 [54] and 4 [55] evaluated icodec when administered to patients with long-standing T2D. ONWARDS 2 results showed that icodec was statistically superior to iDeg in lowering HbA1c and statistically non-inferior with respect to the incidence of hypoglycemia. The results of ONWARDS 2 and 4 confirmed that a once-weekly insulin such as icodec could deliver equivalent or even slightly superior performance relative to a conventional once-daily basal insulin while reducing the number of yearly injections from 365 (daily) to 52 (weekly), more than 85% reduction. Patients in both studies expressed greater treatment satisfaction with icodec relative to the conventional once-daily basal insulins.

Eli Lilly has advanced basal insulin Fc (BIF, insulin efsitora alpha), which consists of single-chain insulin fused to a human IgG2 Fc domain to clinical trials. BIF was shown to have a 17-day half-life and a sustained lowering of fasting blood glucose of five days, with a within-the-week peak-to-trough ratio of 1.14 [56]. The safety and efficacy of BIF was evaluated in a Phase 3 study of T2D patients who had been previously treated with basal insulin. The patients were randomized to receive either daily iDeg or weekly BIF. The study demonstrated non-inferiority of BIF versus iDeg in HbA1c reduction and noted that the hypoglycemia event rates were 25% lower in the BIF cohort [57]. The safety and efficacy of weekly BIF was also compared to once-daily iDeg in insulin-naive patients with T2D. In this study, BIF demonstrated non-inferior HbA1c change from baseline with similar rates of hypoglycemia between treatment groups [58]. In T1D patients, BIF demonstrated non-inferior glycemic control and comparable rates of hypoglycemia relative to iDeg [59]. BIF is currently being evaluated in the comprehensive QWINT (Once Weekly Insulin Therapy) Phase 3 clinical trial program, which is scheduled to be completed in 2024.

8.3 Structure

The IR is a functional dimer, with each protomer containing five extracellular domains: L1, CR, L2, FnIII-1, FnIII-2, and FnIII-3. Additionally, it has a single-pass transmembrane domain and an intracellular kinase domain [60]. Structural studies have been crucial in elucidating the molecular details of insulin action on the IR. The first crystal structure of the insulin-free IR (apo-IR) extracellular domain showed an inverted V-shaped conformation with the membrane-proximal FnIII-3 domains more than 140 Å apart [61]. This conformation corresponds to the inactive state as it prevents direct interaction between the kinase domains, thus hindering the trans-autophosphorylation necessary for activation.

The crystal structure of the L1 domain and the C-terminal helix of the α-chain (α-CT) of IR in complex with insulin revealed how insulin binds to the primary site of IR [62] but did not explain how this binding leads to activation. Recent cryo-EM structures of IR/insulin complexes have shown that insulin binding disrupts the apo-state of IR, inducing a large conformational change from an inverted V to a T-conformation. This brings the FnIII-3 domains closer together, enabling trans-autophosphorylation of the kinase domains [63–67] (Figure 8.2).

The primary insulin-binding site of IR consists of the L1 domain and α-CT. Binding of insulin causes a notable shift of the α-CT in relation to the L1 domain, demonstrating the flexibility of the α-CT is crucial for binding. Impressively, 24 of the 51 insulin residues are involved in Site 1 binding (Figure 8.3). The A chain of insulin mainly contacts the α-CT, while the B chain interacts with both the L1 domain and α-CT using Van der Waals and hydrogen-bonding (Figure 8.3). Insulin bound to Site 1 also contacts the top loops of the FnIII-1′ domain of the neighboring protomer, where specific B chain residues play key roles. This binding method allows one insulin to engage two protomers, stabilizing IR's T-shaped active conformation. The second site is located on one side of a major β sheet of the FnIII-1 domain of IR (Figure 8.3) and is necessary for IR activation; mutation of this site diminishes both insulin binding and IR autophosphorylation [66].

These studies demonstrate that the binding of multiple insulins at two distinct sites on IR is required for physiological IR activation. To clarify the individual contributions of each site to the cooperative formation of the active T-state and the

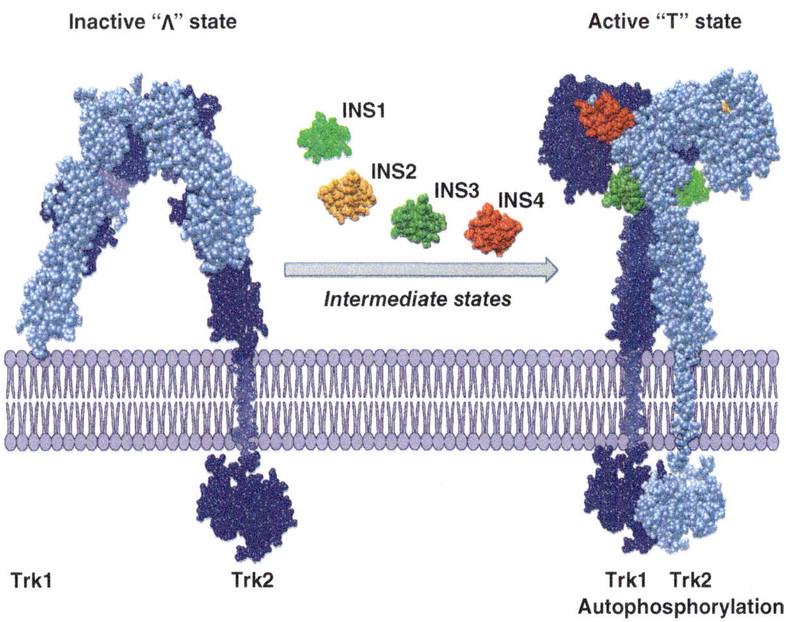

Figure 8.2 The transition from the inactive inverted "^" state to the active "T" state involves the binding of four insulin molecules at both Site 1 and Site 2 of the insulin dimer. Upon binding IR undergoes a conformational transition that brings the tyrosine kinase domains (Trk) into close proximity for phosphorylation.

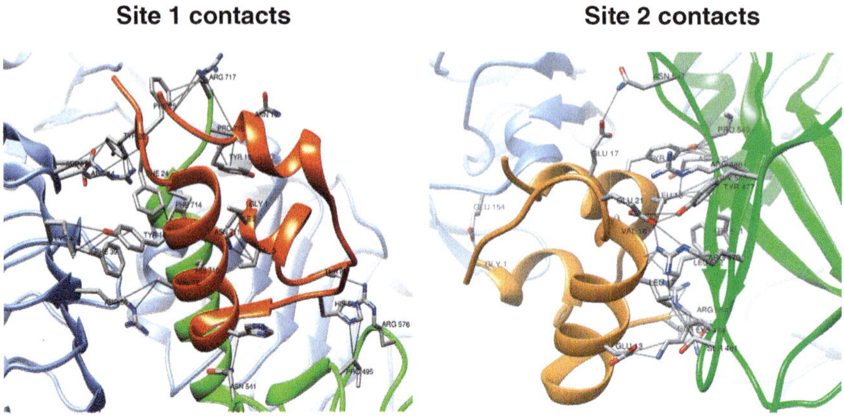

A-chain GIVEQCCTSICSLYQLENYCN

B-chain FVNQHLCGSHLVEALYLVCGERGFFYTPKT

Figure 8.3 Binding interactions of Site 1 and Site 2. Specific residues of the insulin A and B chains are color coded to show the nonoverlapping interactions of insulin with either Site 1 or Site 2, red for Site 1 and orange for Site 2.

physiological necessity of this complex activation mechanism, a series of insulin analogs with point mutations were synthesized to abrogate binding to either Site 1 or Site 2 [68], and structure analysis by cryo-EM was performed.

Insulin binding to Site 1 of apo-IR causes a separation between the L1 domain of one protomer and the FnIII-2′ domain of another within the inverted V-shaped IR dimer. However, due to long interdomain linkers, interactions between the L2 and FnIII-1′ domains remain; suggesting insulin at Site 1 may not fully disrupt IR's autoinhibited state. Cryo-EM analysis shows that insulin binding solely at Site 2 does not alter IR's conformation [68]. When insulin binds at both Site 1 and Site 2, they clash, possibly pushing the Site 2–bound insulin and its domain away, further separating the protomers. This indicates that insulins at Site 1 and Site 2 collectively disrupt IR's autoinhibited state and are required for IR activation.

8.4 Synthesis

Despite a six-decade-long effort, insulin chemistry continues to present a "persistent synthetic challenge" [69]. Moreover, the initial focus on the total synthesis of mammalian insulins has now shifted to broader classes of structurally and biologically diverse insulin-related peptides. Synthesis of other members of the insulin family, such as the relaxins [70], the viral insulin-like peptides [71], and the snail-derived insulins [72], have revealed limitations of classical, insulin-focused approaches. This concern has stimulated the development of novel, more versatile methods, which have been successfully applied to a range of insulin-like structures. These advances have also enabled more extensive exploration of insulin structure–activity relationships, such as the installation of a fourth disulfide, disulfide isosteres, and other synthetically challenging modifications of the insulin pharmacophore. These reports have been reviewed recently by Wade [73], DiMarchi [74], and Liu [75].

Efforts to prepare insulin by chemical synthesis began in the early 1960s, shortly following the landmark elucidation of its primary and secondary structure by Sanger [76]. Methods utilizing solution-phase synthesis were reported in the 1960s by Zahn [77], Katsoyannis [78], and Du [79], followed by the first Boc/Bzl-based solid-phase peptide synthesis (SPPS) by Merrifield and Marglin [80]. This intense interest was stimulated by an eventual need to secure an alternative supply to animal-sourced insulin as well as by plans to conduct structure–activity studies related to this important hormone. These reports are impressive not only in view of the enormous synthetic challenges involved, but also the limitations imposed by the rudimentary analytical and preparative capabilities of this period. In the mid-1970s, Sieber et al. [81] reported a solution-phase approach utilizing an orthogonal cysteine protection scheme to direct regioselective disulfide pairing. The FDA approval of rDNA-derived human insulin in 1983 was a transforming event in insulin history, ensuring a nearly limitless supply of the hormone [82].

Several important semi-synthetic methods were advanced in 1979 and shortly thereafter, such as those of Inouye [83] and Morihara [84], and Markussen [85], which utilized porcine insulin transpeptidation to obtain human insulin and facilitated modification of the B23-B30 octapeptide region. The 1993 total synthesis by Kiso et al. [86] (Figure 8.4) combined Fmoc-based SPPS with directed disulfide bond formation featuring their novel silyl chloride/sulfoxide method to form the CysA6—CysA11 bond. The Kiso general strategy of conducting independent syntheses of the A and B chains, assembly of the heterodimer, followed by regioselective installation of the remaining two disulfides, established an important precedent. A number of subsequently published syntheses differed only in the choice of protecting groups and the specific order of disulfide bond formation [87]. While versatile in scope, methods relying on directed disulfide bond formation tend to be laborious in practice, resulting in modest yields due to the need for multiple high-performance liquid chromatography purification steps as well as poor solubility and compromised handling of the intermediates.

The complications due to poor solubility and handling have been addressed by the incorporation of enzymatically labile polyglutamic acid tags [88] and

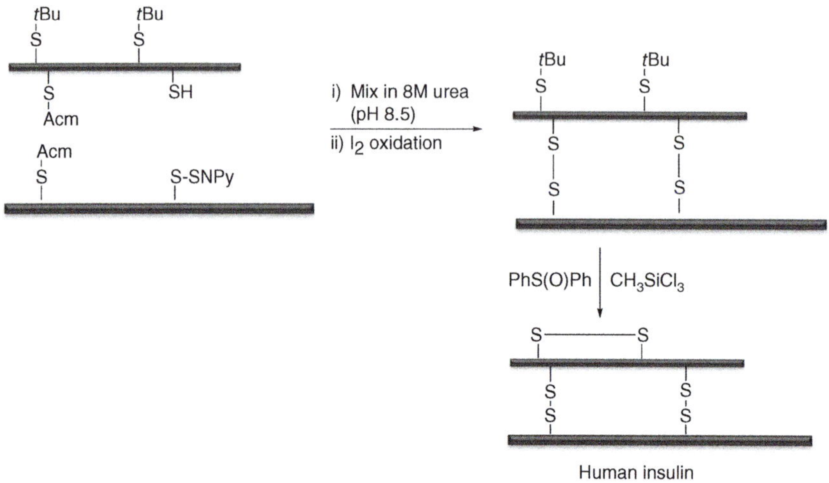

Figure 8.4 Synthesis of human insulin by sequential, regioselective disulfide bond formation. *Source*: Adapted from Ref. [86].

chemically labile poly-lysine tags [89,90]. Various embodiments of this technique utilizing either cationic or anionic tags have been applied by Zaykov [91], Brimble [92], Chou [93], and Bode [94]. A novel and effective technique was introduced by Liu et al., who incorporated iso-acyl dipeptide isosteres at ThrA10:SerA11 and TyrB26:ThrB27, which dramatically improved the handling of both chains [95] (Figure 8.5). Following the formation of the insulin heterodimer, both iso-acyl moieties were induced to undergo an O->N shift under basic conditions, reverting to the native backbone. The incorporation of iso-acyl dipeptides to aid handling was subsequently adopted by several groups [96–99].

The single-chain or biomimetic strategy offers an alternative to the two-chain, directed disulfide method. This involves an SPPS synthesis of a single-chain insulin where the GlyA1 is either directly linked to LysB29 or connected via a short, enzymatically labile surrogate C-peptide. The linear precursor is folded under redox conditions and enzymatically converted to the two-chain hormone. Tofteng [88] evaluated several connecting peptides and solubilizing extensions. Both the connecting and solubilizing peptides were simultaneously removed by the action of *Achromobacter lyticus* (Lys–C) protease to yield desB30 human insulin. A similar approach was used by Zaykov [91] (Figure 8.6), who identified a 49 amino acid precursor “desDi,” with superior folding properties and tolerance for several previously problematic amino acid substitutions. Similar to the Tofteng method, the folded single-chain intermediate was converted to the two-chain bioactive hormone through Lys–C proteolysis. While the biomimetic approach offers important advantages, such as fewer purification steps and acceptable yields, it

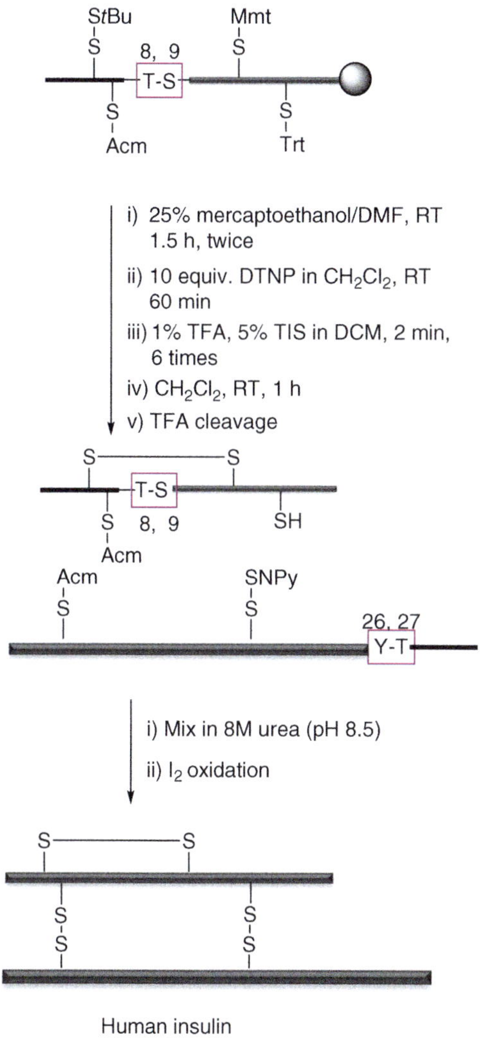

Figure 8.5 A synthetic route to human insulin using iso-acyl dipeptides. *Source*: Adapted from Ref. [95].

is inherently limited in scope. Specifically, the biomimetic approach requires an enzyme, and it is consequently restricted to insulin analogs with a single, unique proteolytic site, typically LysB29, for conversion to the two-chain final product.

The prospect of a chemically removable linker promises to greatly expand the scope of the single-chain approach. This concept was first validated by Busse and Carpenter, who cross-linked the A-chain N-terminus to LysB29 of porcine insulin

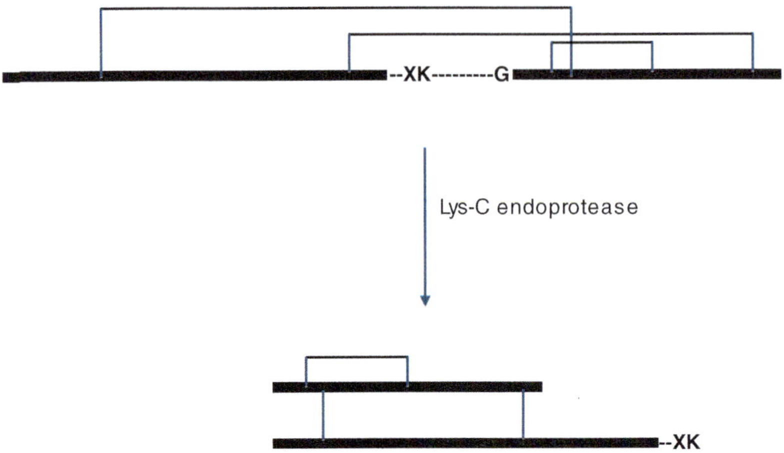

Figure 8.6 Enzymatic conversion of single-chain precursors to two-chain insulin analogs [88,91].

with carbonylbis (methionine-*p*-nitrophenyl ester). The cross-linked insulin was first reduced, then oxidized, and the insulin was liberated through cyanogen bromide treatment [100]. With a resurgence of interest in insulin chemistry over the last two decades, the concept of a chemically labile surrogate linker was revisited. In their synthesis of insulin lispro, the Kent group introduced an oxime-forming ligation linking the N-terminus of the A-chain to the C-terminus of the B chain [101], which facilitated efficient folding and disulfide formation. Despite the use of a non-peptide linker, conversion to two-chain insulin lispro still required treatment with trypsin and carboxypeptidase B. A subsequent report from the Kent group devised a "traceless" chemically labile tether by forming an ester between the γ-carboxyl group of GluA4 and the β-hydroxyl group of ThrB30. The folding, disulfide bond formation, and saponification of the ester proceeded in high yield to give insulin lispro [102,103]. The Bode group applied their α-ketoacid-hydroxylamine (KAHA) ligation to the synthesis of several insulin analogs applying the 5-oxaproline [104,105] and the 4,4-difluoro-5-oxaproline [94] versions of the KAHA chemistry. The method utilized a sulfone C-peptide surrogate, which was excised under basic conditions following folding and disulfide formation. The DiMarchi group recently developed several methods utilizing chemically labile tethers to synthesize insulin and related two-chain peptides. The first employed a dual oxime ligation to link the A- and B-chains in the native (N-terminus of A-chain to C-terminus of B-chain) orientation. Following folding/disulfide formation, the linker was chemically excised in a two-step reaction sequence through a diketopiperazine (DKP) cyclative release at the A-chain junction and ester hydrolysis at the B-chain junction [70,106]. A subsequent report introduced a symmetrical polyethylene glycol linker joining

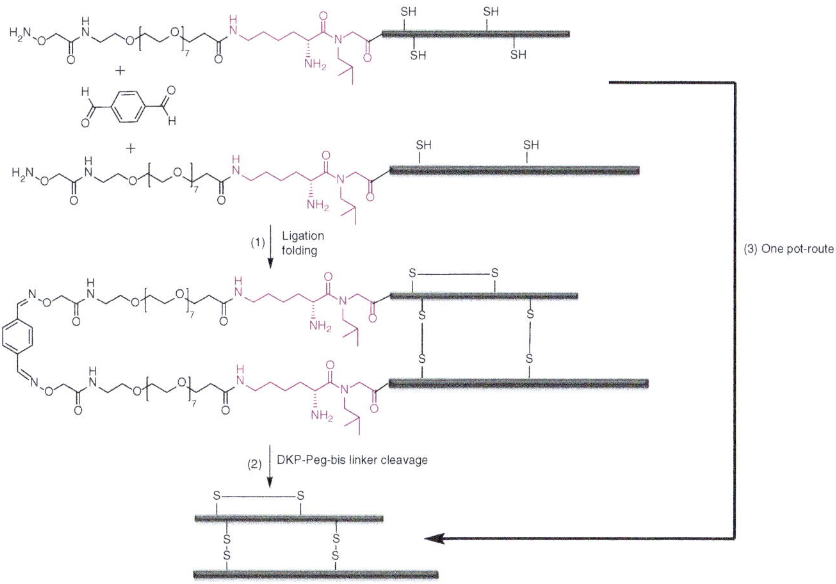

Figure 8.7 Synthesis of disulfide-rich heterodimeric peptides through an auxiliary *N, N*-crosslink. *Source*: Adapted from Ref. [89].

the respective N-termini of the A and B chains in a noncanonical configuration [99] (Figure 8.7). Following folding/disulfide formation, the linker was excised under mild basic conditions in a single step through concomitant DKP formation at both N-termini. A unique and important attribute of this approach is that the installation and removal of the noncanonical linkage do not rely on any specific sequence or side-chain functionality, which allows application to any two-chain peptide.

The decades of synthetic innovation summarized above have enabled detailed, broader interrogation of insulin's structural features, such as the individual disulfides. Examples include replacement of the A6:A11 disulfide with a thioether [107], a methylene thioacetal [108], an olefin [109,110], and di-selenide [111]. Vinther et al. [112] reported an insulin analog with an additional disulfide at A10-B4 with full potency and increased stability. The influence of an additional disulfide was explored by several groups. The DiMarchi group utilized sequential disulfide bond formation to crosslink A10-B4, A10-B5, and A10-B6 positions [98]. The same group employed a novel [3+1] synthetic strategy to construct 20 analogs constrained by a fourth disulfide [113]. This approach allowed a comparison between the activity of the constrained and unconstrained analogs. Xiong et al. prepared several 4-disulfide analogs, notably A22-B22 with full potency and improved aggregation stability [114]. Syntheses of glyco-insulins were also reported by the Wade [115] and DiMarchi groups [116].

The recent synthetic advances have greatly enhanced our ability to conduct structure–function studies of insulin with the promise of improving its therapeutic properties. This expanded synthetic repertoire is additionally applicable to structurally more diverse insulin-related peptides, which will prove essential for exploration of their biology and potential therapeutic uses.

8.5 Aspirational Aims

8.5.1 Continuous Glucose Monitoring

The first- and second-generation prandial and basal insulins have enabled dramatically improved clinical outcomes for T1D and T2D patients. Historically, the primary goal of insulin therapy has been the minimization of hyperglycemia, and the consequent reduction of long-term pathologies such as micro- and macro-vascular complications, neuropathy, and nephropathy. The benefits of tighter glucose control, however, must be weighed against the risk of hypoglycemia. The T1D or T2D patient is challenged with maintaining the precarious balance between avoiding hyperglycemia while minimizing episodes of hypoglycemia. Glucose monitoring can inform and help maintain a personalized insulin regimen and is increasingly considered an essential part of effective disease management for T1D and T2D patients. The introduction of glucose test strips and glucose meters constituted the first step toward enabling glucose self-monitoring [117]. The concept of CGM technology was first used in the late 1990s and this technology has undergone tremendous progress in terms of accuracy, device miniaturization, sensor life, and compatibility with smartphones. This has resulted in broad adoption of CGM devices among mainly T1D patients but also T2D patients on intensive insulin therapy. The recent advances in CGM technology were recently reviewed by Galindo and Aleppo [118]. There are currently two basic types of CGM devices: real-time CGM (rtCGM) and intermittently scanned CGM (isCGM). There are several rtCGM devices on the market, including Dexcom G6 (Dexcom Inc.), Medtronic Guardian Sensor 3 (Medtronic, Inc.), and Senseonics Eversense (Senseonics, Inc.). All three can be interfaced with a smartphone receiver, with the Guardian 3 and Dexcom also compatible with an insulin pump. The Freestyle Libre 3 (Abbott Diabetes Care) is an example of an isCGM device, which differs from the rtCGM devices in that the glucose reading is updated periodically.

The use of rtCGM and isCGM device systems has been shown to substantially simplify and improve diabetes disease management control through reducing glycemic variability, and risk of hyper- and hypoglycemia. The International Study Group on Hypoglycemia recognizes three levels of hypoglycemia: level 1 (<3.9 mmol/L) when symptoms may or may not be present, level 2 (<3.0 mmol/L) when the patient likely experiences reduced cognition, and level 3, severe hypoglycemia, which may include symptoms such as diminished physical function, seizure, or coma [119]. Information regarding an individual's glucose profile is essential to minimizing hypoglycemic episodes. Several studies have demonstrated the favorable impact of CGM on the incidence of hypoglycemic episodes, hospitalizations, work absenteeism, and quality of life. A total of 515 T1D patients in a Belgian study using insulin pumps were enrolled in a program utilizing rtCGM for a 12-month period. The patients experienced reductions in HbA1c from baseline, reduced hospitalizations due to hypoglycemia or ketoacidosis, as well as reduced work absenteeism, and an improved quality of life [120]. The utility of CGM has also been validated in T1D patients taking multiple daily insulin injections. In a clinical trial, 158 patients were randomized to CGM and standard-of-care groups. At 24 weeks, the CGM cohort reported a mean HbA1c reduction of 0.6% versus the control group while also experiencing significantly shorter periods of hypoglycemia [121].

8.5.2 Glucose-responsive Insulins

The advent of a glucose-responsive or glucose-sensitive insulin has become somewhat of a holy grail for insulin discovery and diabetes therapy. An insulin molecule capable of adjusting its activity in real time, in response to glucose levels, could have the potential to dramatically transform diabetes therapy. It would likely minimize or even displace the need for glucose monitoring, insulin pumps, and closed-loop devices. The promise of this technology has attracted enormous interest from academic and industrial groups, and while these have produced several ingenious concepts, none have thus far achieved clinical validation. The basic approaches are summarized below.

8.5.2.1 Polymeric GRI Systems

These systems involve the encapsulation of insulin within glucose-responsive polymeric matrices [122,123]. To confer glucose-responsiveness, polymeric GRI systems include either enzymatic or synthetic glucose-binding elements (GBEs) that stimulate the release of insulin from the matrix upon the binding of glucose [124,125]. GBEs are either co-encapsulated with insulin or embedded within the polymeric matrix [126]. The pore sizes of the polymeric matrices are wide enough to allow the passage of glucose yet narrow enough to retain encapsulated insulin and GBEs [124,125]. Following subcutaneous administration, this system facilitates glucose-responsive release of insulin [122,127].

8.5.2.1.1 ConA GRI Systems

The first polymeric GRI system utilized Concanavalin A (ConA) [128], a lectin metalloprotein that reversibly binds glucose and several other carbohydrates [129]. This GRI system consists of tetrameric ConA-insulin conjugates quadruply tethered to a glycopolymeric matrix [130]. Initially, ConA binds the pendant saccharides of the glycopolymeric matrix, thereby anchoring the ConA-insulin conjugates to the matrix [131,132]. Under hyperglycemic conditions, glucose participates in competitive binding for the ConA binding site against the pendant saccharides of the matrix [125,133]. Upon the binding of glucose to the ConA binding site, the pendant saccharides are displaced, thereby disconnecting the ConA-insulin conjugate from the matrix [133,134] Once released, ConA-insulin tetramers readily dissociate into insulin monomers [134]. Unfortunately, ConA GRI systems have not yet been able to overcome their significant issues with toxicity, mitogenicity, and immunogenicity [135].

8.5.2.1.2 Glucose Oxidase GRI Systems

Another GRI system employs the enzyme glucose oxidase (GOx) as the GBE [126,136]. GOx catalyzes the conversion of glucose to gluconic acid and H_2O_2, releasing H^+ ions in the process [137,138]. GOx GRI systems respond to changes in O_2, H_2O_2, and H^+ ions to control the release of encapsulated insulin from a polymeric matrix [139]. GOx systems that are pH-sensitive may be composed of either basic or acidic polymeric matrices [126]. Basic matrices become ionized and negatively charged under low pH conditions induced by the binding of glucose to GOx. The resulting charge repulsion causes the polymeric matrix to expand and become more porous, thereby releasing insulin from the matrix and into the bloodstream [140]. In the case of acidic matrices, the glucose-stimulated pH decrease ionizes the polymer such that it will have increased solubility or be degraded by hydrolysis [124,141]. The resulting solvation/degradation of the polymeric matrix allows the escape of the cargo insulin and GOx from the matrix and into the blood.

In several GOx systems, particularly those with acidic matrices, the production of reactive H_2O_2 contributes to the destabilization of the matrix [126,139]. GOx and insulin can also be co-encapsulated with a catalase enzyme, forcing the equilibrium of the GOx reaction forward by consuming H_2O_2 thereby producing additional O_2 for consumption by GOx. This addition of catalase has been shown to enhance the glucose sensitivity of GOx systems by amplifying the response to the binding of glucose [139,142].

GOx GRI systems have also been assembled from hypoxia-sensitive hyaluronic polymers derivatized with 2-nitroimidazole [141,143]. Under the hypoxic conditions created by the binding of glucose to GOx, the matrix is enzymatically reduced from a hydrophobic state (2-nitroimidazole) into a hydrophilic state (2-aminoimidazole), increasing the solubility of the matrix [144]. Solvation of the matrix facilitates the release of insulin and GOx into the bloodstream.

8.5.2.1.3 Phenylboronic Acids GRI Systems

Phenylboronic acids (PBAs) participate in the reversible esterification of 1,2- or 1,3-*cis*-diols groups, such as those in glucose [145]. Under physiological conditions,

PBAs exist in an equilibrium between the initial uncharged trigonal planar configuration and the anionic, hydroxyl-bound tetrahedral configuration [146]. PBAs have been used as GBEs embedded directly within the polymeric matrix of GRI systems. The binding affinity of PBAs for glucose is significantly greater in the anionic tetrahedral configuration of PBA as compared to its neutral trigonal planar configurations [147]. Higher blood glucose levels naturally lead to more binding interactions between glucose and the PBAs, leading to a greater density of negatively charged PBAs in the matrix. The resulting charge repulsion causes the polymeric matrix to expand and become more porous, liberating insulin from the matrix and into the bloodstream [123,140]. The hydrophilicity of the negative charges also functions to recruit additional water molecules into the matrix and cause further swelling and insulin release [148]. The status of PBAs as a synthetic GBE presents unique advantages and challenges as compared to enzymatic GBEs such as ConA and GOx. As a synthetic compound, PBAs have improved stability *in vivo* and do not elicit the same immunogenic responses that handicap GRI systems with enzymatic GBEs [149]. However, PBAs struggle to achieve the same rates of glucose responsiveness as their enzymatic counterparts.

8.5.2.2 Non-polymeric GRI Systems

Non-polymeric GRI systems include all GRI systems that do not involve the encapsulation of insulins within a matrix.

8.5.2.2.1 Multihexameric Insulin Complexes

One seminal non-polymeric formulation involves the coordination of natural insulin hexamers into large multihexameric insulin complexes. In this formulation, the exterior of insulin hexamers is adorned with PBAs and polyols, allowing them to interconnect with one another to form a multihexameric structure. At high glucose concentrations, glucose molecules compete with the polyols for PBA binding sites. Upon the binding of glucose to PBA, polyols are displaced from the binding site, and hexamers are thereby released from the multihexameric complex to stimulate glucose uptake into cells [150]. Despite functioning well at glucose concentrations above the physiological range, multihexameric insulin complex formulations have struggled to achieve a glucose sensitivity appropriate for medical use.

8.5.2.2.2 Human Serum Albumin Bioconjugates

The bioconjugation of multihexameric insulins to human serum albumin has been explored as a unique strategy for albumin-bound depots in the blood. In these formulations, insulins are chemically modified to include an aliphatic group for hydrophobic coordination to albumin. This aliphatic group is equipped with a PBA such that insulin will remain bound until the competitive binding of glucose at the PBA group displaces the insulin [151]. The jury is still out as to the ability of such non-covalent conjugates to achieve glucose responsiveness in the physiological range.

8.5.2.2.3 Mannosylated Insulins

GRI systems featuring mannosylated insulins achieve glucose-responsiveness by maintaining balanced binding interactions between human insulin receptor (HIR) and mannose receptor C-Type 1 (MRC1) [152]. At low glucose concentrations, the majority of mannosylated insulin binding interactions occur with MRC1, leading to degradation of the insulin and a dormant state of the therapy. At high glucose concentrations, the majority of mannosylated insulin interactions occur at the HIR, thereby enabling glucose uptake. Mannosylated insulins have been unsuccessful in clinical trials due to high rates of background clearance through MRC1 endocytosis and the significantly low binding affinity of HIR for mannosylated insulins [123,149].

8.5.2.2.4 Macrocycles

Macrocycles have been developed as highly specific GBEs [153,154]. In these GRI systems, macrocycles can be pre-equipped with glucose mimetics in their binding sites and conjugated to insulins. While macrocycle binding sites are occupied by glucose mimetics, bound insulins exist in an inactive conformation [155,156]. Under hyperglycemic conditions, glucose molecules replace glucose mimetics in the macrocycle binding sites through competitive binding, inducing a conformational change in the structure and activating the bound insulin to stimulate glucose uptake [155,156].

8.6 Concluding Thoughts

It is no great exaggeration to state that as a therapeutic entity, insulin has done as much for human health in terms of prolonging and saving lives as virtually any other drug. Much has been accomplished since insulin's discovery: its purification, structural characterization, recombinant production, and elucidation of the signaling mechanism and physiological functions, to name just a few milestones. At this point, we should ask what remains to be done and what specifically are the goals for the next century.

In the near term, several developments promise to improve upon the shortcomings of current insulin therapy. Despite decades of work, the approved insulins are differentiated only by their time action, ranging from prandial to intermediate, to basal. While used as monotherapy or in a combination role as part of a basal-bolus regimen, insulin has greatly improved diabetic outcomes. Nonetheless, the performance falls short of healthy physiological glucose control leaving patients to experience both hyper- and hypoglycemic episodes. The introduction of the two advanced once-weekly insulins, icodec, and efsitora alfa should bring us even closer to achieving near-normal glucose control. Advances in supportive technologies that integrate CGM with pens, pumps, and other automated delivery devices will further improve and simplify T1D and T2D disease management. Ultimately, precise glucose control may be achieved by so-called closed-loop systems capable of delivering precise doses of insulin or glucagon in response to glucose fluctuations.

Notwithstanding the promise of the above technologies, insulin as a chemical entity is almost certain to remain the central focus of innovation and the source of transformative peptide therapies. The next generation of insulins could, for example, include analogs, which function in a glucose-responsive manner similar to that of GLP-1 drugs in T2D patients. We can also envision analogs with a wider therapeutic index, or those that are hepato-selective, or those of higher potency that could substantially reduce production costs. These novel insulins will require further exploration of the insulin structure–activity relationships, much of which will be driven by the molecular insights gained from cryo-EM-based analysis of interaction with its receptor. In addition, we must consider the pragmatic aspects in structural complexity, as large-scale production may not be accessible through existing rDNA methods (such as yeast or *Escherichia coli*) and may require considerable chemical modification or total chemical synthesis.

The past century of insulin research has brought revolutionary advances in science and medicine that have provided enormous benefits to patients. This century is poised to deliver further advances and potentially a transformative leap in the management of insulin-dependent diabetes.

References

1 Rostène, W. and De Meyts, P. Insulin: a 100-year-old discovery with a fascinating history. *Endocr Rev*. 2021;42(5):503–527.
2 Flier, J.S. and Kahn, C.R. Insulin: a pacesetter for the shape of modern biomedical science and the Nobel Prize. *Mol Metab*. 2021;52:101194.
3 Nathan, D.M. Realising the long-term promise of insulin therapy: the DCCT/EDIC study. *Diabetologia*. 2021;64(5):1049–1058.
4 The ORIGIN Trial Investigators. Basal insulin and cardiovascular and other outcomes in dysglycemia. *N Engl J Med*. 2012;367(4):319–328.
5 Stratton, I.M. Association of glycaemia with macrovascular and microvascular complications of type 2 diabetes (UKPDS 35): prospective observational study. *BMJ*. 2000;321(7258):405–412.
6 Holman, R.R., Paul, S.K., Bethel, M.A., Matthews, D.R., and Neil, H.A.W. 10-year follow-up of intensive glucose control in type 2 diabetes. *N Engl J Med*. 2008;359(15):1577–1589.

7 Pop-Busui, R., Boulton, A.J.M., Feldman, E.L., et al. Diabetic neuropathy: a position statement by the American Diabetes Association. *Diabetes Care*. 2017;40(1):136–154.
8 Kilpatrick, E.S., Rigby, A.S., and Atkin, S.L. A1C variability and the risk of microvascular complications in type 1 diabetes. *Diabetes Care*. 2008;31(11):2198–2202.
9 Mathieu, C. Minimising hypoglycaemia in the real world: the challenge of insulin. *Diabetologia*. 2021;64(5):978–984.
10 Khunti, K., Davies, M., Majeed, A., Thorsted, B.L., Wolden, M.L., and Paul, S.K. Hypoglycemia and risk of cardiovascular disease and all-cause mortality in insulin-treated people with type 1 and type 2 diabetes: a cohort study. *Diabetes Care*. 2015;38(2):316–322.
11 Warren, M. and Steel, D. Clinical use of IDegLira: initiation to titration after basal insulin. *Clin Diabetes*. 2020;38(1):62–70.
12 Anderson, J. Improved mealtime treatment of diabetes mellitus using an insulin analogue. *Clin Ther*. 1997;19(1):62–72.
13 Anderson, R.A., Cheng, N., Bryden, N.A., et al. Elevated intakes of supplemental chromium improve glucose and insulin variables in individuals with type 2 diabetes. *Diabetes*. 1997;46(11):1786–1791.
14 Anderson, J.H. Mealtime treatment with insulin analog improves postprandial hyperglycemia and hypoglycemia in patients with non-insulin-dependent diabetes mellitus. *Arch Intern Med*. 1997;157(11):1249.
15 Owens, D.R., Landgraf, W., Frier, B.M., et al. Commencing insulin glargine 100 U/mL therapy in individuals with type 2 diabetes: determinants of achievement of HbA1c goal less than 7.0%. *Diabetes Obes Metab*. 2019;21(2):321–329.
16 Davis, A., Kuriakose, J., and Clements, J.N. Faster insulin aspart: a new bolus option for diabetes mellitus. *Clin Pharmacokinet*. 2019;58(4):421–430.
17 Evans, M., Wilkinson, M., and Giannpolou, A. Fast-acting insulin aspart: the rationale for a new mealtime insulin. *Diabetes Ther*. 2019;10(5):1793–1800.
18 Mathieu, C., Bode, B.W., Franek, E., et al. Efficacy and safety of fast-acting insulin aspart in comparison with insulin aspart in type 1 diabetes (onset 1): a 52-week, randomized, treat-to-target, phase III trial. *Diabetes Obes Metab*. 2018;20(5):1148–1155.
19 Heise, T., Pieber, T.R., Danne, T., Erichsen, L., and Haahr, H. A pooled analysis of clinical pharmacology trials investigating the pharmacokinetic and pharmacodynamic characteristics of fast-acting insulin aspart in adults with type 1 diabetes. *Clin Pharmacokinet*. 2017;56(5):551–559.
20 Bowering, K., Harvey, J., Kolaczynski, J.W., Snyder, J.W., and Bode, B.W. Mealtime fast-acting insulin aspart versus insulin aspart for controlling postprandial hyperglycaemia in people with insulin-resistant type 2 diabetes. *Diabet Med*. 2019;36(6):771–775.

21 Klaff, L., Cao, D., Dellva, M.A., et al. Ultra rapid lispro improves postprandial glucose control compared with lispro in patients with type 1 diabetes: results from the 26-week PRONTO-T1D study. *Diabetes Obes Metab*. 2020;22(10):1799–1807.

22 Leohr, J., Dellva, M.A., Coutant, D.E., et al. Pharmacokinetics and glucodynamics of ultra rapid lispro (URLi) versus Humalog® (Lispro) in patients with type 2 diabetes mellitus: a phase I randomised, crossover study. *Clin Pharmacokinet*. 2020;59(12):1601–1610.

23 Andersen, G., Meiffren, G., Lamers, D., et al. Ultra-rapid BioChaperone Lispro improves postprandial blood glucose excursions vs insulin lispro in a 14-day crossover treatment study in people with type 1 diabetes. *Diabetes Obes Metab*. 2018;20(11):2627–2632.

24 Heise, T., Meiffren, G., Alluis, B., et al. BioChaperone Lispro versus faster aspart and insulin aspart in patients with type 1 diabetes using continuous subcutaneous insulin infusion: a randomized euglycemic clamp study. *Diabetes Obes Metab*. 2019;21(4):1066–1070.

25 Meiffren, G., Herbrand, T., Anastassiadis, E., et al. Better glycaemic control with BioChaperone glargine lispro co-formulation than with insulin lispro Mix25 or separate glargine and lispro administrations after a test meal in people with type 2 diabetes. *Diabetes Obes Metab*. 2019;21(7):1570–1575.

26 Xiong, X., Menting, J.G., Disotuar, M.M., et al. A structurally minimized yet fully active insulin based on cone-snail venom insulin principles. *Nat Struct Mol Biol*. 2020;27(7):615–624.

27 Lovre, D. and Fonseca, V. Benefits of timely basal insulin control in patients with type 2 diabetes. *J Diabetes Complications*. 2015;29(2):295–301.

28 Hanefeld, M., Monnier, L., Schnell, O., and Owens, D. Early treatment with basal insulin glargine in people with type 2 diabetes: lessons from ORIGIN and other cardiovascular trials. *Diabetes Ther*. 2016;7(2):187–201.

29 Inzucchi, S.E., Bergenstal, R.M., Buse, J.B., et al. Management of hyperglycaemia in type 2 diabetes: a patient-centered approach. Position statement of the American Diabetes Association (ADA) and the European Association for the Study of Diabetes (EASD). *Diabetologia*. 2012;55(6):1577–1596.

30 Klein, O., Lynge, J., Endahl, L., Damholt, B., Nosek, L., and Heise, T. Albumin-bound basal insulin analogues (insulin detemir and NN344): comparable time-action profiles but less variability than insulin glargine in type 2 diabetes. *Diabetes Obes Metab*. 2007;9(3):290–299.

31 Porcellati, F., Rossetti, P., Ricci, N.B., et al. Pharmacokinetics and pharmacodynamics of the long-acting insulin analog glargine after 1 week of use compared with its first administration in subjects with type 1 diabetes. *Diabetes Care*. 2007;30(5):1261–1263.

32 Kurtzhals, P., Havelund, S., Jonassen, I., et al. Albumin binding of insulins acylated with fatty acids: characterization of the ligand-protein interaction and correlation between binding affinity and timing of the insulin effect *in vivo*. *Biochem J.* 1995;312(3):725–731.

33 Heise, T. and Pieber TR. Towards peakless, reproducible and long-acting insulins. An assessment of the basal analogues based on isoglycaemic clamp studies. *Diabetes Obes Metab.* 2007;9(5):648–659.

34 Porcellati, F., Rossetti, P., Busciantella, N.R., et al. Comparison of pharmacokinetics and dynamics of the long-acting insulin analogs glargine and detemir at steady state in type 1 diabetes. *Diabetes Care.* 2007;30(10):2447–2452.

35 Heller, S., Koenen, C., and Bode, B. Comparison of insulin detemir and insulin glargine in a basal—bolus regimen, with insulin aspart as the mealtime insulin, in patients with type 1 diabetes: a 52-week, multinational, randomized, open-label, parallel-group, treat-to-target noninferiority trial. *Clin Ther.* 2009;31(10):2086–2097.

36 Goldman, J. and White, J.R. Advances in basal insulin therapy. *J Pharm Technol.* 2016;32(6):260–268.

37 Becker, R.H.A., Dahmen, R., Bergmann, K., Lehmann, A., Jax, T., and Heise, T. New insulin glargine 300 units·mL^{-1} provides a more even activity profile and prolonged glycemic control at steady state compared with insulin glargine 100 units·mL^{-1}. *Diabetes Care.* 2015;38(4):637–643.

38 Riddle, M.C., Bolli, G.B., Ziemen, M., Muehlen-Bartmer, I., Bizet, F., and Home, P.D. New insulin glargine 300 units/mL versus glargine 100 units/mL in people with type 2 diabetes using basal and mealtime insulin: glucose control and hypoglycemia in a 6-month randomized controlled trial (EDITION 1). *Diabetes Care.* 2014;37(10):2755–2762.

39 Yki-Järvinen, H., Bergenstal, R.M., Bolli, G.B., et al. Glycaemic control and hypoglycaemia with new insulin glargine 300 U/ml versus insulin glargine 100 U/ml in people with type 2 diabetes using basal insulin and oral antihyperglycaemic drugs: the EDITION 2 randomized 12-month trial including 6-month extension. *Diabetes Obes Metab.* 2015;17(12):1142–1149.

40 Danne, T., Matsuhisa, M., Sussebach, C., et al. Lower risk of severe hypoglycaemia with insulin glargine 300 U/mL versus glargine 100 U/mL in participants with type 1 diabetes: a meta-analysis of 6-month phase 3 clinical trials. *Diabetes Obes Metab.* 2020;22(10):1880–1885.

41 Jonassen, I., Havelund, S., Hoeg-Jensen, T., Steensgaard, D.B., Wahlund, P.O., and Ribel, U. Design of the novel protraction mechanism of insulin degludec, an ultra-long-acting basal insulin. *Pharm Res.* 2012;29(8):2104–2114.

42 Heise, T., Nosek, L., Bøttcher, S.G., Hastrup, H., and Haahr, H. Ultra-long-acting insulin degludec has a flat and stable glucose-lowering effect in type 2 diabetes. *Diabetes Obes Metab.* 2012;14(10):944–950.

43 Rosenstock, J., Cheng, A., Ritzel, R., et al. More similarities than differences testing insulin glargine 300 units/mL versus insulin degludec 100 units/mL in insulin-naive type 2 diabetes: the randomized head-to-head BRIGHT trial. *Diabetes Care*. 2018;41(10):2147–2154.

44 Bailey, T.S., Pettus, J., Roussel, R., et al. Morning administration of 0.4 U/kg/day insulin glargine 300 U/mL provides less fluctuating 24-hour pharmacodynamics and more even pharmacokinetic profiles compared with insulin degludec 100 U/mL in type 1 diabetes. *Diabetes Metab*. 2018;44(1):15–21.

45 Heise, T., Nørskov, M., Nosek, L., Kaplan, K., Famulla, S., and Haahr, H.L. Insulin degludec: Lower day-to-day and within-day variability in pharmacodynamic response compared with insulin glargine 300 U/mL in type 1 diabetes. *Diabetes Obes Metab*. 2017;19(7):1032–1039.

46 On behalf of the CONCLUDE Study Group, Philis-Tsimikas, A., Klonoff, D.C., et al. Risk of hypoglycaemia with insulin degludec versus insulin glargine U300 in insulin-treated patients with type 2 diabetes: the randomised, head-to-head CONCLUDE trial. *Diabetologia*. 2020;63(4):698–710.

47 Rosenstock, J. and Del Prato, S. Basal weekly insulins: the way of the future! *Metabolism*. 2022;126:154924.

48 Kjeldsen, T.B., Hubálek, F., Hjørringgaard, C.U., et al. Molecular engineering of insulin icodec, the first acylated insulin analog for once-weekly administration in humans. *J Med Chem*. 2021;64(13):8942–8950.

49 Pieber, T.R., Asong, M., Fluhr, G., et al. Pharmacokinetic and pharmacodynamic properties of once-weekly insulin icodec in individuals with type 2 diabetes. *Diabetes Obes Metab*. 2023;25(12):3716–3723.

50 Philis-Tsimikas, A., Bajaj, H.S., Begtrup, K., et al. Rationale and design of the phase 3a development programme (ONWARDS 1–6 trials) investigating once-weekly insulin icodec in diabetes. *Diabetes Obes Metab*. 2023;25(2): 331–341.

51 Rosenstock, J., Bain, S.C., Gowda, A., et al. Weekly icodec versus daily glargine U100 in type 2 diabetes without previous insulin. *N Engl J Med*. 2023;389(4):297–308.

52 Lingvay, I., Asong, M., Desouza, C., et al. Once-weekly insulin icodec vs once-daily insulin degludec in adults with insulin-naive type 2 diabetes: the ONWARDS 3 randomized clinical trial. *JAMA*. 2023;330(3): 228.

53 Bajaj, H.S., Aberle, J., Davies, M., et al. Once-weekly insulin icodec with dosing guide app versus once-daily basal insulin analogues in insulin-naive type 2 diabetes (ONWARDS 5): a randomized trial. *Ann Intern Med*. 2023;176(11):1476–1485.

54 Philis-Tsimikas, A., Asong, M., Franek, E., et al. Switching to once-weekly insulin icodec versus once-daily insulin degludec in individuals with basal insulin-treated type 2 diabetes (ONWARDS 2): a phase 3a, randomised,

open label, multicentre, treat-to-target trial. *Lancet Diabetes Endocrinol.* 2023;11(6):414–425.

55 Mathieu, C., Ásbjörnsdóttir, B., Bajaj, H.S., et al. Switching to once-weekly insulin icodec versus once-daily insulin glargine U100 in individuals with basal-bolus insulin-treated type 2 diabetes (ONWARDS 4): a phase 3a, randomised, open-label, multicentre, treat-to-target, non-inferiority trial. *Lancet.* 2023;401(10392):1929–1940.

56 Heise, T., Chien, J., Beals, J.M., et al. Pharmacokinetic and pharmacodynamic properties of the novel basal insulin Fc (insulin efsitora alfa), an insulin fusion protein in development for once-weekly dosing for the treatment of patients with diabetes. *Diabetes Obes Metab.* 2023;25(4):1080–1090.

57 Frias, J., Chien, J., Zhang, Q., et al. Safety and efficacy of once-weekly basal insulin Fc in people with type 2 diabetes previously treated with basal insulin: a multicentre, open-label, randomised, phase 2 study. *Lancet Diabetes Endocrinol.* 2023;11(3):158–168.

58 Bue-Valleskey, J.M., Kazda, C.M., Ma, C., et al. Once-weekly basal insulin Fc demonstrated similar glycemic control to once-daily insulin degludec in insulin-naive patients with type 2 diabetes: a phase 2 randomized control trial. *Diabetes Care.* 2023;46(5):1060–1067.

59 Kazda, C.M., Bue-Valleskey, J.M., Chien, J., et al. Novel once-weekly basal insulin Fc achieved similar glycemic control with a safety profile comparable to insulin degludec in patients with type 1 diabetes. *Diabetes Care.* 2023;46(5):1052–1059.

60 Lawrence, M.C. Understanding insulin and its receptor from their three-dimensional structures. *Mol Metab.* 2021;52:101255.

61 McKern, N.M., Lawrence, M.C., Streltsov, V.A., et al. Structure of the insulin receptor ectodomain reveals a folded-over conformation. *Nature.* 2006;443(7108):218–221.

62 Menting, J.G., Whittaker, J., Margetts, M.B., et al. How insulin engages its primary binding site on the insulin receptor. *Nature.* 2013;493(7431):241–245.

63 Gutmann, T., Kim, K.H., Grzybek, M., Walz, T., and Coskun, Ü. Visualization of ligand-induced transmembrane signaling in the full-length human insulin receptor. *J Cell Biol.* 2018;217(5):1643–1649.

64 Gutmann, T., Schäfer, I.B., Poojari, C., et al. Cryo-EM structure of the complete and ligand-saturated insulin receptor ectodomain. *J Cell Biol.* 2020;219(1): e201907210.

65 Scapin, G., Dandey, V.P., Zhang, Z., et al. Structure of the insulin receptor–insulin complex by single-particle cryo-EM analysis. *Nature.* 2018;556(7699):122–125.

66 Uchikawa, E., Choi, E., Shang, G., Yu, H., and Bai, X.C. Activation mechanism of the insulin receptor revealed by cryo-EM structure of the fully liganded receptor–ligand complex. *eLife.* 2019;8:e48630.

67 Weis, F., Menting, J.G., Margetts, M.B., et al. The signalling conformation of the insulin receptor ectodomain. *Nat Commun.* 2018;9(1):4420.

68 Jie, L., Junhee, P., John, P.M., et al. Synergistic activation of the insulin receptor via two distinct sites. *Nat Struct Mol Biol.* 2022;29(4):357–368.

69 Moroder L, Musiol H. Insulin—from its discovery to the industrial synthesis of modern insulin analogues. Angew Chem Int Ed. 2017;56(36):10656–10669.

70 Thalluri, K., Kou, B., Yang, X., et al. Synthesis of relaxin-2 and insulin-like peptide 5 enabled by novel tethering and traceless chemical excision. *J Pept Sci.* 2017;23(6):455–465.

71 Altindis, E., Cai, W., Sakaguchi, M., et al. Viral insulin-like peptides activate human insulin and IGF-1 receptor signaling: a paradigm shift for host–microbe interactions. *Proc Natl Acad Sci.* 2018;115(10):2461–2466.

72 Safavi-Hemami, H., Gajewiak, J., Karanth, S., et al. Specialized insulin is used for chemical warfare by fish-hunting cone snails. *Proc Natl Acad Sci.* 2015;112(6):1743–1748.

73 Karas, J.A., Wade, J.D., and Hossain, M.A. The chemical synthesis of insulin: an enduring challenge. *Chem Rev.* 2021;121(8):4531–4560.

74 Brunel, F.M., Mayer, J.P., and DiMarchi, R.D. Recent advances in the chemical synthesis of insulin and related peptides. *Future Med Chem.* 2020;12(8):649–654.

75 Liu, F., Li, P., Gelfanov, V., Mayer, J., and DiMarchi, R. Synthetic advances in insulin-like peptides enable novel bioactivity. *Acc Chem Res.* 2017;50(8):1855–1865.

76 Ryle, A.P., Sanger, F., Smith, L.F., and Kitai, R. The disulphide bonds of insulin. *Biochem J.* 1955;60(4):541–556.

77 Meienhofer, J., Schnabel, E., Bremer, H., et al. Synthesis of insulin chains and their combination to insulin-active preparations. *Z Naturforsch B.* 1963;18:1120–1121.

78 Katsoyannis, P.G., Tometsko, A.M., Zalut, C., and Fukuda, K. Insulin peptides. XV. The synthesis of the A chain of sheep insulin and its combination with synthetic or natural B chain to produce insulin[1,2]. *J Am Chem Soc.* 1966;88(23):5625–5635.

79 Kung, Y.T., Du, Y.C., Huang, W.T., Chen, C.C., and Ke, L.T. Total synthesis of crystalline bovine insulin. *Sci Sin.* 1965;14(11):1710–1716.

80 Marglin, B. and Merrifield, R.B. The synthesis of bovine insulin by the solid phase method[1]. *J Am Chem Soc.* 1966;88(21):5051–5052.

81 Sieber, P., Kamber, B., Hartmann, A., Jöhl, A., Riniker, B., and Rittel, W. Totalsynthese von humaninsulin unter gezielter bildung der disulfidbindungen. vorläufige mitteilung. *Helv Chim Acta.* 1974;57(8):2617–2621.

82 Johnson, I.S. Human insulin from recombinant DNA technology. *Science.* 1983;219(4585):632–637.

83 Inouye, K., Watanabe, K., Morihara, K., et al. Enzyme-assisted semisynthesis of human insulin. *J Am Chem Soc.* 1979;101(3):751–752.

84 Morihara, K., Oka, T., and Tsuzuki, H. Semi-synthesis of human insulin by trypsin-catalysed replacement of Ala-B30 by Thr in porcine insulin. *Nature*. 1979;280(5721):412–413.
85 Markussen, J., Damgaard, U., Pingel, M., Snel, L., Sorensen, A.R., and Sorensen, E. Human insulin (Novo): chemistry and characteristics. *Diabetes Care*. 1983; Suppl 1:4–8.
86 Akaji, K., Fujino, K., Tatsumi, T., and Kiso, Y. Total synthesis of human insulin by regioselective disulfide formation using the silyl chloride-sulfoxide method. *J Am Chem Soc*. 1993;115(24):11384–11392.
87 Liu, F., Luo, E.Y., Flora, D.B., and Mayer, J.P. Concise synthetic routes to human insulin. *Org Lett*. 2013;15(4):960–963.
88 Tofteng, A.P., Jensen, K.J., Schäffer, L., and Hoeg-Jensen, T. Total synthesis of desB30 insulin analogues by biomimetic folding of single-chain precursors. *ChemBioChem*. 2008;9(18):2989–2996.
89 Hossain, M.A., Belgi, A., Lin, F., et al. Use of a temporary "solubilizing" peptide tag for the Fmoc solid-phase synthesis of human insulin glargine via use of regioselective disulfide bond formation. *Bioconjug Chem*. 2009;20(7):1390–1396.
90 Kwak, S.Y., Forbes, B.E., Lee, Y.S., Belgi, A., Wade, J.D., and Hossain, M.A. Solid phase synthesis of an analogue of insulin, A0:R glargine, that exhibits decreased mitogenic activity. *Int J Pept Res Ther*. 2010;16(3):153–158.
91 Zaykov, A.N., Mayer, J.P., Gelfanov, V.M., and DiMarchi, R.D. Chemical synthesis of insulin analogs through a novel precursor. *ACS Chem Biol*. 2014;9(3):683–691.
92 Williams, G.M., Lee, K., Li, X., Cooper, G.J.S., and Brimble, M.A. Replacement of the CysA7–CysB7 disulfide bond with a 1,2,3-triazole linker causes unfolding in insulin glargine. *Org Biomol Chem*. 2015;13(13):4059–4063.
93 Disotuar, M.M., Petersen, M.E., Nogueira, J.M., Kay, M.S., and Chou, D.H.C. Synthesis of hydrophobic insulin-based peptides using a helping hand strategy. *Org Biomol Chem*. 2019;17(7):1703–1708.
94 Baldauf, S., Ogunkoya, A.O., Boross, G.N., and Bode, J.W. Aspartic acid forming α-ketoacid-hydroxylamine (KAHA) ligations with (S)-4,4-difluoro-5-oxaproline. *J Org Chem*. 2020;85(3):1352–1364.
95 Liu, F., Luo, E.Y., Flora, D.B., and Mezo, A.R. A synthetic route to human insulin using isoacyl peptides. *Angew Chem Int Ed*. 2014;53(15):3983–3987.
96 Karas, J.A., Scanlon, D.B., Forbes, B.E., et al. 2-Nitroveratryl as a photocleavable thiol-protecting group for directed disulfide bond formation in the chemical synthesis of insulin. *Chem – Eur J*. 2014;20(31):9549–9552.
97 Patil, N.A., Karas, J.A., Wade, J.D., Hossain, M.A., and Tailhades, J. Rapid photolysis-mediated folding of disulfide-rich peptides. *Chem – Eur J*. 2019;25(36):8599–8603.

98 Wu, F., Mayer, J.P., Gelfanov, V.M., Liu, F., and DiMarchi, R.D. Synthesis of four-disulfide insulin analogs via sequential disulfide bond formation. *J Org Chem*. 2017;82(7):3506–3512.

99 Thalluri, K., Mayer, J.P., Chabenne, J.R., Gelfanov, V., and DiMarchi, R.D. Synthesis of disulfide-rich heterodimeric peptides through an auxiliary N, N-crosslink. *Commun Chem*. 2018;1(1):36.

100 Busse, W.D. and Carpenter, F.H. Synthesis and properties of carbonylbis(methionyl)insulin, a proinsulin analog which is convertible to insulin by cyanogen bromide cleavage. *Biochemistry*. 1976;15(8):1649–1657.

101 Sohma, Y. and Kent, S.B.H. Biomimetic synthesis of lispro insulin via a chemically synthesized "mini-proinsulin" prepared by oxime-forming ligation. *J Am Chem Soc*. 2009;131(44):16313–16318.

102 Sohma, Y., Hua, Q., Whittaker, J., Weiss, M.A., and Kent, S.B.H. Design and folding of [Glu^{A4} (Oβ Thr^{B30})]insulin ("ester insulin"): a minimal proinsulin surrogate that can be chemically converted into human insulin. *Angew Chem Int Ed*. 2010;49(32):5489–5493.

103 Dhayalan, B., Mandal, K., Rege, N., et al. Scope and limitations of Fmoc chemistry SPPS-based approaches to the total synthesis of insulin lispro via ester insulin. *Chem – Eur J*. 2017;23(7):1709–1716.

104 Boross, G.N., Shimura, S., Besenius, M., et al. Facile folding of insulin variants bearing a prosthetic C-peptide prepared by α-ketoacid-hydroxylamine (KAHA) ligation. *Chem Sci*. 2018;9(44):8388–8395.

105 Boross, G.N., Schauenburg, D., and Bode, J.W. Chemoselective derivatization of folded synthetic insulin variants with potassium acyltrifluoroborates (KATs). *Helv Chim Acta*. 2019;102(2):e1800214.

106 Thalluri, K., Kou, B., Gelfanov, V., Mayer, J.P., Liu, F., and DiMarchi, R.D. Biomimetic synthesis of insulin enabled by oxime ligation and traceless "C-peptide" chemical excision. *Org Lett*. 2017;19(3):706–709.

107 Karas, J.A., Patil, N.A., Tailhades, J., et al. Total chemical synthesis of an intra-A-chain cystathionine human insulin analogue with enhanced thermal stability. *Angew Chem Int Ed*. 2016;55(47):14743–14747.

108 Zheng, N., Karra, P., VandenBerg, M.A., et al. Synthesis and characterization of an A6-A11 methylene thioacetal human insulin analogue with enhanced stability. *J Med Chem*. 2019;62(24):11437–11443.

109 Van Lierop, B., Ong, S.C., Belgi, A., et al. Insulin in motion: the A6-A11 disulfide bond allosterically modulates structural transitions required for insulin activity. *Sci Rep*. 2017;7(1):17239.

110 Ong, S.C., Belgi, A., Van Lierop, B., et al. Probing the correlation between insulin activity and structural stability through introduction of the rigid A6–A11 bond. *J Biol Chem*. 2018;293(30):11928–11943.

111 Weil-Ktorza, O., Rege, N., Lansky, S., et al. Substitution of an internal disulfide bridge with a diselenide enhances both foldability and stability of human insulin. *Chem – Eur J*. 2019;25(36):8513–8521.

112 Vinther, T.N., Norrman, M., Ribel, U., et al. Insulin analog with additional disulfide bond has increased stability and preserved activity. *Protein Sci*. 2013;22(3):296–305.

113 Brunel, F.M., Mayer, J.P., Gelfanov, V.M., et al. A disulfide scan of insulin by [3 + 1] methodology exhibits site-specific influence on bioactivity. *ACS Chem Biol*. 2019;14(8):1829–1835.

114 Xiong, X., Blakely, A., Karra, P., et al. Novel four-disulfide insulin analog with high aggregation stability and potency. *Chem Sci*. 2020;11(1):195–200.

115 Hossain, M.A., Okamoto, R., Karas, J.A., et al. Total chemical synthesis of a nonfibrillating human glycoinsulin. *J Am Chem Soc*. 2020;142(3): 1164–1169.

116 Kabotso, D.E.K., Smiley, D., Mayer, J.P., et al. Addition of sialic acid to insulin confers superior physical properties and bioequivalence. *J Med Chem*. 2020;63(11):6134–6143.

117 Clarke, S.F. and Foster, J.R. A history of blood glucose meters and their role in self-monitoring of diabetes mellitus. *Br J Biomed Sci*. 2012;69(2):83–93.

118 Galindo, R.J. and Aleppo, G. Continuous glucose monitoring: the achievement of 100 years of innovation in diabetes technology. *Diabetes Res Clin Pract*. 2020;170:108502.

119 The International Hypoglycaemia Study Group. Glucose concentrations of less than 3.0 mmol/l (54 mg/dl) should be reported in clinical trials: a joint position statement of the American Diabetes Association and the European Association for the Study of Diabetes. *Diabetologia*. 2017;60(1):3–6.

120 Charleer, S., Mathieu, C., Nobels, F., et al. Effect of continuous glucose monitoring on glycemic control, acute admissions, and quality of life: a real-world study. *J Clin Endocrinol Metab*. 2018;103(3):1224–1232.

121 Beck, R.W., Riddlesworth, T., Ruedy, K., et al. Effect of continuous glucose monitoring on glycemic control in adults with type 1 diabetes using insulin injections: the DIAMOND randomized clinical trial. *JAMA*. 2017; 317(4):371.

122 Hoeg-Jensen, T. Review: glucose-sensitive insulin. *Mol Metab*. 2021;46:101107.

123 Disotuar, M.M., Chen, D., Lin, N.P., and Chou, D.H.C. Glucose-responsive insulin through bioconjugation approaches. *J Diabetes Sci Technol*. 2020;14(2):198–203.

124 Rege, N.K., Phillips, N.F.B., and Weiss, M.A. Development of glucose-responsive 'smart' insulin systems. *Curr Opin Endocrinol Diabetes Obes*. 2017;24(4):267–278.

125 Wang, Z., Wang, J., Kahkoska, A.R., Buse, J.B., and Gu, Z. Developing insulin delivery devices with glucose responsiveness. *Trends Pharmacol Sci.* 2020;42(1):31–44.
126 Fuchs, S., Shariati, K., and Ma, M. Stimuli-responsive insulin delivery devices. *Pharm Res.* 2020;37(10):202.
127 Matsumoto, A., Tanaka, M., Matsumoto, H., et al. Synthetic "smart gel" provides glucose-responsive insulin delivery in diabetic mice. *Sci Adv.* 2017;3(11): eaaq0723.
128 Brownlee, M. and Cerami, A. Glycosylated insulin complexed to concanavalin A: biochemical basis for a closed-loop insulin delivery system. *Diabetes.* 1983;32(6):499–504.
129 Cavada, B., Pinto-Junior, V., Osterne, V., and Nascimento, K. ConA-like lectins: high similarity proteins as models to study structure/biological activities relationships. *Int J Mol Sci.* 2018;20(1):30.
130 Lin, K., Yi, J., Mao, X., Wu, H., Zhang, L.M., and Yang, L. Glucose-sensitive hydrogels from covalently modified carboxylated pullulan and concanavalin A for smart controlled release of insulin. *React Funct Polym.* 2019;139: 112–119.
131 Webber, M.J., Anderson, D.G., and Langer, R. Engineering synthetically modified insulin for glucose-responsive diabetes therapy. *Expert Rev Endocrinol Metab.* 2015;10(5):483–489.
132 Yin, R., Tong, Z., Yang, D., and Nie, J. Glucose and pH dual-responsive concanavalin A based microhydrogels for insulin delivery. *Int J Biol Macromol.* 2011;49(5):1137–1142.
133 Kim, J.J. and Park, K. Modulated insulin delivery from glucose-sensitive hydrogel dosage forms. *J Control Release.* 2001;77(1–2):39–47.
134 Yin, R., Bai, M., He, J., Nie, J., and Zhang, W. Concanavalin A-sugar affinity based system: binding interactions, principle of glucose-responsiveness, and modulated insulin release for diabetes care. *Int J Biol Macromol.* 2019;124:724–732.
135 Hadden, J.W., Hadden, E.M., Sadlik, J.R., and Coffey, R.G. Effects of concanavalin A and a succinylated derivative on lymphocyte proliferation and cyclic nucleotide levels. *Proc Natl Acad Sci U S A.* 1976;73(5): 1717–1721.
136 Wohlfahrt, G., Witt, S., Hendle, J., Schomburg, D., Kalisz, H.M., and Hecht, H.J. 1.8 and 1.9 Å resolution structures of the *Penicillium amagasakiense* and *Aspergillus niger* glucose oxidases as a basis for modelling substrate complexes. *Acta Crystallogr D Biol Crystallogr.* 1999;55(5):969–977.
137 Weibel, M.K. and Bright, H.J. The glucose oxidase mechanism. *J Biol Chem.* 1971;246(9):2734–2744.

138 Wong, C.M., Wong, K.H., and Chen, X.D. Glucose oxidase: natural occurrence, function, properties and industrial applications. *Appl Microbiol Biotechnol.* 2008;78(6):927–938.

139 Fuchs, S., Ernst, A.U., Wang, L.H., et al. Hydrogels in emerging technologies for type 1 diabetes. *Chem Rev.* 2021;121(18):11458–11526.

140 Rizwan, M., Yahya, R., Hassan, A., et al. pH sensitive hydrogels in drug delivery: brief history, properties, swelling, and release mechanism, material selection and applications. *Polymers.* 2017;9(4):137.

141 Jana, B.A., Shinde, U., and Wadhwani, A. Synthetic enzyme-based nanoparticles act as smart catalyst for glucose responsive release of insulin. *J Biotechnol.* 2020;324:1–6.

142 Zhao, L., Wang, L., Zhang, Y., et al. Glucose oxidase-based glucose-sensitive drug delivery for diabetes treatment. *Polymers.* 2017;9(12):255.

143 Zhou, X., Wu, H., Long, R., et al. Oral delivery of insulin with intelligent glucose-responsive switch for blood glucose regulation. *J Nanobiotechnol.* 2020;18(1):96.

144 Yu, J., Qian, C., Zhang, Y., et al. Hypoxia and H_2O_2 dual-sensitive vesicles for enhanced glucose-responsive insulin delivery. *Nano Lett.* 2017;17(2):733–739.

145 Cambre, J.N. and Sumerlin, B.S. Biomedical applications of boronic acid polymers. *Polymer.* 2011;52(21):4631–4643.

146 Wang, R., Bian, Z., Zhan, D., Wu, Z., Yao, Q., and Zhang, G. Boronic acid-based sensors for small-molecule reactive species: a review. *Dyes Pigments.* 2021;185 (PA):108885.

147 Springsteen, G. and Wang, B. A detailed examination of boronic acid-diol complexation. *Tetrahedron.* 2002;58:5291–5300.

148 Huang, Q., Wang, L., Yu, H., and Ur-Rahman, K. Advances in phenylboronic acid-based closed-loop smart drug delivery system for diabetic therapy. *J Control Release.* 2019;305:50–64.

149 Wang, J., Wang, Z., Yu, J., Kahkoska, A.R., Buse, J.B., and Gu, Z. Glucose-responsive insulin and delivery systems: innovation and translation. *Adv Mater.* 2020;32(13):1–19.

150 Hoeg-Jensen, T., Ridderberg, S., Havelund, S., et al. Insulins with built-in glucose sensors for glucose responsive insulin release. *J Pept Sci.* 2005;11(6):339–346.

151 Chou, D.H.C., Webber, M.J., Tang, B.C., et al. Glucose-responsive insulin activity by covalent modification with aliphatic phenylboronic acid conjugates. *Proc Natl Acad Sci U S A.* 2015;112(8):2401–2406.

152 Visser, S.A.G., Kandala, B., Fancourt, C., Krug, A.W., and Cho, C.R. A model-informed drug discovery and development strategy for the novel glucose-responsive insulin MK-2640 enabled rapid decision making. *Clin Pharmacol Ther.* 2020;107(6):1296–311.

153 Tromans, R.A., Carter, T.S., Chabanne, L., et al. A biomimetic receptor for glucose. *Nat Chem*. 2019;11(1):52–56.

154 Tromans, R.A., Samanta, S.K., Chapman, A.M, and Davis, A.P. Selective glucose sensing in complex media using a biomimetic receptor. *Chem Sci*. 2020;11(12):3223–3227.

155 Hoeg, J., Thomas, M., Alice, R., et al. Glucose sensitive insulins and uses thereof. WO 2020/058322 A1, 2020.156. Hoeg-Jensen, T., Kruse, T., Brand, C.L. et al. Glucose-sensitive insulin with attenuation of hypoglycaemia. *Nature* 2024;634:944–951. https://doi.org/10.1038/s41586-024-08042-3.

9

Incretins and Development of Incretin-based Therapeutics

Nan Zheng, Xiao-xuan Su, Shengping Zhang and Weijun Shen

Center for Translational Research (CTR), Shenzhen Bay Laboratory, Shenzhen, China

9.1 Introduction

9.1.1 Incretin Hormones

Incretin hormones are gastrointestinal (GI) peptides that are secreted in response to nutrient intake, effectively stimulating insulin release to maintain blood glucose and energy homeostasis. Glucagon-like peptide-1 (GLP-1) and glucose-dependent insulinotropic polypeptide (GIP) are recognized as the primary incretin hormones originating from the upper and lower segments of the GI tract [1]. Given that the incretin affects over half of postprandial insulin secretion, GLP-1 and GIP have emerged as the mainstay in the development of incretin-based therapeutic agents [2].

GLP-1, identified as part of the gene sequence encoding pro-glucagon, is secreted from L cells of the lower intestine and exhibits a gradient in distribution, with lower concentrations in the duodenum and higher concentrations in the ileum, colon, rectum, pancreatic islet α cells, and neurons within the nucleus of the solitary tract. The biologically active form of GLP-1, consisting of 31 amino acids, is currently referred to as $GLP\text{-}1^{[7\text{-}37]}$ [3]. Under physiological conditions, GLP-1 plays a crucial role in maintaining normal glucose tolerance by activating specific GLP-1 receptors (GLP-1Rs), which belong to the class B G-protein-coupled receptors family. These receptors are expressed in various locations, including pancreatic islet β cells, the stomach, small intestine, mucosa, heart, and other cell types [4,5]. GLP-1R primarily signals through

Peptide Science: Chemical Ligation, Lead Generation, and Therapeutic Advances, First Edition. Fa Liu.

the $G\alpha_s$/cAMP pathway, which in turn triggers the glucose-induced secretion of insulin via voltage-dependent calcium channel-induced Ca^{2+} influx. GLP-1R activation exerts a series of biological functions, including glucose-dependent insulin secretion stimulation, glucagon release inhibition, satiety induction, and gastric emptying deceleration [1,6].

GIP, a peptide containing 42 amino acids, is produced within enteroendocrine K cells primarily located in the duodenum and jejunum. The mature GIP peptide is produced through posttranslational processing of the 153-amino acid precursor peptide, known as pro-GIP [7]. GIP functions through binding with GIPR, which is highly expressed in pancreatic β cells to increase cAMP-dependent protein kinase A (PKA) signaling [8].

In fasting human subjects, GLP-1 and GIP maintain basal plasma concentrations, typically within the low picomolar range. Upon nutrient intake, both peptides exhibit a prompt elevation in plasma levels within a few minutes, reaching their peak after approximately one hour, and then gradually reverting to basal concentrations within several hours. This concentration increase can be induced by various nutrients, including carbohydrates, triglycerides, amino acids, and some proteins [9]. The release of GLP-1 and GIP depends on a minimal rate of gastric emptying, which allows nutrients to reach the specific locations of K and L cells within the GI tract. The minimal delivery rate is typically lower for GIP, likely attributed to the direct delivery of nutrients to the L cells displaying distal regions of the gut instead of stimulating GIP-producing K cells in the proximal regions [10]. However, GLP-1 secretion from L cells initiates shortly after nutrient intake, often concurrently with GIP secretion. While the secretion mechanism is well established, a significant degree of inter-individual variability is observed in both GLP-1 and GIP secretion, which may result from other physiological factors [9,11].

9.1.2 Biological Functions of Incretin Hormones

Both GLP-1 and GIP stimulate insulin secretion in a glucose-dependent manner, which results in enhanced cAMP production and subsequent activation of the PKA pathway when binding with their receptors, respectively. However, this pathway alone cannot initiate the release of insulin secretory because the insulinotropic effects of incretin hormones always require a permissive level of hyperglycemia to initiate calcium ion influx [12]. Therefore, rather than directly stimulating insulin secretion, the primary role of incretin hormones is to enhance the insulin secretory responses initiated by hyperglycemia. GLP-1 and GIP collaboratively orchestrate the "incretin effect," characterized by an enhanced insulin secretory response upon oral glucose administration compared to intravenous delivery, even when striving to achieve identical plasma glucose concentrations [13].

Besides, incretin hormones also exert an influence on glucagon release, referred to as glucagonostatic effects. GLP-1 exhibits the capacity to suppress glucagon secretion, which leads to a reduction in hepatic glucose production, whereas GIP has been identified as a stimulator of glucagon secretion, especially when glucose concentrations are lower. Moreover, GLP-1 appears to decrease hepatic glucose output independently of plasma glucagon changes. Given that the liver seemingly lacks direct GLP-1 receptors, this modulation is likely mediated indirectly, potentially through the autonomic nervous system [14,15] (Figure 9.1).

Incretin hormones exhibit additional biological effects that contribute crucial dimensions to their comprehensive range of functions, including regulation of adipose cells, bone metabolism, and the cardiovascular system [2,16]. GLP-1 has been shown to reduce appetite and enhance feelings of satiety by activating GLP-1Rs located in the hypothalamus. Endogenous and exogenous GLP-1R agonists have been consistently shown to impede gastric emptying both at physiological and pharmacological concentrations, which leads to a diminished transfer of nutrients into the intestinal lumen, subsequently resulting in delayed and reduced nutrient absorption that manifests as less significant elevations in glycaemia and triglyceride levels following meals [2]. While a definitive physiological role for GLP-1 within the cardiovascular system remains elusive, it imparts a broad spectrum of advantageous effects in high-risk patient populations, including enhancements in cardiac blood supply, ischemia tolerance, endothelial function, attenuation of inflammatory responses, and modulation of associated cytokines [16]. In adipose tissue metabolism, GIPR is expressed within adipose

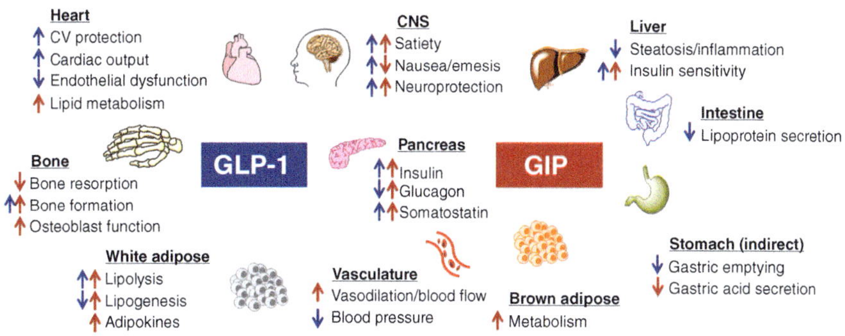

Figure 9.1 Metabolic functions of the incretin hormones GLP-1 and GIP in different organs.

tissue but not GLP-1R. GIP enhances white adipose tissue (WAT) blood flow and stimulates lipoprotein lipase to eliminate chylomicron triglycerides by releasing fatty acids. In this way, GIP primarily upregulates WAT lipid disposal and reduces inflammation [17]. On the contrary, GLP-1 stimulates brown adipose tissue thermogenesis and adipocyte browning "indirectly" through the central nervous system (CNS) hypothalamic AMP-activated protein kinase pathway independent of nutrient intake, leading to augmented energy expenditure [18]. In bone metabolism, GLP-1 improves bone metabolism by enhancing osteoblast proliferation and differentiation while decreasing the accumulation of advanced glycation end products. GIPR signaling also plays a role in stimulating bone formation and inhibiting bone resorption, affecting factors such as osteoclast number, osteoblast function, particularly in association with meal intake [19].

9.1.3 Incretin Hormones in Obesity and Type 2 Diabetes

Interest in the involvement of incretins in obesity has been piqued by the findings that bariatric surgery in obese patients leads to a substantial increase in the pattern of GI hormone secretion. This includes not only GLP-1 and GIP, but also other gut hormones produced in the lower small intestines, such as peptide tyrosine tyrosine (PYY) [20]. The incretin effect has been documented to diminish in obesity, likely due to a reduced contribution of GLP-1 to insulin secretory responses or decreased responsiveness to GIP. It has been observed that with an increase in the body mass index, there is a concurrent reduction in GLP-1 secretion levels, leading to limited benefits from GLP-1, such as reduced appetite and food intake, as well as increased satiety. Conversely, it has been demonstrated that obesity induces hypersecretion of GIP, possibly in an effort to compensate for insulin hypersecretion and mitigate the metabolic consequences of insulin resistance [21].

Type 2 diabetes (T2D) arises from a combination of insulin resistance and the incapacity of the endocrine pancreas to release sufficient insulin in response to hyperglycemia. Hyperglucagonemia represents another facet of the pathophysiology of T2D [22]. Owing to the potential of incretin hormones to enhance insulin secretion and ability of GLP-1 to reduce glucagon levels, there has been substantial interest in uncovering the role of incretin hormones in the pathophysiology of T2D. Analyses suggest no systematic differences in the nutrient-induced secretion of GIP and GLP-1 between healthy and T2D subjects [23]. Nevertheless, the incretin effect in T2D patients is attenuated or absent, primarily due to the significantly diminished efficacy of GIP on the diabetic endocrine pancreas and the marginal physiological role of GLP-1 to elicit insulinotropic and glucagonostatic effect, when compared with healthy subjects [1].

9.2 Development of Novel GLP-1 and GIP-based Therapeutics

Diabetes is the fastest-growing global health issue, impacting ~10% of the world's adult population. T2D constitutes more than 90% of all diabetes cases, although this proportion could potentially be greater, given that approximately one-third of individuals with T2D remain undiagnosed [24]. Meanwhile, obesity stands as a primary contributor to global morbidity and mortality, serving as a catalyst for numerous co-morbidities, including T2D, nonalcoholic fatty liver disease, hypertension, hypercholesterolemia, cardiovascular disease (CVD), and cancer [25]. For decades, there has been a persistent and urgent demand for pharmaceuticals capable of effectively regulating body fat, managing excess weight, and combating obesity. Despite the substantial advancement in scientific knowledge of molecular systems of energy homeostasis, only bariatric surgery thus far offers a weight loss exceeding 20%. The development of novel pharmaceutical agents targeting the GLP-1R has ushered in a new era of opportunity for achieving substantial medical weight loss at tolerable doses [26]. More recently, the identification of unimolecular peptides that concurrently act on the GLP-1R and other hormonal receptors as multi-receptor co-agonists has demonstrated clinical weight reduction that was previously unattainable through other medical interventions [14].

9.2.1 GLP-1R Mono-agonists

Native GLP-1 is unsuitable for therapeutic application primarily because of its short circulating half-life ($t_{1/2}$ < two minutes), which is largely attributed to a rapid cleavage by dipeptidyl peptidase-4 (DPP-4) at the site of residue Ala^8 (or Ala^2 in other analogs) [27]. Various strategies have been employed to extend the half-life of native GLP-1 and improve patient compliance, leading to the development of pharmacological agents that efficaciously combat obesity and T2D. These strategies encompass sequence modifications, chemical alterations, and fusion techniques, which are frequently employed in tandem to deliver heightened stability [28].

Exendin-4, a natural peptide derived from Gila monster *Heloderma* venom, exhibited similar pharmacodynamic characteristics but relatively prolonged pharmacokinetic profiles, with a half-life of ~2.4 hours due to its resistance to DPP-4 proteolysis (residue Gly^8 instead of Ala^8). Although synthetic exendin-4, also known as exenatide, was the first GLP-1R agonist approved for treating T2D, it requires twice-daily injection to maintain active circulating concentrations, resulting in patient discomfort and inconvenience [29,30]. Moreover, as a natural toxin, exenatide has immunogenicity concerns with the potential to elicit allergic reactions upon administration. The development of antibodies against exenatide was reported in its clinical trials. However, reductions in HbA1c levels were significant in participants with or without antibodies, and the presence of antibodies did not correlate with reported rates of GI adverse effects [30].

Building on exenatide, lixisenatide removed one proline and incorporated a six-Lys C-terminal tail, affording an extended half-life and a fourfold increase in binding affinity to GLP-1R. The synergy between its high binding affinity and reduced dissociation rate effectively prolongs the duration of its pharmacological effects, allowing for a once-daily administration schedule [31,32].

Fatty acid derivatization entails chemically coupling a fatty acid to the amino acid side chains of a peptide. This imparts reversible binding to albumin, a multifunctional and abundant protein crucial for transporting endogenous fatty acids and other substances. Additionally, fatty acid-derivatized peptides can spontaneously oligomerize at the site of subcutaneous injection, delaying absorption due to their amphiphilicity and resulting in a sustained release effect [28].

Liraglutide is synthesized by the addition of a C16-palmitoyl fatty acid to the side chain of residue Lys^{20}, along with the substitution of Lys^{28} with Arg^{28}. These modifications enhance oligomerization in the injection site and promote non-covalent binding affinity to albumin during circulation. Liraglutide is suitable for once-daily subcutaneous administration, with a maximum concentration achieved at 9–13 hours (t_{max}) and a half-life of 13 hours [33]. In comparison to exenatide and lixisenatide, liraglutide demonstrates a more substantial improvement in reducing HbA1c levels and fasting blood glucose levels. Moreover, it exhibits fewer side effects than exenatide when administered twice a day. Antibody formation was also observed in liraglutide clinical trials [34]. However, a meta-analysis indicated a lower level of immunogenicity with liraglutide compared to exenatide administered twice daily. Notably, the immunogenicity of liraglutide did not impact its effectiveness in glycemic control [35].

The above GLP-1 analogs characterized by short-lived peaks in plasma drug concentrations following each injection with intermittent periods of near-zero concentrations are defined as short-acting GLP-1 agonists. However, long-acting GLP-1 agonists that support once-weekly therapies are particularly desirable for an ideal regimen and are closely associated with enhanced convenience, treatment adherence, and improved quality of life. Moreover, they provide fewer daunting experiences for individuals managing frequently chronic conditions. In addition, various technologies, such as C-terminal amidation, fatty acid derivatization, PEGylation, sequence alterations, and amino acid substitutions, have been employed with GLP-1R peptide agonists to facilitate once-weekly treatment. Particularly, this is accomplished by replacing the DPP-4 labile Ala8 with uncleavable noncanonical amino acids, including α-aminoisobutyric acid (Aib), 1-aminocyclobutyl-1-carboxylic acid (Ac4c), and D-serine [28].

These pivotal technologies contributed in conjunction to the development of the novel GLP-1R peptide agonist, semaglutide, which is approved for once-weekly subcutaneous treatment of T2D. Structurally, semaglutide incorporates three critical modifications to native GLP-1, endowing its prolonged pharmacokinetics (PK): the replacement of Ala^8 with Aib^2, the substitution of Lys^{28} with Arg^{28}, and the acylation of Lys^{20} with a C18-diacid. This acylation results in a much tighter binding to albumin compared to the C16-palmitoyl fatty acid in liraglutide. Beyond its successful application in treating T2D, semaglutide has shown efficacy in terms of body weight reduction and was approved for the treatment of obesity [36,37]. Moreover, the use of semaglutide has demonstrated cardiovascular benefits by improving the health status of heart failure (HF) patients, leading to a significant reduction in emergency department visits and hospitalizations related to HF [38]. More recently, semaglutide has proven effective in slowing the progression of chronic kidney diseases and reducing the risk of kidney failure-related mortality [39]. In addition to metabolic diseases, semaglutide has even entered clinical development for Alzheimer's disease treatment, exhibiting promising results in memory function improvement and the reduction of phospho-tau accumulation [40,41].

Ecnoglutide (Sciwind Biosciences), a novel long-acting cAMP-biased GLP-1R agonist, exhibits suppressed GLP-1R internalization and consequently demonstrates superior glucose and body weight control in comparison to semaglutide in multiple preclinical animal models. β-Arrestins mediated receptor internalization is considered a pivotal factor in the *in vivo* clearance of GLP-1 peptides. Ecnoglutide retains the full natural amino acid sequence with Val^2 in place of Ala^2 and is acylated closer to the C-terminus at Lys^{30} side chain with a C18-diacid. These structural modifications endow ecnoglutide with comparable cAMP activation potency while significantly reducing receptor internalization in contrast to semaglutide, which was associated with its prolonged retention of the GLP-1R on the cell surface and continued signaling over an extended time. In a Phase 1 clinical study, ecnoglutide was found to be generally safe and well tolerated, with PK properties supporting once-weekly subcutaneous treatment. Ecnoglutide is currently being evaluated in Phase 3 clinical trials for the treatment of T2D and obesity in China [42].

Conjugation of peptides to larger proteins, such as an immunoglobulin Fc fragment, produces "Peptibodies," which prove to be an efficient strategy to extend circulating half-life. The Fc domain enhances the *in vivo* circulation duration of hybrid proteins by interacting with FcRn, which has the ability to bind to the Fc domain within the acidic endosomal environment and prevent Fc domain from degradation. The complex subsequently dissociates to release peptibodies when cycled back to neutralize the pH environment outside the cell. Upon administration, these Fc-conjugated analogs therefore achieve effective circulating concentrations relatively early, thereby initiating the reduction of plasma glucose shortly after treatment commencement [43].

Dulaglutide is a once-weekly GLP-1R agonist approved for the treatment of T2D. It is a recombinant peptide analog of IgG4-Fc covalently linked GLP-$1^{[7\text{-}37]}$ where Ala^8 is replaced by Gly^8 to resist DPP-4 degradation, and Gly^{36} is substituted for Arg^{36} to avoid a potential T-cell epitope interaction. The structure of dulaglutide confers improved solubility, reduced immunogenicity, and a lower rate of renal clearance [44].

Human serum albumin is well recognized as another potential carrier for therapeutic peptides due to its extended circulation half-life, capacity for widespread distribution in the body, and low immunogenicity. This approach has proven successful in the development of long-acting peptide and protein drugs. Albiglutide, for instance, is a recombinant protein consisting of two repeated copies of GLP-1 analogs fused to human albumin. This molecule features an Ala^8 to Gly^8 substitution in both copies, enhancing resistance to DPP-4. Albiglutide possesses an *in vivo* half-life of six to eight days and is approved for once-weekly administration [45].

9.2.2 GIPR Mono-agonists

The contribution of GIPR to overall metabolic effects is ambiguous in that both agonists and antagonists of the GIPR show efficacy, at least for weight loss therapy [46]. One hypothesis is that GIPR activation is associated with strong β-arrestin recruitment that may lead to subsequent internalization of the GIP/GIPR complex, which causes "GIPR desensitization" [47]. Although unclear in mechanism, several long-acting GIPR mono-agonists, such as LY3537021 (Eli Lilly) and ZP6590 (Zealand Pharma), have been developed and tested in clinical trials. In the Phase 1 study, once-weekly repeated dosing of a LY-353702 induced weight loss without GI adverse effects, suggesting that GIP agonism can contribute to weight loss effects [48]. Nevertheless, unlike GLP-1R mono-agonist, the weight reduction effect of long-acting GIPR mono-agonist is limited, even with improved safety profiles. Therefore, strategies in co-formulation GLP-1R and GIPR mono-agonists are under development, like ZP6590 with semaglutide, to achieve an antiemetic effect along with ideal body weight reduction.

9.2.3 GLP-1R/GIPR Dual Agonists

Pharmacologic engagement of the GLP-1R is a proven strategy for treating hyperglycemia in T2D and reducing body weight. The concept of synergistically stimulating both GLP-1R and GIPR holds greater promise as it has the potential to significantly increase insulin secretion and improve insulin sensitivity. In addition, GIPR activation simultaneously is reported to decrease the GI adverse effect associated with GLP-1R agonists, thus increasing the tolerability of GLP-1R agonists at higher doses. Recent research endeavors have delved into the exploration of co-activating GIPR as a strategy to amplify the therapeutic advantages associated with GLP-1R activation. Early investigations have confirmed that GLP-1 and GIP share sufficient structural homology, enabling the development of a unimolecular peptide sequence with high-potency agonistic effects at both receptors [49].

RG7697 was the first engineered GIPR/GLP-1R dual agonist by merging the GIP sequence with the exendin-4-derived helix cap tail and Aib^2 substitution for DPP-4 resistance and was subsequently acylated with a C16-palmitoyl fatty acid at the C-terminus for PK extension. RG7697 had balanced activity for both human GLP-1R ($EC_{50} = 5$ pM) and GIPR ($EC_{50} = 3$ pM) and was suitable for daily dosing but suffered from severe dose-limiting GLP-1R-mediated GI adverse effects in both health and T2D subjects. Unfortunately, although the effects of RG7697 were promising, the changes in glycemic control and body weight were not statistically different from liraglutide at the single dose tested [50].

Tirzepatide was the second developed but the first once-weekly GIPR/GLP-1R dual agonist advanced to clinical studies and was ultimately approved by the FDA in 2022 [51]. It is a 39-amino-acid peptide developed by specific residue alterations in the native GIP sequence to induce agonistic effects at GLP-1R. The core structure is accompanied by a C20-diacid moiety at the side chain of residue K^{20}, which provides PK profiles suitable for once-weekly dosing. The acylation moiety also contributes to its interaction with both receptors while reducing GLP-1R β-arrestin recruitment, thereby leading to reduced internalization [52]. In contrast to RG7697, tirzepatide is a biased dual agonist, which exhibits a binding affinity comparable to native GIP for GIPR but ~11-fold lower when compared to native GLP-1 for GLP-1R [53]. High doses of tirzepatide stimulate insulin secretion and reduce hyperglycemia in both *Gipr*$^{-/-}$ and *Glp1r*$^{-/-}$ mice, providing evidence of its *in vivo* activity at both receptors. Compared to semaglutide, once-weekly tirzepatide treatment demonstrated superior outcomes in terms of HbA1c reduction and body weight loss, while maintaining similar safety profiles [54].

Given the success of tirzepatide, other once-weekly GIPR/GLP-1R dual agonists with different characteristics have been developed, and some of them are currently in clinical trials. These promising agents include VK2735 (Viking Therapeutics), SCO-094 (Scohia Pharma, Inc.), CT-388 (Carmot Therapeutics), DR10627 (Doer Biologics), and others (Table 9.1).

Table 9.1 Sequence representation of native hormones and novel GLP-1R mono-, GLP-1R/GIPR dual agonists.

Name	Sequence	Modifications
Native hormone		
GLP-1[7-37]	HAEGTFTSDVSSYLEGQAAKEFIAWLVKGRG-*OH*	
GIP	YAEGTFISDYSIAMDKIHQQDFVNWLLAQKGKKNDWKHNITQ-*OH*	
GLP-1R mono-agonist		
Exenatide	HGEGTFTSDLSKQMEEEAVRLFIEWLKNGGPSSGAPPPS-NH_2	
Lixisenatide	HGEGTFTSDLSKQMEEEAVRLFIEWLKNGGPSSGAPPS-[K_6]-NH_2	
Liraglutide	HAEGTFTSDVSSYLEGQAA***J***EFIAWLVRGRG-*OH*	***J***: [K-γE-C^{16}monoacid]
Semaglutide	HXEGTFTSDVSSYLEGQAA***J***EFIAWLVRGRG-*OH*	X: Aib; ***J***: [K-$AEEA_2$-γE-C^{18}diacid]
Ecnoglutide	HVEGTFTSDVSSYLEEQAAREFI***J***WLVRGRG-*OH*	***J***: [K-$AEEA_2$-γE-C^{18}diacid]
Dulaglutide	HGEGTFTSDVSSYLEEQAAKEFIAWLVKGGG-Spacer-Fc	
Albiglutide	HGEGTFTSDVSSYLEGQAAKEFIAWLVKGRG-Albumin	
GLP-1R/GIPR dual agonist		
RG7697	YXEGTFTSDYSIYLDKQAAXEFVNWLLAGGPSSGAPPPS***J***-NH_2	X: Aib; ***J***: [K-C^{16}monoacid]
Tirzepatide	YXEGTFTSDYSI**X**LDKIAQ***J***AFVQWLIAGGPSSGAPPPS-NH_2	X: Aib; ***J***: [K-$AEEA_2$-γE-C^{20}diacid]

9.3 Development of Multi-receptor Agonists Targeting Incretin and Other Hormonal Receptors

9.3.1 Gut-brain Hormonal Regulation of Glucose and Energy Homeostasis

The multifactorial nature of the physiological control of metabolism, energy homeostasis, and body weight, combined with evidence demonstrating that gastric bypass surgery increases the secretion of GLP-1 and several other gut-derived hormones, suggests that the development of medications combining GLP-1R activity with other hormonal stimuli may offer more vibrant clinical outcomes beyond single GLP-1R activation [55]. Consequently, as a promising new concept for pharmaceutical intervention, various combinations of multi-receptor agonists that target GLP-1R and complementary metabolically related hormonal receptors are currently advancing through clinical trials involving participants with diabetes, obesity, and nonalcoholic steatosis hepatitis (NASH) [56,57] (Figure 9.2).

Glucagon (GCG), a 29-residue hormone synthesized within α cells, collaborates with insulin in the regulation of blood glucose levels. GCG, as the principal counter-regulatory hormone to insulin, serves to promote liver gluconeogenesis, enhancing glycogen hydrolysis and increasing glucose output [58]. Additionally, GCG plays a pivotal role in diverse metabolic pathways, which influence lipid metabolism, energy storage, and amino acid metabolism. Dysregulation in GCG function may lead to a range of pathological conditions, such as insulin resistance

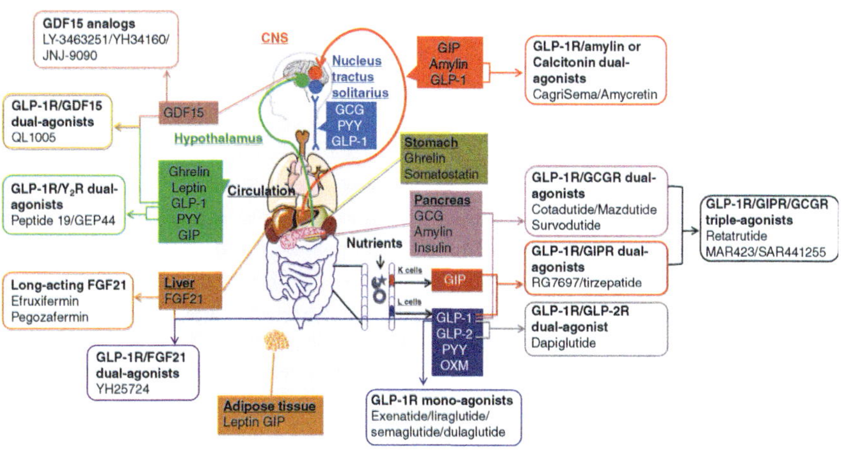

Figure 9.2 Complementary physiological functions of incretins and other gut hormones in regulating glucose and energy homeostasis.

in T2D, glucagonoma-accelerating amino acid turnover, and abnormal triglyceride lipolysis [59].

Glucagon-like peptide-2 (GLP-2) is a gut-derived hormone that shares high homology with GLP-1. This 33-amino acid peptide serves as a selective agonist for GLP-2R over GLP-1R, albeit some studies have hinted at partial functional overlap between GLP-1 and GLP-2. GLP-2 plays a pivotal role in upholding the structural and functional integrity of the intestines by stimulating mucosal epithelial cell proliferation and inhibiting its apoptosis, enhancing nutrient absorption, increasing intestinal blood flow, reducing gut permeability and motility, and attenuating intestinal inflammation [60].

Amylin and calcitonin belong to the calcitonin family of peptides together with calcitonin gene-related peptide and adrenomedullin, which participate in diverse physiological regulations, including bone metabolism, appetite control, nutrient uptake, and vasodilation [61]. Amylin is a 37-amino acid pancreatic hormone secreted alongside insulin from β cells and has roles in satiety, gastric emptying, and glucagon secretion. Amylin controls the satiety signal via direct action on amylin receptors in the brain. Among all three subtypes of amylin receptors, namely Amy1R, Amy2R, and Amy3R, it is believed that Amy1R is associated with lipid metabolism, while Amyl3R is responsible for glycemic regulation. Amylin is also an efficacious inhibitor of gastric emptying, further suppressing glucagon secretion via central mechanisms. Calcitonin is a 32-amino-acid peptide secreted by the C cells of the thyroid gland. Calcitonin exerts a primary hypocalcemia effect via inhibition of osteoclasts and promotion of renal calcium excretion. Recent studies have also indicated that calcitonin possesses metabolism benefits such as affecting insulin sensitivity, reducing the gastric emptying rate, and promoting satiation.

PYY is a gut-derived peptide hormone co-secreted from the enteroendocrine L cells of the intestines in response to nutrient intake. $PYY^{[3-36]}$, derived from full-length PYY through rapid DPP-4 cleavage, is the primary active form in circulation [62]. $PYY^{[3-36]}$ selectively activates the neuropeptide Y2 receptor (Y_2R), which is part of the NPYR family consisting of four different members: Y_1, Y_2, Y_4, and Y_5. Activation of NPY_2R in the hypothalamus and intestinal afferent vagus nerve terminals leads to appetite suppression, consequently inhibiting food intake and promoting weight loss [63].

Fibroblast growth factor 21 (FGF21) is a 209-amino acid hormone induced by starvation and stress, primarily distributed in the liver, pancreas, and adipose tissue. FGF21 interacts with the FGF receptor FGFR1c and a co-receptor protein, β-klotho, to form a stable heterotrimeric FGF21/β-Klotho/FGFR complex, which activates downstream receptor tyrosine kinase activities. Endogenous FGF21 plays pivotal roles in various physiological functions, including angiogenesis, mitogenesis, pattern formation, cell differentiation, metabolic regulation, and tissue injury repair. Administration of recombinant human FGF21 in preclinical animal models effectively regulates insulin sensitivity, energy expenditure, and lipid metabolism. Notably, FGF21 exhibits the potential to ameliorate hepatic steatosis, highlighting its multifaceted therapeutic utility in addressing metabolic disorders [64].

Growth differentiation factor 15 (GDF15), also known as macrophage inhibitory cytokine 1, is a 25-kDa disulfide-linked homodimeric protein belonging to the transforming growth factor-β superfamily. Physiologically, GDF15 is secreted at a low concentration but is upregulated in response to cell stress. GDF15 specifically binds with the brain-stem-restricted glial-derived neurotrophic factor-family receptor α-like (GFRAL) complexed with its co-receptor RET, through which it reduces food intake and lowers body weight. These unique characteristics have sparked interest in exploring its therapeutic potential for obesity, T2D, and CVD [65].

In the realm of pharmaceutical innovation, diverse chemical methodologies have been developed to create multi-receptor agonists. These strategies include novel peptide modification approaches such as residue alteration for sequences with high homology, truncation, and hybridization of minimal but functionally essential segments into one peptide chimera, or direct conjugation of individual unrelated peptide hormones with optimal linkers. These methods collectively stand as pivotal techniques in the design and synthesis of multi-receptor agonists, aiming to harness synergistic therapeutic effects from multiple hormone pathways (Figure 9.3).

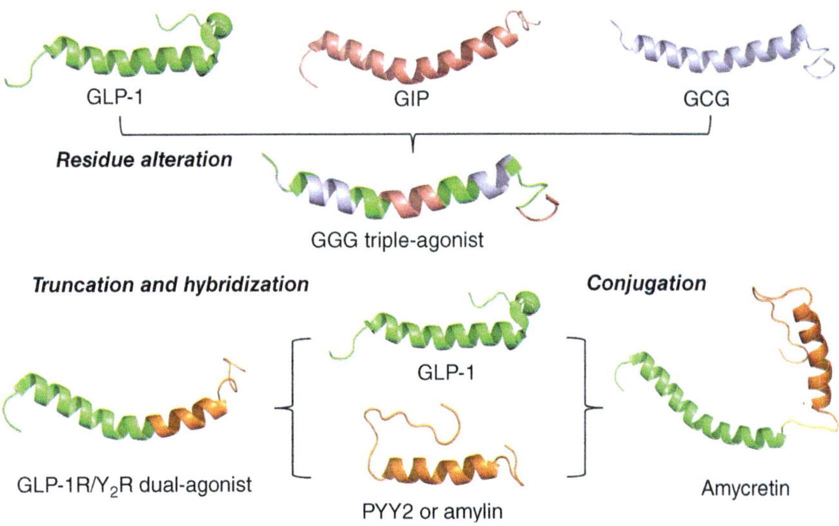

Figure 9.3 Chemical strategies of unimolecular multi-receptor targeting agonists.

9.3.2 Multi-receptor Agonists Targeting GLP-1R and Other Hormonal Receptors

The following context will assess the translational potential of these pioneering multi-receptor agonist developments, including merging incretins with GCG, GLP-2, amylin and calcitonin, PYY2, FGF21, and GDF15.

9.3.2.1 GLP-1R/GCGR Dual Agonists

It is commonly accepted that obesity is a consequence of a chronic imbalance of energy intake and energy expenditure, which results in excess fat storage. Recent advancements have elucidated the beneficial effects of GCG receptor activation in relation to energy expenditure, lipid metabolism, cardiorenal protection, and satiety regulation, thereby leading to interest in its potential for obesity treatment [59]. Subsequently, simultaneous activation of the GLP-1R and glucagon receptor (GCGR) is believed to target both appetite suppression and energy expenditure enhancement, which mimics the complementary effects of diet and exercise on weight loss [66]. This concept has been demonstrated with the peptide hormone oxyntomodulin (OXM), an endogenous GLP-1R/GCGR dual agonist that reduces body weight and food intake while enhancing energy expenditure in humans. Early clinical research has demonstrated that repeated subcutaneous injections (three times per day) of OXM to obesity subjects rapidly suppressed energy intake and significantly reduced body weight in addition to concurrent changes in leptin and adiponectin, suggesting a reduction in adiposity. Although the outcomes of studies employing native OXM in rodents and humans appeared promising, its short *in vivo* half-life rendered it an impractical choice for clinical use [67].

Cotadutide, modified from OXM sequence with C16-palmitoyl fatty acid acylated to the side chain of residue K^{10}, was the first once-daily GLP-1R/GCGR dual agonist advanced to clinical trials. In an early Phase 2a study in adults with T2D, daily treatment with cotadutide demonstrated optimal weight loss and glycemic control but was associated with dose-dependent GI adverse events, especially nausea and vomiting, in common with the GLP-1R agonists [68] (Table 9.2).

SAR425899, another once-daily GLP-1R/GCGR dual agonist, was modified with a D-form Ser^2 substitution to reduce DPP-4 cleavage. A Phase 1a study of SAR425899 revealed a dose-dependent weight loss in overweight patients. However, its Phase 2 trials were halted due to excessive rates of GI adverse events [69,70].

Although unimolecular GLP-1R/GCGR dual agonists were able to leverage the anorectic and antidiabetic effects of GLP-1 with the ability of GCG to increase energy expenditure, it is crucial to emphasize that the diabetogenic consequences of GCGR agonism must be counterbalanced by the insulinotropic impacts of GLP-1 [71]. Typically, this guides the molecular design of next-generation

Table 9.2 Sequence representation of native hormones and novel multi-agonists targeting GLP-1R with other hormonal receptors.

Name	Sequence	Modifications
Native hormone		
GCG	HSQGTFTSDYSKYLDSRRAQDFVQ WLMNT-*OH*	
OXM	HSQGTFTSDYSKYLDSRRAQDFVQWLMNTKRNKNNIA-*OH*	
GLP-2	HADGSFSDEMNTILDNLAARDFINWLIQTKITD-*OH*	
GLP-1R/GCGR dual agonist		
Cotadutide	HSQGTFTSD***J***SEYLDSERARDFVAWLEAGG-*OH*	***J***: [K-γE-C^{16}monoacid]
SAR425899	HsQGTFTSDLSKQ***J***ESKAAQDFIEWLKAGGPSSGAPPPS-NH_2	***J***: [K-γE-C^{16}monoacid]
Mazdutide	HXQGTFTSDYSKYLDEKKA***J***EFVEWLLEGGPSG-NH_2	X: Aib; ***J***: [K-$AEEA_2$-γE-C^{20}diacid]
Survodutide	HXQGTFTSDYSKyLDERAAKDFI***J***wlesa-NH_2	X: Ac4c; ***J***: [k-GGsGsG-γE-C^{18}diacid]
Pemvidutide	HXQGTFTSDYSKYLDE***J***AAKEFIQWLLQT-NH_2	X: Aib; ***J***: [K-^{D}gluco-1-C^{18}monoacid]
GLP-1R/GIPR/GCGR triple agonist		
Peptide 20	HXQGTFTSD***J***SKYLDERAAQDFVQWLLDGGPSSGAPPPS-NH_2	X: Aib; ***J***: [K-γE-C^{16}monoacid]
SAR441255	HXHGTFTSDLSKL***J***EEQRQXEFIEWLKAaGPPSXKPPPK-NH_2	X: Aib; ***J***: [K-γE-γE-C^{16}monoacid]
Retatrutide	YXQGTFTSDYSIULDK***J***AQXAFIEYLLEGGPSSGAPPPS-NH_2	X: Aib; U: αMeL; ***J***: [K-$AEEA_2$-γE-C^{20}diacid]
Peptide 16	HXHGTFTSDYSIYLE***J***KYAXEFVQWLLEGGPSSGAPPPS-NH_2	X: Aib; ***J***: [K-εLys_2-γE-C^{20}diacid]
GLP-1R/GLP-2R dual agonist		
Dapiglutide	HXEGSFTSELATILD***J***QAARDFIAWLIQHKITD-*OH*	X: Aib; ***J***: [K-γE-C^{18}diacid]

GLP-1R/GCGR dual agonists toward a relatively higher potency for GLP-1R compared to GCGR, also known as "balanced" agonism, mimicking that of OXM. Based on this conclusion, several modified once-weekly GLP-1R/GCGR dual agonists have been developed and advanced into clinical trials.

Mazdutide (IBI362 or LY-3305677) is a once-weekly OXM analog that features a cleavage-resistant residue substitution (Aib^2) and a C20-fatty-diacyl moiety to extend *in vivo* half-life. Mazdutide has been engineered to achieve balanced GLP-1R and GCGR potency in order to prevent glucagon receptor-induced hyperglycemia while preserving the desired benefits of reducing HbA1c levels and body weight. Besides weight loss and glycemic control, mazdutide elicited robust improvement in lipid profiles, including significant reduction of total cholesterol, LDL cholesterol, and triglycerides levels, together with decreased ALT and AST levels. These results confer comprehensive metabolic benefits to patients with T2D and supported further explorations for its application in NASH [72,73].

Survodutide (BI456906) is a potent GLP-1R/GCGR dual agonist containing a C18-diacid moiety at the side chain of residue D-form Lys^{24} to support once-weekly dosing. Structurally, survodutide features several unnatural mutations, namely an $Ac4c^2$ substitution to prevent DPP-4 cleavage; a D-form [Wlesa] C-terminal sequence to enhance stability; and a [GGsGsG] spacer on the side chain. In a Phase 2 study involving nondiabetic subjects living with an average BMI of 37, survodutide led to weight loss of up to 19% over 46 weeks. Furthermore, survodutide also led to improvements in cardiometabolic health, such as reduced blood pressure and waist circumference [74].

Pemvidutide (ALT-801) is a long-acting balanced GLP-1R/GCGR dual agonist that is in development for the treatment of obesity and NASH but not for T2D, as its effect on glucose is neutral. Pemvidutide is developed by the proprietary EuPort™ technology, which significantly increases the circulating half-life of the peptide to support once-weekly dosing and may slow down the entry of pemvidutide into the bloodstream for improved tolerability. EuPort™ is a platform technology consisting of a fatty acid-conjugated sugar moiety that can be coupled to the peptides. Pemvidutide also has beneficial effects on insulin resistance by increasing fat mobilization within the liver. In a Phase 1 clinical study, pemvidutide demonstrated striking reductions in body weight and significant reductions in liver fat content, serum lipids, and markers of liver inflammation, which together may potentially provide additional cardioprotective effects [75].

9.3.2.2 GLP-1R/GIPR/GCGR Triple Agonists

GLP-1R agonism contributes to weight reduction by suppressing appetite and enhancing insulin secretion and sensitivity, thereby aiding in glycemic control. GIP amplifies the therapeutic benefits of GLP-1R agonists by mitigating common GI adverse effects. In addition to its crucial role in maintaining euglycemia, GCG also regulates energy expenditure and hepatic lipid metabolism. The physiological and pharmacological functions of GIP and GCG logically appear as suitable complements to the beneficial effects of GLP-1. Given the high degree of sequence homology among these three peptides, the concept of a unimolecular GLP-1R/GIPR/GCGR triple agonist could potentially establish a new standard for pharmaceutical interventions in T2D, obesity, and even NASH. As a result, several GIPR/GLP-1R/GCGR triple-agonists have demonstrated superior metabolic benefits along with minimized side effects arising from this orchestrated agonism, advancing them into clinical trials [76].

MAR423 (Peptide 20), the first highly potent unimolecular GLP-1R/GIPR/GCGR triple agonist, evolved from residue alterations based on OXM and exenatide. Namely, substitutions at Glu^{16}, Arg^{17}, Gln^{20}, Leu^{27}, and Asp^{28} were made to improve GCGR potency, enhance solubility, reinforce secondary structure, and improve chemical stability. Additionally, the acyl moiety of C^{16}-palmitoyl chain and a γ-Glu spacer facilitates albumin binding to support once-daily injection [77]. In a Phase 1 clinical study, MAR423 effectively reversed glucose dysregulation without adverse effects and demonstrated significant reductions in body weight and HbA1c level. Moreover, it improved glucose tolerance and helped preserve pancreatic islet architecture in diabetic fatty Zucker rat models [78].

SAR441255 represents another unimolecular GLP-1R/GIPR/GCGR triple agonist designed for once-daily subcutaneous injections in humans. Besides the Aib^2 substitution to prevent DPP-4 cleavage, four residues were mutated in the exendin tail (Pro^{32}, Aib^{34}, Lys^{35}, and Lys^{39}) to prevent aggregation problems at acidic pH in the presence of phenolic preservatives. Further residue changes including Gln^3, Leu^{13}, Aib^{20}, and ${}^{d}Ala^{29}$ were implemented to achieve high and balanced activation at each receptor and enhance chemical stability in buffer formulations. Similar to MAR423, SAR441255 is acylated on K^{14} with two γ-Glu spacers and a C^{16}-palmitoyl moiety to extend its *in vivo* half-life via albumin binding. Preclinical studies of SAR441255 conducted in mouse and cynomolgus monkey models of obesity and T2DM demonstrated better effectiveness in reducing body weight and enhancing metabolic parameters than a dual GLP1R/GCGR dual agonist. A Phase 1 clinical study revealed that SAR441255 significantly improved glycemic control during a mixed-meal tolerance test and exhibited good tolerability [76].

A "next-generation" GLP-1R/GIPR/GCGR triple agonist (Peptide 16) was developed by DiMarchi et al. based on the once-daily GLP-1R/GIPR dual-agonist NN0090-2746. The optimized peptide employs a protraction strategy permitting once-weekly human dosing. A dual mutation of His^1 and His^3 shows significant improvement in GCGR potency. Residue changes involving Lys^{17} and Tyr^{18} also improved GCGR potency with a slight reduction in GLP-1R potency. The acylation of Lys^{16} with εLys-εLys-γE spacer and a C^{20}-diacid moiety provides attractive PK profiles for potential once-weekly administration compared to semaglutide.

This optimized triple agonist normalizes body weight in diet-induced obese (DIO) mice and enhances energy expenditure in a manner superior to that of semaglutide and tirzepatide [79].

Retatrutide (LY-3437943) is a 39-amino acid peptide engineered from the human GIP sequence to achieve triple agonism at GCGR, GIPR, and GLP-1R [80]. Structurally, retatrutide incorporates three noncanonical amino acid residues Aib^2, Aib^{20}, and αMeL^{13}. Aib^2 provides stability against DPP-4 cleavage, whereas Aib^{20} contributes to optimal GIP activity, the PK profile, and overall developability. αMeL^{13} was found to enhance both glucagon and GIP activity and help maintain helicity in the meanwhile. Conjugation with a C^{20}-diacid moiety at Lys^{17} facilitated strong albumin binding and extended its half-life for sustained pharmacological effects, which enables once-weekly treatment. Early-phase clinical studies demonstrated promising outcomes in individuals with T2D where retatrutide exhibited robust reductions in both glucose levels and body weight [81]. Notably, a Phase 2 study evidenced a substantial mean weight reduction of up to 24.2% after the 48-week treatment duration. These findings suggest its significant potential for further clinical development [82]. As it advances into Phase 3 trials, a more comprehensive understanding of its long-term efficacy, safety profile, and potential clinical application is expected.

9.3.2.3 GLP-1R/GLP-2R Dual Agonists

GLP-2 agonism proves to be a practical means to enhance intestinal repair and attenuates inflammation in preclinical inflammatory bowel disease (IBD) models, making GLP-2 analogs attractive candidates for IBD therapy. The first GLP-2R agonist approved for clinical use is Teduglutide, a DPP-4 cleavage-resistant analog of GLP-2. Teduglutide has gained approval for the treatment of short bowel syndrome (SBS), a severe condition arising from nutrient malabsorption when the small intestine is either shortened or damaged [83]. Glepaglutide is a long-acting GLP-2 receptor agonist in clinical development for the treatment of SBS [84]. A stapled long-acting GLP-2R agonist, Peptide 10, was reported to exhibit over 10-fold extended half-life than teduglutide and improved intestinotrophic and anti-inflammatory effects in mouse models of DSS-induced colitis [85].

In SBS and IBD, intestinal insufficiency and failure result in secretion deficiency of both GLP-1 and GLP-2, which are further associated with accelerated transit and epithelial surface loss. Therefore, it is reasonable that additional benefit can be generated from a combination of GLP-1 and GLP-2 agonism, with the aim to enlarge intestinal surface area while decelerating intestinal transit. Dapiglutide (Zealand Pharma A/S) has been developed as a GLP-1R/GLP-2R dual agonist by merging the sequences of GLP-1 and GLP-2. It is currently undergoing Phase 2 clinical trials for the treatment of obesity [86]. Previous clinical trials have shown an average weight reduction of 4.5% in healthy participants after a four-week, once-weekly regimen, and its side effects align with those of GLP-1 mono-agonists. Furthermore, studies in rodent models have suggested that dapiglutide significantly promotes intestinal growth and improves intestinal barrier function through GLP-2R agonism, as evidenced by increased mucosal height and intestinal length. The activation of GLP-2R synergizes with GLP-1 agonism, slowing GI transit time and facilitating nutrient absorption. Therefore, the various benefits of dapiglutide for intestinal health also pave the way for its potential therapeutic application for SBS [87].

9.3.2.4 GLP-1 Co-agonist with Amylin and/or Calcitonin

The inherent fibrillation propensity and short half-life of native human amylin pose a major challenge for its clinical application. Modified amylin analogs have been developed by sequence alteration and/or half-life extension technologies. Pramlintide, which is a non-fibrillogenic amylin analog, was approved for the management of Type 1 diabetes and T2D as an adjunct to preprandial insulin therapy in patients with inadequate glycemic control of insulin therapy [88]. Meanwhile, inspired by natural salmon calcitonin (sCT), which exhibits unmatched AmyR and calcitonin receptor (CTR) agonism, the concept of dual amylin and calcitonin receptor agonists (DACRAs) have emerged as novel approaches to achieve better glycemic control. KBP-336 was developed as a long-acting, once-weekly DACRA with the ability to prolong activation of the receptor [89]. In preclinical studies, KBP-336 showed superior long-term efficacy on both glucose and weight reduction, supporting the relevance of the agonism balance and highlighting KBP-336 as a promising agent for the treatment of obesity and T2D (Table 9.3).

Amylin/calcitonin and GLP-1 mediate feedback control of food intake by seemingly distinct, yet overlapping, mechanisms. The combination application of exenatide and an amylin analog effectively suppressed food intake in rhesus monkeys, which is significantly greater than those generated by the individual dose [90]. Thereby, the combination treatment with amylin and GLP-1R agonists represents a new anti-obesity treatment concept in humans [91]. CagriSema, a co-formulation medication consisting of the long-acting amylin analog cagrilintide and the GLP-1R agonist semaglutide, which can be administered subcutaneously once a week, has progressed to Phase 3 clinical trials. In a Phase 2 trial for T2D, subjects in the CagriSema group had a mean reduction in HbA1c from baseline of −2.18% and a mean weight loss of 15.6%, which was significantly better than that of semaglutide alone (−1.79% and −5.1%) [92]. Similarly, the combined use of KBP-336 and semaglutide significantly improved weight loss compared to either monotherapy alone, as evidenced by the reduction in adiposity at the study end. All treatments improved glucose tolerance [93].

Table 9.3 Novel amylin and calcitonin analogs.

Name	Sequence	Modifications
Amylin and analogs		
hAmylin	K[CNTATC]ATQRLANFLVHSSNNFGAILSSTNVGSNTY-NH_2	[]: disulfide
Pramlintide	K[CNTATC]ATQRLANFLVHSSNNFGPILPPTNVGSNTY-NH_2	[]: disulfide
Cagrilintide	***J***K[CNTATC]ATQRLAEFLRHSSNNFGPILPPTNVGSNTP-NH_2	[]: disulfide; ***J***: [γE-C^{20}diacid]
Calcitonin and DACRA analogs		
hCT	[CGNLSTC]MLGTYTQDFNKFHTFPQTAIGVGAP-NH_2	[]: disulfide
sCT	[CSNLSTC]VLGKLSQELHKLQTYPRTNTGSGTP-NH_2	[]: disulfide
KBP-336	Ac-[CASLSTC]XLGRLSAELHK***J***ATYPKTDVGANAP-NH_2	[]: disulfide; X: Aib; ***J***: [K-$AEEA_2$-γE-C^{18}diacid]

Based on the success of CagriSema, a "next-generation" unimolecular long-acting amylin and GLP-1 analog by directly hybridizing two peptides via a linker, named as amycretin (NN9487), was developed by Novo Nordisk, which is now in Phase 1 clinical trials for obesity.

9.3.2.5 GLP-1R/Y_2R Co-agonists

The secretion of both PYY and GLP-1 in serum was significantly decreased in obese individuals but recovered after bariatric surgery, suggesting that the elevation of both peptide levels simultaneously reduces appetite and food intake. Therefore, mimicking the biological effects of PYY and GLP-1 after bariatric surgery presents a novel approach for treating obesity and diabetes. Considering the lability of $PYY^{[3-36]}$ to proteolysis, the protracted $PYY^{[3-36]}$ derivatives, such as nisotirostide (Eli Lilly), Peptide 22 (Calibr-Scripps), and Peptide 5 (Novo Nordisk), have been developed and demonstrated remarkable performance in sustained weight loss in animal models [94,95]. More intriguingly, the combined use of long-acting Y_2R and GLP-1R agonist has been proved to act synergistically to achieve improved efficacy in terms of food intake suppression, weight loss, and improved glycemic control, thereby highlighting the necessity of developing potential unimolecular co-agonists targeting GLP-1R, Y_2R, and others if possible [96]. Another once-weekly long-acting PYY analog, NNC0165-1875, in combination use with semaglutide, is currently being evaluated in Phase 2 studies for the treatment of T2D and obesity [97] (Table 9.4).

Several co-agonists targeting GLP-1R, Y2R, and other receptors (e.g. GCGR) have been reported thus far, and their effects in combating obesity and diabetes have been investigated in rodent-based models. A unimolecular GLP-1R/Y2R dual agonist, Peptide 19, was developed by fusing the N-terminal sequence of GLP-1 with truncated C-terminal $PYY^{[24-36]}$. Peptide 19, which shares the same length as GLP-1, exhibited potent activity against both Y2R and GLP-1R, with a twofold and fivefold reduction in activity relative to native $PYY^{[3-36]}$ and GLP-1, respectively. Subsequent *in vivo* studies of Peptide 19 in rodent models suggested its superior capacity to reduce food intake when compared to the anorectic effects when individual GLP-1 and PYY were applied [96].

GEP44 is another potent GLP-1R/Y_2R dual agonist, which is a chimeric peptide designed by fusing the N-terminal sequence of exenatide with the C-terminus of $PYY^{[25-36]}$. Daily repeated administration of GEP44 in animal models resulted in significant body weight loss and improved blood glucose control, outperforming the effect of exenatide. Notably, albeit with over 20-fold lower GLP-1R potency, the dual-agonism of Y_2R and GLP-1R by GEP44 was able to prevent potential GI adverse effects associated with GLP-1R agonism, revealing an additional therapeutic benefit [98]. Beyond appetite suppression, further investigation of GEP44 also suggested its potential therapeutic application in reducing opioid-seeking behavior.

Table 9.4 Novel GLP-1R/Y2R dual agonists and GLP1R/CGR/Y2R triple agonist.

Name	Sequence	Modifications
PYY[3-36] and long-acting analogs		
PYY[3-36]	IKPEAPGEDASPEELNRYYASLRHYLNLVTRQRY-NH_2	
Nisotirostide	PKPE***J***PGEDASPEEWQRYYAELRHYLNWLTRQRY-NH_2	***J***: [K-Ahx-E-γE-C^{18}diacid]
Peptide 22	PKPEAPGKDASPEEWNRYYA***J***LRHYLN***J***LTRQRY-NH_2	***J-J***: [Cys-Cys]-C^{18}diacid-staple
Peptide 5	XRPE***J***PGEDASPEELQRYYASLRHYLNWVTRQ[^{Me}R]Y-NH_2	X: 3-Mba; ***J***: [K-AEEA$_2$-γE-Sulphonic-C^{18}acid]
GLP-1R/Y$_2$R dual agonists		
Peptide 19	HXEGTFTSDVSSYLEGQALRHYINWLTRQRY-NH_2	X: Aib
GEP44	HsQGTFTSDLSKYLEEEAVREFIAWLKNGGPSRHYLNLVTRQRY-NH_2	
Peptide 6q	[HXEGTYTNDVTEYLEE***J***AAKEFIEWLIKGK**C**-NH_2]—Mal-AEEA$_4$—[PKPEAPGKDASPEEWNYYADLRHYLNWLTRQRY-NH_2]	X: Aib; ***J***: [K-γE-C^{16}monoacid]
GLP-1R/GCGR/Y2R triple agonist		
Peptide 3b	[HsQGTYTNDVSKY***J***DSRRAQDFIEWLKNGGPSSGAPPPS**C**-NH_2]—Mal-AEEA$_4$—[PKPEAPGKDASPEEWNRYYADLRHYLNWLTRQRY-NH_2]	***J***: [K-γE-C^{16}monoacid]

Moreover, a series of long-acting GLP-1R/Y_2R dual agonists and GLP-1R/GCGR/Y_2R triple agonists by directly linking the C-terminus of GLP-1/GGG peptide derivatives to the N-terminus of PYY[3-36] via various lengths of spacers. Most of these peptide conjugates retained similar affinity and potency on GLP-1R and Y_2R compared to their parent mono-agonists [99,100]. Among them, peptide 6q, which contains a C16 palmitoyl acylated motif to achieve protracted *in vivo* half-life, was found to have enhanced effects on long-term glycemic control and body weight reduction compared to liraglutide in db/db and DIO mice models. Notably, the administration of peptide 6q in high-fat diet-induced (NASH) mice led to a significant reduction in hepatic triglyceride and total cholesterol levels and could also reverse hepatic steatosis and fibrosis [99]. Consequently, these data provide supporting evidence for Y2 and GLP-1R co-agonists as promising agents for combating hepatic steatosis.

9.3.2.6 GLP-1/FGF21 Dual Agonists

Several engineered FGF21 analogs for enhanced pharmacokinetic profiles, such as LY2405319, PF-05231023, and AKR-0010, have been developed and demonstrated remarkable effects on body weight reduction and glycemic control in animal models. However, clinical trials in humans have yielded variable results, with significant improvements in lipid profiles but less pronounced effects on glucose regulation and weight reduction [64]. Therefore, most FGF21 analogs currently advanced to clinical trials, including efruxifermin (Akero Therapeutics), pegozafermin (89bio, Inc.), pegbelfermin (BMS), BOS-580 (Boston Pharmaceuticals), are primarily focused on NASH treatment by simultaneously addressing the liver fibrosis and managing the underlying causes such as insulin resistance, dyslipidemia, and excessive body weight, which result in excess fat deposition in the liver [101–104].

Recent investigations have demonstrated that FGF21 and GLP-1 exhibit interactions in hepatic glucose metabolism and regulation within hepatocytes. GLP-1 analogs can stimulate hepatic FGF21 production. Furthermore, while GLP-1 primarily reduces body weight by inhibiting energy intake, FGF21 has been shown to stimulate energy expenditure, indicating that these two hormones have complementary effects on body weight control [105]. These findings collectively suggest that GLP-1 and FGF21 could potentially have synergistic functions.

An innovative fusion protein, consisting of modified N-terminal GLP-1 and C-terminal FGF21, linked via an elastin-linker polypeptide (ELP) moiety serving as a sustained release module, was developed as a long-acting therapeutic agent for the treatment of T2D. In animal studies, weekly administration of this GLP-1-ELP-FGF21 fusion protein resulted in a significantly more robust reduction in body weight and improvements in glycemic control compared to what could be achieved with either agonist administered independently. Additionally, this unimolecular GLP-1/FGF21 dual agonist offers the benefits of reducing the required injection concentration and viscosity, making it a more readily administered therapeutic agent with enhanced stability [106].

Fc-conjugation proves to be a practical strategy to generate GLP-1/FGF21 dual agonists. Early research involving IgG4-Fc conjugation of GLP-1 agonists with an FGF21 variant bearing enhanced β-Klotho (KLB) binding affinity led to the discovery of GLP-1-Fc-FGF21 D1 (HEC Pharma). *In vivo* studies have revealed GLP-1-Fc-FGF21 D1 manifested more robust and sustained reductions in glucose levels and body weight, along with improving liver function, serum, and hepatic lipid profiles, compared to the independent application of Dulaglutide and FGF21 [105]. Therefore, this novel GLP-1R/FGFR dual agonist exhibits substantial promise for clinical applications in the treatment of T2D, obesity, and NASH.

YH25724 is another Fc-conjugated long-acting GLP-1R/FGFR dual agonist now in clinical trial. In preclinical studies, YH25724 exhibited a potent and synergistic glucose-lowering effect and body weight reduction in rodent models, which is superior to dulaglutide and FGF21 monotherapy. Noteworthily, remarkable antisteatotic, anti-inflammatory, and anti-fibrotic effects were consistently observed after administration of YH25724 in mouse models, supporting its therapeutic potential for NASH treatment [107].

9.3.2.7 GLP-1/GDF15 Dual Agonists

GDF15 has lately garnered interest in obesity treatment due to its beneficial effect on energy homeostasis discovered in animal studies. In light of the short *in vivo* half-life of native GDF15, a few long-acting GDF15 analogs (such as LY-3463251, YH34160, and JNJ-9090) have been developed by fusing GDF15 with HSA or IgG-Fc to extend its half-life [108,109]. Although a remarkable reduction of food intake and body weight was observed in obese rodents and nonhuman primates after administration, a lack of significant effect on the body weight loss in overweight human cases poses a new challenge for leveraging these long-acting GDF15 derivatives in anti-obesity therapy [65].

Intriguingly, a synergistic effect has been observed in rodent models treated with a combination of GDF15 and GLP-1R agonists. This combination amplifies body weight reduction due to the distinct mechanisms through which GDF15 and GLP-1R agonists concurrently suppress appetite. Inspired by this result, a novel GLP-1R and GFRAL dual agonist QL1005 was constructed by fusing GLP-1 and GDF15 analogs via a peptide linker, which was followed by conjugation with a fatty acid for half-life protraction [110]. In DIO mice, QL1005 exhibited a superior capability of lowering body weight, food intake, fasting glucose, and triglyceride than the efficacy of semaglutide or long-acting GDF15 standalone.

Notably, these metabolic effects stemmed from the activity of both GLP-1 and GDF15 in an individual pathway-balanced fashion, which echoes the beneficiary complementary effects previously observed between these two agonists. Furthermore, QL1005 was also more efficacious in reducing body weight and lowering glucose than semaglutide in studies involving cynomolgus monkeys, and only limited incidence of GI adverse effects was observed during the experiment [110]. Given this evidence, this long-acting dual GLP-1/GDF15 molecule may present a new opportunity to address the clinical ineffectiveness of GDF15 monotherapy, but further clinical evaluation would still be needed to prove the rationale of this novel therapeutic strategy.

9.4 Summary and Perspective

The rising prevalence of obesity and T2D has underscored the need for the development of novel and more effective methods to attain proficient glycemic control and weight loss. Patients with obesity and T2D are frequently at high risk of CVD and are affected by comorbidities that complicate the evaluation of drug safety. Besides diet and lifestyle modifications, bariatric surgeries result in effective alleviation of obesity and T2D and highlight the therapeutic inspiration of gut-brain axis regulation strategy to provide more patient-friendly potentials.

Over the past decades, in-depth investigations into incretin pharmacology concerning glucose and energy regulation have illuminated the novel development of incretin-based therapeutics. Thereby, GLP-1R agonists have become well established as an efficacious treatment option for patients with T2D and obesity and provide both effective glycemic control and weight management with well-characterized safety profiles. Meanwhile, regulation of energy balance is a complex process involving multiple central and peripheral signaling pathways in different tissues and organs. The significance of combined actions involving incretin and other regulatory peptide hormones in metabolic control is gradually becoming evident. The recent precedent-setting clinical results with tirzepatide and retatrutide inspire confidence for the future development of unimolecular multi-receptor targeting peptide approaches as well as novel combination therapies involving additional peptide hormones like GLP-2, PYY, amylin, FGF21, et al. Although thorough clinical studies are still required to determine the sustainability, safety, and translational benefits, these advancements have the potential to redefine the treatment paradigm for individuals with metabolic disorders, offering new prospects for effective body management and enhanced quality of life.

References

1 Campbell, J.E. and Drucker, D.J. Pharmacology, physiology, and mechanisms of incretin hormone action. *Cell Metab*. 2013;17(6):819–837.

2 Nauck, M.A. and Meier, J.J. Incretin hormones: their role in health and disease. *Diabetes Obes Metab*. 2018;20 (Suppl 1):5–21.

3 Bell, G.I., Sanchez-Pescador, R., Laybourn, P.J., and Najarian, R.C. Exon duplication and divergence in the human preproglucagon gene. *Nature*. 1983;304(5924):368–371.

4 Gasbjerg, L.S., Helsted, M.M., Hartmann, B., et al. Separate and combined glucometabolic effects of endogenous glucose-dependent insulinotropic polypeptide and glucagon-like peptide 1 in healthy individuals. *Diabetes*. 2019;68(5):906–917.

5 de Graaf, C., Song, G., Cao, C., et al. Extending the structural view of class B GPCRs. *Trends Biochem Sci*. 2017;42(12):946–960.

6 Marzook, A., Tomas, A., and Jones, B. The interplay of glucagon-like peptide-1 receptor trafficking and signalling in pancreatic beta cells. *Front Endocrinol (Lausanne)*. 2021;12:678055.

7 Takeda, J., Seino, Y., Tanaka, K., et al. Sequence of an intestinal cDNA encoding human gastric inhibitory polypeptide precursor. *Proc Natl Acad Sci USA*. 1987;84(20):7005–7008.

8 Yuan, L., Zhang, J., Guo, J.-H., et al. DAla2-GIP-GLU-PAL protects against cognitive deficits and pathology in APP/PS1 mice by inhibiting neuroinflammation and upregulating cAMP/PKA/CREB signaling pathways. *J Alzheimers Dis*. 2021;80(2):695–713.

9 Nauck, M.A. and Meier, J.J. The incretin effect in healthy individuals and those with type 2 diabetes: physiology, pathophysiology, and response to therapeutic interventions. *Lancet Diabetes Endocrinol*. 2016;4(6):525–536.

10 Schirra, J., Katschinski, M., Weidmann, C., et al. Gastric emptying and release of incretin hormones after glucose ingestion in humans. *J Clin Invest*. 1996;97(1):92.

11 Nauck, M.A., El-Ouaghlidi, A., Gabrys, B., et al. Secretion of incretin hormones (GIP and GLP-1) and incretin effect after oral glucose in first-degree relatives of patients with type 2 diabetes. *Regul Pept*. 2004;122(3):209–217.

12 Watkins, J.D., Koumanov, F., and Gonzalez, J.T. Protein- and calcium-mediated GLP-1 secretion: a narrative review. *Adv Nutr*. 2021;12(6):2540–2552.

13 Drucker, D.J. and Holst, J.J. The expanding incretin universe: from basic biology to clinical translation. *Diabetologia*. 2023;66(10):1765–1779.

14 Drucker, D.J. Mechanisms of action and therapeutic application of glucagon-like peptide-1. *Cell Metab*. 2018;27(4):740–756.

15 Hare, K.J., Vilsbøll, T., Asmar, M., Deacon, C.F., Knop, F.K., and Holst, J.J. The glucagonostatic and insulinotropic effects of glucagon-like peptide 1 contribute equally to its glucose-lowering action. *Diabetes*. 2010;59(7):1765–1770.

16 Hammoud, R. and Drucker, D.J. Beyond the pancreas: contrasting cardiometabolic actions of GIP and GLP1. *Nat Rev Endocrinol*. 2023;19(4):201–216.

17 Conceição-Furber, E., Coskun, T., Sloop, K.W., and Samms, R.J. Is glucagon receptor activation the thermogenic solution for treating obesity? *Front Endocrinol (Lausanne)*. 2022;13:868037.

18 Hropot, T., Herman, R., Janez, A., Lezaic, L., and Jensterle, M. Brown adipose tissue: a new potential target for glucagon-like peptide 1 receptor agonists in the treatment of obesity. *Int J Mol Sci*. 2023;24(10):8592.

19 Xie, B., Chen, S., Xu, Y., et al. The impact of glucagon-like peptide 1 receptor agonists on bone metabolism and its possible mechanisms in osteoporosis treatment. *Front Pharmacol*. 2021;12:697442.

20 Akalestou, E., Miras, A.D., Rutter, G.A., and le Roux, C.W. Mechanisms of weight loss after obesity surgery. *Endocr Rev*. 2022;43(1):19–34.

21 Beaudry, J.L., Kaur, K.D., Varin, E.M., et al. Physiological roles of the GIP receptor in murine brown adipose tissue. *Mol Metab*. 2019;28:14–25.

22 Petersen, M.C. and Shulman, G.I. Mechanisms of insulin action and insulin resistance. *Physiol Rev*. 2018;98(4):2133–2223.

23 Seino, Y., Fukushima, M., and Yabe, D. GIP and GLP-1, the two incretin hormones: similarities and differences. *J Diabetes Investig*. 2010;1(1–2):8.

24 Zheng, Y., Ley, S.H., and Hu, F.B. Global aetiology and epidemiology of type 2 diabetes mellitus and its complications. *Nat Rev Endocrinol*. 2018;14(2):88–98.

25 Bhupathiraju, S.N. and Hu, F.B. Epidemiology of obesity and diabetes and their cardiovascular complications. *Circ Res*. 2016;118(11):1723–1735.

26 Graaf, C.D., Donnelly, D., Wootten, D., et al. Glucagon-like peptide-1 and its class B G protein-coupled receptors: a long march to therapeutic successes. *Pharmacol Rev*. 2016;68(4):954.

27 Mentlein, R., Gallwitz, B., and Schmidt, W.E. Dipeptidyl-peptidase IV hydrolyses gastric inhibitory polypeptide, glucagon-like peptide-1(7-36)amide, peptide histidine methionine and is responsible for their degradation in human serum. *Eur J Biochem*. 1993;214(3):829–835.

28 Lamers, C. Overcoming the shortcomings of peptide-based therapeutics. *Future Drug Discov*. 2022;4(2):FDD75.

29 Raufman, J.P., Singh, L., Singh, G., and Eng, J. Truncated glucagon-like peptide-1 interacts with exendin receptors on dispersed acini from guinea pig pancreas. Identification of a mammalian analog of the reptilian peptide exendin-4. *J Biol Chem*. 1992;267(30):21432–21437.

30 Folli, F. and Guardado Mendoza, R. Potential use of exenatide for the treatment of obesity. *Expert Opin Investig Drugs*. 2011;20(12):1717–1722.
31 Petersen, A.B., Knop, F.K., and Christensen, M. Lixisenatide for the treatment of type 2 diabetes. *Drugs Today (Barc)*. 2013;49(9):537–553.
32 Anderson, S.L. and Trujillo, J.M. Lixisenatide in type 2 diabetes: latest evidence and clinical usefulness. *Ther Adv Chronic Dis*. 2016;7(1):4.
33 Knudsen, L.B., Nielsen, P.F., Huusfeldt, P.O., et al. Potent derivatives of glucagon-like peptide-1 with pharmacokinetic properties suitable for once daily administration. *J Med Chem*. 2000;43(9):1664–1669.
34 Alruwaili, H., Dehestani, B., and le Roux, C.W. Clinical impact of liraglutide as a treatment of obesity. *Clin Pharmacol*. 2021;13:53–60.
35 Buse, J.B., Garber, A., Rosenstock, J., et al. Liraglutide treatment is associated with a low frequency and magnitude of antibody formation with no apparent impact on glycemic response or increased frequency of adverse events: results from the liraglutide effect and action in diabetes (LEAD) trials. *J Clin Endocrinol Metab*. 2011;96(6):1695–1702.
36 Wilding, J.P.H., Batterham, R.L., Calanna, S., et al. Once-weekly semaglutide in adults with overweight or obesity. *N Engl J Med*. 2021;384(11):989.
37 Wadden, T.A., Bailey, T.S., Billings, L.K., et al. Effect of subcutaneous semaglutide vs placebo as an adjunct to intensive behavioral therapy on body weight in adults with overweight or obesity: the STEP 3 randomized clinical trial. *JAMA*. 2021;325(14):1403–1413.
38 Harris, E. Semaglutide improves heart failure and cardiovascular disease. *JAMA*. 2023;330(12):1127.
39 Mann, J.F.E., Hansen, T., Idorn, T., et al. Effects of once-weekly subcutaneous semaglutide on kidney function and safety in patients with type 2 diabetes: a post-hoc analysis of the SUSTAIN 1-7 randomised controlled trials. *Lancet Diabetes Endocrinol*. 2020;8(11):880–893.
40 Salameh, T.S., Rhea, E.M., Talbot, K., and Banks, W.A. Brain uptake pharmacokinetics of incretin receptor agonists showing promise as Alzheimer's and Parkinson's disease therapeutics. *Biochem Pharmacol*. 2020;180:114187.
41 Wang, Z.-J., Li, X.-R., Chai, S.-F., et al. Semaglutide ameliorates cognition and glucose metabolism dysfunction in the 3xTg mouse model of Alzheimer's disease via the GLP-1R/SIRT1/GLUT4 pathway. *Neuropharmacology*. 2023;240:109716.
42 Guo, W., Xu, Z., Zou, H., et al. Discovery of ecnoglutide—a novel, long-acting, cAMP-biased glucagon-like peptide-1 (GLP-1) analog. *Mol Metab*. 2023;75:101762.
43 Cavaco, M., Castanho, M.A.R.B., and Neves, V. Peptibodies: an elegant solution for a long-standing problem. *Biopolymers*. 2017;110:e23095.

44 Gerstein, H.C., Colhoun, H.M., Dagenais, G.R., et al. Dulaglutide and cardiovascular outcomes in type 2 diabetes (REWIND): a double-blind, randomised placebo-controlled trial. *Lancet*. 2019;394(10193):121–130.

45 Blair, H.A. and Keating, G.M. Albiglutide: a review of its use in patients with type 2 diabetes mellitus. *Drugs*. 2015;75(6):651–663.

46 Campbell, J.E. Targeting the GIPR for obesity: to agonize or antagonize? Potential mechanisms. *Mol Metab*. 2021;46:101139.

47 Killion, E.A., Chen, M., Falsey, J.R., et al. Chronic glucose-dependent insulinotropic polypeptide receptor (GIPR) agonism desensitizes adipocyte GIPR activity mimicking functional GIPR antagonism. *Nat Commun*. 2020;11(1):4981.

48 Knop, F.K., Urva, S., Rettiganti, M., et al. 56-OR: a long-acting glucose-dependent insulinotropic polypeptide receptor agonist shows weight loss without nausea or vomiting. *Diabetes*. 2023;72(Suppl 1):56-OR.

49 Samms, R.J., Coghlan, M.P., and Sloop, K.W. How may GIP enhance the therapeutic efficacy of GLP-1? *Trends Endocrinol Metab*. 2020;31(6):410–421.

50 Frias, J.P., Bastyr, E.J., Vignati, L., et al. The sustained effects of a dual GIP/GLP-1 receptor agonist, NNC0090-2746, in patients with type 2 diabetes. *Cell Metab*. 2017;26(2):343.

51 Syed, Y.Y. Tirzepatide: first approval. *Drugs*. 2022;82(11):1213–1220.

52 Jastreboff, A.M., Aronne, L.J., Ahmad, NN, et al. Tirzepatide once weekly for the treatment of obesity. *N Engl J Med*. 2022;387(3):205–216.

53 Willard, F.S., Douros, J.D., Gabe, M.B., et al. Tirzepatide is an imbalanced and biased dual GIP and GLP-1 receptor agonist. *JCI Insight*. 2020;5(17).

54 Frías, J.P., Davies, M.J., Rosenstock, J., et al. Tirzepatide versus semaglutide once weekly in patients with type 2 diabetes. *N Engl J Med*. 2021;385(6):503–515.

55 Lyons, S.A. and Beaudry, J.L. Synergistic combinations of gut- and pancreas-hormone-based therapies: advancements in treatments for metabolic diseases. *Endocrinology*. 2023;164(11):bqad153.

56 Zanchi, D., Depoorter, A., Egloff, L., et al. The impact of gut hormones on the neural circuit of appetite and satiety: a systematic review. *Neurosci Biobehav Rev*. 2017;80:457–475.

57 Steinert, R.E., Feinle-Bisset, C., Asarian, L., Horowitz, M., Beglinger, C., and Geary, N. Ghrelin, CCK, GLP-1, and PYY(3-36): secretory controls and physiological roles in eating and glycemia in health, obesity, and after RYGB. *Physiol Rev*. 2017;97(1):411–463.

58 Al-Massadi, O., Fernø, J., Diéguez, C., Nogueiras, R., and Quiñones, M. Glucagon control on food intake and energy balance. *Int J Mol Sci*. 2019;20(16):3905.

59 Capozzi, M.E., D'Alessio, D.A., and Campbell, J.E. The past, present, and future physiology and pharmacology of glucagon. *Cell Metab*. 2022;34(11):1654–1674.

60 Kounatidis, D., Vallianou, N.G., Tsilingiris, D., et al. Therapeutic potential of GLP-2 analogs in gastrointestinal disorders: current knowledge, nutritional aspects, and future perspectives. *Curr Nutr Rep*. 2022;11(4):618–642.
61 Mathiesen, D.S., Lund, A., Holst, J.J., Knop, F.K., Lutz, T.A., and Bagger, J.I. Therapy of endocrine disease: amylin and calcitonin—physiology and pharmacology. *Eur J Endocrinol*. 2022;186(6).
62 Adrian, T.E., Ferri, G.L., Bacarese-Hamilton, A.J., Fuessl, H.S., Polak, J.M., and Bloom, S.R. Human distribution and release of a putative new gut hormone, peptide YY. *Gastroenterology*. 1985;89(5):1070–1077.
63 Batterham, R.L. and Bloom, S.R. The gut hormone peptide YY regulates appetite. *Ann N Y Acad Sci*. 2003;994:162–168.
64 Geng, L., Lam, K.S.L., and Xu, A. The therapeutic potential of FGF21 in metabolic diseases: from bench to clinic. *Nat Rev Endocrinol*. 2020;16(11):654–667.
65 Wang, D., Day, E.A., Townsend, L.K., Djordjevic, D., Jørgensen, S.B., and Steinberg, G.R. GDF15: emerging biology and therapeutic applications for obesity and cardiometabolic disease. *Nat Rev Endocrinol*. 2021;17(10):592–607.
66 Pocai, A., Carrington, P.E., Adams, J.R., et al. Glucagon-like peptide 1/glucagon receptor dual agonism reverses obesity in mice. *Diabetes*. 2009;58(10):2258–2266.
67 Wynne, K., Field, B.C., and Bloom, S.R. The mechanism of action for oxyntomodulin in the regulation of obesity. *Curr Opin Investig Drugs*. 2010;11(10):1151–1157.
68 Ambery, P., Parker, V.E., Stumvoll, M., et al. MEDI0382, a GLP-1 and glucagon receptor dual agonist, in obese or overweight patients with type 2 diabetes: a randomised, controlled, double-blind, ascending dose and phase 2a study. *Lancet*. 2018;391(10140):2607–2618.
69 Schiavon, M., Visentin, R., Göbel, B., et al. Improved postprandial glucose metabolism in type 2 diabetes by the dual glucagon-like peptide-1/glucagon receptor agonist SAR425899 in comparison with liraglutide. *Diabetes Obes Metab*. 2021;23(8):1795–1805.
70 Corbin, K.D., Carnero, E.A., Allerton, T.D., et al. Glucagon-like peptide-1/glucagon receptor agonism associates with reduced metabolic adaptation and higher fat oxidation: a randomized trial. *Obesity (Silver Spring)*. 2023;31(2):350–362.
71 Hope, D.C.D., Vincent, M.L., and Tan, T.M.M. Striking the balance: GLP-1/glucagon co-agonism as a treatment strategy for obesity. *Front Endocrinol (Lausanne)*. 2021;12:735019.
72 Ji, L., Gao, L., Jiang, H., et al. Safety and efficacy of a GLP-1 and glucagon receptor dual agonist mazdutide (IBI362) 9 mg and 10 mg in Chinese adults

with overweight or obesity: a randomised, placebo-controlled, multiple-ascending-dose phase 1b trial. *EClinicalMedicine*. 2022;54:101691.

73 Jiang, H., Pang, S., Zhang, Y., et al. A phase 1b randomised controlled trial of a glucagon-like peptide-1 and glucagon receptor dual agonist IBI362 (LY3305677) in Chinese patients with type 2 diabetes. *Nat Commun*. 2022;13(1):3613.

74 Zimmermann, T., Thomas, L., Baader-Pagler, T., et al. BI 456906: discovery and preclinical pharmacology of a novel GCGR/GLP-1R dual agonist with robust anti-obesity efficacy. *Mol Metab*. 2022;66:101633.

75 Nestor, J.J., Parkes, D., Feigh, M., Suschak, J.J., and Harris, M.S. Effects of ALT-801, a GLP-1 and glucagon receptor dual agonist, in a translational mouse model of non-alcoholic steatohepatitis. *Sci Rep*. 2022;12(1):6666.

76 Finan, B. and Douros, J.D. GLP-1/GIP/glucagon receptor triagonism gets its try in humans. *Cell Metab*. 2022;34(1):3–4.

77 Finan, B., Yang, B., Ottaway, N., et al. A rationally designed monomeric peptide triagonist corrects obesity and diabetes in rodents. *Nat Med*. 2015;21(1):27–36.

78 Zhao, F., Zhou, Q., Cong, Z., et al. Structural insights into multiplexed pharmacological actions of tirzepatide and peptide 20 at the GIP, GLP-1 or glucagon receptors. *Nat Commun*. 2022;13(1):1057.

79 Knerr, P.J., Mowery, S.A., Douros, J.D., et al. Next generation GLP-1/GIP/glucagon triple agonists normalize body weight in obese mice. *Mol Metab*. 2022;63:101533.

80 Coskun, T., Urva, S., Roell, W.C., et al. LY3437943, a novel triple glucagon, GIP, and GLP-1 receptor agonist for glycemic control and weight loss: from discovery to clinical proof of concept. *Cell Metab*. 2022;34(9):1234.

81 Urva, S., Coskun, T., Loh, M.T., et al. LY3437943, a novel triple GIP, GLP-1, and glucagon receptor agonist in people with type 2 diabetes: a phase 1b, multicentre, double-blind, placebo-controlled, randomised, multiple-ascending dose trial. *Lancet*. 2022;400(10366):1869–1881.

82 Jastreboff, A.M., Kaplan, L.M., Frías, J.P., et al. Triple-hormone-receptor agonist retatrutide for obesity – a phase 2 trial. *N Engl J Med*. 2023;389(6):514–526.

83 Wilhelm, S.M., Lipari, M., Kulik, J.K., and Kale-Pradhan, P.B. Teduglutide for the treatment of short bowel syndrome. *Ann Pharmacother*. 2014;48(9):1209–1213.

84 Naimi, R.M., Hvistendahl, M., Enevoldsen, L.H., et al. Glepaglutide, a novel long-acting glucagon-like peptide-2 analogue, for patients with short bowel syndrome: a randomised phase 2 trial. *Lancet Gastroenterol Hepatol*. 2019;4(5):354–363.

85 Yang, P.-Y., Zou, H., Lee, C., et al. Stapled, long-acting glucagon-like peptide 2 analog with efficacy in dextran sodium sulfate induced mouse colitis models. *J Med Chem*. 2018;61(7):3218–3223.

86 Globel Newswire by notified. Zealand Pharma Presents Data from Phase 1 Trial of Dapiglutide at the 82nd Annual American Diabetes Association Scientific Sessions and Announces Dapiglutide to Move into Phase 2 Trial for Obesity. https://www.globenewswire.com/news-release/2022/06/06/2457233/0/en/Zealand-Pharma-Presents-Data-from-Phase-1-Trial-of-Dapiglutide-at-the-82nd-Annual-American-Diabetes-Association-Scientific-Sessions-and-Announces-Dapiglutide-to-Move-into-Phase-2-T.html. [cited 2023 Nov 2].

87 Reiner, J., Berlin, P., Held, J., et al. Dapiglutide, a novel dual GLP-1 and GLP-2 receptor agonist, attenuates intestinal insufficiency in a murine model of short bowel. *JPEN J Parenter Enteral Nutr*. 2022;46(5):1107–1118.

88 Singh-Franco, D., Robles, G., and Gazze, D. Pramlintide acetate injection for the treatment of type 1 and type 2 diabetes mellitus. *Clin Ther*. 2007;29(4):535–562.

89 Larsen, A.T., Mohamed, K.E., Sonne, N., et al. Does receptor balance matter? - Comparing the efficacies of the dual amylin and calcitonin receptor agonists cagrilintide and KBP-336 on metabolic parameters in preclinical models. *Biomed Pharmacother*. 2022;156:113842.

90 Bello, N.T., Kemm, M.H., Ofeldt, E.M., and Moran, T.H. Dose combinations of exendin-4 and salmon calcitonin produce additive and synergistic reductions in food intake in nonhuman primates. *Am J Physiol Regul Integr Comp Physiol*. 2010;299(3):R945–R952.

91 Liberini, C.G., Koch-Laskowski, K., Shaulson, E., et al. Combined Amylin/GLP-1 pharmacotherapy to promote and sustain long-lasting weight loss. *Sci Rep*. 2019;9(1):8447.

92 Frias, J.P., Deenadayalan, S., Erichsen, L., et al. Efficacy and safety of co-administered once-weekly cagrilintide 2·4 mg with once-weekly semaglutide 2·4 mg in type 2 diabetes: a multicentre, randomised, double-blind, active-controlled, phase 2 trial. *Lancet*. 2023;402(10403):720–730.

93 Thorsø Larsen, A., Karsdal, M.A., and Henriksen, K. Treatment sequencing using the dual amylin and calcitonin receptor agonist KBP-336 and semaglutide results in durable weight loss. *Eur J Pharmacol*. 2023;954:175837.

94 Lear, S., Pflimlin, E., Zhou, Z., et al. Engineering of a potent, long-acting NPY2R agonist for combination with a GLP-1R agonist as a multi-hormonal treatment for obesity. *J Med Chem*. 2020;63(17):9660–9671.

95 Chen, W., Binbin, G., Lidan, S., Qiang, Z., and Jing, H. Evolution of peptide YY analogs for the management of type 2 diabetes and obesity. *Bioorg Chem*. 2023;140:106808.

96 Østergaard, S., Paulsson, J.F., Kjaergaard Gerstenberg, M., and Wulff, B.S. The design of a GLP-1/PYY dual acting agonist. *Angew Chem Int Ed Engl*. 2021;60(15):8268–8275.

97 Müller, T.D., Blüher, M., Tschöp, M.H., and DiMarchi, R.D. Anti-obesity drug discovery: advances and challenges. *Nat Rev Drug Discov*. 2022;21(3):201–223.

98 Milliken, B.T., Elfers, C., Chepurny, O.G., et al. Design and evaluation of peptide dual-agonists of GLP-1 and NPY2 receptors for glucoregulation and weight loss with mitigated nausea and emesis. *J Med Chem*. 2021;64(2):1127–1138.

99 Yang, Q., Tang, W., Sun, L., et al. Design of xenopus GLP-1-based long-acting dual GLP-1/Y2 receptor agonists. *J Med Chem*. 2022;65(20):14201–14220.

100 Yuan, Y., Yan, Z., Lao, Q., et al. Discovery of a potent and long-acting xenopus GLP-1-based GLP-1/glucagon/Y2 receptor triple agonist. *Eur J Med Chem*. 2023;247:115036.

101 Harrison, S.A., Ruane, P.J., Freilich, B.L., et al. Efruxifermin in non-alcoholic steatohepatitis: a randomized, double-blind, placebo-controlled, phase 2a trial. *Nat Med*. 2021;27(7):1262–1271.

102 Loomba, R., Sanyal, A.J., Kowdley, K.V., et al. Randomized, controlled trial of the FGF21 analogue pegozafermin in NASH. *N Engl J Med*. 2023;389(11):998.

103 Luo, Y., Decato, B.E., Charles, E.D., et al. Pegbelfermin selectively reduces secondary bile acid concentrations in patients with non-alcoholic steatohepatitis. *JHEP Rep*. 2022;4(1):100392.

104 ClinicalTrials.gov. A study of BOS-580 in obese subjects at risk for, or with biopsy-confirmed, nonalcoholic steatohepatitis (NASH. https://clinicaltrials.gov/study/NCT04880031. [cited 2023 Nov 2].

105 Pan, Q., Lin, S., Li, Y., et al. A novel GLP-1 and FGF21 dual agonist has therapeutic potential for diabetes and non-alcoholic steatohepatitis. *EBioMedicine*. 2021;63:103202.

106 Gilroy, C.A., Capozzi, M.E., Varanko, A.K., et al. Sustained release of a GLP-1 and FGF21 dual agonist from an injectable depot protects mice from obesity and hyperglycemia. *Sci Adv*. 2020;6(35):eaaz9890.

107 Hong, H.N., Kim, J.H., Choi, H.H., et al. YH25724, a novel long-acting GLP-1/FGF21 dual agonist lowers both non-alcoholic fatty liver disease activity score and fibrosis stage in a diet-induced obese mouse model of biopsy-confirmed non-alcoholic steatohepatitis. *J Hepatol*. 2017;66:S16–S17.

108 Benichou, O., Coskun, T., Gonciarz, M.D., et al. Discovery, development, and clinical proof of mechanism of LY3463251, a long-acting GDF15 receptor agonist. *Cell Metab*. 2023;35(2):274.

109 Lim, S., Kim, D.-H., Yang, J., et al. 235-LB: YH34160, a novel long-acting GDF15 fusion protein, exerts potent and sustained body weight loss in rodent obesity models. *Diabetes*. 2022;71(Suppl 1):235–LB.

110 Zhang, Y., Zhao, X., Dong, X., et al. Activity-balanced GLP-1/GDF15 dual agonist reduces body weight and metabolic disorder in mice and non-human primates. *Cell Metab*. 2023;35(2):287.

10

Peptide-based Therapeutics for Inflammatory and Autoimmune Diseases

Sunay V. Chankeshwara, Werngard Czechtizky and Wu Su

Medicinal Chemistry, Research and Early Development, Respiratory and Immunology, BioPharmaceuticals R&D, AstraZeneca, Gothenburg, Sweden

10.1 Introduction

Inflammation is an essential physiological defense mechanism against infection and tissue damage [1]. The inflammatory response includes activating biological mechanisms like the origin of pro-inflammatory mediators, either produced or secreted, such as chemokines, cytokines, interleukins (IL), and growth factors [2]. The regulated inflammatory process is crucial for maintaining or restoring homeostasis in damaged biological compartments. However, excessive inflammation, which sometimes lasts for years, causes significant damage to tissues and organs [3]. Increasing evidence has shown that inflammatory dysregulation is closely associated with many chronic inflammatory diseases and, notably, autoimmune diseases, including rheumatoid arthritis, inflammatory bowel disease (IBD), and systemic lupus erythematosus (SLE) [4].

Over the last decades, critical focus and important progress have been made in gathering insights and building an understanding of the pathogenic mechanisms involved in chronic inflammatory diseases. Inflammatory stimuli like pathogen-derived molecules, damaged cell products, toxins, or allergens trigger inflammatory reactions, leading to the release of cell-derived mediators such as IL, cytokines, tumor necrosis factor-alpha (TNF-α),

Peptide Science: Chemical Ligation, Lead Generation, and Therapeutic Advances, First Edition. Fa Liu.

prostaglandins, nitric oxide, and leukotrienes [5]. Several key research findings have identified cellular and molecular targets, as well as soluble molecules, leading to the creation of specific tools to block, neutralize, or modulate key pathways that are disrupted in chronic inflammatory diseases [6]. For example, using monoclonal antibodies (mAb) to inhibit cytokines or their receptors has effectively reduced chronically high cytokine signaling and uncontrolled cytokine effects [7].

The currently available biological therapies include a panel of mAb directed to cytokines. The discovery of mAb and their introduction in clinical practice for the use directed against the pro-inflammatory cytokine TNF-α, e.g. infliximab, adalimumab, golimumab, and certolizumab pegol, revolutionized chronic inflammatory disease management. Further, these anti-TNF-α drugs induce and maintain clinical remission and mucosal healing, which reduces the need for corticosteroids, hospitalization, and surgery. This improved the health-related quality of life of chronic inflammatory disease patients massively [7,8]. Unfortunately, anti-TNF-α therapy has been associated with infectious complications, and a high rate of primary and secondary non-responsiveness limits clinical efficacy. Further, the production of antibodies against the TNF-α antagonists neutralizes the drugs and accelerates their clearance or induces reactions that consequently lead to loss of therapeutic response. Therefore, new therapies equipped with better safety profiles and specific targeting mechanisms of inflammation were clinically approved for practice, e.g. vedolizumab and ustekinumab, with many others currently being studied [9].

Small molecules targeting intracellular components have also been developed and approved for clinical use (such as IKK/NFkb inhibitors and JAK inhibitors) [10]. Unfortunately, some small molecule inhibitors are associated with high toxicity, low selectivity, and a wide range of undesired side effects, including blood–brain barrier crossing and the generation of toxic molecules after metabolization. This can make them inadequate in prolonged therapy [11].

Synthetic peptides constitute highly attractive tools in the panel of emerging therapeutics for inflammatory diseases [12]. Peptides, with respect to small molecules, usually possess higher potency and specificity for their targets, mainly because of their chemical and consequent biological selection processes. The last two decades witnessed the emergence of innovative technologies as well as the expansion of peptide libraries and scaffolds enabled by peptide arrays and biosensors that have opened up a new perspective [13]. Enhanced peptide array capability led to the construction of hundreds and thousands of peptides, which are not only screened in high-throughput fashion but also new reactivities have been characterized by completely novel strategies. In comparison to therapeutic fusion proteins and mAb, peptides are uniquely positioned to offer many advantages with their cost of production and purification and the possibility to scale up drug substances under good manufacturing practice conditions. Furthermore, peptides bearing natural or non-natural modifications and peptide mimics of protein or non-protein antigens (DNA, RNA, carbohydrates) can be designed and may potentially replace native antigens. Thus, among alternative strategies developed to treat patients with chronic inflammatory diseases, immune intervention based on synthetic peptides occupies a place of choice [14].

Several endogenous peptides secreted during inflammatory response and others deriving from interacting regions of protein complexes triggered by inflammatory mediators showed anti-inflammatory properties and might be engaged in developing new therapies [15]. According to the 2022–2030 forecast reports compiled by Growth Plus Report, the global market size of the ant-inflammatory peptide is expected to reach US$139.4 billion by 2030 [16]. In this chapter, inspired by a strong research drive and the rising demand in the field of anti-inflammatory peptides, we review the peptide candidates in current clinical trials for inflammatory disease as leading indications, their targets, peptide origins, pharmacology, delivery routes, and current status.

10.2 Literature Search and Methods

Literature sources included Cortellis, SciFinder, PubMed, and Google Scholar databases (date of last search: 18 September 2023). The keywords and filters included in the search were "anti-inflammatory peptides" or "autoinflammatory disease" or "dermatitis" or "fibrosis" or" gastrointestinal inflammation" or "genitourinary tract inflammation" or "granulomatosis" or "musculoskeletal system inflammation" or "ulcerative colitis" or "ocular and orbital inflammation" or "systemic inflammatory response syndrome" or "peptide" or "clinical trials" or "cardiovascular disease" or "dermatological disease" or "immune disorder" or "inflammatory disease" or "respiratory disease" or "inflammation and autoimmune disease" or "systemic lupus erythematosus" or "LSALT PEPTIDE" or "EA 230" or "Larazotide" or "AT-1001" or "Forigerimod" or "Lupuzor" or "IPP-201101," or "P140" or "systemic lupus erythematosus" or "B27PD" or "SP-16 SERPIN-like Peptide" or "SP-16a SERPIN-Like, Small Peptide Agonist" or "AMTX-100" or "JNJ-2113" or "Jusvinza" or "HSP60" or "RLS-0071" or "GLP-1" or "Optiquel" or "JNJ2113."

10.3 Representative Anti-inflammatory Peptides in Clinical Trials

10.3.1 AMTX-100

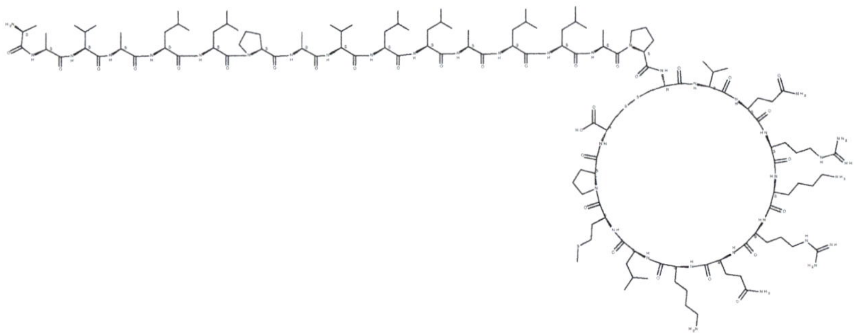

AMTX-100 is a novel 28-amino acid synthetic cyclic polypeptide (formerly cSN50.1) [17–21]. It is bioengineered by combining the hydrophobic domain of the human fibroblast growth factor-4 and a cell-permeable cyclized motif of the nuclear localization sequence of the nuclear factor kappa B. AMTX-100 (cSN50.1) was initially discovered by Jacek Hawiger at Vanderbilt University [17,22] and later licensed to the biopharmaceutical company Amytrx Therapeutics, Inc. The Food and Drug Administration (FDA) approved AMTX-100 for testing as a potential therapeutic for mild-to-moderate eczema [18–21].

Cytokines, TNF-α, interferon-gamma (IFN-γ), and IL-1 are mediators of localized and systemic responses. Activating genes encoding cytokines involves sequestering cytoplasmic nuclear translocation of stress-responsive transcription factors (SRTFs) and mobilization to the nucleus following cellular stimulation. AMTX-100 is a novel targeted agent that targets two cytoplasmic nuclear transport shuttles, importin α5 and importin β1. This nuclear transport checkpoint inhibitor (NTCI) competitively inhibits transcription factors [16], such as activator protein 1 (AP-1) and STAT1, thereby reducing the production of pro-inflammatory cytokines and chemokines. Thus, NTCI affects downstream SRTF-regulated genes.

One of the most complex and chronic inflammatory skin disorders is atopic dermatitis (AD), which is directly linked to dysregulation of the immune system. The development of AD and skin barrier dysfunction is associated with the involvement of Type 2 cytokines, IL-17, and IL-22 [19,20]. The traditional management of AD includes topical steroids as a standard of care to reduce inflammation. However, second-line agents and nonsteroidal alternatives, including topical calcineurin inhibitors and phosphodiesterase inhibitors, act as immunosuppressants and inhibit T-cell activation, decreasing the expression of pro-inflammatory cytokines. The current therapeutic agent and more than 40 topical treatments in the pipeline are dominated by small molecules and antibody therapeutics.

AMTX-100 is considered as first-in-class and best-in-class peptide therapeutics representing a targeted approach (T-cell transcription factor NFAT modulator; AP-1 transcription factor modulator; STAT-1 modulator) that attenuates the intracellular molecular machinery in immune cells and suppresses inflammation originating from the gene activation process. Amytrx is initiating the second part of the Phase 1/2 trial (NCT04313400), investigating AMTX-100 CF3 in adults with mild-to-moderate AD. Data from the first part of the trial – Phase 1/2a with safety and exploratory endpoints – showed that half of the patients achieved total or almost total clearance of AD lesions in the 7-day treatment protocol. Currently, it is in a two-part, Phase 1/2 clinical trial (NCT04313400) in the United States sponsored by AmbioPharm Inc. and Amytrx Therapeutics, Inc., for topically applied AMTX-100 safety and efficacy in adult patients with mild-to-moderate AD [18,21,23].

10.3.2 B27PD

B27PD (also known as HLA-B27, Optiquel®) is a small peptide identified from the sequence of human leukocyte antigen (HLA)-B molecules [24–26]. The peptide sequences comprised the residues from 125 to 138 [24,27]. It mimics the retinal S-antigen, which is responsible for ocular inflammation [28].

Specific organs, e.g. the inner eye and the inner ear, are regarded as immune-privileged and essential for the survival of the individual and/or the species. These survival organs are often threatened by autoimmune diseases and exposure to cross-reactive antigens, e.g. autoimmune uveitis.

"Antigenic mimicry" refers to the similarity between antigens that causes cross-reactivity with immune receptors, while similar antigen epitopes are referred to as "mimotopes." The postulation believes the initiation of autoimmune diseases is caused by antigenic mimicry, while the similarity of auto-antigens and mimotopes from infectious agents misguide immune cells to combat against self. Antigenic mimicry is postulated to be the cause and initiating event of autoimmune diseases when mimotopes from infectious agents look similar to autoantigens and thus confuse immune cells to combat against self erroneously.

The HLA system is also called the major histocompatibility complex (MHC) in humans. It is an essential part of the immune system, which is controlled by genes located on chromosome 6 that help code for proteins on the cell surface to present antigenic peptides to the T-cell receptor (TCR) on T cells that differentiate between self and nonself.

The recognition of peptides presented in MHC Class I and II can only be done by TCRs. Further, TCRs of CD8+ and CD4+ T helper cells recognize antigen peptides on MHC Class I and II, respectively. Further, MHC Class I molecules binds to shorter peptide sequences containing 8–9 amino acids, while MHC Class II molecules can bind longer peptides.

Therefore, HLA molecules play a pivotal role in immune recognition, and several autoimmune diseases are associated with certain HLA molecules with their antigen presentation. HLA-B27 was the first HLA molecule discussed in the context of antigenic mimicry and autoimmunity. Specific HLA alleles are linked to some autoimmune disorders – for example, ankylosing spondylitis and reactive arthritis to HLA-B27. Further, HLA- HLA-B27-associated uveitis is commonly found in uveitis worldwide.

The peptide sequence representing B27PD offers specificity for several HLA-B antigens, excluding HLA-A or -C. B27PD and retinal S-antigen-peptide PDSAg share five discontinuous amino acid homologies that serve as anchor positions for HLA Class I1 antigen presentation. Using peptide B27PD, the earlier Phase 1 studies on patients with therapy-refractive uveitis had promising outcomes. However, the later Phase 1/2 trial (NCT01195948) with oral B27PD peptide with patients under concomitant immunosuppression lacked efficacy [28,29].

10.3.3 EA-230

EA-230 is a synthetic linear tetrapeptide (AQGV) identified from the human chorionic gonadotropin (hCG), also called the "pregnancy hormone" [30–34]. Immunomodulatory effects of hCG are well documented and originate from the b-loop of hCG, which is abundantly present throughout pregnancy. Several oligopeptides derived from the b-loop of hCG are studied for immunomodulation; a specific tetrapeptide, Ala—Glu—Gly—Val (later named EA-230), effective on attenuation of systemic inflammation, thus reducing organ failure and mortality in hemorrhage models. The mechanism of action of EA-230 is still unclear; however, the adaptation of the maternal immune system during the pregnancy for tolerating a semi-allogeneic fetus and protection from invading pathogens inspired the immunomodulatory strategies to combat inflammation [30–32].

In a preclinical study on renal injury models, EA-230 significantly attenuated systemic inflammation by the attenuation of circulating levels of IL-6/10, IFN-γ, and TNF-α. Further, EA-230 promotes the preservation of renal function and increases survival by reducing neutrophil influx, lowering the expression of the leukocyte adhesion molecule E-selectin in renal tissue [33].

A Phase 1/2a trial study conducted by Radboud University and Exponential Biotherapies Inc. in Europe (NCT02629874) confirmed its immunomodulatory effects, and an optimal dose was established. Owing to all-natural amino acid residues and short sequence, EA-230 peptide exhibited a short elimination half-life and a large volume of distribution. In the highest dosage group, EA-230 peptide showed a favorable safety profile. Further, a Phase 2 clinical study (NCT03145220) was conducted to test the effects of EA-230 on the systemic inflammatory response following on-pump cardiac surgery. EA-230 was administered during surgery by infusion of 90 mg/kg/h over two hours as the highest dose. The study concluded that EA-230 was safe in patients undergoing on-pump cardiac surgery. It did not modulate IL-6 plasma concentrations but appeared to exert beneficial renal and cardiovascular effects and shorten in-hospital length of stay. Efficacy results were more pronounced in patients with longer surgery duration and thus longer study drug infusion duration [34].

10.3.4 Lupuzor

Lupuzor™ (also known as Forigerimod or P140, IPP-201101; Cephalon CEP-33457, Rigerimod; SyB-1001; SyBL-1001 peptide) is a 21-amino-acid fragment of U1 small nuclear ribonucleoprotein (snRNP) that is phosphorylated at Ser11 (RIHMVYSKR-{pSer}-GKPRGYAFIEY) [35,36]. Lupuzor™ was licensed to Cephalon Inc. in February 2009. In late 2011, following the acquisition of Cephalon by Teva Pharmaceuticals, ImmuPharma PLC regained all product rights to the drug.

P140 is a spliceosomal peptide of U1 snRNP that is suggested to prevent the proliferation of CD4+ T cells, thus promoting tolerance. It appeared to decrease IL-10 secretion and anti-dsDNA levels. The P140 peptide can potently inhibit autophagy and target SLE, a chronic, life-threatening autoimmune, inflammatory disease with a pattern of areas and remission.

As a CD4+ T-cell signaling modulator in major indications such as chronic inflammatory demyelinating polyneuropathy (CIDP) and SLE. Other known actions of lupuzor are T-lymphocyte modulator and immunomodulator. P140, a phosphopeptide, functions by binding to some residues on the heat shock protein 8 (HSPA8)-nonbinding domain. It modulates the immune system by binding to HSPA8, inhibiting the trafficking of the autophagosome to the lysosome. It has a novel action distinct and upstream from approved drugs. Unlike other therapies, it is not an immunosuppressant. It "normalizes" what is otherwise a hyperactive immune state [37].

P140 as Lupuzor™ received fast-track designation from the FDA in an orphan drug category. The current delivery route is a subcutaneous formulation. ImmuPharm PLC is developing a pharmacologically improved version of P140 to provide therapies with improved administration modalities while maintaining P140 as the active moiety.

The emerging research on P140 has revealed new indications and shared the common cause at the mechanistic level of the cell. Several preclinical studies have confirmed P140 activity in asthma (acute and chronic), gout, and periodontitis.

P140 is under clinical development, with the potential to be a novel first-line drug therapy for the treatment of lupus by Immupharma and is currently in the Phase 2/3 trial process for CIDP [35].

ImmuPharma completed multisite Phase 3 trial studies (NCT03427151 and NCT02504645) of repeated administration of a 200 μg dose of ipp-201101 plus standard of care in patients with SLE. ImmuPharma PLC is preparing to move into an adaptive Phase 2/3 international pivotal trial in H2–2023 [38].

10.3.5 SP16

SP16 is a novel synthetic short linear peptide developed by Serpin Pharma for treating autoimmune and inflammatory diseases [39–43]. It is derived from the carboxyl terminus (corresponding to residues 364–380, VKFNKPFVFL-Nle-IEQNTK) of an endogenous serine protease of α1-antitrypsin (AAT). It contains recognition sequences for the low-density lipoprotein receptor-related protein 1 (LRP1). SP16, as an LRP1 agonist, mimics the LRP1 binding portion of the AAT, leading to LRP1 activation, which modulates inflammation and promotes cell survival. The SP16 contains the 5-amino acid sequence homology with endogenous serpin AAT; however, the peptide does not need to bind to a serine protease to become active.

The LRP1, or cluster of differentiation 91, is a multifunctional transmembrane receptor that plays a role in lipoprotein endocytosis, signal transduction, and modulates the inflammatory response to various noxious stimuli. Thus, regulation and multifunction by LRP1 affects cytokine secretion, phagocytosis, and migration of immune system cells. LRP1 receptor is ubiquitously expressed in different organs and under various pathophysiological conditions and, thereby, inflammation linked to atherosclerosis, neurodegeneration, and organ fibrosis, for example, of the lung or liver. As a scavenger receptor, LRP1 internalizes endogenous ligands and harmful substances such as pathogens, debris, and apoptotic cells, thereby limiting the inflammatory response and helping to control tissue damage. Additionally, as a scaffold receptor, LRP1 modulates the activity of other membrane proteins, including integrins and receptor tyrosine kinases. Consequently, targeting LRP1 to modulate the inflammatory response enhances its protective functions.

These peculiar properties enable LRP1 to couple extracellular microenvironments and intracellular signaling and response. Serine protease inhibitors are endogenous proteins, including AAT, α2-macroglobulin, and antithrombin III, known as SERPINs, which inactivate proteolytic enzymes released during inflammatory responses. SERPINS and other ligands have been found to transduce intracellular signals through actions on LRP1. In a Phase 1 trial (NCT03651089) conducted at Virginia Commonwealth University in the United States, sponsored by Serpin Pharma, the peptide showed no safety concerns, no off-target effects, no cardiac liability, no humoral immune response, and no drug-specific toxicity. Based on pre-clinical research and LRP1 multifunction, the SP16 peptide can potentially explore a wide range of chronic and acute inflammatory diseases involving cytokines and cell signaling pathways (JAK/STAT) [42,43].

10.3.6 JNJ2113

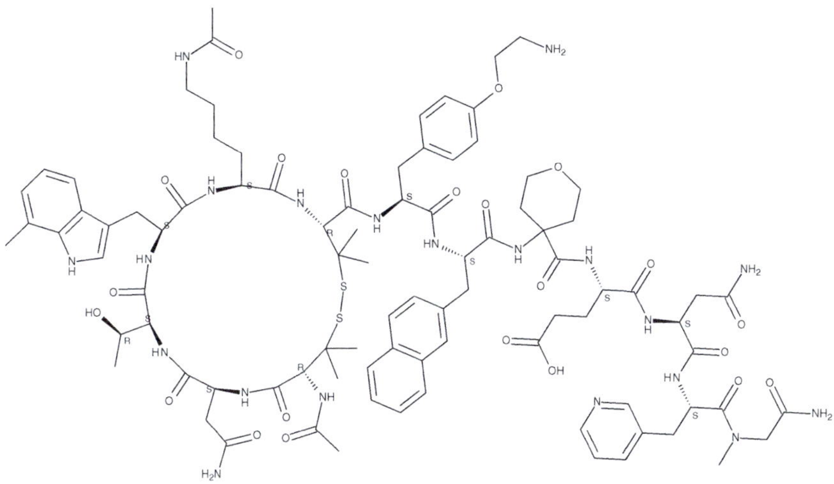

Source: Ref. [44]/Russell Publishing Limited.

JNJ-2113 is a novel and high-affinity IL-23R antagonist peptide absorbed with oral dosing [44]. IL-23 is critical in regulating the cell regulation pathways and mechanisms involved in inflammation. Specifically, it enhances the T helper Type 17 (Th17) expansion that potentiates myriads of the inflammatory autoimmune responses. Along with the IL-23 and IL-17, a new axis formed through Th17 cells, further initiating the various pathogenic responses associated with immune activation and chronic inflammation. Therefore, targeting IL-23 as a receptor or IL-23 axis holds great potential for therapeutic modalities for treating a range of autoimmune diseases that include psoriasis, IBD (Crohn's disease), RA, and multiple sclerosis.

Psoriasis is a common immune-mediated chronic epithelial inflammatory Type 1 cytokine-related skin disease characterized by abnormal keratinocyte proliferation and differentiation and infiltrates of activated memory T cells with high IFN-γ expression.

Remarkably, the effect of anti-human IL-23 mAb was as efficacious as anti-TNF mAb, which is currently the standard therapy for psoriasis. A recent study further expands the importance of blocking the IL-23 pathway on psoriasis' epithelial and immunological components. While IFN-γ is one of the major cytokines in psoriasis, IL-12 and IL-23 have also been specifically implicated in the disease's development due to their crucial roles in linking the innate and adaptive immune responses. Functionally, IL-12 induces and sustains Th1 immune responses, leading to the secretion of IFN-α and the homing of T cells to the skin. However, IL-23 activates macrophages and maintains chronic autoimmune inflammation via regulating T memory cells and the induction of IL-17. Serum level of IL-23 was negatively correlated with the disease duration. IL-23 was higher in the earlier lesion. This might suggest that it could be a crucial early mediator in the induction of psoriatic lesions.

IL-23 can be targeted by using an antibody against IL-12/IL-23; nevertheless, it would be much more helpful to design drugs that target the IL-23p19 or IL-23 receptor, so inhibiting IL-23 without modifying the effects of IL-12.

JNJ-2113 is a disulfide-bridged macrocyclic peptide consisting of 13 amino acids with a single-digit picomolar affinity for the IL-23 receptor. As a potent and the first oral IL-23 antagonist, JNJ-2113 selectively inhibits IL-23 signaling in human T cells, resulting in the inhibition of inflammatory cytokine production. JNJ-2113 (formerly PN-235) was discovered and is being developed according to the license and collaboration agreement between Protagonist Therapeutics and Janssen Biotech, Inc. In 2017, the license and collaboration agreement established between the Protagonist and Janssen Biotech, Inc. enabled the companies to collaborate to discover and develop next-generation compounds that ultimately led to JNJ-2113. According to the agreement, including a broad range of therapeutic indications, the worldwide rights to further develop JNJ-2113 in Phase 2 clinical trials and beyond up to commercialization of compounds derived from the research are retained by Janssen.

10.3.7 Larazotide

Larazotide acetate, also known as AT-1001, is a synthetic peptide consisting of eight amino acids (sequence: GGVLVQPG) derived from the C-terminal of the Vibrio cholerae protein zonula occludens toxin [45–48]. It acts as an anti-zonulin receptor inhibitor, reducing zonulin-mediated increases in intestinal barrier permeability. Larazotide acetate also helps rearrange various tight junction proteins and cytoskeletal actin filaments to enhance tight junction assembly. Additionally, it suppresses the enzyme myosin light chain kinase, reducing tension on cytoskeletal actin filaments and promoting tight junction closure. Overall, larazotide acetate restores the integrity of tight junctions, preventing the intestinal permeation of gliadin, an immunopathogenic substance from gluten metabolism [46,48].

Preclinically, larazotide acetate has demonstrated positive effects by enhancing tight junction integrity and reducing paracellular permeability in various *in vitro* and *in vivo* models of celiac disease (CD). Clinically, it has been tested for tolerability in two Phase 1 trials involving healthy volunteers. Single and multiple doses of larazotide acetate, ranging from 0.25 to 36 mg, did not cause any severe drug-related side effects, and no participants withdrew from the studies. The most common mild side effect reported was a headache. These Phase 1 trials confirmed the safety of larazotide acetate in healthy volunteers [47].

Following this, the safety and efficacy of larazotide acetate have been evaluated in multiple doses across four randomized placebo-controlled trials in CD patients, both with and without gluten challenge. In these trials, larazotide acetate proved to be more effective than placebo in reducing gastrointestinal symptoms in gluten-challenged patients, with a well-tolerated safety profile.

The analysis of an independent statistician involved in the ongoing Phase 3 trial, CedLara, has concluded the ongoing requirement to add a substantial number of new patients to the existing trial to get scientifically meaningful results. Meters decided that the required number of additional patients was too large, so the trial could not continue [45].

10.3.8 LSALT

Dipeptidase 1 (DPEP1) function as a physical adhesion receptor for neutrophil sequestration, regardless of its enzymatic activity [49–54]. Genetic removal or peptide blocking of DPEP1 significantly reduces neutrophil recruitment to the lungs and liver, leading to improved survival in endotoxemia models.

Arch Biopartners discovered a novel peptide called LSALT (LSALTPSPSWLKYKAL), which has been shown to inhibit dipeptidase 1, essential for the recruitment of leucocytes in the lungs and liver [54]. LSALT peptide is a specific DPEP1-binding peptide and nonenzymatic DPEP1 inhibitor. LSALT peptide binds specifically to DPEP1 with no detectable binding to DPEP2 or DPEP3. LSALT peptide significantly reduces neutrophil recruitment to the lungs and liver and improves survival in endotoxemia models. LSALT peptide blocks neutrophil adhesion to peritubular capillaries and reduces inflammatory monocyte recruitment to the kidney after IRI [53].

According to Arch Biopartners' announcement, in a recently completed human trial on healthy normal volunteers in Australia, the daily dose of LSALT peptide has been safely increased to 20 mg per day [51]. LSALT peptide is their lead drug candidate for treating inflammation in the lungs, liver, and kidneys. The company plans to use this new maximum dose in designing a Phase 2 trial targeting cardiac surgery-associated acute kidney injury (CS-AKI).

10.3.9 RLS-0071

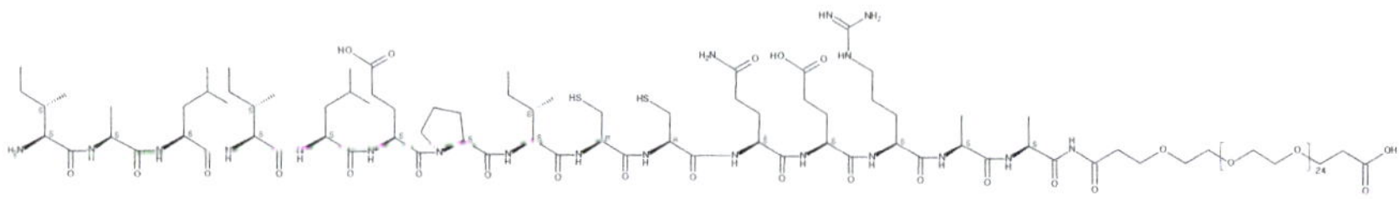

RLS-0071 is a dual-targeting peptide (IALILEPICCQERAA-dPEG24) derived from the coat protein of human astrovirus Type 1 [55–60]. It regulates humoral and cellular inflammation by inhibiting neutrophil effectors, such as myeloperoxidase and the formation of neutrophil extracellular traps (NETosis). RLS-0071 blocks the activation of the classical complement pathway at C1 and inhibits the enzymatic activity of myeloperoxidase, preventing the generation of hypochlorous acid and NETosis. *In vitro* studies have demonstrated that this peptide inhibitor of complement C1 (PIC1) binds to the collagen-like region of the initiator molecule of the classical pathway, C1q. PIC1 binding to C1q blocks activation of the associated serine proteases (C1s—C1r—C1r—C1s) and subsequent downstream complement activation. PIC1 inhibits classical pathway-mediated complement activation in ABO incompatibility *in vitro* and inhibits classical pathway activation *in vivo* in rats.

This peptide is being developed for the treatment of neonatal hypoxic-ischemic encephalopathy and neutrophilic pulmonary diseases [59].

10.3.10 GLP-1 and Antidiabetic GLP-1R Agonists

The posttranslational processing of proglucagon results in the production of glucagon-like peptide-1 (GLP-1), a peptide that is 30–31 amino acids long. This peptide is produced by intestinal enteroendocrine L cells and neurons in the brainstem's nucleus of the solitary tract in response to nutritional and inflammatory stimuli. GLP-1 and its receptor agonists (GLP-1 RAs) have been shown to be effective as adjuvant therapies for Type 2 diabetes (T2D) mellitus, obesity control, and other metabolic diseases [61–63].

Costanzo et al. suggested the GLP-1 agonist semaglutide as a possible treatment for obese patients with T2DM and psoriasis [62]. The observations were supported by a 4- to 10-month treatment of semaglutide added to metformin that unexpectedly improved skin lesions of plaque psoriasis with the expected improved outcome of glycemic parameters and decreased body weight. Recently, Mehdi et al. reported that GLP-1 has been recognized for its anti-inflammatory properties, which protect cardiovascular and nervous system cells by reducing inflammation and apoptosis [61].

In a clinical study (NCT03449654), Zobel et al. investigated the anti-inflammatory effects of liraglutide, a GLP-1 receptor agonist, in individuals with T2D. They found that liraglutide indirectly influenced the expression of inflammatory genes in peripheral blood mononuclear cells (PBMCs). However, since this effect was not observed in PMCs and THP1 cells, it suggests that liraglutide's impact on inflammatory gene expression in PBMCs is complex and warrants further investigation [64,65].

People with chronic kidney disease (CKD) often experience systemic and chronic inflammation, which is closely linked to the progression of CKD and can lead to cardiovascular disease. Additionally, CKD is commonly found in individuals with both Type 1 diabetes (T1D) and T2D. Also, CKD is common in people with diabetes, both T1D and T2D. Semaglutide was also investigated in another Phase 3 clinical study (NCT03819153; FLOW) in T2D and CKD. Novo Nordisk terminated this study early due to strong efficacy signals and expects to share FLOW's readout in 2024 [66].

Recently, a Phase 3 study (NCT05891496) evaluated the impact of semaglutide on the immune system and other biological processes in individuals with Alzheimer's disease. This multisite study, covering Europe, the United States, and Canada, covers many conditions, including inflammatory disease, with the dosing of once-weekly subcutaneously administered controlled-release formulation. The primary endpoint completion was May 2024, and the study end date is June 2025 [67].

10.4 Summary and Discussion

This review summarizes anti-inflammatory peptides in clinical trials in the last 5 years. As exemplified above, several peptides are entirely natural or analogs, bioengineered from a pharmacological understanding of the inflammation pathway and synthetically modified to enable multiple delivery routes for various indications. The overview also shows that anti-inflammatory peptides represent a successful approach for several inflammation pathways that regulate diverse physiological events and related dysfunctions.

Although the preclinical research on anti-inflammatory peptides is ever-increasing, the vast potential of anti-inflammatory peptides still needs to be explored. Thanks to access to natural sources and advancement in drug delivery sciences, robust synthetic methodologies enabled medicinal chemists to explore peptide scaffold designs with an improved pharmacological profile beyond injectable dosing to topical and oral delivery of peptides. The opportunity for therapeutic anti-inflammatory peptide development is vast due to the association of excessive and uncontrolled inflammation in multiple disease indications. The success rate of anti-inflammatory peptides in the clinic with regard to efficacy and market positioning remains to be followed in the next 3–5 years.

References

1 Monteleone, G., Moscardelli, A., Colella, A., Marafini, I., and Salvatori S. Immune-mediated inflammatory diseases: common and different pathogenic and clinical features. *Autoimmun Rev.* 2023;22(10):103410.

2 Izvolskaia, M., Sharova, V., and Zakharova, L. Perinatal inflammation reprograms neuroendocrine, immune, and reproductive functions: profile of cytokine biomarkers. *Inflammation.* 2020;43(4):1175–1183.

3 Beland, L.C., Markovinovic, A., Jakovac, H., et al. Immunity in amyotrophic lateral sclerosis: blurred lines between excessive inflammation and inefficient immune responses. *Brain Commun.* 2020;2(2):fcaa124.

4 Sarrand, J. and Soyfoo, M.S. Involvement of epithelial-mesenchymal transition (EMT) in autoimmune diseases. *Int J Mol Sci.* 2023;24(19):14481.

5 Tamamoto-Mochizuki, C., Santoro, D., Saridomikelakis, M.N., et al. Update on the role of cytokines and chemokines in canine atopic dermatitis. *Vet Dermatol.* 2023;35:25.

6 Muth, K.N., Rech, J., Losch, F.O., and Hoerning, A. Reversing the inflammatory process-25 years of tumor necrosis factor-alpha inhibitors. *J Clin Med.* 2023;12(15):5039.

7 Briani, C. and Visentin, A. Therapeutic monoclonal antibody therapies in chronic autoimmune demyelinating neuropathies. *Neurotherapeutics.* 2022;19(3):874–884.

8 Feldmann, M. What is the mechanism of action of anti-tumour necrosis factor-alpha antibody in rheumatoid arthritis? *Int Arch Allergy Immunol.* 1996;111(4):362–365.

9 Gisbert, J.P. and Chaparro, M. Anti-TNF agents and new biological agents (Vedolizumab and Ustekinumab) in the prevention and treatment of postoperative recurrence after surgery in Crohn's disease. *Drugs.* 2023;83(13):1179–1205.

10 Li, T., Yang, X., Zhu, J., et al. Current application status and structure-activity relationship of selective and non-selective JAK inhibitors in diseases. *Int Immunopharmacol.* 2023;122:110660.

11 Yang, V., Kragstrup, T.W., McMaster, C., et al. Managing cardiovascular and cancer risk associated with JAK inhibitors. *Drug Saf.* 2023;46:1049.

12 La Manna, S., Di Natale, C., Florio, D., and Marasco, D. Peptides as therapeutic agents for inflammatory-related diseases. *Int J Mol Sci.* 2018;19(9):2714.

13 Zhang, K., Tang, Y., Chen, Q., and Liu, Y. The screening of therapeutic peptides for anti-inflammation through phage display technology. *Int J Mol Sci.* 2022;23(15):8554.

14 Briand, J.P. and Muller, S. Emerging peptide therapeutics for inflammatory autoimmune diseases. *Curr Pharm Des.* 2010;16(9):1136–1142.

15 Dadar, M., Shahali, Y., Chakraborty, S., et al. Anti-inflammatory peptides: current knowledge and promising prospects. *Inflamm Res.* 2019;68(2):125–145.

16 Report, G.P. Anti-inflammatory peptides market size worth US$ 139.4 billion by 2030. https://finance.yahoo.com/news/anti-inflammatory-peptides-market-size-064200806.html. [cited 2023 Aug 01].

17 Liu, X.Y., Robinson, D., Veach, R.A., et al. Peptide-directed suppression of a pro-inflammatory cytokine response. *J Biol Chem.* 2000;275(22):16774–16778.

18 Amytrx therapeutics announces open enrollment for phase 2b clinical trial of AMTX-100 CF3, a promising anti-inflammatory peptide therapeutic for atopic dermatitis. https://www.biospace.com/article/releases/amytrx-therapeutic-s-amtx-100-cf3-moves-to-phase-ii-clinical-trial-for-atopic-dermatitis/. [cited 2023 June 08].

19 Kleinman, E., Laborada, J., Metterle, L., and Eichenfield, L.F. What's new in topicals for atopic dermatitis? *Am J Clin Dermatol.* 2022;23(5):595–603.

20 Liu, Y., Zienkiewicz, J., Qiao, H., et al. Genomic control of inflammation in experimental atopic dermatitis. *Sci Rep.* 2022;12(1):18891.

21 Therapeutics, A. AMTX-100, the next generation of anti-inflammatory medicines. https://www.amytrx.com. [cited 2023 Sep 14].

22 University, V. Vanderbilt University anti-inflammatory therapy licensed to Amytrx is now in clinical trials for eczema treatment. https://news.vanderbilt.edu/2020/10/27/vanderbilt-university-anti-inflammatory-therapy-licensed-to-amytrx-is-now-in-clinical-trials-for-eczema-treatment/. [cited 2023 Sep 14].

23 Amytrx Therapeutics, I. Amytrx works with ambiopharm to manufacture its phase I/II atopic dermatitis asset; august bioservices, tergus pharma tasked for topical formulation. https://www.linkedin.com/pulse/amytrx-works-ambiopharm-manufacture-its-phase-iii-/. [cited 2023 Sep 14].

24 Wildner, G., Hünig, T., and Thurau, S.R. Orally induced, peptide-specific gamma/delta TCR+ cells suppress experimental autoimmune uveitis. *Eur J Immunol.* 1996;26(9):2140–2148.

25 Likelihood of approval and phase transition success rate model – optiquel in non-infectious uveitis. https://www.globaldata.com/store/report/optiquel-in-non-infectious-uveitis-loa-and-ptsr-drug-analysis-18405/.

26 Optiquel by enzo biochem for non-infectious uveitis: likelihood of approval. https://www.pharmaceutical-technology.com/data-insights/?wpsolr_fq%5B1%5D=categories%3AData%20Insights. [cited 2023 Sep 14].

27 Wildner, G. and Thurau, S.R. Cross-reactivity between an HLA-B27-derived peptide and a retinal autoantigen peptide: a clue to major histocompatibility complex association with autoimmune disease. *Eur J Immunol.* 1994;24(11):2579–2585.

28 Wildner, G. and Diedrichs-Möhring, M. Molecular mimicry and uveitis. *Front Immunol.* 2020;11:580636.

29 Wildner, G. Antigenic mimicry—the key to autoimmunity in immune privileged organs. *J Autoimmun.* 2023;137:102942.

30 van Groenendael, R., Aarnoutse, R., Kox, M., van Eijk, L., and Pickkers, P. Pharmacokinetics, safety and tolerability of the novel β-hCG derived immunomodulatory compound, EA-230. *Br J Clin Pharmacol.* 2019;85(7):1572–1584.

31 van Groenendael, R., Beunders, R., Hemelaar, P., et al. Safety and efficacy of human chorionic gonadotropin hormone-derivative EA-230 in cardiac surgery patients: a randomized double-blind placebo-controlled study. *Crit Care Med.* 2021;49(5):790–803.

32 van Groenendael, R., Beunders, R., Hofland, J., et al. The safety, tolerability, and effects on the systemic inflammatory response and renal function of the human chorionic gonadotropin hormone-derivative EA-230 following on-pump cardiac surgery (the EASI study): protocol for a randomized, double-blind, placebo-controlled phase 2 study. *JMIR Res Protoc.* 2019;8(2):e11441.

33 van Groenendael, R., Beunders, R., Kox, M., van Eijk, L.T., and Pickkers, P. The human chorionic gonadotropin derivate EA-230 modulates the immune response and exerts renal protective properties: therapeutic potential in humans. *Semin Nephrol.* 2019;39(5):496–504.

34 van Groenendael, R., Kox, M., Leijte, G., et al. A randomized double-blind, placebo-controlled clinical phase IIa trial on safety, immunomodulatory effects and pharmacokinetics of EA-230 during experimental human endotoxaemia. *Br J Clin Pharmacol.* 2019;85(7):1559–1571.

35 LUPUZOR™ for Lupus. https://www.immupharma.co.uk/pipeline/autoimmunity-inflammation/. [cited 2023 Sep 13].

36 Forigerimod, (Rigerimod) also known as Lupuzor, CEP-3345 for treatment of systemic lupus erythematosus (SLE). https://newdrugapprovals.org/tag/forigerimod/. [cited 2023 Sep 18].

37 Wang, F., Bonam, S.R., Schall, N., et al. Blocking nuclear export of HSPA8 after heat shock stress severely alters cell survival. *Sci Rep.* 2018;8(1):16820.

38 Phase 2/3 adaptive clinical trial of P140 (Lupuzor™) in Lupus to commence following FDA meeting. https://www.immupharma.co.uk/phase-2-3-adaptive-clinical-trial-of-p140-lupuzor-in-lupus-to-commence-following-fda-meeting/. [cited 2023 Sep 27].

39 Pipeline: target indications for SP16. https://serpinpharma.com. [cited 2023 Sep 14].

40 Soler, Y., Rodriguez, M., Austin, D., Gineste, C., Gelber, C., and El-Hage, N. SERPIN-derived small peptide (SP16) as a potential therapeutic agent against HIV-induced inflammatory molecules and viral replication in cells of the central nervous system. *Cells*. 2023;12:632.

41 Van Tassell, B.W., Wohlford, G.F., Del Buono, M.G., et al. Safety, tolerability, and effects of a single subcutaneous administration of SP16 – a SERPIN-like, small peptide agonist of the low-density lipoprotein–like receptor 1– on the acute inflammatory response in patients with ST-segment elevation myocardial infarction. *J Cardiovasc Pharmacol*. 2022;80(5):672.

42 Wang, Z., Martellucci, S., Enoo, A., Austin, D., Gelber, C., and Campana W. α1-antitrypsin derived SP16 peptide demonstrates efficacy in rodent models of acute and neuropathic pain. *FASEB J*. 2022;36:e22093.

43 Wohlford, G., Buckley, L., Kadariya, D., et al. A phase 1 clinical trial of SP16, a first-in-class anti-inflammatory LRP1 agonist, in healthy volunteers. PLOS One. 2021;16:e0247357.

44 Novel oral therapy JNJ-2113 shows potential in plaque psoriasis. https://www.europeanpharmaceuticalreview.com/news/184233/novel-oral-therapy-jnj-2113-shows-potential-in-plaque-psoriasis/. [cited 2023 Sep 15].

45 9 meters discontinues phase 3 clinical trial for potential celiac disease drug larazotide. https://celiac.org/about-the-foundation/featured-news/2022/06/9-meters-discontinues-phase-3-clinical-trial-for-potential-celiac-disease-drug-larazotide/. [cited 2023 June 21].

46 Larazotide—an overview. https://www.sciencedirect.com/topics/medicine-and-dentistry/larazotide. [cited 2023 Sep 13].

47 Hoilat, G.J., Altowairqi, A.K., Ayas, M.F., et al. Larazotide acetate for treatment of celiac disease: a systematic review and meta-analysis of randomized controlled trials. *Clin Res Hepatol Gastroenterol*. 2022;46(1):101782.

48 Slifer, Z.M., Krishnan, B.R., Madan, J., and Blikslager, A.T. Larazotide acetate: a pharmacological peptide approach to tight junction regulation. *Am J Physiol Gastrointest Liver Physiol*. 2021;320(6):G983-g9.

49 Targeting inflammation with LSALT peptide. https://archbiopartners.com/our-science/metablok/. [cited 2023 Aug 20].

50 LSALT peptide vs. placebo to prevent ARDS and acute kidney injury in patients infected with SARS-CoV-2 (COVID-19).

51 Arch biopartners submits IND application to the FDA for metablok (LSALT Peptide). https://drug-dev.com/arch-biopartners-submits-ind-application-to-the-fda-for-metablok-lsalt-peptide/. [cited 2023 Aug 28].

52 Arch biopartners announces positive results of dose escalation human trial for LSALT peptide. https://www.biospace.com/article/releases/arch-biopartners-announces-positive-results-of-dose-escalation-human-trial-for-lsalt-peptide/. [cited 2023 Aug 28].

53 Lau, A., Rahn, J.J., Chappellaz, M., et al. Dipeptidase-1 governs renal inflammation during ischemia reperfusion injury. *Sci Adv.* 2022;8(5):eabm0142.
54 Choudhury, S.R., Babes, L., Rahn, J.J., et al. Dipeptidase-1 is an adhesion receptor for neutrophil recruitment in lungs and liver. *Cell.* 2019;178(5): 1205– 1221.e17.
55 ReAlta life sciences receives FDA clearance for phase 2 trial of RLS-0071 in hospitalized patients with steroid-refractory acute graft-versus-host disease. https://finance.yahoo.com/news/realta-life-sciences-receives-fda-130000310.html. [cited 2023 Sep 06].
56 Goss, J., Hair, P., Kumar, P., et al. RLS-0071, a dual-targeting anti-inflammatory peptide - biomarker findings from a first in human clinical trial. *Transl Med Commun.* 2023;8(1):1.
57 Goss, J.M., Hair, P., Kumar, P.S., Thienel, U., Cunnion, K., and Ellis A. RLS-0071 moderated elevated myeloperoxidase level and activity in an asymptomatic subject in clinical trial RLS-0071-101. *Am J Case Rep.* 2023;24:e939803.
58 Kenji, C., Parvathi, K., Neel, K., Ulrich, T., and Matthew, F. RLS-0071-101 a study in healthy human volunteers demonstrates safety, complement and inflammatory effector target engagement activity of this novel peptide (P12-7.002). *Neurology.* 2022;98(Suppl 18):1416.
59 Rickman, J. ReAlta doses first patient in phase 2 STAR study of RLS-0071 for HIE in newborns. https://realtalifesciences.com/news/realta-doses-first-patient-in-phase-2-star-study-of-rls-0071-for-hie-in-newborns/. [cited 2023 Oct 17].
60 Sciences, R.L. ReAlta life sciences' RLS-0071 demonstrates excellent safety profile and confirmed target engagement in first-in-human phase 1 clinical trial. [cited 2023 Sep 14].
61 Mehdi, S.F., Pusapati, S., Anwar, M.S., et al. Glucagon-like peptide-1: a multi-faceted anti-inflammatory agent. *Front Immunol.* 2023;14:1148209.
62 Costanzo, G., Curatolo, S., Busà, B., Belfiore, A., and Gullo, D. Two birds one stone: semaglutide is highly effective against severe psoriasis in a type 2 diabetic patient. *Endocrinol Diabetes Metab Case Rep.* 2021; 2021.
63 Pang, J., Feng, J.N., Ling, W., and Jin, T. The anti-inflammatory feature of glucagon-like peptide-1 and its based diabetes drugs-therapeutic potential exploration in lung injury. *Acta Pharm Sin B.* 2022;12(11):4040–4055.
64 Ripa, R.S., Zobel, E.H., von Scholten, B.J., et al. Effect of liraglutide on arterial inflammation assessed as [(18)F]FDG uptake in patients with type 2 diabetes: a randomized, double-blind, placebo-controlled trial. *Circ Cardiovasc Imaging.* 2021;14(7):e012174.
65 Zobel, E.H., Ripa, R.S., von Scholten, B.J., et al. Effect of liraglutide on vascular inflammation evaluated by [(64)Cu]DOTATATE. *Diagnostics (Basel).* 2021;11(8):1431.

66 Rossing, P., Baeres, F.M.M., Bakris, G., et al. The rationale, design and baseline data of FLOW, a kidney outcomes trial with once-weekly semaglutide in people with type 2 diabetes and chronic kidney disease. *Nephrol Dial Transplant.* 2023;38(9):2041–2051.

67 A research study looking at the effect of semaglutide on the immune system and other biological processes in people with Alzheimer's disease. https://www.novotrials.com/trials-conditions/all-trials-v2/NN6535-7519.html. [cited 2023 Oct 23].

11

Radioligand as a Unique Rising Modality for Cancer Diagnosis and Treatment

Ming-Jin Jheng[#], Xiaozhu Wang[#], Pengyun Li[#] and Fa Liu

Full-Life Technologies Limited, Shanghai, China
[#]These authors contributed equally to this manuscript.

11.1 Radioligand Introduction

A radionuclide is an unstable nuclide possessing high energy that constantly undergoes decay processes toward its stable state. During decay, the excess energy is released as gamma rays, positrons, or alpha- or beta-particles. This radiation emission has enabled various medical applications of radionuclides, which can be traced back to centuries ago and have been continuously evolving. The most notable applications include medical imaging for disease diagnosis and radiation therapy for cancer treatment. The well-established single-photon emission computed tomography (SPECT) and positron emission tomography (PET) are medical imaging technologies that detect gamma rays and positrons, respectively [1]. In parallel, modern cancer radiotherapies utilize alpha- or beta-particle emitting radionuclides to kill tumor cells. The demonstrated clinical benefits and the recent advancements in radioisotope technology have been synergistically promoting the rapid development of nuclear medicine and radiopharmaceuticals [2].

A small set of radionuclides, such as iodine and radium, when given intravenously as an inorganic salt solution, can selectively accumulate in target tissues. The well-established examples are the long-time used radioactive iodine (^{131}I) for the treatment of hyperthyroidism and thyroid cancer [3], and the 2013 FDA-approved radium-223 (^{223}Ra) chloride solution (Xofigo®) for the management of adults with castration-resistant prostate cancer (CRPC) with symptomatic bone

Peptide Science: Chemical Ligation, Lead Generation, and Therapeutic Advances, First Edition. Fa Liu.

metastases and no known visceral metastases [4]. ^{131}I is taken up by the iodide transporter of the thyroid, leading to its enrichment in the thyroid glands, while ^{223}Ra, similar to calcium, can be absorbed by bone cells, particularly in areas of bone metastases where bone remodeling and vascular supply are more active. However, despite the success of the direct application of radioactive iodine and radium as drugs, most radionuclides cannot actively engage tumor lesions as inorganic salt forms, and their medical applications require targeted delivery to the tumor lesions.

The cancer-associated biomarker can potentially serve as an anchor for targeted radiotherapy. A ligand that recognizes such a biomarker, when conjugated with a radionuclide, can enrich the radionuclide at the tumor. Such a molecule, consisting of a tumor cell surface biomarker recognizing domain and a radionuclide, is termed a radioligand. When constructing a radioligand, halogen radionuclides such as fluorine and astatine can be directly labeled onto the binding domain via covalent bonds. In most cases, a chelating group is necessitated, which undergoes complexation with metallic radionuclides via coordination bonds, resulting in a radioligand that is typically stable enough for pharmaceutical applications [5]. The radioligand precursor that is to be complexed with metallic radionuclides is also called a vector.

A radioligand bearing an imaging radionuclide such as gallium-68 (^{68}Ga) or indium-111 (^{111}In) can be used as a diagnostic agent whose *in vivo* biodistribution can be visualized through PET and SPECT technologies to detect possible tumor lesions. A radioligand bearing a therapeutic radionuclide such as lutetium-177 (^{177}Lu), yttrium-90 (^{90}Y), or actinium-225 (^{225}Ac) is expected to have a biocompatible biodistribution profile similar to its imaging counterpart and can therefore serve as a tumor-killing drug candidate. In some cases, a therapeutic radionuclide also emits gamma rays that are detectable by SPECT, allowing it to simultaneously serve as an imaging agent. In general, the sequential application of a diagnostic radioligand for cancer detection and a therapeutic radioligand for treatment is applied in the clinical practices. This process is termed as "theranostics." "Being theranostic" is one of the most distinct and advantageous features of radioligand therapy compared to all other drug modalities, where the tumor lesion is first seen and then treated and the treatment response can be timely and precisely monitored [6–9].

A significant amount of radioligands, including FDA-approved drugs such as Lutathera and Pluvicto, comprise a peptide as the tumor-binding domain, making them essentially peptide-radionuclide conjugates or peptide-drug conjugates (PDCs) [10]. The primary objective of this chapter is to provide a review of this unique PDC modality with a focus on its application in cancer diagnosis and treatment. We will discuss the successful examples, including the FDA-approved theranostic applications in SSTR2 and PSMA-expressing cancers, and the recent exciting development of fibroblast activation protein inhibitor (FAPi)-based radioligand as novel cancer imaging agents. We will also survey emerging radioligand targets, including folate receptor-α, gastrin-releasing peptide receptor (GRPR), neurotensin receptor 1 (NTSR1), Glypican-3 (GPC-3), and carbonic anhydrase 9 (CA9). In addition, we will cover novel vector modalities and new radionuclides that are being actively explored for radioligand applications. In the end, we will share our perspectives on the future opportunities and challenges in the development of nuclear medicine and radiopharmaceuticals toward the ambition of releasing its full potential.

11.2 Radionuclides and Chelators Developed for Medical Imaging and Cancer Treatment

11.2.1 Radionuclides Being Developed as Radioligands

SPECT and PET are two major diagnostic imaging techniques used in nuclear medicine. A SPECT system utilizes a gamma camera to collect gamma-ray signals emitted by γ-emitting radionuclides such as ^{99m}Tc, ^{123}I, ^{67}Ga, or ^{111}In and converts them into electrical signals, which are subsequently processed by an online computer to construct 3D images. In a PET system, a cylindrical array of detectors is used to detect the coincidence events from the two 511-keV photons emitted in opposite directions after the annihilation of a positron from a β^+ emitter interacting with an electron in the medium. The commonly used PET radionuclides are ^{11}C, ^{13}N, ^{15}O, ^{18}F, ^{64}Cu, ^{68}Ga, and ^{82}Rb. Some widely used radioisotopes for SPECT and PET imaging are summarized in Table 11.1.

SPECT only provides one piece of spatial information per decay event, whereas PET imaging provides two. Consequently, PET generally has a higher spatial resolution and requires shorter scan time than SPECT. However, SPECT is currently more widely used, attributed to the facts that the SPECT radiotracers (radionuclides) are more broadly available at lower cost, and the SPECT instruments are more established in hospitals.

Targeted radiotherapy can typically be achieved with high-energy β^-emitters (e.g. ^{67}Cu, ^{131}I, ^{90}Y, ^{153}Sm, ^{161}Tb, ^{177}Lu, ^{186}Re, ^{188}Re, and ^{198}Au) and α-emitters (e.g. ^{223}Ra, ^{212}Pb, and ^{225}Ac). Low-energy Auger electrons (e.g. ^{111}In) have also been studied in neuroendocrine tumor (NET) patients attracted by its nano-micrometer radiation distance while high linear transfer energy (LET) that can be lethal to cancer cells while sparing surrounding normal cells. In general, the therapeutic effects of radiopharmaceuticals rely on the tissue-destructive power of ionizing radiation [12]. Some widely used radionuclides for therapy are summarized in Table 11.2.

Table 11.1 Widely used radionuclides for SPECT and PET imaging.

Radionuclide	Daughter isotopes	$t_{1/2}$	Imaging modality
^{11}C	^{11}B	20.4 min	PET (β^+)
^{18}F	^{18}O	109.7 min	PET (β^+)
^{64}Cu	^{64}Ni	12.8 h	PET (β^+)
^{67}Ga	^{67}Zn	78.2 h	SPECT (γ, 184 keV)
^{68}Ga	^{68}Zn	68 min	PET (β^+)
^{82}Rb	^{82}Kr	1.273 min	PET (β^+)
^{99m}Tc	^{99}Tc	6.01 h	SPECT (γ, 140.5 keV)
^{111}In	^{111}Cd	67.4 h	SPECT (γ, 171 keV)
^{123}I	^{123}Te	13.22 h	SPECT (γ, 157 keV)
^{89}Zr	^{89m}Y	78.41 h	PET (β^+)

Source: Adapted from [11].

Table 11.2 Widely used radionuclides for therapy.

Radionuclide	Daughter isotopes	$t_{1/2}$	Emitted particles	Avg. energy	Range (mm)
^{64}Cu	^{64}Zn	12.8 h	β	573 keV	~ 2.1
^{67}Cu	^{67}Zn	61.9 h	β	141 keV	0.26
^{90}Y	^{90}Zr	64.1 h	β	932 keV	4.0
^{131}I	^{131}Xe	8.06 d	β	192 keV	0.6–2
^{161}Tb	^{161}Dy	6.89 d	β	154 keV	0.29
^{177}Lu	^{177}Hf	6.7 d	β	133 keV	0.23
^{186}Re	^{186}Os	3.8 d	β	349 keV	1.1
^{188}Re	^{188}Os	17 h	β	764 keV	3.1
^{224}Ra	^{220}Rn	3.6 d	α	5.45 MeV	~ 0.06
^{225}Ac	^{209}Bi	9.9 d	α; β	5.8–8.4 MeV[a]	~ 0.05–0.08[a]
^{211}At	^{207}Pb	7.2 h	α	6.79 MeV	~ 0.06
^{212}Pb	^{208}Pb	10.6 h	α; β	6.1–8.8 MeV[a]	~ 0.05–0.08[a]

Source: Adapted from[11].
[a] α-particle energy and range.

11.2.2 Radiometal Chelators Being Utilized in Radioligands

Among all the radionuclides being developed for radiopharmaceuticals, a number of them have drawn great interest due to multiple advantages they possess, including long half-lives, straightforward labeling procedures, convenient source, and sometimes combined imaging and therapeutic capabilities [12]. The complexation (also called labeling) of radionuclides to form radioligands requires a chelator, which is covalently attached to the binding domain. As shown in Figure 11.1, most radionuclide chelators used in radiopharmaceuticals are "bifunctional agents" consisting of two chemical moieties: a strong metal-binding unit to sequester the radionuclide and a chemically reactive group for bioconjugation to the binding domain that recognizes the target receptor [13,14]. Generally, an ideal radionuclide chelator agent should possess the following properties [14,15]:

a. Thermodynamic stability and kinetic inertness: The complex should be sufficiently stable *in vitro* (drug product formulation stability) and *in vivo* (drug *in vivo* stability), with kinetic inertness more important than thermodynamic

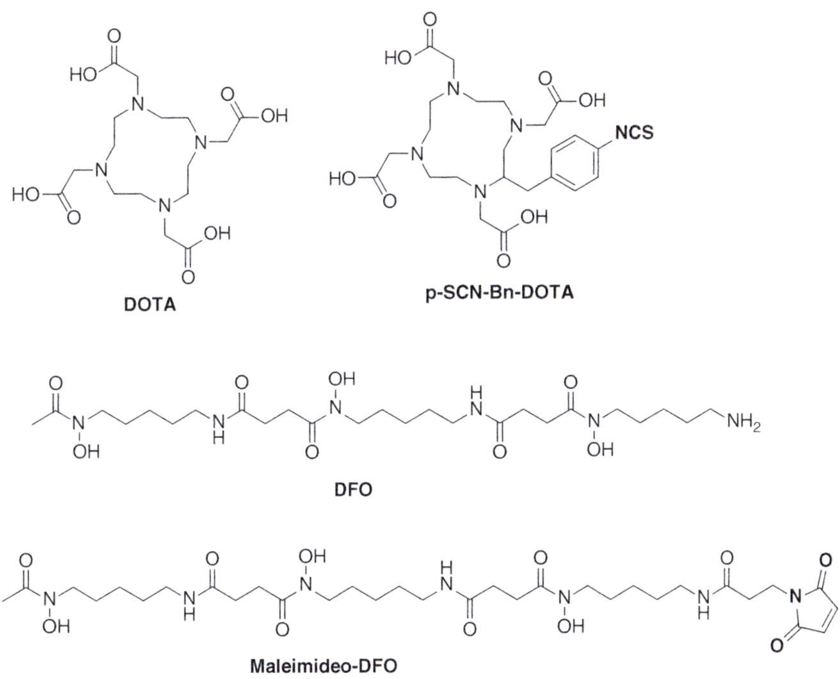

Figure 11.1 The chemical structures of metallic radionuclide chelators: DOTA, DFO, and their bifunctional version: *p*-SCN-Bn-DOTA, maleimideo-DFO.

stability. Low *in vivo* inertness will lead to complex decomposition and the release of free radionuclides, deteriorating imaging quality and causing normal organ toxicity.

b. Fast complexation kinetics: The complexation process should take place quickly at low temperatures with low concentrations of the chelator (vector) and radionuclide. This is especially important for short half-lived radionuclides and chemically or biophysically labile vectors such as antibodies.
c. Bioconjugation capability: The chelators should have a chemically reactive group (e.g. activated esters, isothiocyanates, terminal alkynes, azides, or free primary amines) that can form covalent linkages with the binding domain without affecting the complexation behavior of the chelating unit.
d. Synthetic accessibility: The synthesis of the chelator should be straightforward and can be scaled up at low cost.

11.3 Notable Radioligands Developed for Cancer Diagnosis and Treatment

11.3.1 SSTR2-targeting Theranostic Radioligands for SSTR2-overexpressed Cancers

Somatostatin and its receptor family (SSTR), consisting of five G protein-coupled receptors (GPCRs), play important roles in modulating the secretion of several essential hormones. The abnormality of this signaling pathway has been found to be associated with various diseases, including cancers. Among the five family receptors, it has been established that SSTR2 is more abundantly and broadly overexpressed in various cancers, including NETs, breast and colon adenocarcinomas, and pancreatic tumors. The overexpression of SSTR2 in multiple cancers makes it a viable tumor cell surface receptor for targeted therapies [16].

Somatostatin, a peptide consisting of 14 amino acids and 1 disulfide bond, functions as a high-affinity universal native ligand for the SSTR family receptors. Somatostatin is too short-lived *in vivo* as a drug; therefore, its truncated synthetic analogs have been extensively developed aiming for better drug-like properties. The marketed somatostatin analogs include octreotide (Sandostatin®) (FDA approval in 1988), lanreotide (Somatuline® Depot), and pasireotide (Signifor LAR®). It is worth noting that although all three analogs bind to SSTR2 with high affinity, octreotide and lanreotide are selective for SSTR2, whereas pasireotide also binds tightly to SSTR1, SSTR3, and SSTR5. The on-label indications of these drugs include acromegaly, gastroenteropancreatic neuroendocrine tumors (GEP-NETs), or Cushing's disease [17–19].

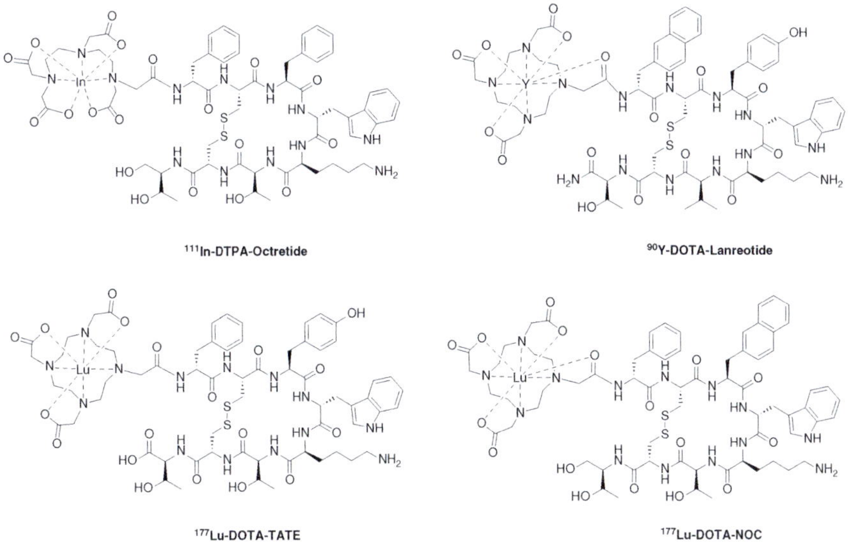

Figure 11.2 The chemical structures of key SSTR2 targeting radioligands.

The well-established somatostatin medicinal chemistries provided solid starting points for developing radioligands for SSTR2-overexpressing tumors (Figure 11.2). The clinical applications began in the mid-1990s with the FDA approval of ^{111}In-DTPA-octreotide (Octreoscan™) for the diagnostic imaging of NETs [20,21]. ^{111}In-DTPA-octreotide was also attempted for the treatment of NETs, theoretically relying on DNA damage induced by the Auger energy emitted from ^{111}In [22]. However, only limited antitumor activities of ^{111}In-DTPA-octreotide was observed in NET patients, which subsequently directed the field toward stronger radionuclides for therapeutic purposes, such as the β-emitters ^{90}Y and ^{177}Lu.

In designing therapeutic SSTR2-targeting radioligands, the macrocyclic chelator DOTA was used instead of the acyclic DTPA of Octreoscan™, enabling stable complexation with ^{90}Y and ^{177}Lu. The radioligand construction started with appending DOTA directly to the N-termini of the well-established binders such as octreotide and lanreotide. The binders were then further evolved with the development of several important octreotide derivatives, including Tyr3-octreotide (TOC), Tyr3,Thr8-octreotide (TATE), and Nal3-octreotide (NOC). DOTA-TATE exhibits approximately fivefold higher affinity for SSTR2 compared to DOTA-TOC, while DOTA-NOC not only binds to SSTR2 strongly but also exhibits significant affinity for SSTR3 and SSTR5 [18]. The clinical development of these diverse set of radioligands shed important insights into the human translation from their preclinical profiles to the biodistribution and antitumor activities in NET patients.

^{90}Y and ^{177}Lu-SSTR2 radioligands demonstrated significantly superior efficacy than ^{111}In-DTPA-octreotide. The antitumor activity of ^{90}Y and ^{177}Lu-SSTR2 radioligands appeared to be comparable; however, ^{177}Lu demonstrated more favorable safety profiles. This is most likely due to that ^{90}Y has significantly longer path length in soft tissue than ^{177}Lu (12 mm vs. 2 mm). The longer path length was theorized to provide greater efficacy for large tumors; unfortunately, it also resulted in excess radiation in kidney and bone marrow, leading to high toxicity [18,23]. The collective clinical data supported Lu-177 as a β-emitter that provides promising efficacy and an acceptable safety window, which eventually led to the FDA approval of ^{177}Lu-DOTATATE (Lutathera®) in 2018 [24]. Several other β-emitter-based SSTR2 radioligands have also reached the late-phase clinical developments, including ^{177}Lu-DOTATOC at Phase 3 and ^{67}Cu-SARTATE at Phase 2 [25–27].

Along with the development of SSTR2-targeting therapeutic radioligands, the diagnostic partners have also been continuously matured. The first FDA approval was ^{111}In-DTPA-octreotide (^{111}In-pentertreotide, Octreoscan™) in 1994. It was a SPECT imaging agent where ^{111}In decays by electron capture emitting high-energy gamma photos, and the imaging resolution was suboptimal. In comparison, the PET imaging acquired through positron-emitting radionuclides has significantly superior resolution and becomes a preferred imaging modality. The first SSTR2 PET imaging agent approved by FDA was ^{68}Ga-DOTATATE (NetSpot®) in 2016 and then ^{68}Ga-DOTATOC in 2019 (approved in Europe in 2016) under various trade names, including TASOtoc® and TOCscan® [28–30]. In addition to ^{68}Ga imaging, other PET imaging radionuclides have also been developed; among these, ^{64}Cu-DOTATE (Detectnet®) was approved by FDA in 2020 [31]. ^{68}Ga has a decay half-life of 68 minutes, and its clinical imaging is typically acquired 40–90 minutes after dosing. In comparison, ^{64}Cu has a decay half-life of 12.7 hours, which not only provides logistic flexibility during the hot drug manufacturing but also enables the imaging of later time points such as 24 hours or even 48 hours, should such needs arise. Another ^{64}Cu imaging agent, ^{64}Cu-SARTATE where a bicyclic chelator sarcophagine (SAR) is used instead of DOTA, is currently under Phase 2 trial studies [32].

Further development around SSTR2 radioligands has been centered on improving therapeutic outcome and broadening therapeutic indications. Lutathera (^{177}Lu-DOTATATE) is currently approved for GEP-NETs, which only represent half of the total NET incidents. In addition, the objective radiographic response rate (ORR) of Lutathera in its registrational trial (NETTER1) was 18%, and there is no treatment option after Lutathera [33]. Separately, SSTR2 is also confirmed to be overexpressed in small cell lung cancer (SCLC) and other tumor types, and SSTR2 radiotherapy has shown certain antitumor activity in the treatment of these cancers, however still at a very early development stage [34]. Collectively, there are still sizable opportunities for SSTR2-targeted radioligand therapy. The programs under development include ^{225}Ac-DOTATATE (Phase 3), ^{212}Pb-DOTAMTATE (Phase 3), ^{212}Pd-VMT-a-NET (Phase 1), ^{177}Lu-DOTA-EB-TATE (Phase 1), and ^{177}Lu-Satoreotide (Phase 1) [35–38]. The former three programs develop α-emitting radionuclides whose tumor-killing effort is considered as 2–3 orders of magnitudes stronger than β-emitter ^{177}Lu, the latter two programs focus on binders with further optimized distribution profiles highlighted by higher tumor uptake and longer tumor retention. Satoreotide is a SSTR2 antagonist whose blood clearance kinetics is comparable to DOTATATE, while EB-TATE has a significantly slower blood clearance, both of which demonstrated superior tumor uptake than DOTATATE in NET patients. Provided that the toxicity is manageable, all five programs can potentially improve the therapeutic outcome and expand the indication of SSTR2-targeted radiotherapies from GEP-NETs to other NET subtypes, SCLC, and other SSTR2-postive indications whose tumors are SSTR2-medium to low expressed and/or resistant or refractory to Lutathera.

11.3.2 PSMA-targeting Theranostic Radioligands for Prostate Cancer

Prostate cancer (PC) is the most common cancer in men worldwide with about 1.4 million new cases and nearly 0.4 million deaths annually. The traditional treatments, including surgical resection, external radiotherapy, endocrine therapy, and chemotherapy, have been offering significant clinical benefits to the patients. However, most patients under androgen deprivation therapies develop resistance and progress to the stage of castration-resistant (CR) PC; in addition, more than 80% of CRPC patients will eventually advance to metastatic CRPC (mCRPC) [39,40], a stage that still lacks effective treatments [41].

It has been well established that prostate-specific membrane antigen (PSMA) is overexpressed in more than 80% of prostate cancer patients, and it has been extensively pursued as a radiopharmaceutical target for both diagnostic and therapeutic applications. PSMA is a cell surface-anchored glutamate carboxypeptidase whose substrates include folate and the neuropeptide *N*-acetyl-L-aspartyl-L-glutamate. The medicinal chemistry efforts of developing PSMA inhibitors started later in 1990 when both substrate analogs and transition state analogs were established [42,43]. The pharmacophore of the former consists of a urea bond while the latter bears a phosphonate group, both serving as a glutamate peptide bond surrogate that is nonhydrolyzable by PSMA (Figure 11.3). In addition to short peptidomimetic based inhibitors, antibody-based PSMA binders have

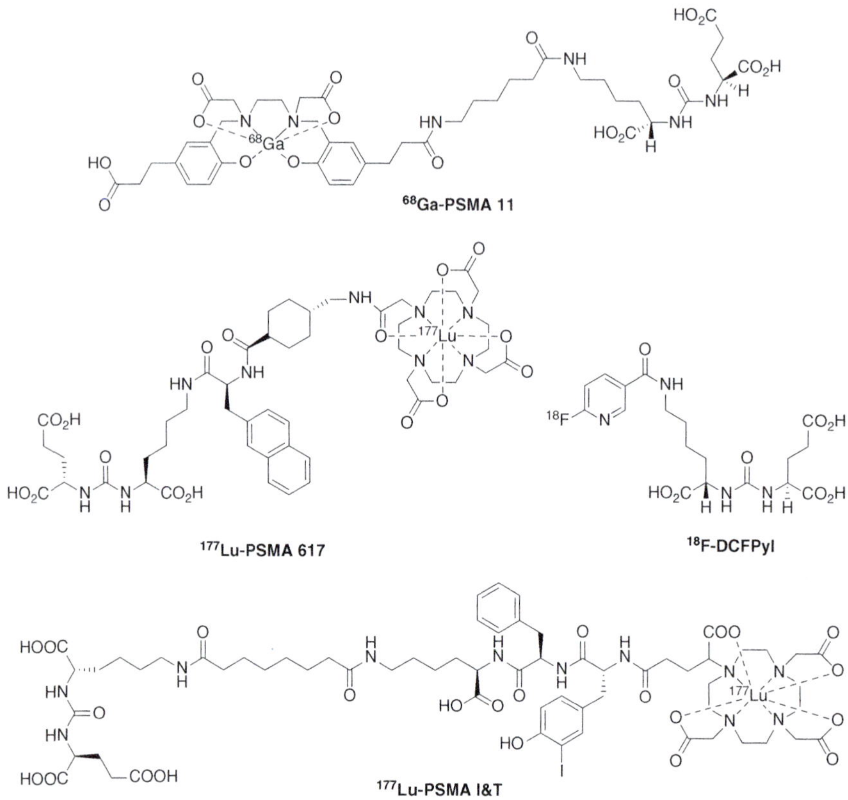

Figure 11.3 The chemical structures of key PSMA radioligands.

also been reported, including full-length antibody and the truncated versions such as minibody and nanobody. The binding region of the antibody binders is typically outside of the enzymatic pocket where the peptidomimetic inhibitor binds [7,44].

In the past nearly 20 years, the development of PSMA radioligands has primarily centered on the peptidomimetic inhibitor-derived ones. However, in fact, the initial clinical studies investigated the full-length antibody J591 conjugated with ^{111}In for imaging and with ^{90}Y and ^{177}Lu for therapy [45,46]. In these early trails, the SPECT imaging of J591 radioligands was successful in detecting tumor lesions in prostate cancer patients. However, the inherent slow clearance rate of full-length antibody, with about 70% injected radioactivity remaining within the body after 7 days following administration, led to significant hematological toxicities challenging their further development paths. Similar to SSTR2-radioligand therapy, ^{90}Y-J591 caused more severe toxicity than ^{177}Lu-J591 attributed to the longer tissue penetration distance of ^{90}Y radiation [46].

With stark difference from full-length antibody, peptidomimetic based PSMA radioligands can be quickly cleared from the body. In the case of ^{177}Lu-vipivotide tetraxetan (PSMA617), 70% of the injected radioactivity was excreted within 24 hours after administration. At the same time, their small hydrodynamic size also enables fast tumor uptake, which typically reaches its maximal level within the first few hours after dosing [47,48]. Most importantly, the uptake of peptidomimetic PSMA radioligands at tumor lesions is often sufficiently high to achieve therapeutic effects. These favorable systemic clearance and tumor uptake profiles have supported the therapeutic development of ^{177}Lu-peptidomimetic PSMA radioligands. In 2022, ^{177}Lu-PSMA617 (Pluvicto) was approved by the US FDA and by European Commission for the treatment of patients with PSMA-positive mCRPC who have been previously treated with androgen receptor pathway inhibition and taxane-based chemotherapy. In the pivotal "VISION" trial, patients receiving Pluvicto plus Standard of Care (SOC) demonstrated a median radiographic progression-free survival (rPFS) of 8.7 months and a median overall survival (OS) of 15.3 months. The control group was SOC alone, which provided an rPFS of 4.3 months and a median OS of 11.3 months [49]. A similar PSMA radioligand, ^{177}Lu-PNT2002 (PSMA I&T), also completed its Phase 3 trial (SPLASH) late 2023, where the ^{177}Lu-PNT2002 group demonstrated a rPFS of 9.5 months versus 6.0 months of the androgen receptor pathway inhibitor (ARPI) treated control group, the OS of SPLASH trial is immature to be concluded yet; nevertheless, it is reasonably safe to conclude that the efficacy and safety profile of ^{177}Lu-PNT2002 is not significantly distinguished from Pluvicto [50].

Along with the progress of PSMA radiotherapies, the diagnostic partners have been developed in parallel. The diagnostic radioligands have mostly utilized the urea-based PSMA peptidomimetic inhibitors as the binding domain; however, a range of radioisotopes have been explored including ^{11}C, ^{99m}Tc, ^{123}I, ^{125}I, ^{68}Ga, ^{18}F, and ^{64}Cu. Many years of preclinical and clinical efforts have led to the approval of Illuccix®/Locametz® (^{68}Ga-gozetotide/PSMA11 injection kit) and Pylarify® (^{18}F-Piflufolastat), all in 2022 [51–54]. Both approvals are PET imaging agents; however, the radioisotope of Illuccix®/Locametz® is ^{68}Ga with a decay $T_{1/2}$ of 68 minutes and the one of Pylarify® is ^{18}F whose $T_{1/2}$ is 110 minutes. The clinical imaging efficacy is comparable between Illuccix®/Locametz® and Pylarify®; however, their production scale and shipping logistics are quite different. Pylarify® is manufactured using cyclotron-generated ^{18}F, which supports large-scale production, and the $T_{1/2}$ of ^{18}F enables shipping distance of hundreds of miles. In comparison, Illuccix® is manufactured by both cyclotron and generator-produced ^{68}Ga on site due to the shorter life of

^{68}Ga; in addition, the Ge/Ga generator has limited daily production that one generator can only supply a few patient scans daily. In addition to ^{18}F and ^{68}Ga, ^{64}Cu PET imaging is also currently being actively developed including ^{64}Cu-PSMA I&T and ^{64}Cu-SAR-bisPSMA [55,56]. The apparent advantage of ^{64}Cu is its longer $T_{1/2}$ of 12 hours which offers not only manufacturing and logistics flexibility but also the possibility of imaging later time points such as 24 or 48 hours. In addition, ^{64}Cu can potentially work as a "true theranostic" pair with its therapeutic partner ^{67}Cu (trail number: NCT04868604).

Similar to the trend of SSTR2-radioligands, the further development of PSMA-radioligands has now heavily focused on next generation therapies. The fact is that despite the promising clinical results of Pluvicto and ^{177}Lu-PSMA I&T, however, only about 30% of patients developed a radiographic response [49]. In addition, the clinical data has also clearly indicated that the outcome of ^{177}Lu-PSMA-617 treatment correlates positively with PSMA expression level, and the efficacy was very limited on low PSMA expressing patients [57]. Furthermore, the significant accumulation of the radioactivity in salivary glands and kidney has become critical dose-limiting factors that ruled out the possibility of driving better efficacy by higher dosing level [58]. The clinical validation and well-recognized unmet needs together warrant the development of next generation therapy aiming for better efficacy and potentially larger safety window where both new vectors and radioisotopes are being investigated.

In addition to ^{177}Lu-PSMA I&T, there are additional multiple ^{177}Lu-PSMA therapy programs in development, which include ^{177}Lu-CTT-1403, ^{177}Lu-EB-PSMA-617, ^{177}Lu-Ludotadipep, ^{177}Lu-rhPSMA-10.1, ^{177}Lu-PSMA-DGUL and ^{177}Lu-TLX-591 (full-length antibody) [59–64]. All aim to increase the radioactivity uptake at the tumor lesions through a novel PSMA binder which typically has longer blood circulation $T_{1/2}$ therefore managing hematological toxicity is critical toward their clinical successes. Other β-emitter-based therapies in development include ^{161}Tb-PSMA I&T and ^{67}Cu-SAR-bisPSMA [65,66]. ^{161}Tb emits stronger but shorter radiation than ^{177}Lu projecting potentially better efficacy and safety window. SAR-bisPSMA vector consisting of a homodimeric PSMA binding domain which has demonstrated higher tumor uptake and longer retention while maintaining similar blood-clearance kinetic to PSMA-617. Both programs hold promise as the next-generation β-emitter-based PSMA therapy.

α-Emitter-based PSMA therapy has been currently dominated by ^{225}Ac. Initial clinical experience with ^{225}Ac was established on two late stage mCRPC patients who were contraindicated for treatment with ^{177}Lu-PSMA-617 or progressed after treatment with ^{177}Lu-PSMA-617 where ^{225}Ac-PSMA-617 achieved a promising clinical response in both patients [67]. In a later retrospective study, ^{225}Ac-PSMA-617 treatment experience was analyzed for 73 consecutive patients with histologically proven mCRPC that relapsed after initial therapy, 70% of subjects responded to therapy [68]. However, the promising clinical results were associated with a significantly increased incidence of xerostomia which was less pronounced for ^{177}Lu-PSMA therapies, and salivary glands have become one of the key dose-limiting organs for ^{225}Ac-PSMA therapies [69–71]. Currently, the peptidomimetic-based ^{225}Ac programs at clinical stage include ^{225}Ac-PSMA I&T, ^{225}Ac-PSMA 617, ^{225}Ac-PSMA R2, ^{225}Ac-PSMA-Trillium, and ^{225}Ac-FL-020 [72–75]. In addition, the full-length antibody binder J591 is being re-developed as ^{225}Ac-J591 (^{225}Ac-CONV01-a), largely driven by its higher tumor retention and potentially less salivary glands accumulation therefore less xerostomia incidents [76]. Other α-emitters in development include ^{212}Pd-NG001, ^{212}Pb-ADVC001 and ^{227}Th-BAY-2315497 (full-length antibody), all are currently at the stage of early clinical development [77–79]. Overall, α-emitter-based PSMA therapies have clearly demonstrated stronger anti-tumor activity, their ultimate clinical success is dependent on the therapeutic windows addressing key dose-limiting organs including bone marrow, salivary glands, and kidney.

11.4 Emerging Targets

The clinical validation of the efficacy of radioligand therapy for solid tumor treatment through Lutathera and Pluvicto and their commercial success have set the stage for the next wave of radioligand therapies, where the targets beyond SSTR2 and PSMA, and indications beyond NETs and mCRPC, have been actively explored.

11.4.1 FAP-targeting Radioligands as Pan-tumor Diagnostic Agents and Potentially Therapeutic Drugs

Solid tumors are composed of tumor cells and their ecosystem tumor stroma (also called tumor microenvironment, TME). Tumor stroma includes diverse types of immune cells, cancer-associated fibroblasts (CAFs), endothelial cells, and various additional tissue-resident cell types, and extracellular matrix (ECM), blood vessels, cytokines, and other noncellular components. CAFs are a major component of the tumor stroma and produce a majority of extracellular components, including growth factors, cytokines, and ECM, and play a pivotal role in tumor proliferation, metastasis, neo angiogenesis, ECM remodeling, immunosuppression, and chemo-resistance [80–82].

Fibroblast activation protein-α (FAP) is a type II transmembrane protease closely related to proline-specific serine proteases such as prolyl oligopeptidase (PREP) and the dipeptidyl peptidase 4, 8, and 9. FAP is a specific biomarker of CAFs and is highly expressed in more than 90% of epithelial tumors, but has low to undetectable expression in healthy adult tissues, which makes it an attractive pan-tumor target for cancer diagnosis and treatment.

The first clinical report of FAP-targeted tumor imaging was made by Welt et al. in 1994. In this study, ^{131}I-labeled murine monoclonal antibody F19 (^{131}I-mAbF19, anti-FAP) was administered intravenously to 17 patients with hepatic metastases from colorectal carcinoma. The smallest lesion visualized by SPECT imaging 3 to 5 days after administration was 1 cm in diameter [83]. Later, sibrotuzumab, a humanized version of mAb F19, was used to explore FAP-targeted therapy but failed in early Phase 2 trial due to lack of efficacy [84].

Radiolabeled antibodies have slow blood clearance because of their long half-life, resulting in higher background signals, an unfavorable tumor-to-healthy tissue signal ratio, lower focal detection sensitivity, and higher overall radiation exposure, so small-molecule and peptide FAP inhibitors have attracted tremendous attention lately. Early work on FAP inhibitors (FAPIs) focused on boronic acid derivatives, which exhibited affinity for members of the DPP subfamily, but were not very specific for FAP [85]. Lately, Veken et al. at the University of Antwerp developed a series of FAPIs based on an *N*-(4-quinolinoyl)-Gly-(2-cyanopyrrolidine) scaffold and identified a highly potent, selective, and orally bioavailable FAP inhibitor, UAMC1110 [86–89]. Subsequently, 122–312 MBq of ^{68}Ga-FAPI-04 was studied in 80 patients with 28 different tumor entities (Figure 11.4). It was found that ^{68}Ga-FAPI-04 had

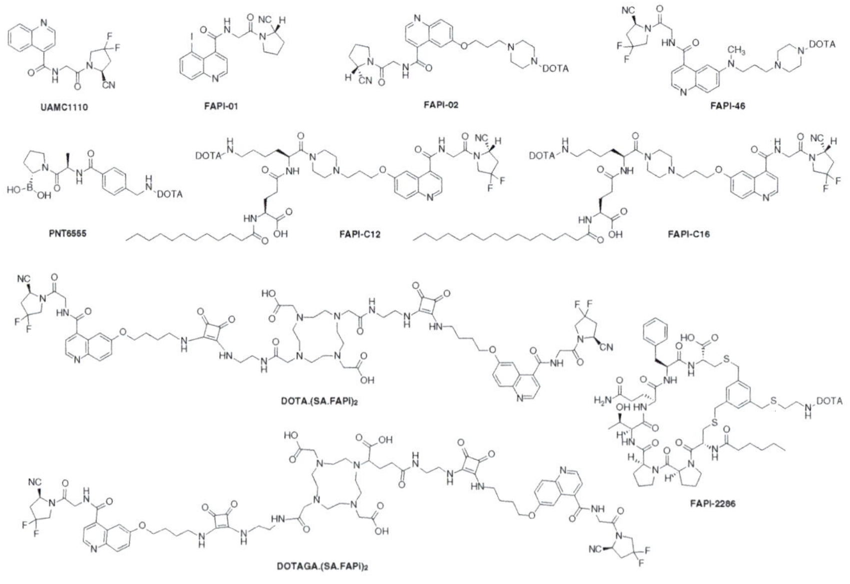

Figure 11.4 The chemical structures of key FAPI and FAP-targeting radioligand vectors.

the highest uptake in sarcoma, cholangiocarcinoma, esophageal, and breast cancer, while the lowest uptake was observed in pheochromocytoma, renal cell carcinoma, differentiated thyroid cancer, adenoid cystic carcinoma, and gastric cancer. The tumor-to-background ratios were more than threefold in the intermediate uptake group and more than sixfold in the high uptake group, which may provide new applications for noninvasive tumor characterization and staging examinations for prevalent tumor entities [90].

FAPI conjugated with the macrocyclic DOTA chelator can be labeled with the PET imaging radionuclide ^{68}Ga, and therapeutic radionuclides such as ^{177}Lu, ^{90}Y, and ^{225}Ac, making these radioligand vectors promising candidates for both PET imaging and targeted radiotherapy. However, short intratumoral $T_{1/2}$ was found to be one of the major limitations of FAPI radioligand therapies. Several strategies were developed to enhance the tumor uptake and prolong retention while minimizing nontarget tissue accumulation. One approach is to increase the avidity of FAP binding moiety through multimerization. Millul et al. developed ^{177}Lu-FAPI-46-F1D, which is a dimer of FAPI-46. Moon et al. started from a highly potent FAP inhibitor UAMC1110 and developed FAPI homodimers with two squaramide-coupled FAP inhibitor conjugates DOTA.(SA.FAPi)$_2$ and DOTAGA.(SA.FAPi)$_2$. These dimers demonstrated higher FAP affinity *in vitro*, and increased tumor accumulation and prolonged retention time *in vivo*; however, a higher radiation dose in nontargeted organs and overall background activity was also observed, suggesting higher safety concerns [91,92]. Another strategy is to introduce an albumin-binding moiety to prolong the blood circulation time of the FAPI radioligand. Zhang et al. prepared FAPI-C12 and FAPI-C16 by conjugating lauric acid (C12) and palmitic acid (C16) to FAPI-04, respectively. Whole-body SPECT imaging of ^{177}Lu-FAPI-C12 and ^{177}Lu-FAPI-C16 showed longer circulation times and significantly higher tumor uptake and retention compared to ^{177}Lu-FAPI-04. ^{177}Lu-FAPI-C16 had the higher tumor uptake at both 24 and 72 hours ($11.22 \pm 1.18\%$ ID/g and $6.50 \pm 1.19\%$ ID/g, respectively) than ^{177}Lu-FAPI-C12 ($7.54 \pm 0.97\%$IA/g and $2.62 \pm 0.65\%$IA/g, respectively). In comparison, ^{177}Lu-FAPI-04 had significantly lower tumor uptake of $1.24 \pm 0.54\%$ ID/g at 24 hours. At a dose of 29.6 MBq of ^{177}Lu-FAPI-C16, significant tumor volume inhibition was observed, with a median survival of 28 days, much longer than that of ^{177}lu-FAPI-04 treated group (10 days). However, long circulation time also increased normal tissue uptake, resulting in worsened tumor-to-background ratios. The same group also prepared TEFAPI-06 and TEFFAPI-07 by introducing two well-studied albumin binders: a 4-(*p*-iodophenyl) butyric acid moiety (TEFAPI-06) and a truncated Evans Blue moiety (EB) (TEFAPI-07), respectively. When compared with ^{177}Lu-FAPI-04, both ^{177}Lu-TEFAPI-06 and ^{177}Lu-TEFAPI-07 demonstrated enhanced tumor uptake and prolonged retention time in imaging and biodistribution studies, along with remarkable tumor growth inhibitions in patient-derived xenograft (PDX) tumors with negligible side effects [93,94].

3B-Pharmaceuticals in Germany developed a novel class of cyclic peptide-based FAP-targeting radioligand vectors. FAP-2286, bearing DOTA as the chelator, exhibits great binding affinity, selectivity, and plasma stability. Comparative imaging studies demonstrated that both ^{68}Ga-FAP-2286 and ^{68}Ga-FAPI-46 were rapidly enriched within the HEK-FAP xenografts with predominantly renal elimination, and similar tumor accumulation was observed with 9.8 versus 9.3% ID/g at 0.5 hour and 10.8 versus 9.2% ID/g at 3 hours. The tumor uptake of ^{177}Lu-FAP-2286 was 21.1% ID/g at 3 hours. Importantly, it also demonstrated durable retention, with 16.4% ID/g at 72 hours. In contrast, ^{177}Lu-FAPI-46 had a slightly lower tumor uptake at 3 hours of 15.3% ID/g, but washed out quickly to 1.6% ID/g at 72 hours. It is worth noting that higher kidney accumulation was observed for ^{177}Lu-FAP-2286 over ^{177}Lu-FAPI-46 (2.2% ID/g vs. 0.7% ID/g at 3 hours, 0.6% ID/g vs. 0.2% ID/g at 72 hours). ^{177}Lu-FAP-2286 exhibited antitumor activity in FAP-expressing HEK293 tumors and sarcoma patient-derived xenografts, without radioligand-related weight loss observed. In addition, ^{177}Lu-FAP-2286 maintained longer growth suppression over ^{177}Lu-FAPI-46. These results have supported the clinical development of ^{68}Ga-FAP-2286 for imaging and ^{177}Lu-FAP-2286 for therapy for a broad range of tumor types [95].

More recently, Poplawski et al. developed PNT6555 based on the boronic acid derivative *N*-(pyridine-4-carbonyl)-D-Ala-boroPro. PNT6555 was proven to be a highly potent and selective inhibitor of FAP. ^{68}Ga-PNT6555 showed rapid renal clearance and continuous accumulation in tumors, resulting in high-contrast PET images of tumors at 60 minutes. ^{177}Lu-PNT6555 exhibited selective uptake in tumors 4 hours after a single dose of injection, and the tumor retention time is over 168 hours. In therapeutic studies, ^{177}Lu-PNT6555 exhibited dose-dependent tumor growth delay, and animal survival at well-tolerated doses [95].

In summary, FAP-targeted imaging radioligands including PET and SPECT have been proven to be promising pan-cancer diagnostic agents for a broad range of tumor types. However, FAP expression is also increased in several nonmalignant diseases where active tissue remodeling is involved, including hepatic fibrosis [96], idiopathic pulmonary fibrosis [97], and rheumatoid arthritis [98]. Thus, PET and SPECT imaging interpretation with FAP radioligands should be interpreted with caution to avoid false positives. In comparison to the success of FAPI-based radioligands in cancer diagnosis, the applications in targeted cancer therapies remain to be further validated, where FAP-2286 and PNT6555 are among the promising vectors for such clinical investigations.

11.4.2 Gastrin-releasing Peptide Receptor

The gastrin-releasing peptide receptor, also known as the bombesin receptor subtype 2, is a GPCR found in various organs, including the gastrointestinal tract and pancreas [99,100]. The activation of GRPR triggers multiple physiological processes, such as the regulation of exocrine and endocrine secretion. Over the past few decades, GRPR expression has been identified in various cancer types, including prostate cancer and breast cancer [101,102].

Dalm et al. described a potent GRPR antagonist-based radioligand, ^{177}Lu-NeoBOMB1, for the treatment of prostate cancer (Figure 11.5) [103]. In a radioligand binding assay, NeoBOMB1 demonstrated a half-maximal inhibitory concentration (IC_{50}) of 1 nM [103, 104]. Biodistribution studies using ^{177}Lu-NeoBOMB1 in PC3 tumor-bearing mice with doses of 10 pmol or 200 pmol confirmed GRPR-driven tumor uptake; however, they also revealed on-target uptake by healthy pancreas. Dosimetry calculations indicated that the higher mass dose (200 pmol) afforded a higher tumor radiation dose (581 vs. 435 mGy/MBq) and a lower pancreas radiation dose (265 vs. 1393 mGy/MBq) when compared to the mass dose of 10 pmol. Verhoeven et al. evaluated the efficacy and safety of ^{177}Lu-NeoBOMB1 in a PC3 model [105]. Treatment of PC3 tumors with ^{177}Lu-NeoBOMB1 at dose range from 30 MBq to 60 MBq demonstrated anti-tumor activities. Neither significant weight loss nor histopathological finding was observed in animals during the study and the follow-up period.

Damiana et al. compared ^{177}Lu-NeoBOMB1 with another antagonist radioligand, ^{177}Lu-RM2, in the preclinical studies [106]. They found that both radioligands showed high stability in PBS and human serum; however, their stability in mouse serum was significantly poorer. In addition, the binding affinity of both radioligands was similar, both in the nanomolar range, and the tumor uptake was also comparable at all time points. However, the normal organ uptake differed

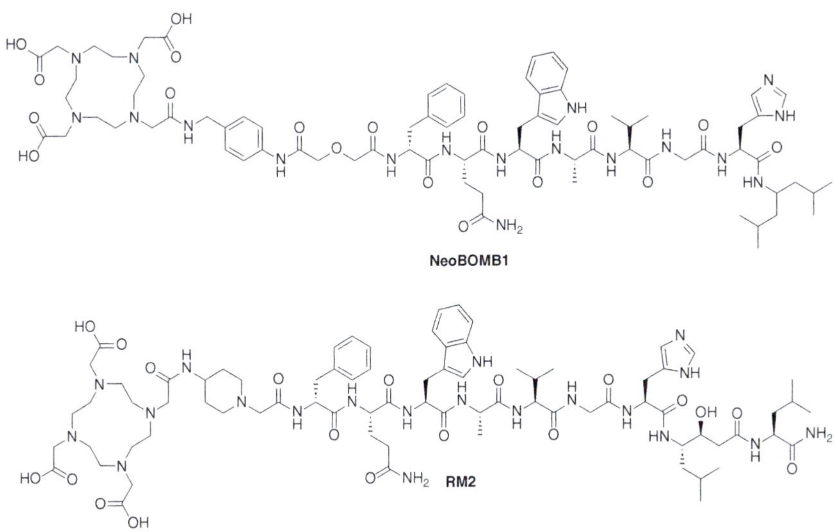

Figure 11.5 The chemical structures of GRPR-targeting radioligand vectors NeoBOMB1 and RM2.

between the two, leading to different tumor-to-normal organ ratios. ^{177}Lu-RM2 had more favorable tumor-to-blood, tumor-to-liver, tumor-to-pancreas, and tumor-to-gastrointestinal tract ratios, but a lower tumor-to-kidney ratio. The tumor-to-kidney ratios of ^{177}Lu-RM2 increased from 3.09 at 1-hour post injection to 11.6 at 24 hours and 24.8 at 72 hours. Similarly, the tumor-to-pancreas ratios of ^{177}Lu-RM2 increased from 0.644 at 1-hour post injection to 29.3 at 24 hours and 59.4 at 72 hours.

Kurth et al. reported the first-in-human dosimetry of ^{177}Lu-RM2 in a group of mCRPC patients who had inadequate PSMA expression or were refractory to ^{177}Lu-PSMA-617 therapy [107]. The study began with ^{68}Ga-RM2-PET/CT scan of 35 patients, followed by ^{177}Lu-RM2 therapeutic study in 4 patients. Increased uptake in tumor lesions and pancreas was noted within the first hour. Mean absorbed organ doses were 1.08 ± 0.44 Gy/GBq in the pancreas, 0.35 ± 0.14 Gy/GBq in the kidneys, 0.05 ± 0.02 Gy/GBq in the liver, 0.07 ± 0.02 Gy/GBq in the gallbladder wall, 0.10 ± 0.06 Gy/GBq in the spleen, and 0.02 ± 0.01 Gy/GBq in the red bone marrow. The mean dose for tumor lesions was 6.20 ± 3.00 Gy/GBq. For the therapeutic doses, a mean activity of 4.5 ± 0.9 GBq was administered, and all patients tolerated the treatment well without notable side effects.

11.4.3 Neurotensin Receptor 1 (NTSR1)

Neurotensin receptors (NTSRs) mediate the effects of the neuropeptide hormone neurotensin, functioning as a neuromodulator in the central nervous system and acting peripherally as a para- and endocrine factor in the gastrointestinal system [108]. NTSRs are primarily expressed in the central nervous system, as well as in the intestinal and hepatic systems [109]. Among the three known NTSR subtypes, NTSR1 is the principal mediator of neurotensin signaling due to its sub-nanomolar affinity for its natural ligand. NTSR1 expression has been linked to neoplastic progression in colon carcinoma, breast cancer, and pancreatic adenocarcinoma [110–113]. Autoradiographic studies have shown that NTSR1 is expressed in pancreatic adenocarcinoma, with densities ranging from 10^3 to 10^4 dpm/mg in both primary tumors and metastases [113,114]. Similar receptor densities are found in Ewing's sarcoma, SCLC, and medullary thyroid carcinoma, although some individual tumors show receptor densities below 10^3 dpm/mg [115]. These receptor densities are comparable to those of SSTR2 in NETs. Collectively, it is rationalized that NTSR1-targeting radioligand therapy for NTR1-expressing tumors may be as effective as Lutathera for NETs.

Schulz et al. described a potent NTSR1 antagonist-based radioligand vector, 3BP-227, for the treatment of colon carcinoma (Figure 11.6) [116,117]. The IC_{50} of ^{nat}Lu-3BP-227 was determined to be 5.95 nM in an *in vitro* functional calcium assay (antagonism mode) and 0.59 nM in a competitive radioligand binding assay [118].

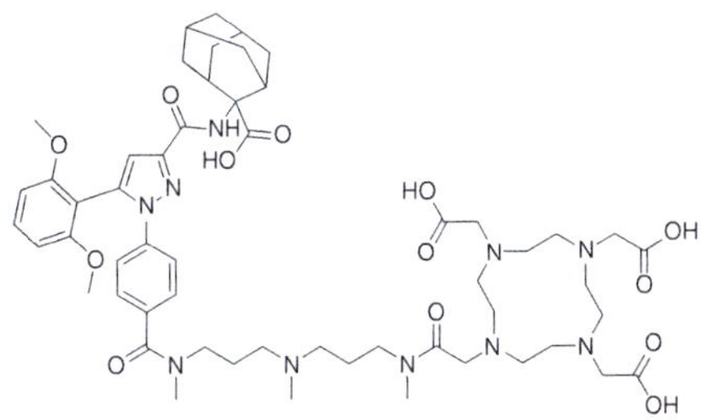

Figure 11.6 The chemical structure of the NTSR1-targeting radioligand vector 3BP-227.

A biodistribution study of ^{177}Lu-3BP-227 in HT29 tumor-bearing mice showed high tumor-specific uptake with low background uptake. Tumor-to-kidney ratios rose from 5.2 ± 2.5 at 3 hours after injection to 6.3 ± 4.3 at 20 hours and 15.9 ± 19.3 at 69 hours after injection. Tumor-to-blood ratios rose from 2.8 ± 0.5 at 3 hours after injection to 44.2 ± 13.5 at 20 hours and 85.0 ± 50.0 at 69 hours after injection [116]. Based on these data, the human dosimetry prediction indicated that a cumulative patient dose of 50 GBq of ^{177}Lu-3BP-227 would result in approximately 6 Gy radiation dose to the kidneys, well below the 23 Gy threshold currently established for kidneys. The efficacy study of ^{177}Lu-3BP-227 in an HT-29 xenograft mouse model with 110 MBq/mouse or 165 MBq/mouse doses demonstrated significant tumor growth inhibition of 55 or 88% respectively, and the mice tolerated the treatment well, as indicated by body weight changes, functional kidney biomarkers, and histological examinations.

Baum et al. reported the first clinical results of ^{177}Lu-3BP-227 for NTSR1-targeted radioligand therapy where six patients with metastatic ductal pancreatic adenocarcinoma, who had exhausted all other treatment options, were treated [119]. It was found that ^{177}Lu-3BP-227 was well tolerated by all patients; the most severe adverse event was reversible grade 2 anemia. Dosimetry calculations identified the kidney as the dose-limiting organ, and a cumulative dose of 20.8 GBq delivered 22 Gy radiation dose to the kidney. One patient achieved a partial response and experienced significant improvements in symptoms and quality of life.

Recently, Liu et al. described a potent NTSR1 radioligand, ^{177}Lu-FL-091, for the treatment of NTSR1-positive tumors [120]. In head-to-head studies, ^{177}Lu-FL-091 demonstrated tighter binding affinity *in vitro* and higher tumor uptake, faster systemic clearance and greater antitumor activities *in vivo* when compared to ^{177}Lu-3BP-227.

11.4.4 Glypican 3 (GPC3)

Glypican 3 (GPC3) is a membrane-bound heparan sulfate proteoglycan primarily involved in embryonic development, with its expression levels significantly decreasing after birth [121]. GPC3 is overexpressed in approximately 70–80% of hepatocellular carcinomas (HCCs) but is not commonly found in normal adult tissues [122]. In addition to HCC, GPC3 expression has been observed in other adult and pediatric cancers, including lung adenocarcinoma, squamous cell carcinoma, embryonal tumors, testicular germ cell tumors, and liposarcoma [123].

Lin et al. recently disclosed a potent and selective cyclic peptide-based GPC3 radioligand vector, DOTA-RYZ-GPC3 (RAYZ-8009), for the treatment of HCC [123]. It was reported that the equilibrium dissociation constants (K_D) of ^{177}Lu-RAYZ-8009 to recombinant human, mouse, and cynomolgus monkey GPC3 were 2.1, 1.4, and 4.0 nM, respectively, and the K_D in the cell-based binding assay using HepG2 cells was 10.8 nM [123,124]. The binding of RAYZ-8009, when conjugated with various stable isotopes (^{175}Lu, ^{139}La, ^{63}Cu, or ^{69}Ga), to HepG2 cells was comparable. ^{177}Lu-RAYZ-8009 demonstrates specificity for GPC3 with no cross-reactivity to other human glypican family proteins (GPC1, 2, 4, 5, and 6) and showed rapid internalization into HepG2 cells at 37 °C. The plasma pharmacokinetic half-life of RAYZ-8009 in mice was determined to be 0.3 hour, and the biodistribution study of ^{177}Lu-RAYZ-8009 in HepG2 tumor-bearing mice demonstrated GPC3-specific tumor uptake and low background accumulation. Tumor-to-kidney ratios increased from 1.3 at 1-hour post injection to 3.7 at 24 hours, 11.3 at 48 hours, and 15.0 at 96 hours. Human dosimetry prediction indicated the kidneys as the dose-limiting organ. In an efficacy study using the same mouse model, ^{177}Lu-RAYZ-8009 at 37 MBq/mouse or ^{225}Ac-RAYZ-8009 at 0.0111, 0.0185, or 0.037 MBq/mouse all exhibited significant tumor growth inhibition. No significant body weight changes or clinical signs of stress were observed in any group.

11.4.5 P-cadherin (CDH3)

P-cadherin (placental cadherin; CDH3) is a calcium-dependent cell-cell adhesion protein belonging to the classical cadherin superfamily, which is essential for cell-cell adhesion and tissue integrity. It consists of five extracellular cadherin repeats (EC1–EC5), a transmembrane region, and a highly conserved cytoplasmic tail [125]. P-cadherin mRNA is overexpressed in a wide range of cancers, including lung, bladder, cervix, esophagus, head and neck, stomach, pancreas, rectum, colon, and others. High cell surface expression of P-cadherin protein, as detected via immunohistochemistry (IHC), has been confirmed in triple-negative breast cancer (TNBC), esophageal squamous cell carcinoma (ESCC), and head and neck squamous cell carcinoma (HNSCC) [126]. Specifically, 37% of TNBC (7/19), 68% of ESCC (15/22), and 30% of HNSCC (6/20) tumors exhibited greater than 2+ P-cadherin staining. P-cadherin staining was also observed in normal tissues, including the epidermis and ducts of sweat glands in the skin, mammary ducts, bronchial epithelium and ducts of bronchial glands, surface epithelium of the tongue, esophagus, ureter, bladder, and vagina, and glandular epithelium of the prostate gland [127]. The density of endogenous P-cadherin was analyzed by quantitative flow cytometry (QFACS) in cancer cell lines, with the highest expression levels ranging between 10^5 and 10^6 receptors per cell. Notable examples include HCT-116 (colorectal; receptor density of 114,000), NCI-H1650 (adenocarcinoma; receptor density of 368,000), and DU-145 (prostate; receptor density of 231,000) [128]. Given the favorable expression profile of P-cadherin, radioligand approaches have been developed to target P-cadherin, including full-length antibody-based radioligand ^{90}Y-21101 and peptide-based vector 3BP-6121.

11.5 Emerging Binder Modalities as Radioligand Vector

Various molecular modalities have been explored as the radioligand vector, including small molecules, peptides, monoclonal antibodies and their fragments, single domain antibodies, DARPins, oligonucleotides, liposomal vectors, and others. Among these, in addition to peptides, small molecules and antibody-based vectors have been investigated clinically while the rest remain to be tested in humans.

11.5.1 Small Molecules

Small molecules are typically synthetic compounds with molecular weight less than 1000 Da. The small molecular size often provides the advantages of quick target engagement, fast systemic clearance, and deep tissue penetration [129,130].

Most small molecule-based radioligand vectors are derived from enzyme inhibitors. In addition to PSMA and FAP discussed earlier, poly(ADP-ribose)polymerase (PARP) and carbonic anhydrase IX (CA-9) inhibitors have also been broadly explored.

PARPs are DNA repair enzymes that are found to be elevated in the cytoplasm and nucleus of tumor cells due to their high proliferation rate and replicative stress. Several PARP1 inhibitors, which disrupt tumor cell replication by blocking DNA repair, have been approved in the past decade, including olaparib, talazoparib, and rucaparib. When converted to radioligand, PARP inhibitors are usually labeled with radiohalogens rather than radiometals to maintain good membrane permeability (Figure 11.7) [131–133]. The gamma ray and auger-emitting ^{123}I labeled PARPi, ^{123}I-MAPI, showed a survival benefit in mice with glioblastoma multiforme [134]. ^{211}At-MM4, an α-radionuclide radioligand derived from rucaparib, also demonstrated a durable antitumor response and an increase in median survival in mice with neuroblastoma tumor [135].

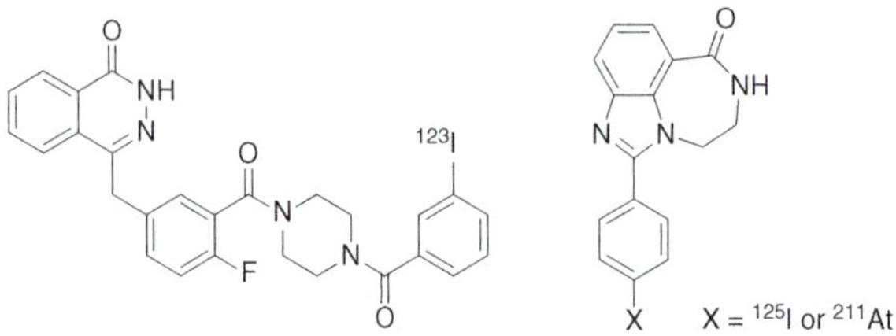

Figure 11.7 The chemical structures of key PARP radioligands.

Currently, the clinical application of PARPi-based radioligands has been limited by the nonoptimal biodistribution profiles with most showing a low tumor uptake and fast tumor clearance. Improvements toward these directions are necessary to ensure successful clinical translation [131,136].

Carbonic anhydrase IX (CA-9) is a cell-surface glycoprotein that is induced by hypoxia and associated with most solid tumors and gastrointestinal mucosa [137]. Small molecules such as sulfonamides, coumarins, and dithiocarbamates are established CA inhibitors [138–140], among which the aromatic sulfonamides are the most explored scaffold as radioligands (Figure 11.8). A variety of aromatic sulfonamides derivatives such as benzenesulfonamide (BSA), acetazolamide (AAZ), and imidazothiadiazole sulfonamide (IS) have been labeled with diagnostic isotopes (^{99m}Tc, ^{18}F, ^{68}Ga, ^{64}Cu, and ^{111}In) as radioligand imaging tracers [141]. While most of the ^{99m}Tc-labeled tracer showed low tumor uptake and high background signal [142–145], ^{99m}Tc-PHC-102, an acetazolamide-based radioligand, demonstrated a high tumor uptake and retention of 22% ID/g at 3 hours and 20% ID/g at 6 hours in the renal cell carcinoma SKRC-52 mouse model [146]. The favorable preclinical profile warranted a Phase 1 study with 5 patients where reasonable tumor lesion uptake was observed in three CA-9-positive RCC patients and the uptake was very low for two CA-9-negative patients [147]. This study demonstrated the potential of CA-9 radioligand imaging in primary and metastatic RCC patients. Similar results were obtained by ^{18}F-VM4-037, a ^{18}F-labeled benzenesulfonamide

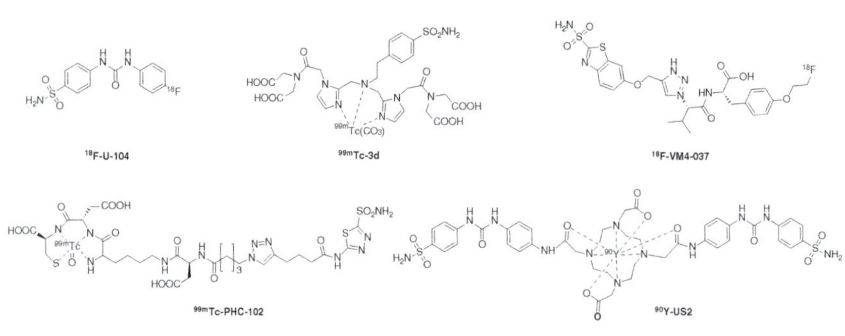

Figure 11.8 The chemical structures of key CA-9 radioligands.

derivative which reached Phase 2 clinical trials [148]. In general, the high kidney accumulation presented a major challenge for their clinical development for RCC.

Only a few CA-9-targeting small molecule-based therapeutic radioligands have been reported. ^{90}Y-US2, a ^{90}Y-labeled ureidosulfonamide derivative, was tested in a CA-9 high-expressing tumor mouse model, where a significant tumor growth delay was observed in ^{90}Y-US2-treated group [149]. The clinical potential of therapeutic CA-9 radioligands remained to be demonstrated.

In addition to the synthetic small molecule inhibitors, small molecule biomolecules and their analogs have also been explored for radioligand imaging and therapy where vitamins and phospholipid ether analogs have drawn the most attention. Vitamins are often more demanded by the tumor cells than the normal ones and therefore have potential to be developed as tumor-targeting radioligands [150]. Folic acid, also known as vitamin B9, is the most extensively studied vitamin radioligand [151]. Folate receptors were found to be overexpressed in many solid epithelial tumors, including breast, cervical, colorectal, and ovarian carcinomas. A number of folic acid-based radioligands have been reported where one of the key challenges was the high on-target kidney uptake [152–155]. It was found that the incorporation of albumin-binding motifs resulted in meaningfully improved tumor-to-kidney ratio [156].

11.5.2 Antibody and Antibody Fragments

The first set of monoclonal antibody-based radioligands was reported between the late 1940s and early 1950s by Pressman [157,158]. Since then, a number of antibody-based radioligand vectors have been developed for diagnosis (with ^{111}In and ^{89}Zr) and therapy (with ^{131}I, ^{90}Y, ^{177}Lu, and ^{225}Ac) [159] (Figure 11.9). The notable examples include Zevalin and Bexxar, both of which target CD20 for the treatment of patients with CD20-positive, non-Hodgkin's lymphoma – Zevalin is ^{90}Y-ibritumomab tiuxetan while Bexxar is ^{131}I-tositumomab.

The long plasma PK half-life of full-length antibodies, typically ranging from days to 1 to 2 weeks, enables high tumor uptake but also causes prolonged exposure of blood and other normal organs to radiation. For diagnostic applications, it requires long waiting time between dosing the radioligand to the patient and SPECT or PET scanning; for therapy applications, it causes safety concerns, in particular, bone marrow toxicity. To address these limitations, various antibody fragments have been developed, which retain high binding affinity similar to full-length antibody but have a much smaller molecular size, resulting in significantly faster blood clearance kinetics. The structure and sizes of antibody fragments are summarized in Figure 11.9 [159].

The fast blood clearance of antibody fragments is a double-edged sword. On one hand, the antibody fragment-based diagnostic radioligands behave similarly

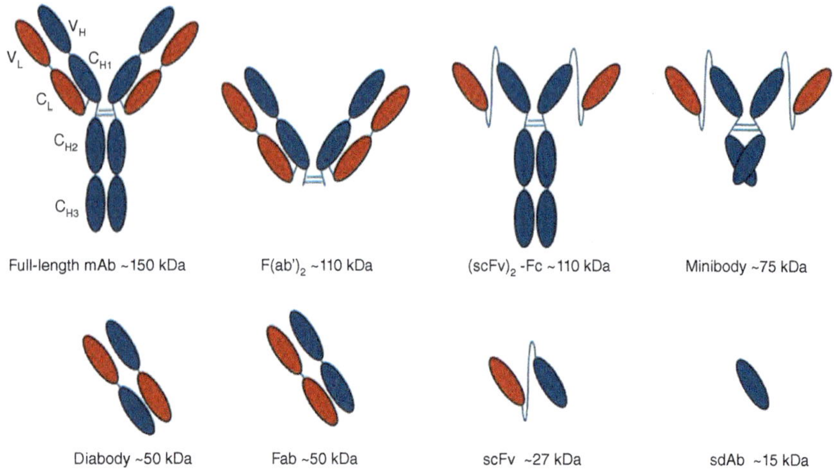

Figure 11.9 The anatomies of IgG1 antibody and its fragments and their respective molecular weight.

to peptide imaging agents, allowing scans to be performed shortly after dosing. On the other hand, they typically provide low and transient tumor uptake and high kidney accumulation.

Recently, several promising antibody fragment-based radioligands have been reported in preclinical imaging and therapy applications. ^{64}Cu-labeled anti-CD8 cys-diabody has demonstrated promise in imaging CD8-postive T-cell density [160]. An ^{131}I-labeled human epidermal growth factor receptor 2 (HER2)-targeting sdAb, ^{131}I-SGMIB-VHH_1028, demonstrated excellent tumor growth inhibition [161]. Clinically, in a Phase 1 study, ^{89}Zr-IAB2M, a prostate-specific membrane antigen (PSMA) targeting minibody, was confirmed to provide good tumor lesion visualization for mCRPC patients [162].

11.5.3 Engineered Scaffold Protein

Since the late 1990s, the concept of engineered scaffold proteins emerged, which provides alternatives to antibody and its fragments [163,164]. A number of scaffold proteins have been extensively studied, including affibody derived from streptococcal protein A, monobody (adnectin) from human fibronectin, anticalin proteins from natural lipocalin proteins, albumin-binding domain-derived affinity proteins (ADAPTs) originally from streptococcal protein G, and designed ankyrin repeat proteins (DARPins) from ankyrin [165].

These scaffold proteins, with a molecular weight range of 6–18 KDa, not only share favorable properties with antibody fragments but also offer great biophysical stability and production feasibility. Short sequences like affibody can be routinely chemically synthesized and derivatized [166–171].

DARPin consists of four to six repeating sequences, each containing one β-turn and two antiparallel α-helices. G3, a HER2-targeting DARPin ^{99m}Tc radioligand, has been studied in a Phase 1 study for HER2+ breast cancer imaging [172–174]. This diagnostic radioligand was confirmed to be able to detect HER2+ tumor lesions with high contrast. However, similar to the small-sized antibody fragments, it accumulates very highly in kidney, preventing its application to therapy. Co-dosing with kidney-protecting agents such as lysine, gelofusine, fructose, or maleate was not sufficiently effective in reducing kidney accumulation [175].

Several approaches have been explored to improve the pharmacokinetics profile of DARPin, including introducing an albumin-binding domain, large unstructured polypeptides such as PAS (polypeptide containing proline, alanine, and serine) or XTEN (polypeptide containing proline, alanine, serine, glycine, and glutamate), and a long polyethylene glycol (PEG) [176–178]. Although high tumor uptake can be achieved by these approaches, the associated long blood circulation therefore raises concerns about bone marrow safety, presenting a major challenge for their clinical development as therapeutic drugs [168].

11.5.4 Oligonucleotide Aptamers

Aptamers are short, single-stranded oligonucleotides (DNA or RNA) that can recognize the targeting receptors. The binding affinity of the aptamer, screened by Systematic Evolution of Ligands by Exponential Enrichment (SELEX) technology, can reach the same range of an antibody (nM–pM) [179,180]. With a relatively small size (10–20 KDa), aptamers can be chemically synthesized and derivatized, and they are sufficiently stable under radiolabeling conditions. Similar to small-sized protein radioligand vectors, aptamer-based radioligands can offer favorable profiles such as deep tumor penetration, fast tumor uptake, and quick systemic clearance [181].

^{99m}Tc-NXZ1909 was one of the early aptamer-based radioligands tested for inflammation imaging [182]. Recently, a number of aptamer-based diagnostic radioligands labeled with ^{18}F, ^{68}Ga, ^{64}Cu, ^{99m}Tc, ^{188}Re, and ^{89}Zr have been explored in cancer or inflammation imaging [183–191].

Chemical modifications with non-oligonucleotide building blocks such as sugar, unnatural nucleotides, and phosphodiester bonds have been utilized to improve the stability and pharmacokinetic profiles [192–195]; however, several main challenges remain to be addressed, including instability *in vivo*, low tumor uptake, and rapid renal clearance. The typical reported tumor uptake was below 1% ID/g [196–199].

Recently, the first-in-human pharmacokinetics study of an aptamer radioligand, ^{68}Ga-NOTA-SGC8, targeting SGC8 protein tyrosine kinase 7, was reported where fast renal clearance and high kidney accumulation were observed, which was consistent with preclinical findings. The study provides valuable human data on aptamer-based radioligands, also suggesting that significant optimization is necessary prior to the clinical use of such vector [200].

11.6 Perspectives: Promises and Challenges

The well-established clinical benefits of SSTR2 and PSMA targeted radioligands and their commercial successes have been aggressively promoting further development of such modality for solid tumor indications. There are multiple directions that are concurrently being explored, including alternative radioisotopes, vectors with new molecular formats, and targets beyond SSTR2 and PSMA.

Therapeutic radioisotopes have evolved from the historical radioactive iodine to the currently well-accepted ^{68}Ga and ^{18}F for diagnosis and ^{177}Lu for therapy. Alternative PET imaging radioisotopes that possess longer $T_{1/2}$ such as ^{64}Cu and ^{89}Zr are of great interest for further diagnostic applications. For therapeutic applications, α-particle-emitting radioisotopes such as ^{225}Ac, ^{212}Pb, and ^{211}At have emerged as potential next-generation treatments. In addition, ^{161}Tb is being considered as an attractive β-emitter alterative to ^{177}Lu, as it delivers stronger radiation within a shorter range compared to ^{177}Lu. Separately, the element-equivalent theranostic pairs, including ^{203}Pb (SPECT)/^{212}Pb (α-emitter) and ^{64}Cu (PET)/^{67}Cu (β-emitter), are gaining attractions due to the chemical – and therefore biodistribution – equivalence of the diagnostic agent and therapeutic drug.

Peptide, as a molecular format of the vector, has demonstrated great success in SSTR2 and PSMA radioligand development attributed to its high specificity, tight binding affinity, and rapid *in vivo* clearance kinetics. However, it is estimated that only 10–15% of the drug targets are druggable by peptides, so there is a significant need to explore alternative binders to access these peptide "undruggable" radioligand targets. Currently, various formats of antibodies, DARPins, affibodies, and other protein binders are being actively explored as they often provide high-affinity binding to the targets that are undruggable by peptides. However, their development path toward a therapeutic radioligand is currently hampered by either the slow systemic clearance such as full-length Mab or extensive accumulation in the kidney such as nanobody. Efforts addressing such limitations are underway, and any breakthrough has the potential to drastically increase our accessibility to RDC target space.

Lastly, the increased versatility of the radioisotopes and binder molecular formats has been enabling the exploration of the indications outside of NETs and prostate cancer and the targets beyond SSTR2 and PSMA. The next milestones where radioligand therapy demonstrates clinical proof of concept (POC) in the new solid tumor targets will be essential to further support and sustain the long-term development interest of this field.

Overall, we firmly believe that the development of the radioligand field will follow a trajectory of ADC field, however, with a steeper slope and potentially larger scale. We envision that clinical POC for new targets will be achieved within the next five years potentially with multiple radioisotopes and vector molecular formats, which will eventually lead to new approvals and bring unmet clinical benefits to the patients.

References

1 Crisan, G., Moldovean-Cioroianu, N.S., Timaru, D.G., Andries, G., Cainap, C., and Chis, V. Radiopharmaceuticals for PET and SPECT imaging: a literature review over the last decade. *Int J Mol Sci*. 2022;23(9):5023.

2 Sgouros, G., Bodei, L., McDevitt, M.R., and Nedrow, J.R. Radiopharmaceutical therapy in cancer: clinical advances and challenges. *Nat Rev Drug Discov*. 2020;19(9):589–608.

3 Ciarallo, A. and Rivera J. Radioactive iodine therapy in differentiated thyroid cancer: 2020 update. *AJR Am J Roentgenol*. 2020;215(2):285–291.

4 Parker, C., Nilsson, S., Heinrich, D., et al. Alpha emitter radium-223 and survival in metastatic prostate cancer. *N Engl J Med*. 2013;369(3):213–223.

5 Tornesello, A.L., Buonaguro, L., Tornesello, M.L., and Buonaguro, F.M. New insights in the design of bioactive peptides and chelating agents for imaging and therapy in oncology. *Molecules*. 2017;22(8):1282.

6 Turner, J.H. Philosophy of cancer theranostics. *Cancer Biother Radiopharm*. 2023;38(1):1–7.

7 Jang, A., Kendi, A.T., and Sartor, O. Status of PSMA-targeted radioligand therapy in prostate cancer: current data and future trials. *Ther Adv Med Oncol*. 2023;15:17588359231157632.

8 Aloj, L., Attili, B., Lau, D., et al. The emerging role of cell surface receptor and protein binding radiopharmaceuticals in cancer diagnostics and therapy. *Nucl Med Biol*. 2021;92:53–64.

9 Refardt, J., Hofland, J., Kwadwo, A., et al. Theranostics in neuroendocrine tumors: an overview of current approaches and future challenges. *Rev Endocr Metab Disord*. 2021;22(3):581–594.

10 Alas, M., Saghaeidehkordi, A., and Kaur, K. Peptide-drug conjugates with different linkers for cancer therapy. *J Med Chem*. 2021;64(1):216–232.

11 Kilbourn, M.R. and Scott, P.J.H. *Handbook of Radiopharmaceuticals: Radiochemistry and Applications*. Wiley; 2003.

12 Brechbiel, M.W. Bifunctional chelates for metal nuclides. *Q J Nucl Med Mol Imaging*. 2008;52(2):166–173.
13 Milenic, D.E., Brady, E.D., and Brechbiel, M.W. Antibody-targeted radiation cancer therapy. *Nat Rev Drug Discov*. 2004;3(6):488–499.
14 Notni, J., Hermann, P., Havlickova, J., et al. A triazacyclononane-based bifunctional phosphinate ligand for the preparation of multimeric 68Ga tracers for positron emission tomography. *Chemistry*. 2010;16(24):7174–7185.
15 Bartholoma, M.D., Louie, A.S., Valliant, J.F., and Zubieta, J. Technetium and gallium derived radiopharmaceuticals: comparing and contrasting the chemistry of two important radiometals for the molecular imaging era. *Chem Rev*. 2010;110(5):2903–2920.
16 Weckbecker, G., Raulf, F., Stolz, B., and Bruns, C. Somatostatin analogs for diagnosis and treatment of cancer. *Pharmacol Ther*. 1993;60(2):245–264.
17 Janecka, A., Zubrzycka, M., and Janecki, T. Somatostatin analogs. *J Pept Res*. 2001;58(2):91–107.
18 Dong, C., Liu, Z., and Wang, F. Peptide-based radiopharmaceuticals for targeted tumor therapy. *Curr Med Chem*. 2014;21(1):139–152.
19 Gomes-Porras, M., Cardenas-Salas, J., and Alvarez-Escola, C. Somatostatin analogs in clinical practice: a review. *Int J Mol Sci*. 2020;21(5):1682.
20 Maini, C.L., Tofani, A., Venturo, I., et al. Somatostatin receptor imaging in small cell lung cancer using 111In-DTPA-octreotide: a preliminary study. *Nucl Med Commun*. 1993;14(11):962–968.
21 Oliaro, A., Filosso, P.L., Bello, M., et al. Use of 111In-DTPA-octreotide scintigraphy in the diagnosis of neuroendocrine and non-neuroendocrine tumors of the lung. Preliminary results. *J Cardiovasc Surg (Torino)*. 1997;38(3):313–315.
22 Bomanji, J.B. and Papathanasiou, N.D. (1)(1)(1)In-DTPA(0)-octreotide (Octreoscan), (1)(3)(1)I-MIBG and other agents for radionuclide therapy of NETs. *Eur J Nucl Med Mol Imaging*. 2012; 39 (Suppl 1):S113–S125.
23 Imhof, A., Brunner, P., Marincek, N., et al. Response, survival, and long-term toxicity after therapy with the radiolabeled somatostatin analogue [90Y-DOTA]-TOC in metastasized neuroendocrine cancers. *J Clin Oncol*. 2011;29(17):2416–2423.
24 Hennrich, U., Kopka, K. Lutathera((R)): the first FDA- and EMA-approved radiopharmaceutical for peptide receptor radionuclide therapy. *Pharmaceuticals (Basel)*. 2019;12(3):114.
25 Wang, L.F., Lin, L., Wang, M.J., and Li, Y. The therapeutic efficacy of 177Lu-DOTATATE/DOTATOC in advanced neuroendocrine tumors: a meta-analysis. *Medicine (Baltimore)*. 2020;99(10):e19304.
26 Cullinane, C., Jeffery, C.M., Roselt, P.D., et al. Peptide receptor radionuclide therapy with (67)Cu-CuSarTATE is highly efficacious against a somatostatin-positive neuroendocrine tumor model. *J Nucl Med*. 2020;61(12):1800–1805.

27 Kwekkeboom, D.J. and Krenning, E.P. Peptide receptor radionuclide therapy in the treatment of neuroendocrine tumors. *Hematol Oncol Clin North Am.* 2016;30(1):179–191.
28 Nguyen, N.C., Moon, C.H., Muthukrishnan, A., and Furlan, A. 68Ga-DOTATATE PET/MRI for neuroendocrine tumors: a pictorial review. *Clin Nucl Med.* 2020;45(9):e406–e410.
29 Hennrich, U. and Benesova, M. [(68)Ga]Ga-DOTA-TOC: the first FDA-approved (68)Ga-radiopharmaceutical for PET imaging. *Pharmaceuticals (Basel).* 2020;13(3):38.
30 Graham, M.M., Gu, X., Ginader, T., Breheny, P., and Sunderland, J.J. (68) Ga-DOTATOC imaging of neuroendocrine tumors: a systematic review and metaanalysis. *J Nucl Med.* 2017;58(9):1452–1458.
31 Delpassand, E.S., Ranganathan, D., Wagh, N., et al. (64)Cu-DOTATATE PET/CT for imaging patients with known or suspected somatostatin receptor-positive neuroendocrine tumors: results of the first U.S. prospective, reader-masked clinical trial. *J Nucl Med.* 2020;61(6):890–896.
32 Hicks, R.J., Jackson, P., Kong, G., et al. (64)Cu-SARTATE PET imaging of patients with neuroendocrine tumors demonstrates high tumor uptake and retention, potentially allowing prospective dosimetry for peptide receptor radionuclide therapy. *J Nucl Med.* 2019;60(6):777–785.
33 Strosberg, J.R., Caplin, M.E., Kunz, P.L., et al. (177)Lu-dotatate plus long-acting octreotide versus high-dose long-acting octreotide in patients with midgut neuroendocrine tumours (NETTER-1): final overall survival and long-term safety results from an open-label, randomised, controlled, phase 3 trial. *Lancet Oncol.* 2021;22(12):1752–1763.
34 Lehman, J.M. Somatostatin receptor 2 targeting in small cell lung carcinoma: perspectives. Oncotarget. 2019;10(46):4727–4730.
35 Delpassand, E.S., Tworowska, I., Esfandiari, R., et al. Targeted alpha-emitter therapy with (212)Pb-DOTAMTATE for the treatment of metastatic SSTR-expressing neuroendocrine tumors: first-in-humans dose-escalation clinical trial. *J Nucl Med.* 2022;63(9):1326–1333.
36 Michler, E., Kastner, D., Brogsitter, C., et al. First-in-human SPECT/CT imaging of [(212)Pb]Pb-VMT-alpha-NET in a patient with metastatic neuroendocrine tumor. *Eur J Nucl Med Mol Imaging.* 2024;51(5):1490–1492.
37 Thakur, S., Daley, B., Millo, C., et al. (177)Lu-DOTA-EB-TATE, a radiolabeled analogue of somatostatin receptor type 2, for the imaging and treatment of thyroid cancer. *Clin Cancer Res.* 2021;27(5):1399–1409.
38 Wild, D., Gronbaek, H., Navalkissoor, S., et al. A phase I/II study of the safety and efficacy of [(177)Lu]Lu-satoreotide tetraxetan in advanced somatostatin receptor-positive neuroendocrine tumours. *Eur J Nucl Med Mol Imaging.* 2023;51(1):183–195.

39 Cheng, Q., Butler, W., Zhou, Y., et al. Pre-existing castration-resistant prostate cancer-like cells in primary prostate cancer promote resistance to hormonal therapy. *Eur Urol.* 2022;81(5):446–455.
40 Yamada, Y. and Beltran, H. The treatment landscape of metastatic prostate cancer. *Cancer Lett.* 2021;519:20–29.
41 Parker, C., Castro, E., Fizazi, K., et al. Prostate cancer: ESMO clinical practice guidelines for diagnosis, treatment and follow-up. *Ann Oncol.* 2020;31(9):1119–1134.
42 Wustemann, T., Haberkorn, U., Babich, J., and Mier, W. Targeting prostate cancer: prostate-specific membrane antigen based diagnosis and therapy. *Med Res Rev.* 2019;39(1):40–69.
43 Al Saffar, H., Chen, D.C., Delgado, C., et al. The current landscape of prostate-specific membrane antigen (PSMA) imaging biomarkers for aggressive prostate cancer. *Cancers (Basel).* 2024;16(5):939.
44 Niaz, M.O., Sun, M., Ramirez-Fort, M.K., and Niaz, M.J. Review of lutetium-177-labeled anti-prostate-specific membrane antigen monoclonal antibody J591 for the treatment of metastatic castration-resistant prostate cancer. *Cureus.* 2020;12(2):e7107.
45 Vallabhajosula, S., Kuji, I., Hamacher, K.A., et al. Pharmacokinetics and biodistribution of 111In- and 177Lu-labeled J591 antibody specific for prostate-specific membrane antigen: prediction of 90Y-J591 radiation dosimetry based on 111In or 177Lu? *J Nucl Med.* 2005;46(4):634–641.
46 Vallabhajosula, S., Goldsmith, S.J., Kostakoglu, L., Milowsky, M.I., Nanus, D.M., and Bander, N.H. Radioimmunotherapy of prostate cancer using 90Y- and 177Lu-labeled J591 monoclonal antibodies: effect of multiple treatments on myelotoxicity. *Clin Cancer Res.* 2005;11(19 Pt 2):7195s–7200s.
47 Benesova, M., Bauder-Wust, U., Schafer, M., et al. Linker modification strategies to control the prostate-specific membrane antigen (PSMA)-targeting and pharmacokinetic properties of DOTA-conjugated PSMA inhibitors. *J Med Chem.* 2016;59(5):1761–1775.
48 Benesova, M., Schafer, M., Bauder-Wust, U., et al. Preclinical evaluation of a tailor-made DOTA-conjugated PSMA inhibitor with optimized linker moiety for imaging and endoradiotherapy of prostate cancer. *J Nucl Med.* 2015;56(6):914–920.
49 Sartor, O., de Bono, J., Chi, K.N., et al. Lutetium-177-PSMA-617 for metastatic castration-resistant prostate cancer. *N Engl J Med.* 2021;385(12):1091–1103.
50 Hansen, A.R., Probst, S., Tutrone, R.F., et al. Efficacy and safety of 177Lu-PNT2002 prostate-specific membrane antigen (PSMA) therapy in metastatic castration resistant prostate cancer (mCRPC): initial results from SPLASH. ESMO. 2022;33(Suppl 7):S1185.

51 Szabo, Z., Mena, E., Rowe, S.P., et al. Initial evaluation of [(18)F]DCFPyL for prostate-specific membrane antigen (PSMA)-targeted PET imaging of prostate cancer. *Mol Imaging Biol*. 2015;17(4):565–574.

52 Ferreira, G., Iravani, A., Hofman, M.S., and Hicks, R.J. Intra-individual comparison of (68)Ga-PSMA-11 and (18)F-DCFPyL normal-organ biodistribution. *Cancer Imaging*. 2019;19(1):23.

53 Sallam, M., Nguyen, N.T., Sainsbury, F., Kimizuka, N., Muyldermans, S., and Benesova-Schafer, M. PSMA-targeted radiotheranostics in modern nuclear medicine: then, now, and what of the future? *Theranostics*. 2024;14(8):3043–3079.

54 Keam, S.J. Piflufolastat F 18: diagnostic first approval. *Mol Diagn Ther*. 2021;25(5):647–656.

55 Lee, C.H., Lim, I., Woo, S.K., et al. The feasibility of (64)Cu-PSMA I&T PET for prostate cancer. *Cancer Biother Radiopharm*. 2022;37(6):417–423.

56 Lengyelova, E., Wong, V., Lenzo, N., Parker, M., and Emmett, L. 64Cu-SAR-bisPSMA (PROPELLER) positron emission tomography (PET) imaging in patients with confirmed prostate cancer. *J Clin Oncol*. 2023; 41(16_suppl):5039.

57 Seifert, R., Seitzer, K., Herrmann, K., et al. Analysis of PSMA expression and outcome in patients with advanced prostate cancer receiving (177)Lu-PSMA-617 radioligand therapy. *Theranostics*. 2020;10(17):7812–7820.

58 Jones, W., Griffiths, K., Barata, P.C., and Paller, C.J. PSMA theranostics: review of the current status of PSMA-targeted imaging and radioligand therapy. *Cancers (Basel)*. 2020;12(6):1367.

59 Ling, X., Latoche, J.D., Choy, C.J., et al. Preclinical dosimetry, imaging, and targeted radionuclide therapy studies of Lu-177-labeled albumin-binding, PSMA-targeted CTT1403. *Mol Imaging Biol*. 2020;22(2):274–284.

60 Zang, J., Fan, X., Wang, H., et al. First-in-human study of (177)Lu-EB-PSMA-617 in patients with metastatic castration-resistant prostate cancer. *Eur J Nucl Med Mol Imaging*. 2019;46(1):148–158.

61 Shin, D., Ha, S., O, J.H., et al. A single dose of novel PSMA-targeting radiopharmaceutical agent [(177)Lu]Ludotadipep for patients with metastatic castration-resistant prostate cancer: phase I clinical trial. *Cancers (Basel)*. 2022;14(24):6225.

62 Bundschuh, R.A., Pfob, C.H., Wienand, G., Dierks, A., Kircher, M., and Lapa, C. 177 Lu-rhPSMA-10.1 induces tumor response in a patient with mCRPC after PSMA-directed radioligand therapy with 177 Lu-PSMA-I&T. *Clin Nucl Med*. 2023;48(4):337–338.

63 Suh, M., Cheon, G.J., Kang, K.W., et al. Phase I/II clinical trial of 177Lu-DGUL in metastatic castration-resistant prostate cancer patients: preliminary results of safety, dosimetry, and efficacy. *J Nucl Med*. 2023;64(Suppl 1):823.

64 Antonini, P., Meyrick, D., and Hayward, C. 177Lu-DOTA-TLX591 safety, biodistribution and dosimetry study (ProstACT-SELECT). *J Clin Oncol.* 2022;40(16_suppl):TPS5109.

65 Tschan, V.J., Busslinger, S.D., Bernhardt, P., et al. Albumin-binding and conventional PSMA ligands in combination with (161)Tb: biodistribution, dosimetry, and preclinical therapy. *J Nucl Med.* 2023;64(10):1625–1631.

66 Johnson, G., Lengyelova, E., Nordquist, L.T., et al. SECuRE: a dose escalation/expansion study to assess the anti-tumor efficacy of 67Cu-SAR-bisPSMA in patients with metastatic castrate-resistant prostate cancer. *J Clin Oncol.* 2024;42(4_suppl). Abstract#: TPS246.

67 Kratochwil, C., Bruchertseifer, F., Giesel, F.L., et al. 225Ac-PSMA-617 for PSMA-targeted alpha-radiation therapy of metastatic castration-resistant prostate cancer. *J Nucl Med.* 2016;57(12):1941–1944.

68 Sathekge, M., Bruchertseifer, F., Vorster, M., et al. Predictors of overall and disease-free survival in metastatic castration-resistant prostate cancer patients receiving (225)Ac-PSMA-617 radioligand therapy. *J Nucl Med.* 2020;61(1):62–69.

69 Feuerecker, B., Tauber, R., Knorr, K., et al. Activity and adverse events of actinium-225-PSMA-617 in advanced metastatic castration-resistant prostate cancer after failure of lutetium-177-PSMA. *Eur Urol.* 2021;79(3):343–350.

70 Alam, M.R., Singh, S.B., Thapaliya, S., Shrestha, S., Deo, S., and Khanal, K. A review of 177Lutetium-PSMA and 225Actinium-PSMA as emerging theranostic agents in prostate cancer. *Cureus.* 2022;14(9):e29369.

71 Ling, S.W., de Blois, E., Hooijman, E., van der Veldt, A., and Brabander, T. Advances in (177)Lu-PSMA and (225)Ac-PSMA radionuclide therapy for metastatic castration-resistant prostate cancer. *Pharmaceutics.* 2022;14(10):2166.

72 Sathekge, M.M., Lawal, I.O., Bal, C., et al. Actinium-225-PSMA radioligand therapy of metastatic castration-resistant prostate cancer (WARMTH Act): a multicentre, retrospective study. *Lancet Oncol.* 2024;25(2):175–183.

73 Zacherl, M.J., Gildehaus, F.J., Mittlmeier, L., et al. First clinical results for PSMA-targeted alpha-therapy using (225)Ac-PSMA-I&T in advanced-mCRPC patients. *J Nucl Med.* 2021;62(5):669–674.

74 Zitzmann-Kolbe, S., Hesterman, J., Cole, P.E., et al. Actinium-225 -PSMA-trillium (BAY 3563254): preclinical evaluation and clinical imaging study of a novel 225Ac-labeled PSMA-targeting small molecule triad for the treatment of mCRPC. *Cancer Res.* 2024;84(7_Suppl). Abstract#: ND09.

75 Liu, F., Zhang, J., Yang, J., et al. 225Ac-FL-020 is a novel PSMA-targeting radionuclide drug conjugate (RDC) with superior in vivo anti-tumor activity. *Cancer Res.* 2024;84(6_Suppl). Abstract#: 6023.

76 Tagawa, S.T., Sun, M., Sartor, A.O., et al. Phase I study of 225Ac-J591 for men with metastatic castration-resistant prostate cancer (mCRPC). *J Clin Oncol.* 2021;39(15_suppl). Abstract#: 5015.

77 Stenberg, V.Y., Larsen, R.H., Ma, L.W., et al. Evaluation of the PSMA-binding ligand (212)Pb-NG001 in multicellular tumour spheroid and mouse models of prostate cancer. *Int J Mol Sci*. 2021;22(9):7815.

78 Griffiths, M.R., Pattison, D.A., Latter, M., et al. First-in-human (212)Pb-PSMA-targeted alpha-therapy SPECT/CT imaging in a patient with metastatic castration-resistant prostate cancer. *J Nucl Med*. 2024;65(4):664.

79 Hammer, S., Hagemann, U.B., Zitzmann-Kolbe, S., et al. Preclinical efficacy of a PSMA-targeted thorium-227 conjugate (PSMA-TTC), a targeted alpha therapy for prostate cancer. *Clin Cancer Res*. 2020;26(8):1985–1996.

80 Anderson, N.M. and Simon, M.C. The tumor microenvironment. *Curr Biol*. 2020;30(16):R921–R925.

81 de Visser, K.E. and Joyce, J.A. The evolving tumor microenvironment: from cancer initiation to metastatic outgrowth. *Cancer Cell*. 2023;41(3):374–403.

82 Zhao, Y., Shen, M., Wu, L., et al. Stromal cells in the tumor microenvironment: Accomplices of tumor progression? *Cell Death Dis*. 2023;14(9):587.

83 Welt, S., Divgi, C.R., Scott, A.M., et al. Antibody targeting in metastatic colon cancer: a phase I study of monoclonal antibody F19 against a cell-surface protein of reactive tumor stromal fibroblasts. *J Clin Oncol*. 1994;12(6):1193–1203.

84 Hofheinz, R.D., al-Batran, S.E., Hartmann, F., et al. Stromal antigen targeting by a humanised monoclonal antibody: an early phase II trial of sibrotuzumab in patients with metastatic colorectal cancer. *Onkologie*. 2003;26(1):44–48.

85 Adams, S., Miller, G.T., Jesson, M.I., Watanabe, T., Jones, B., and Wallner, B.P. PT-100, a small molecule dipeptidyl peptidase inhibitor, has potent antitumor effects and augments antibody-mediated cytotoxicity via a novel immune mechanism. *Cancer Res*. 2004;64(15):5471–5480.

86 Jansen, K., Heirbaut, L., Verkerk, R., et al. Extended structure-activity relationship and pharmacokinetic investigation of (4-quinolinoyl)glycyl-2-cyanopyrrolidine inhibitors of fibroblast activation protein (FAP). *J Med Chem*. 2014;57(7):3053–3074.

87 Ryabtsova, O., Jansen, K., Van Goethem, S., et al. Acylated Gly-(2-cyano) pyrrolidines as inhibitors of fibroblast activation protein (FAP) and the issue of FAP/prolyl oligopeptidase (PREP)-selectivity. *Bioorg Med Chem Lett*. 2012;22(10):3412–3417.

88 Loktev, A., Lindner, T., Mier, W., et al. A tumor-imaging method targeting cancer-associated fibroblasts. *J Nucl Med*. 2018;59(9):1423–1429.

89 Lindner, T., Loktev, A., Altmann, A., et al. Development of quinoline-based theranostic ligands for the targeting of fibroblast activation protein. *J Nucl Med*. 2018;59(9):1415–1422.

90 Kratochwil, C., Flechsig, P., Lindner, T., et al. (68)Ga-FAPI PET/CT: tracer uptake in 28 different kinds of cancer. *J Nucl Med*. 2019;60(6):801–805.

91 Millul, J., Koepke, L., Haridas, G.R., Sparrer, K.M.J., Mansi, R., and Fani, M. Head-to-head comparison of different classes of FAP radioligands designed to increase tumor residence time: monomer, dimer, albumin binders, and small molecules vs peptides. *Eur J Nucl Med Mol Imaging*. 2023;50(10):3050–3061.

92 Moon, E.S., Ballal, S., Yadav, M.P., et al. Fibroblast activation protein (FAP) targeting homodimeric FAP inhibitor radiotheranostics: a step to improve tumor uptake and retention time. *Am J Nucl Med Mol Imaging*. 2021;11(6):476–491.

93 Chakravarty, R., Song, W., Chakraborty, S., and Cai, W. Fibroblast activation protein (FAP)-targeted radionuclide therapy: Which ligand is the best? *Eur J Nucl Med Mol Imaging*. 2023;50(10):2935–2939.

94 Zhang, P., Xu, M., Ding, J., et al. Fatty acid-conjugated radiopharmaceuticals for fibroblast activation protein-targeted radiotherapy. *Eur J Nucl Med Mol Imaging*. 2022;49(6):1985–1996.

95 Zboralski, D., Hoehne, A., Bredenbeck, A., et al. Preclinical evaluation of FAP-2286 for fibroblast activation protein targeted radionuclide imaging and therapy. *Eur J Nucl Med Mol Imaging*. 2022;49(11):3651–3667.

96 Levy, M.T., McCaughan, G.W., Abbott, C.A., et al. Fibroblast activation protein: a cell surface dipeptidyl peptidase and gelatinase expressed by stellate cells at the tissue remodelling interface in human cirrhosis. *Hepatology*. 1999;29(6):1768–1778.

97 Acharya, P.S., Zukas, A., Chandan, V., Katzenstein, A.L., and Pure, E. Fibroblast activation protein: a serine protease expressed at the remodeling interface in idiopathic pulmonary fibrosis. *Hum Pathol*. 2006;37(3):352–360.

98 Bauer, S., Jendro, M.C., Wadle, A., et al. Fibroblast activation protein is expressed by rheumatoid myofibroblast-like synoviocytes. *Arthritis Res Ther*. 2006;8(6):R171.

99 Gonzalez, N., Moody, T.W., Igarashi, H., Ito, T., and Jensen, R.T. Bombesin-related peptides and their receptors: recent advances in their role in physiology and disease states. *Curr Opin Endocrinol Diabetes Obes*. 2008;15(1):58–64.

100 Guo, M., Qu, X., and Qin, X.Q. Bombesin-like peptides and their receptors: recent findings in pharmacology and physiology. *Curr Opin Endocrinol Diabetes Obes*. 2015;22(1):3–8.

101 Gugger, M. and Reubi, J.C. Gastrin-releasing peptide receptors in non-neoplastic and neoplastic human breast. *Am J Pathol*. 1999;155(6):2067–2076.

102 Markwalder, R. and Reubi, J.C. Gastrin-releasing peptide receptors in the human prostate: relation to neoplastic transformation. *Cancer Res*. 1999;59(5):1152–1159.

103 Dalm, S.U., Bakker, I.L., de Blois, E., et al. 68Ga/177Lu-NeoBOMB1, a novel radiolabeled GRPR antagonist for theranostic use in oncology. *J Nucl Med*. 2017;58(2):293–299.

104 Nock, B., Kaloudi, A., Lymperis, E., et al. [68Ga]NeoBomb1, a new potent GRPR-antagonist for PET imaging - preclinical and first clinical evaluation in prostate cancer. *J Nucl Med.* 2016;57(supplement 2):583.

105 Verhoeven, M., Haeck, J., de Blois, E., et al. The balance between the therapeutic efficacy and safety of [(177)Lu]Lu-NeoB in a preclinical prostate cancer model. *Mol Imaging Biol.* 2024;26(1):114–123.

106 Damiana, T.S.T., Paraiso, P., de Ridder, C., Stuurman, D., Seimbille, Y., and Dalm, S.U. Side-by-side comparison of the two widely studied GRPR radiotracers, radiolabeled NeoB and RM2, in a preclinical setting. *Eur J Nucl Med Mol Imaging.* 2023;50(13):3851–3861.

107 Kurth, J., Krause, B.J., Schwarzenbock, S.M., Bergner, C., Hakenberg, O.W., and Heuschkel, M. First-in-human dosimetry of gastrin-releasing peptide receptor antagonist [(177)Lu]Lu-RM2: a radiopharmaceutical for the treatment of metastatic castration-resistant prostate cancer. Eur J Nucl Med Mol Imaging. 2020;47(1):123–135.

108 Vincent, J.P., Mazella, J., and Kitabgi, P. Neurotensin and neurotensin receptors. *Trends Pharmacol Sci.* 1999;20(7):302–309.

109 Johnson, L.R. *Physiology of the Gastrointestinal Tract*, Two Volume Set. Burlington, MA: Elsevier Science; 2012.

110 Dupouy, S., Viardot-Foucault, V., Alifano, M., et al. The neurotensin receptor-1 pathway contributes to human ductal breast cancer progression. *PLoS One.* 2009;4(1):e4223.

111 Gui, X., Liu, S., Meng, Z., and Gao, Z.-H. Neurotensin receptor 1 (NTSR1) overexpression in breast carcinomas is common and independent of ER/PR/Her2 expression. *J Cancer Ther.* 2013; 4(7A):12–17.

112 Gui, X., Guzman, G., Dobner, P.R., and Kadkol, S.S. Increased neurotensin receptor-1 expression during progression of colonic adenocarcinoma. *Peptides.* 2008;29(9):1609–1615.

113 Korner, M., Waser, B., Strobel, O., Buchler, M., and Reubi, J.C. Neurotensin receptors in pancreatic ductal carcinomas. *EJNMMI Res.* 2015;5:17.

114 Reubi, J.C., Waser, B., Friess, H., Buchler, M., and Laissue, J. Neurotensin receptors: a new marker for human ductal pancreatic adenocarcinoma. *Gut.* 1998;42(4):546–550.

115 Reubi, J.C., Waser, B., Schaer, J.C., and Laissue, J.A. Neurotensin receptors in human neoplasms: high incidence in Ewing's sarcomas. *Int J Cancer.* 1999;82(2):213–218.

116 Schulz, J., Rohracker, M., Stiebler, M., et al. Proof of therapeutic efficacy of a (177)Lu-labeled neurotensin receptor 1 antagonist in a colon carcinoma xenograft model. *J Nucl Med.* 2017;58(6):936–941.

117 Schulz, J., Rohracker, M., Stiebler, M., et al. Comparative evaluation of the biodistribution profiles of a series of nonpeptidic neurotensin receptor-1 antagonists reveals a promising candidate for theranostic applications. *J Nucl Med*. 2016;57(7):1120–1123.

118 Osterkamp, F., Smerling, C., Reineke, U., and Ungewiß, C.H. Neurotensin receptor ligands. European patent application EP20120008208. 2012.

119 Baum, R.P., Singh, A., Schuchardt, C., et al. (177)Lu-3BP-227 for neurotensin receptor 1-targeted therapy of metastatic pancreatic adenocarcinoma: first clinical results. *J Nucl Med*. 2018;59(5):809–814.

120 Liu, F., Yang, J., Zhang, J., Wong, N., Tang, Z., and Cai, W. Development of 177Lu-FL-091 for the treatment of NTSR1-positive cancers. *J Nucl Med*. 2024; 65(Suppl. 2). Abstract#: 241875.

121 Filmus, J., Capurro, M., and Rast, J. Glypicans. *Genome Biol*. 2008;9(5):224.

122 Moek, K.L., Fehrmann, R.S.N., van der Vegt, B., de Vries, E.G.E., and de Groot, D.J.A. Glypican 3 overexpression across a broad spectrum of tumor types discovered with functional genomic mRNA profiling of a large cancer database. *Am J Pathol*. 2018;188(9):1973–1981.

123 Lin, F., Clift, R., Ehara, T., et al. Peptide binder to glypican-3 as a theranostic agent for hepatocellular carcinoma. *J Nucl Med*. 2024;65(4):586–592.

124 Clift, R., Lin, F., Ehara, T., et al. GPC3 targeted radiopharmaceutical therapy for HCC: preclinical characterization of a novel peptide binder. EASL. 2023. Poster number: P01-03; Abstract number: 279; Pages in the abstract book: 178.

125 Paredes, J., Correia, A.L., Ribeiro, A.S., Albergaria, A., Milanezi, F., and Schmitt, F.C. P-cadherin expression in breast cancer: a review. *Breast Cancer Res*. 2007;9(5):214.

126 Sheng, Q., D'Alessio, J.A., Menezes, D.L., et al. PCA062, a P-cadherin targeting antibody-drug conjugate, displays potent antitumor activity against P-cadherin-expressing malignancies. *Mol Cancer Ther*. 2021;20(7):1270–1282.

127 Shimoyama, Y., Hirohashi, S., Hirano, S., et al. Cadherin cell-adhesion molecules in human epithelial tissues and carcinomas. *Cancer Res*. 1989;49(8):2128–2133.

128 Root, A.R., Cao, W., Li, B., et al. Development of PF-06671008, a highly potent anti-P-cadherin/Anti-CD3 bispecific DART molecule with extended half-life for the treatment of cancer. *Antibodies (Basel)*. 2016;5(1):6.

129 Huck, B.R., Kotzner, L., and Urbahns, K. Small molecules drive big improvements in immuno-oncology therapies. *Angew Chem Int Ed Engl*. 2018;57(16):4412–4428.

130 Adams, J.L., Smothers, J., Srinivasan, R., and Hoos, A. Big opportunities for small molecules in immuno-oncology. *Nat Rev Drug Discov*. 2015;14(9):603–622.

131 Nguyen, N.T., Pacelli, A., Nader, M., and Kossatz, S. DNA repair enzyme poly(ADP-ribose) polymerase 1/2 (PARP1/2)-targeted nuclear imaging and radiotherapy. *Cancers (Basel)*. 2022;14(5):1129.

132 Zhou, D., Chen, H., Mpoy, C., et al. Radiosynthesis and evaluation of talazoparib and its derivatives as PARP-1-targeting agents. *Biomedicines.* 2021;9(5):565.

133 Destro, G., Chen, Z., Chan, C.Y., et al. A radioiodinated rucaparib analogue as an Auger electron emitter for cancer therapy. *Nucl Med Biol.* 2023; 116–117: 108312.

134 Pirovano, G., Jannetti, S.A., Carter, L.M., et al. Targeted brain tumor radiotherapy using an Auger emitter. *Clin Cancer Res.* 2020;26(12):2871–2881.

135 Makvandi, M., Lee, H., Puentes, L.N., et al. Targeting PARP-1 with alpha-particles is potently cytotoxic to human neuroblastoma in preclinical models. *Mol Cancer Ther.* 2019;18(7):1195–1204.

136 Ambur Sankaranarayanan, R., Florea, A., Allekotte, S., et al. PARP targeted Auger emitter therapy with [(125)I]PARPi-01 for triple-negative breast cancer. *EJNMMI Res.* 2022;12(1):60.

137 Pastorekova, S. and Gillies, R.J. The role of carbonic anhydrase IX in cancer development: links to hypoxia, acidosis, and beyond. *Cancer Metastasis Rev.* 2019;38(1–2):65–77.

138 Angapelly, S., Sri Ramya, P.V., Angeli, A., Supuran, C.T., and Arifuddin, M. Sulfocoumarin-, coumarin-, 4-sulfamoylphenyl-bearing indazole-3-carboxamide hybrids: synthesis and selective inhibition of tumor-associated carbonic anhydrase isozymes IX and XII. *ChemMedChem.* 2017;12(19):1578–1584.

139 Carta, F., Aggarwal, M., Maresca, A., et al. Dithiocarbamates strongly inhibit carbonic anhydrases and show antiglaucoma action in vivo. *J Med Chem.* 2012;55(4):1721–1730.

140 Maresca, A., Carta, F., Vullo, D., and Supuran, C.T. Dithiocarbamates strongly inhibit the beta-class carbonic anhydrases from Mycobacterium tuberculosis. *J Enzyme Inhib Med Chem.* 2013;28(2):407–411.

141 Chen, K.T. and Seimbille, Y. New developments in carbonic anhydrase IX-targeted fluorescence and nuclear imaging agents. *Int J Mol Sci.* 2022;23(11):6125.

142 Dubois, L., Lieuwes, N.G., Maresca, A., et al. Imaging of CA IX with fluorescent labelled sulfonamides distinguishes hypoxic and (re)-oxygenated cells in a xenograft tumour model. *Radiother Oncol.* 2009;92(3):423–428.

143 Akurathi, V., Dubois, L., Lieuwes, N.G., et al. Synthesis and biological evaluation of a 99mTc-labelled sulfonamide conjugate for in vivo visualization of carbonic anhydrase IX expression in tumor hypoxia. *Nucl Med Biol.* 2010;37(5):557–564.

144 Lu, G., Hillier, S.M., Maresca, K.P., et al. Synthesis and SAR of novel Re/99mTc-labeled benzenesulfonamide carbonic anhydrase IX inhibitors for molecular imaging of tumor hypoxia. *J Med Chem.* 2013;56(2):510–520.

145 Nakai, M., Pan, J., Lin, K.S., et al. Evaluation of (99 m)Tc-sulfonamide and sulfocoumarin derivatives for imaging carbonic anhydrase IX expression. *J Inorg Biochem*. 2018;185:63–70.

146 Krall, N., Pretto, F., Mattarella, M., Muller, C., and Neri, D. A 99mTc-labeled ligand of carbonic anhydrase IX selectively targets renal cell carcinoma in vivo. *J Nucl Med*. 2016;57(6):943–949.

147 Kulterer, O.C., Pfaff, S., Wadsak, W., et al. A microdosing study with (99 m) Tc-PHC-102 for the SPECT/CT imaging of primary and metastatic lesions in renal cell carcinoma patients. *J Nucl Med*. 2021;62(3):360–365.

148 Turkbey, B., Lindenberg, M.L., Adler, S., et al. PET/CT imaging of renal cell carcinoma with (18)F-VM4-037: a phase II pilot study. *Abdom Radiol (NY)*. 2016;41(1):109–118.

149 Iikuni, S., Ono, M., Watanabe, H., Shimizu, Y., Sano, K., and Saji, H. Cancer radiotheranostics targeting carbonic anhydrase-IX with (111)In- and (90) Y-labeled ureidosulfonamide scaffold for SPECT imaging and radionuclide-based therapy. *Theranostics*. 2018;8(11):2992–3006.

150 Shahrokhi, P., Farahani, A.M., and Tamaddondar, M. Radiolabeled vitamins as the potential diagnostic probes for targeted tumor imaging. *Bioorg Chem*. 2022;122:105717.

151 Russell-Jones, G., McTavish, K., McEwan, J., Rice, J., and Nowotnik, D. Vitamin-mediated targeting as a potential mechanism to increase drug uptake by tumours. *J Inorg Biochem*. 2004;98(10):1625–1633.

152 Low, P.S. and Kularatne, S.A. Folate-targeted therapeutic and imaging agents for cancer. *Curr Opin Chem Biol*. 2009;13(3):256–562.

153 Xia, W. and Low, P.S. Folate-targeted therapies for cancer. *J Med Chem*. 2010;53(19):6811–6824.

154 Muller, C. Folate based radiopharmaceuticals for imaging and therapy of cancer and inflammation. *Curr Pharm Des*. 2012;18(8):1058–1083.

155 Muller, C., Guzik, P., Siwowska, K., Cohrs, S., Schmid, R.M., and Schibli, R. Combining albumin-binding properties and interaction with pemetrexed to improve the tissue distribution of radiofolates. *Molecules*. 2018;23(6):1465.

156 Benesova, M., Guzik, P., Deberle, L.M., et al. Design and evaluation of novel albumin-binding folate radioconjugates: systematic approach of varying the linker entities. *Mol Pharm*. 2022;19(3):963–973.

157 Pressman, D. and Keighley, G. The zone of activity of antibodies as determined by the use of radioactive tracers; the zone of activity of nephritoxic antikidney serum. *J Immunol*. 1948;59(2):141–146.

158 Pressman, D. and Korngold, L. The in vivo localization of anti-Wagner-osteogenic-sarcoma antibodies. *Cancer*. 1953;6(3):619–623.

159 Vivier, D., Sharma, S.K., and Zeglis, B.M. Understanding the in vivo fate of radioimmunoconjugates for nuclear imaging. *J Labelled Comp Radiopharm*. 2018;61(9):672–692.

160 Seo, J.W., Tavare, R., Mahakian, L.M., et al. CD8(+) T-cell density imaging with (64)Cu-labeled cys-diabody informs immunotherapy protocols. *Clin Cancer Res.* 2018;24(20):4976–4987.

161 Feng, Y., Meshaw, R., McDougald, D., et al. Evaluation of an (131)I-labeled HER2-specific single domain antibody fragment for the radiopharmaceutical therapy of HER2-expressing cancers. *Sci Rep.* 2022;12(1):3020.

162 Pandit-Taskar, N., O'Donoghue, J.A., Ruan, S., et al. First-in-human imaging with 89Zr-Df-IAB2M anti-PSMA minibody in patients with metastatic prostate cancer: pharmacokinetics, biodistribution, dosimetry, and lesion uptake. *J Nucl Med.* 2016;57(12):1858–1864.

163 Nygren, P.A. and Uhlen, M. Scaffolds for engineering novel binding sites in proteins. *Curr Opin Struct Biol.* 1997;7(4):463–469.

164 Skerra, A. Engineered protein scaffolds for molecular recognition. *J Mol Recognit.* 2000;13(4):167–187.

165 Gebauer, M. and Skerra, A. Engineered protein scaffolds as next-generation therapeutics. *Annu Rev Pharmacol Toxicol.* 2020;60:391–415.

166 Shilova, O.N. and Deyev, S.M. DARPins: promising scaffolds for theranostics. *Acta Naturae.* 2019;11(4):42–53.

167 Garousi, J., Orlova, A., Frejd, F.Y., and Tolmachev, V. Imaging using radiolabelled targeted proteins: radioimmunodetection and beyond. *EJNMMI Radiopharm Chem.* 2020;5(1):16.

168 Zahnd, C., Kawe, M., Stumpp, M.T., et al. Efficient tumor targeting with high-affinity designed ankyrin repeat proteins: effects of affinity and molecular size. *Cancer Res.* 2010;70(4):1595–1605.

169 Goldstein, R., Sosabowski, J., Livanos, M., et al. Development of the designed ankyrin repeat protein (DARPin) G3 for HER2 molecular imaging. *Eur J Nucl Med Mol Imaging.* 2015;42(2):288–301.

170 Pluckthun, A. Designed ankyrin repeat proteins (DARPins): binding proteins for research, diagnostics, and therapy. *Annu Rev Pharmacol Toxicol.* 2015;55:489–511.

171 Deyev, S., Vorobyeva, A., Schulga, A., et al. Comparative evaluation of two DARPin variants: effect of affinity, size, and label on tumor targeting properties. *Mol Pharm.* 2019;16(3):995–1008.

172 Bragina, O., Chernov, V., Schulga, A., et al. Phase I trial of (99 m)Tc-(HE)(3)-G3, a DARPin-based probe for imaging of HER2 expression in breast cancer. *J Nucl Med.* 2022;63(4):528–535.

173 Zahnd, C., Wyler, E., Schwenk, J.M., et al. A designed ankyrin repeat protein evolved to picomolar affinity to Her2. *J Mol Biol.* 2007;369(4):1015–1028.

174 Theurillat, J.P., Dreier, B., Nagy-Davidescu, G., et al. Designed ankyrin repeat proteins: a novel tool for testing epidermal growth factor receptor 2 expression in breast cancer. *Mod Pathol.* 2010;23(9):1289–1297.

175 Altai, M., Garousi, J., Rinne, S.S., Schulga, A., Deyev, S., and Vorobyeva, A. On the prevention of kidney uptake of radiolabeled DARPins. *EJNMMI Res.* 2020;10(1):7.

176 Steiner, D., Merz, F.W., Sonderegger, I., et al. Half-life extension using serum albumin-binding DARPin(R) domains. *Protein Eng Des Sel.* 2017;30(9):583–591.

177 Brandl, F., Merten, H., Zimmermann, M., Behe, M., Zangemeister-Wittke, U., and Pluckthun, A. Influence of size and charge of unstructured polypeptides on pharmacokinetics and biodistribution of targeted fusion proteins. *J Control Release.* 2019;307:379–392.

178 Brandl, F., Busslinger, S., Zangemeister-Wittke, U., and Pluckthun, A. Optimizing the anti-tumor efficacy of protein-drug conjugates by engineering the molecular size and half-life. *J Control Release.* 2020;327:186–197.

179 Tuerk, C. and Gold, L. Systematic evolution of ligands by exponential enrichment: RNA ligands to bacteriophage T4 DNA polymerase. *Science.* 1990;249(4968):505–510.

180 Ellington, A.D. and Szostak, J.W. In vitro selection of RNA molecules that bind specific ligands. *Nature.* 1990;346(6287):818–822.

181 Kelly, L., Maier, K.E., Yan, A., and Levy, M. A comparative analysis of cell surface targeting aptamers. *Nat Commun.* 2021;12(1):6275.

182 Charlton, J., Sennello, J., and Smith, D. In vivo imaging of inflammation using an aptamer inhibitor of human neutrophil elastase. *Chem Biol.* 1997;4(11):809–816.

183 Li, J., Zheng, H., Bates, P.J., et al. Aptamer imaging with Cu-64 labeled AS1411: preliminary assessment in lung cancer. *Nucl Med Biol.* 2014;41(2):179–185.

184 Wang, L., Jacobson, O., Avdic, D., et al. Ortho-stabilized (18) F-Azido click agents and their application in PET imaging with single-stranded DNA aptamers. *Angew Chem Int Ed Engl.* 2015;54(43):12777–12781.

185 Wu, X., Liang, H., Tan, Y., et al. Cell-SELEX aptamer for highly specific radionuclide molecular imaging of glioblastoma in vivo. *PLoS One.* 2014;9(6):e90752.

186 Jacobson, O., Yan, X., Niu, G., et al. PET imaging of tenascin-C with a radiolabeled single-stranded DNA aptamer. *J Nucl Med.* 2015;56(4):616–621.

187 Park, J.Y., Lee, T.S., Song, I.H., et al. Hybridization-based aptamer labeling using complementary oligonucleotide platform for PET and optical imaging. *Biomaterials.* 2016;100:143–151.

188 de Sousa Lacerda, C.M., Ferreira, I.M., Dos Santos, S.R., et al. (1→3)-beta-D-glucan aptamers labeled with technetium-99 m: Biodistribution and imaging in experimental models of bacterial and fungal infection. *Nucl Med Biol.* 2017;46:19–24.

189 Ferreira, I.M., de Sousa Lacerda, C.M., Dos Santos, S.R., et al. Detection of bacterial infection by a technetium-99 m-labeled peptidoglycan aptamer. *Biomed Pharmacother.* 2017;93:931–938.

190 Santos, S.R.D., de Sousa Lacerda, C.M., Ferreira, I.M., et al. Scintigraphic imaging of Staphylococcus aureus infection using (99 m)Tc radiolabeled aptamers. *Appl Radiat Isot.* 2017;128:22–27.

191 Cheng, S., Jacobson, O., Zhu, G., et al. PET imaging of EGFR expression using an (18)F-labeled RNA aptamer. *Eur J Nucl Med Mol Imaging*. 2019;46(4):948–956.

192 Varmira, K., Hosseinimehr, S.J., Noaparast, Z., and Abedi, S.M. An improved radiolabelled RNA aptamer molecule for HER2 imaging in cancers. *J Drug Target*. 2014;22(2):116–122.

193 Summerton, J. and Weller, D. Morpholino antisense oligomers: design, preparation, and properties. *Antisense Nucleic Acid Drug Dev*. 1997;7(3):187–195.

194 Park, J.Y., Cho, Y.L., Chae, J.R., et al. Gemcitabine-incorporated G-quadruplex aptamer for targeted drug delivery into pancreas cancer. *Mol Ther Nucleic Acids*. 2018;12:543–553.

195 Eder, M., Schafer, M., Bauder-Wust, U., et al. 68Ga-complex lipophilicity and the targeting property of a urea-based PSMA inhibitor for PET imaging. *Bioconjug Chem*. 2012;23(4):688–697.

196 Song, W., Song, Y., Li, Q., Fan, C., Lan, X., and Jiang, D. Advances in aptamer-based nuclear imaging. *Eur J Nucl Med Mol Imaging*. 2022;49(8):2544–2559.

197 Filippi, L., Bagni, O., and Nervi, C. Aptamer-based technology for radionuclide targeted imaging and therapy: a promising weapon against cancer. *Expert Rev Med Devices*. 2020;17(8):751–758.

198 Gijs, M., Aerts, A., Impens, N., Baatout, S., and Luxen, A. Aptamers as radiopharmaceuticals for nuclear imaging and therapy. *Nucl Med Biol*. 2016;43(4):253–271.

199 Zhong, Y., Wu, P., He, J., Zhong, L., and Zhao, Y. Advances of aptamer-based clinical applications for the diagnosis and therapy of cancer. *Discov Med*. 2020;29(158):169–180.

200 Ding, D., Zhao, H., Wei, D., et al. The first-in-human whole-body dynamic pharmacokinetics study of aptamer. *Research (Wash D C)*. 2023;6:0126.

12

Peptide-small Molecule Drug Conjugates for Cancer, Metabolic Diseases, and Beyond

Rongjun He, Sumeet Singh and Adam Mezo

Neurocrine Biosciences, Inc., San Diego, CA, USA

12.1 Introduction

Peptides are a powerful class of medicines used for the treatment of major human diseases [1–4]. Owing to their larger interaction surface as compared to traditional small molecules, peptides often exhibit exquisite potency and specificity on their respective drug targets. Since their endogenous metabolism breakdown products are amino acids, peptides also exhibit minimal off-target adverse effects compared to small molecules. Those benefits have been exemplified by the discovery of insulin in 1921 for the treatment of diabetes [5,6], the discovery of octreotide for the treatment of neuroendocrine cancers [7,8], and more recently the development of incretin-based peptide drugs semaglutide and tirzepatide for the treatment of type 2 diabetes (T2D) and obesity [9].

Currently, more than 60 peptide-based drugs are being sold for the treatment of many human diseases, with more than a hundred in various drug development stages and significantly more in preclinical discovery research [1]. The global peptide drug market was valued at more than US$40 billion in 2022 and is forecast to grow at a yearly rate of 6.3% from 2023 to 2030 to reach over seventy billion dollars [10]. The most active disease areas in peptide drug discovery are oncology, metabolism, infectious disease, and cardiovascular disease, but virtually all disease areas can be impacted by peptides.

Recent frontiers in peptide drug discovery are orally bioavailable peptides through formulation and delivery technologies [11,12], cell-penetrating peptides through structural constraints [13,14], dual- or triple-acting peptides through

Peptide Science: Chemical Ligation, Lead Generation, and Therapeutic Advances, First Edition. Fa Liu.

hybridization [15–17], and targeted drug delivery by peptide small molecule drug conjugation [18–21]. Peptide-small molecule conjugation not only offers organ- and tissue-specific drug delivery as precision medicines, but also empowers the virtues of peptide and small molecule pharmacology in a combined manner. The maturity of solid-phase and solution-phase synthesis methods, coupled with established orthogonal conjugation chemistry, has enabled the successful large-scale production of peptide-drug conjugates, facilitating their translation to clinical studies. In this review, we focus on prominent examples of peptide-small molecule conjugates developed in the fields of oncology and metabolism (Table 12.1). Through analysis of each example, case by case, we illustrate how peptide-small molecule drug conjugates are designed, characterized, and progressed to demonstrate superior efficacy and safety as life-changing medicines.

12.2 Peptide-drug Conjugates for Cancer

12.2.1 PEN-221

Somatostatin (also named growth hormone-inhibiting hormone) is an endocrine peptide hormone produced by hypothalamic neurons, pancreatic delta cells, and gastrointestinal paracrine cells [7,35]. Its major function is to inhibit the secretion of hormones in our body, including growth hormone, thyroid-stimulating hormone, and hormones derived from pituitary, pancreatic, and gastrointestinal cells. Given its hormone inhibitory functions, somatostatin exhibits many direct and indirect effects to suppress the growth and differentiation of cancer cells [7,35]. Somatostatin analogs such as octreotide, lanreotide, and pasireotide have been used for decades in the clinic for the treatment of acromegaly, as well as hormone-dependent tumors such as pancreatic and vasoactive intestinal peptide-secreting tumors [7,36]. Somatostatin exhibits its biological functions by binding and activating a family of five somatostatin receptor subtypes (SSTR1-5), a group of class A G protein-coupled receptor's, that are expressed in the brain, pituitary, adrenals, pancreas, heart, gastrointestinal tract, and many other tissues [7,37]. Somatostain receptor 2 (SSTR2) is reported to be overexpressed in many tumors and undergoes ligand-induced internalization [7,37], which renders SSTR2 a potential target for intracellular delivery of cytotoxic agents to tumor cells [38].

PEN-221 is a peptide-drug conjugate between SSTR2 targeting somatostatin analog [Tyr3,Cys8]octreotate and potent cytotoxic payload DM1, via a disulfide-based linker through the C-terminus of the peptide (Figure 12.1) [22,23]. PEN-221 binds the SSTR2 receptor potently with a Ki of 0.05 nM, and selectivity of >420-fold over SSTR1, 3, 4,5. PEN-221 activates the SSTR2

Table 12.1 Representative examples of peptide-small molecule drug conjugates (PDC) in oncology and metabolic diseases.

PDC	Target	Peptide	Linker	Payload	Indication	Status	References
PEN-221	SSTR2	[Tyr3,Cys8]octreotate	Disulfide	Mertansine (DM1)	Neuroendocrine cancer	Phase 2 completed	[22,23]
BT5528	EphA2	Bicyclic peptide	Val-Cit	Monomethyl auristatin E (MMAE)	Urothelial and ovarian cancers	Phase 2 ongoing	[24,25]
G202	PSMA	Tetrapeptide	Amide	Thapsigargin	Prostate cancer and other solid tumors	Phase 2 completed	[26,27]
ANG1005	LPR1	Trevatide	Ester	Paclitaxel	Breast cancer and other solid tumors	Phase 3 ongoing	[28,29]
TH1902	SORT1	TH19P01	Ester	Docetaxel	Triple-negative breast cancer and other solid tumors	Phase 1 ongoing	[30]
GLP-1-estrogen	GLP-1 receptor	GLP-1 analog	Ether	17β-estradiol	Metabolic diseases	Preclinical	[31]
GLP-1-tesaglitazar	GLP-1 receptor	GLP-1 analog	Amide	Tesaglitazar	Metabolic diseases	Preclinical	[32]
Glucagon-T3	Glucagon receptor	Glucagon analog	Amide	3,3,5-triiodothyronine (T3)	Metabolic diseases	Preclinical	[33]
NPY1-tesaglitazar	NPY1R	NPY1 analog	Amide	Tesaglitazar	Metabolic diseases	Preclinical	[34]

SSTR2, somatostain receptor 2; EphA2, ephrin receptor A2; PSMA, prostate-specific membrane antigen; LPR1, low-density lipoprotein receptor-related protein 1; SORT1, sortilin; GLP-1, glucagon-like peptide-1; NPY1R, neuropeptide Y receptor type 1.

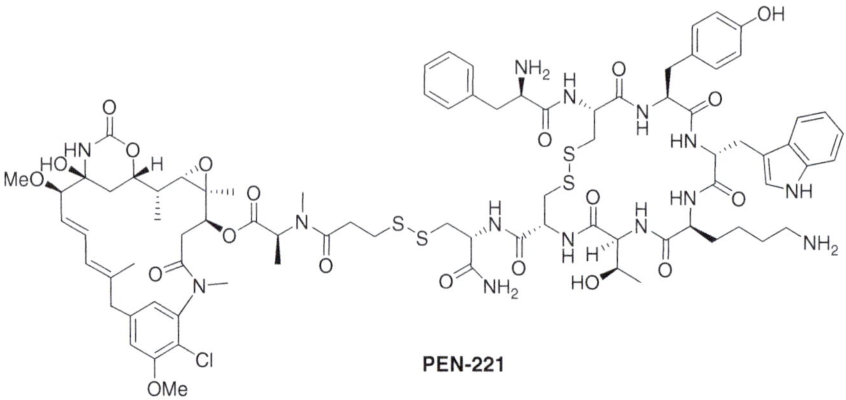

PEN-221

Figure 12.1 Chemical structure of PEN-221, a peptide-drug conjugate between SSTR2-targeting somatostatin analog [Tyr3,Cys8]octreotate and cytotoxic payload DM1.

receptor in CHO cells with an EC_{50} of 0.51 nM, similar or more potent than native peptide ligand somatostatin-28 (EC_{50}:1.08 nM) [22,23]. Upon receptor activation, PEN-221 is internalized in SSTR2-expressing human cancer cells such as NCI-H524, NCI-H69, and HCC-33. The disulfide bond is cleaved under reducing conditions in these cancer cells to release payload DM1, which inhibits cellular proliferation. SSTR2 receptor internalization was observed *in vivo* with an NCI-H524 xenograft tumor mouse model, in which PEN-221 resulted in rapid accumulation and sustained retention of toxic payload DM1 in the tumor tissue up to 100 hours post injection. Additionally, PEN-221 exhibited potent and durable tumor inhibition in mice with SSTR2-expressing xenograft tumors derived from NCI-H524 and HCC-33 small cell lung carcinoma cells, but not in the SSTR2 negative Calu-6 xenograft model, suggesting that *in vivo* efficacy was SSTR2-mediated. The efficacy of PEN-221 was dose-dependent, with tumor inhibition at doses as low as 0.33 mg/kg and considerably lower than the mouse maximum tolerated dose of 2.0 mg/kg, suggesting a reasonable therapeutic index. In a Phase 1/2a clinical trial in patients with advanced gastrointestinal neuroendocrine tumor, PEN-221 was well tolerated given intravenously every 3 weeks at a dose of 8.8 mg/m^2 and possessed a plasma half-life of about 4.5 hours [39]. Of the 26 patients who were evaluable for a response, 23 patients (88.5%) had stable disease, and 10 patients (38%) had target lesion shrinkage, with a median progression-free survival of 9 months.

12.2.2 BT5528

Ephrin receptors (Eph) are an important class of receptor tyrosine kinases, characterized by an extracellular domain, a single transmembrane domain, and an intracellular domain [40–42]. To date, 14 Eph receptors and 8 related ligands (ephrins) have been identified in humans, regulating important functions including tissue patterning, neuronal targeting, and blood vessel development [42,43]. Particularly, EphA2 is found to be overexpressed in several solid tumors and plays a role in carcinogenesis and tumor progression, including prostate, lung, esophageal, colorectal, cervical, ovarian, breast, and skin cancers [42,43]. High expression in cancerous tissue but limited expression in normal tissue makes EphA2 an ideal target for drug delivery to tumor tissues. BT5528 is a bicyclic peptide-drug conjugate made up of an EphA2-binding bicyclic peptide and auristatin toxin MMAE via a Val-Cit dipeptide linker, which can be cleaved by intracellular cathepsin B and other proteases in cancer cells (Figure 12.2) [24,25]. BT5528 has a strong binding affinity to human EphA2 (Kd = 0.9 nM), with cross-reactivity to mouse, rat, and cynomolgus monkey receptors, and high selectivity (> 5000-fold) over closely related human and rodent EphA homologs [24,25]. BT5528, but not nonbinder controls, showed high binding affinity to HT-1080 cells that express high levels of EphA2. BT5528 exhibits high metabolic stability in rat, cynomolgus monkey, and human plasma and human hepatocytes well as low plasma protein binding (fraction unbound [fu] 0.32–0.49). *In vivo*, BT5528 demonstrated a peptide-like pharmacokinetic profile across species with short terminal half-lives: 0.4 hours (mouse), 0.3 hours (rat), and 0.6 hours (nonhuman primate) with a low rate of payload release in plasma. In the mouse model of PC3 tumor xenograft, BT5528 was dosed intravenously at 0.5 mg/kg and gave rise to transient MMAE levels in plasma and persistent exposure in tumor samples for up to 48 hours. In subsequent efficacy studies in the HT-1080, PC3, and NSCLC patient-derived xenograft mouse models, BT5528, dosed once weekly, showed remarkable tumor suppression up to 28 days and significantly prolonged survival in comparison to vehicle controls. Those antitumor effects were not observed from control peptides with either noncleavable linkers or nonbinding bicyclic peptides, suggesting that the effect *in vivo* was the result of

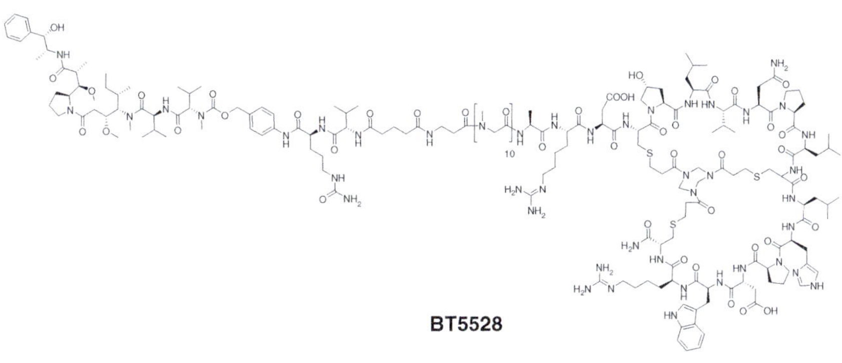

BT5528

Figure 12.2 Chemical structure of BT5528, a peptide-drug conjugate between EphA2-binding bicyclic peptide and auristatin toxin MMAE.

target-mediated payload delivery. BT5528 is currently being evaluated in multicenter, open-label Phase 1/2 clinical trials in solid tumors expressing EphA2 [44]. From the Phase 1 interim results, BT5528 exhibited antitumor activity with two partial responses out of two urothelial cancer patients and an 80% disease control rate in five EphA2-positive ovarian cancer patients. Some mild and transient neutropenia was observed at a dose of 8.5 mg/m^2 when administered every week. A Phase 2 expansion study is ongoing in urothelial and ovarian cancers as well as a basket of other solid tumor types.

12.2.3 G202

Prostate-specific membrane antigen (PSMA) is a type 2 transmembrane carboxypeptidase expressed in healthy prostate tissue [45,46]. PSMA expression is significantly upregulated in prostate cancer and most solid tumors and is positively associated with disease progression and metastasis [45,46], rendering PSMA a highly attractive target for prostate cancer and other solid tumors. Thapsigargin is a highly potent toxin that induces cellular apoptosis by binding the sarco/endoplasmic reticulum calcium ATPase (SERCA), an important regulator of calcium gradient across cytosolic and reticulum compartments [47,48]. Thapsigargin kills both cancer and normal cells, limiting its potential as a chemotherapeutic agent. Conjugation of thapsigargin to a PSMA binding tetra gamma glutamic acid peptide yields a peptide-drug conjugate G202 (mipsagargin) that delivers thapsigargin to tumor tissue via PSMA (Figure 12.3) [26,27]. G202 is unable to cross the cell membrane to reach SERCA due to the peptide's multiple negative charges. However, the tetra gamma glutamic acid peptide is a substrate of PSMA and is processed by PSMA to restore thapsigargin that can cross the cell membrane and kill prostate cancer cells [26,27]. In cell proliferation assays, G202 was reported to be 57-fold more potent in PSMA-expressing human prostate cancer cells than in PSMA-negative cells, suggesting target-mediated cytotoxicity. In mouse xenograft models with human prostate and breast cancer cells, G202 exhibited potent antitumor activities but much-reduced systemic toxicity. In a

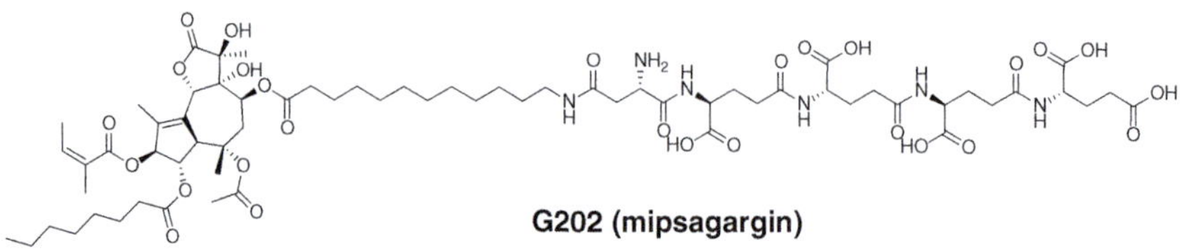

Figure 12.3 Chemical structure of G202 (mipsagargin), a peptide-drug conjugate between PSMA binding tetra gamma glutamic acid peptide and small molecule toxin thapsigargin.

Phase 1 clinical trial with PSMA-positive cancer patients, G202 was found to have a favorable pharmacokinetic profile and was well tolerated at doses up to 88 mg/m^2 (or 2.4 mg/kg) [26]. Several Phase 2 clinical trials were completed in patients with prostate cancer and some other solid tumors, including renal cell carcinoma, hepatocellular carcinoma, and glioblastoma [49].

12.2.4 ANG1005

Paclitaxel is a chemotherapeutic drug widely used for the treatment of various human cancers [50,51]. However, paclitaxel has limited efficacy in primary and metastatic brain tumors due to its low permeability in crossing the blood-brain barrier and efflux by P-glycoprotein pump. The low-density lipoprotein receptor-related protein 1 (LRP1) is a transmembrane protein expressed on the surface of capillary endothelial cells at the blood–brain barrier, which can mediate transcytosis [52]. The importance of LRP1 for cancer drug discovery is that its expression is significantly upregulated in brain tumors, i.e. glioma [52,53]. Angiopep-2 is a peptide that binds to LRP1, and when conjugated to paclitaxel, yields the peptide-drug conjugate ANG1005 (Figure 12.4) (also named GRN1005) that delivers paclitaxel to the brain through LRP1-mediated transcytosis [54]. Interestingly, three molar equivalents of paclitaxel are conjugated to a single peptide, which increases the amount of paclitaxel being delivered [28,29]. In cell proliferation assays, ANG1005 exhibited excellent cytotoxicity against many human cancer cells, including glioblastoma, lung, and ovarian cancer. The peptide-mediated paclitaxel delivery was demonstrated to be LPR1 dependent,

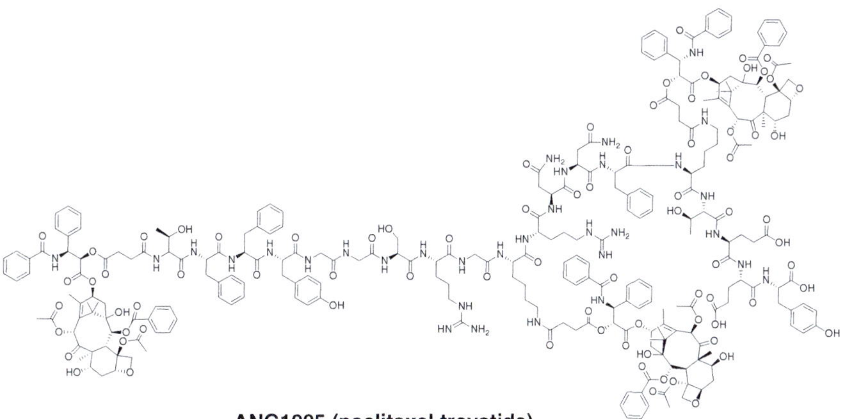

ANG1005 (paclitaxel trevatide)

Figure 12.4 Chemical structure of ANG1005 (paclitaxel trevatide), a peptide-drug conjugate between LRP1-binding peptide angiopep-2 and small molecule toxin paclitaxel.

which would otherwise impair therapeutic efficacy. In xenografted mouse models with glioblastoma and lung carcinoma cells, ANG1005 greatly suppressed tumor growth in comparison to vehicle control and significantly prolonged survival [54]. In a Phase 1 clinical trial in patients with glioma, ANG1005 exhibited a favorable pharmacokinetic profile with a plasma half-life of 3.6 hours and was well tolerated with a toxicity similar to that observed by paclitaxel [28,29]. In a Phase 2 study in patients with breast cancer and brain metastases, ANG1005 demonstrated a median survival time of 8 months, in comparison to 4 months with standard treatment or 2 months without treatment [55]. ANG1005 has also been evaluated in several other Phase 2 studies, including recurrent high-grade glioma, non-small cell lung, and brain metastases [56]. Currently, ANG1005 is in a Phase 3 clinical trial against leptomeningeal disease from breast cancer [57].

12.2.5 TH1902

SORT1, also known as sotilin, the neurotensin receptor 3 or 95-kDa RAP binding protein (Gp95), is a transmembrane receptor described as a scavenger receptor [58]. SORT1 plays important roles in binding, internalizing, and intracellular trafficking of peptide and protein ligands, including neurotensin, progranulin, and apolipoprotein E [59–61]. SORT1's expression has been upregulated in many types of human cancers, including ovarian, endometrial, breast, pancreas, and melanoma. Studies have suggested that SORT1-mediated drug delivery could be useful in anti-cancer therapy [59–61]. TH1902 (sudocetaxel zendusortide) is a peptide-drug conjugate composed of SORT1 binding peptide TH19P01 and small molecule toxin docetaxel in a 1-to-2 molar ratio via ester bond, which is cleaved by esterases or under acidic conditions in the endosome to release the payload docetaxel (Figure 12.5) [30]. TH19P01, when labeled with AlexaFluor-488, showed a binding constant (KD) of 0.32 μM in MDA-MB-231 breast cancer cells. Using Alexa488-labeled TH19P01 as a cell binding tracer, TH19P01 and TH1902 blocked the tracer's cellular uptake, with inhibition constants (Ki) at 0.84 ± 0.35 and 0.55 ± 0.06 μM, respectively. These Ki values suggest that TH1902, with two equivalents of docetaxel has a comparable binding affinity to the unconjugated TH19P01 peptide. In cell assays with SORT1-expressing triple-negative breast cancer cell lines MDA-MB-231 and HCC-70, TH1902 exhibited potent inhibition of cell proliferation, with IC_{50} at sub-nM range. A similar level of anti-proliferative activity was observed in other cancer cell lines such as ES-2, SKOV3, A2780, and AN3-CA [62]. In contrast, TH19P01 had no inhibition up to 1 μM in these cancer cells [30]. TH1902 exhibited rapid plasma clearance in mice after a single i.v. bolus dose, with an elimination half-life of 1.44 hours. In several xenografted mouse models, including triple-negative breast cancer MDA-MB-231, HCC-70, ovarian cancer ES-2, SKOV3, and A2780, as well as endometrial cancer AN3-CA, TH1902

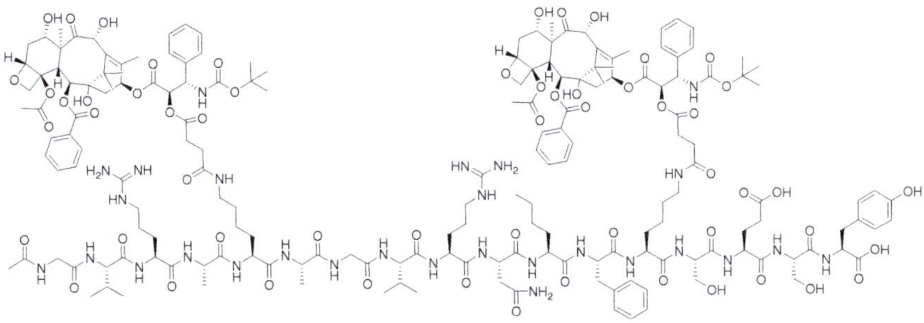

TH1902 (sudocetaxel zendusortide)

Figure 12.5 Chemical structure of TH1902 (sudocetaxel zendusortide), a peptide-drug conjugate between SORT1 binding peptide TH19P01 and small molecule toxin docetaxel.

exhibited potent antitumor efficacy at doses of 8.75–35 mg/kg, as a single agent and in combination with the chemotherapeutic agent carboplatin. In a Phase 1 dose-escalating study, the maximum tolerated dose was established to be 360 mg/m^2. A subsequent expansion study was done at 300 mg/m^2 with dosing every 3 weeks, and five of six patients (83%) with ovarian cancer had either a partial response ($n = 1$) or stable disease ($n = 4$). In the triple-negative breast cancer population, three out of four patients (75%) had the best overall response of stable disease, with one patient having stable disease for at least four cycles and a continued clinical benefit up to 24 weeks [63]. The current Phase 1 study is still ongoing for dose optimization (Part 3) and basket expansion (Part 4).

12.3 Peptide-drug Conjugates for Metabolic Diseases

12.3.1 Glucagon-like Peptide-1-estrogen

GLP-1 receptor agonistic peptides such as exenatide, lixisenatide, liraglutide, dulaglutide, and semaglutide are a powerful class of medicines for the treatment of T2D and obesity with proven cardiovascular benefits [63]. GLP-1 receptor agonism is combined with additional mechanisms to produce GLP-1/glucose-dependent insulinotropic polypeptide (GIP) dual agonist tirzepatide and GLP-1/GIP/glucagon triple agonist retatrutide, which offer additional benefits in the treatment of metabolic diseases. The search for polypharmacological agents that engage the GLP-1 receptor and additional mechanisms continues, with a vision to provide further enhanced metabolic benefits beyond body weight lowering. GLP-1-estrogen conjugation is one such example to combine GLP-1 and a steroid hormone for the treatment of T2D and obesity (Figure 12.6) [31,64]. Estrogens are a group of

GLP-1-estrogen

Figure 12.6 Chemical structure of GLP-1-estrogen, a peptide-drug conjugate between GLP-1 receptor agonist and steroid hormone 17β-estradiol.

steroid hormones that have beneficial effects in enhancing insulin signaling, reducing appetite, and increasing energy expenditure [65–68]. However, estrogens, when used chronically, have increased the risk of cancer in gynecological tissues and increased the risk of cardiovascular diseases [69]. Tissue-targeted estrogens that selectively function in metabolic organs such as the liver, adipose, pancreas, and hypothalamus, but not in reproductive organs such as the ovaries, uterus, and breast could be an approach to harness estrogen's therapeutic benefits while avoiding its undesired toxicity [70,71]. A GLP-1- estrogen conjugate was designed and studied *in vitro* and *in vivo*. The conjugate is linked by an ether bond between a lysine residue at position 40 of GLP-1 and 17β-estradiol, where the ether bond is likely cleaved by lysosomal enzymes upon internalization to release 17β-estradiol [31]. This conjugate has a similar level of activity as the native GLP-1 hormone in cell-based assays, suggesting that conjugation to 17β-estradiol had no adverse impact on activity. The conjugate is stable in human plasma against proteolytical degradation, reducing the possibility of premature release during circulation. When dosed in diet-induced obese (DIO) mice, the conjugate demonstrated greater therapeutic effects than GLP-1 or estrogen alone in reversing obesity, hyperglycemia, and dyslipidemia. Importantly, estrogen-associated gynecological toxicity and oncogenicity were not observed, suggesting that conjugation to GLP-1 can prevent estrogen's adverse effects. Mechanistically, the conjugate exhibited anorexigenic effects from GLP-1's central action and estrogenic action in the dorsal raphe nuclei in female mice, which suppressed binge-like eating behavior [72]. The conjugate also activates both GLP-1 and estrogen receptors in the supra-mammillary nucleus in rats, resulting in greater effects in reducing food intake and body weight [64]. Furthermore, the conjugate increased insulin sensitivity and glucose homeostasis in non-diabetic mice [73], attenuated hyperphagia, and protected β cell failure in New Zealand obese mice [74]. More recently, the conjugate was demonstrated to increase β cell survival and regeneration in mice and protect human β cells against cytokine-induced dysfunction [75]. In all these studies, the metabolic benefits offered by the conjugate were superior to what GLP-1 alone or an untargeted combination provided, which supports future translational studies in a clinical setting for diabetes and obesity.

12.3.2 GLP-1-tesaglitazar

Given the promising results obtained from the GLP-1-estrogen conjugate in combining the therapeutic benefits but avoiding the toxic effects of estrogen, GLP-1 was also conjugated to peroxisome proliferator-activated receptor (PPAR) agonist tesaglitazar. PPARs are nuclear transcription factors that play important roles in regulating energy, glucose, and lipid metabolism [76]. All three subtypes, including PPARα, PPARγ, and PPARβ/δ, have been studied extensively during the past two decades, with many PPARα agonists (fibrates) and PPARγ agonists (thiazolidinediones) being approved by the Food and Drug Administration for the treatment of hypertriglyceridemia and T2D [77]. Tesaglitazar is a PPARα/γ dual agonist, which showed greater efficacy relative to selective PPARγ agonists in improving glucose and lipid metabolism [78]. However, side effects such as weight gain, increased serum levels of creatinine, and decreased glomerular filtration caused its discontinuation in development [79]. It was proposed that tesaglitazar conjugation to GLP-1 could synergize both molecules' insulin-sensitizing effects while avoiding tesaglitazar's body weight gain and adverse kidney effects by altering its drug distribution [32]. Tesaglitazar was covalently linked to a GLP-1 peptide through the sidechain of the C-terminal lysine 40 residue with a gamma glutamic acid spacer (Figure 12.7). GLP1-Tesaglitazar exhibited a similar level *in vitro* activity as compared to unconjugated GLP-1 in Gαs recruitment and cAMP production in HEK293 cells transfected with the human GLP-1 receptor. GLP-1-tesaglitazar treatment resulted in GLP-1 receptor internalization, indicating its intracellular drug delivery capability for tesaglitazar. In a diet-induced obese (DIO) mouse

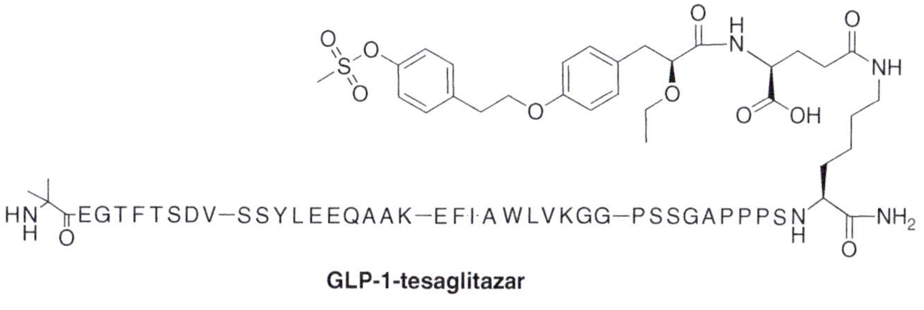

Figure 12.7 Chemical structure of GLP-1-tesaglitazar, a peptide-drug conjugate between GLP-1 receptor agonist and a PPARα/γ dual-agonist tesaglitazar.

study, 14-day treatment with GLP-1-tesaglitazar at the high dose level (50 nmol/kg per day) was able to decrease food intake and body weight more than that from dose-matched GLP-1, tesaglitazar, or a fixed combination of the two agents. Treatment of GLP-1-tesaglitazar also decreased circulating insulin levels and fasted blood glucose levels, effects that were not observed by treatment of GLP-1 or tesglitazar alone, suggesting increased insulin sensitivity from the conjugate. The increased insulin sensitivity and improved glucose handling were also observed at a much-reduced dose (as low as 0.5 nmol/kg) in an acute study in diet-induced obesity (DIO) mice, suggesting its superior efficacy by combining GLP-1 and tesaglitazar pharmacology, and linking the molecules as a conjugate. Importantly, the conjugate had no effects on food intake, body weight, and blood glucose in GLP-1 receptor knockout DIO mice, suggesting GLP-1 receptor-mediated metabolic effects. The current example of GLP-1-mediated delivery of a nuclear acting PPARα/ɣ dual-agonist demonstrated again that peptide-drug conjugation can achieve superior therapeutic efficacy while mitigating the potential adverse effects associated with small-molecule drugs.

12.3.3 Glucagon-T3

Glucagon is a pancreatic hormone that acts on the liver to stimulate glucose production and release [80]. Glucagon also promotes lipolysis and thermogenesis in the liver and offers metabolic benefits, including body weight lowering, lipid lowering and cardiovascular protection [80]. Glucagon's therapeutic potential has been demonstrated by GLP-1/GIP/Glucagon triple agonist retatrutide in clinical trials, which exhibited superior efficacy relative to GLP-1 monoagonist semaglutide or GLP-1/GIP dual agonist tirzepatide in lowering body weight [17]. Glucagon primarily signals in the liver, which expresses a high level of glucagon receptor, suggesting that glucagon can be utilized for liver-targeted drug delivery.

Thyroid hormones, produced by the thyroid gland, play a central role in regulating metabolism [81]. Thyroid hormones' therapeutic potential in metabolic diseases is attributed to their ability to increase energy expenditure, fat oxidation, and cholesterol metabolism in the liver [82,83]. However, thyroid hormones also have many adverse effects, including increased heart rate, muscle wasting, low bone density, and osteoporosis [84]. Liver-targeted thyromimetics have revealed that it is possible to impact hepatic lipids and atherogenic lipoproteins without associated adverse effects [85]. Therefore, targeted thyroid hormone action in the liver and/or adipose tissues constitutes a viable approach to harness the metabolic benefits without stimulating its adverse effects from the cardiovascular system. The glucagon receptor is almost exclusively expressed in the liver, suggesting that conjugation of thyroid hormone and glucagon should be able to confine thyroid's action in the liver and to provide combined effects in reducing body weight [33].

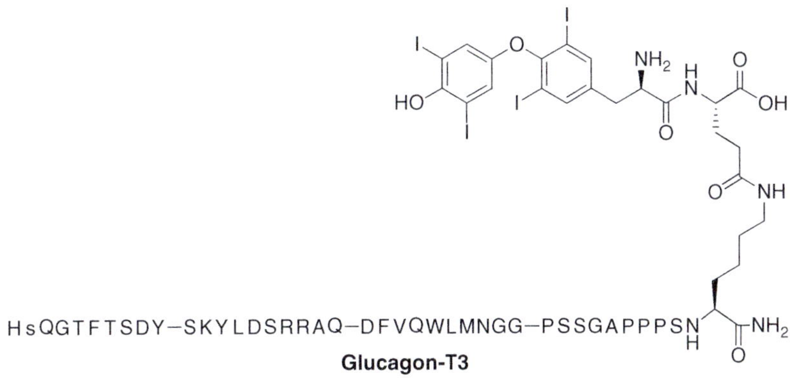

Figure 12.8 Chemical structure of glucagon-T3, a peptide-drug conjugate between glucagon receptor agonist and a thyroid hormone 3,3,5-triiodothyronine (T3).

T3, the most bioactive form of thyroid hormone, was conjugated to a glucagon peptide via a noncleavable ether bond (Figure 12.8) [33]. In cell-based functional assays, the conjugate preserved glucagon activity at its receptor with EC_{50} of 50 pM. In a mouse model of nonalcoholic steatohepatitis, a form of metabolic liver disease, the conjugate demonstrated great benefits in reducing body weight and liver fat, decreasing plasma cholesterol and triglyceride levels, thereby showing great therapeutic potential. A tissue analysis indicated that T3 was enriched in the liver, where the glucagon receptor is highly expressed, but not in the pancreas or heart, where glucagon receptor expression is limited, suggesting glucagon receptor-mediated T3 delivery. Furthermore, the metabolic benefits were the result of glucagon and T3 combined pharmacology in the liver, as the metabolic benefits were lost in glucagon receptor knockout mice and with control conjugates devoid of glucagon or T3 activity. It is worth noting that T3 counteracted glucagon's diabetogenic effects while glucagon lessened T3-induced plasma cholesterol elevation. Hence, glucagon and thyroid hormone conjugation may provide a powerful and safe option for the treatment of multiple metabolic syndromes, including hyperglycemia, obesity, and fatty liver diseases.

12.3.4 NPY1-tesaglitazar

Obesity disorder is associated with an imbalance in food intake and energy expenditure that leads to the accumulation of fat in adipose tissues. The surplus fat stored in newly formed adipocytes leads to an increase in the number of cells (hyperplasia), or in preexisting adipocytes, causing the cells to enlarge (hypertrophy) [86]. The hyperplasia is not considered a metabolic risk as the number of adipocytes is predominantly determined during childhood and the early adolescent stage, and a consistent level is maintained throughout adulthood. In contrast, adipocyte hypertrophy, induced by a persistent surplus of energy, can lead to a substantial increase in size, reaching several hundred micrometers in diameter [87]. As discussed earlier, PPARα/γ agonists (i.e. tesaglitazar) are known for their remarkable anti-diabetic capabilities; they facilitate adipogenesis and contribute to the formation of smaller, metabolically healthy adipocytes [88,89]. Nevertheless, these molecules also regulate diverse processes in other cell types, which leads to some of their major side effects, such as weight gain, increased serum levels of creatinine, congestive heart failure, and decreased glomerular filtration [79]. Thus, targeting a PPARα/γ agonist to adipocytes presents a promising strategy for the development of safer insulin-sensitizing drugs.

The human neuropeptide Y (NPY) receptor family comprises four functional receptors, Y1R, Y2R, Y4R, and Y5R. Among these, NPY1R is highly expressed in adipocytes and overexpressed in the adipose tissue of obese patients, where it leads to enhanced cell proliferation and the induction of food intake upon activation by NPY [90]. Thus, NPY1R antagonism has been proposed as a promising strategy for selective targeting of adipocytes for treating obesity.

A peptide-drug conjugate was synthesized, consisting of NPY1R binding peptide [F7, P34]-NPY, tesaglitazar, and a cleavable peptide GFLG linker (Figure 12.9). The GFLG tetrapeptide is a widely used linker that is cleaved by lysosomal enzyme cathepsin B at the site between F and L and after C-terminal G to release the payload [91]. The conjugate was tested in cell-based assays and leptin receptor-deficient mice (db/db) mice to study its anti-diabetic potential [34]. In NPY1R-expressing HEK293 and murine 3T3-L1 cell-based assays, the conjugate activated PPARα/γ by peptide-mediated internalization and released tesaglitazar into adipocytes, which led to enhanced adipogenesis and adiponectin expression. *In vivo* investigations conducted in db/db mice demonstrated that the conjugate was effective in impeding the progression of diabetes. This was achieved through the reduction of plasma-free fatty acids and triglycerides, as well as the enhancement of plasma adiponectin levels, which is comparable to the effectiveness of

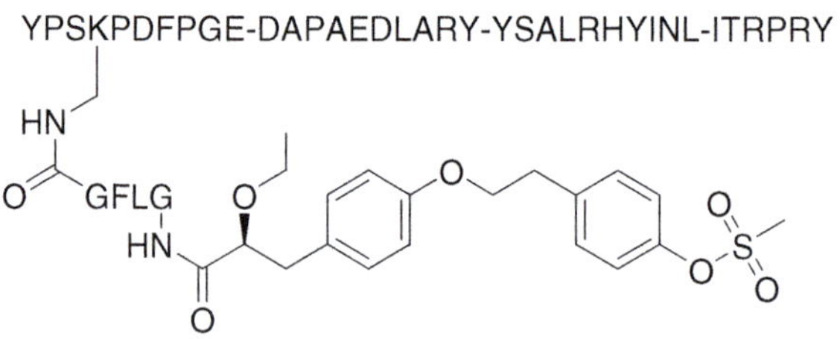

NPY1-tesaglitazar

Figure 12.9 Chemical structure of NPY1-tesaglitazar, a peptide-drug conjugate between NPY1R agonist and a PPARα/γ dual agonist tesaglitazar.

systemically administered tesaglitazar. Consequently, the conjugate emerges as a promising pharmaceutical candidate with the potential to mitigate the known side effects associated with nonselective PPARα/γ agonists in the treatment of T2D.

12.4 Peptide-drug Conjugates for Other Diseases

To date, most peptide-drug conjugates are reported in the oncology and metabolic fields. However, the potential of peptide-drug conjugates goes well beyond and has been reported in other therapeutic areas. For instance, RAB-001 is a peptide-drug conjugate between an α4β1 integrin receptor binding peptide (LLP2A) and a small molecule alendronate [92,93]. The conjugate can recruit mesenchymal stem cells to the bone surface and stimulate bone formation, and it is currently in development for osteonecrosis, osteoporosis, rheumatoid arthritis, and fracture healing. A class of cystine-dense peptides (CDPs) was found to accumulate rapidly in the cartilage of the knees, ankles, hips, shoulders, and intervertebral discs after administration. Conjugation of CDP-11R peptide with the steroid triamcinolone acetonide alleviated joint inflammation in a rat model of rheumatoid arthritis without steroid-associated toxicity [94]. Last, a synovial homing peptide (CKSTHDRLC) was identified to have high synovial endothelium affinity and was conjugated to the highly effective but toxic anti-inflammatory agent sinomenine [95]. The conjugate, as a potential treatment for rheumatoid arthritis, exhibited increased therapeutic efficacy and decreased toxicity through tissue-specific drug delivery.

12.5 Challenges in Peptide-drug Conjugate Design

Despite the great progress in this field, challenges remain for peptide-drug conjugates to become a mainstream therapeutic class like antibody-drug conjugates. The obstacles are numerous, and they pertain to both the chemical and biological aspects of the strategy. In this review, we will only call out a few of the challenges. First, the typical inherent high potency of peptides and typical weaker potency of small molecules generate a deficiency in the transport capacity of the macromolecule relative to the required effective concentration of the small molecules. However, if the therapeutic objective is to enhance the therapeutic index by an order of magnitude via peptide-dependent drug distribution, this is more easily achieved by enriching the pharmacological action at certain tissues. Another challenge of appreciable complexity is the immature nature of the collective knowledge of intracellular biochemistry. While a peptide ligand receptor-binding event can internalize a ligand that carries a small molecule pharmacophore, it is less clear how to transport the entity to the intracellular site where biological action is required. Future research aimed at understanding the limitations and possible solutions to these potential obstacles will help the future of this relatively new modality.

12.6 Summary

Peptide-small molecule conjugation represents a powerful strategy to achieve increased therapeutic efficacy and safety. Most peptide-drug conjugates have been discovered and developed in cancer drug discovery, where peptides have unique advantages as highly potent and tumor-penetrating vehicles to deliver therapeutic drugs to cancer cells. Currently, there are more than a dozen molecules in various stages of clinical development against different types of human cancers.

Peptide-drug conjugates have also found applications in metabolic diseases. Metabolic peptides, given their clinically proven pharmacology, can provide combined or synergistic therapeutic effects with powerful small molecules but with far less off-target toxicity and, therefore, warrant further translational study. Last, peptide-drug conjugates are finding use in emerging therapeutic areas such as osteoporosis and rheumatoid arthritis with autoimmune origin, with notable examples reported in the public domain. With all the significant recent progress in this field, we are very optimistic for the future of peptide-drug conjugates and their place amongst other more mature therapeutic modalities to provide patients with safe and efficacious new medicines to treat unmet medical needs.

References

1 Lau, J.L. and Dunn, M.K. Therapeutic peptides: historical perspectives, current development trends, and future directions. *Bioorg Med Chem.* 2018;26(10):2700–2707.
2 Wang, L., Wang, N., Zhang, W., et al. Therapeutic peptides: current applications and future directions. *Signal Transduct Target Ther.* 2022;7(1):48.
3 Henninot, A., Collins, J.C., and Nuss, J.M. The current state of peptide drug discovery: back to the future? *J Med Chem.* 2018;61(4):1382–1414.
4 Fosgerau, K. and Hoffmann, T. Peptide therapeutics: current status and future directions. *Drug Discov Today.* 2015;20(1):122–128.
5 Liu, F., Li, P., Gelfanov, V., Mayer, J., and DiMarchi, R. Synthetic advances in insulin-like peptides enable novel bioactivity. *Acc Chem Res.* 2017;50(8):1855–1865.

6 Mayer, J.P., Zhang, F., and DiMarchi, R.D. Insulin structure and function. *Pept Sci.* 2007;88(5):687–713.
7 Keskin, O. and Yalcin, S. A review of the use of somatostatin analogs in oncology. *OncoTargets Ther.* 2013;6:471–483.
8 Kaltsas, G.A., Papadogias, D., Makras, P., and Grossman, A.B. Treatment of advanced neuroendocrine tumours with radiolabelled somatostatin analogues. *Endocr Relat Cancer.* 2005;12(4):683–699.
9 Lau, J., Bloch, P., Schäffer, L., et al. Discovery of the once-weekly glucagon-like peptide-1 (GLP-1) analogue semaglutide. *J Med Chem.* 2015;58(18):7370–7380.
10 Research, G.V. Peptide therapeutics market size, share & trends analysis report by application, by type (generic, innovative), by type of manufacturer, by route of administration, by synthesis technology, by region, and segment forecasts, 2023–2030 [Internet]. [cited 2023 Sep 17]. Available from: https://www.grandviewresearch.com/industry-analysis/peptide-therapeutics-market.
11 Drucker, D.J. Advances in oral peptide therapeutics. *Nat Rev Drug Discov.* 2020;19(4):277–289.
12 Tran, H. and ElSayed, M.E.H. Progress and limitations of oral peptide delivery as a potentially transformative therapy. *Expert Opin Drug Deliv.* 2022;19(2):163–178.
13 Xie, J., Bi, Y., Zhang, H., et al. Cell-penetrating peptides in diagnosis and treatment of human diseases: from preclinical research to clinical application. *Front Pharmacol.* 2020;11:697.
14 Reissmann, S. and Filatova, M.P. New generation of cell-penetrating peptides: Functionality and potential clinical application. *J Pept Sci.* 2021;27(5):e3300.
15 Knerr, P.J., Mowery, S.A., Douros, J.D., et al. Next generation GLP-1/GIP/glucagon triple agonists normalize body weight in obese mice. *Mol Metab.* 2022;63:101533.
16 Coskun, T., Urva, S., Roell, W.C., et al. LY3437943, a novel triple glucagon, GIP, and GLP-1 receptor agonist for glycemic control and weight loss: from discovery to clinical proof of concept. *Cell Metab.* 2022;34(9):1234–1247.
17 Jastreboff, A.M., Kaplan, L.M., Frías, J.P., et al. Triple–hormone-receptor agonist retatrutide for obesity—a phase 2 trial. *N Engl J Med.* 2023;389(6):514–526.
18 He, R., Finan, B., Mayer, J.P., and DiMarchi, R.D. Peptide conjugates with small molecules designed to enhance efficacy and safety. *Molecules.* 2019;24(10):1855.
19 Cooper, B.M., Iegre, J., Donovan, D.H.O., Halvarsson, M.Ö., and Spring, D.R. Peptides as a platform for targeted therapeutics for cancer: peptide–drug conjugates (PDCs). *Chem Soc Rev.* 2020;50(3):1480–1494.
20 Fu, C., Yu, L., Miao, Y., Liu, X., Yu, Z., and Wei M. Peptide–drug conjugates (PDCs): A novel trend of research and development on targeted therapy, hype or hope? *Acta Pharm Sin B.* 2023;13(2):498–516.

21 Lindberg, J., Nilvebrant, J., Nygren, P.Å., and Lehmann, F. Progress and future directions with peptide-drug conjugates for targeted cancer therapy. *Molecules.* 2021;26(19):6042.
22 White, B.H., Whalen, K., Kriksciukaite, K., et al. Discovery of an SSTR2-targeting maytansinoid conjugate (PEN-221) with potent activity in vitro and in vivo. *J Med Chem.* 2019;62(5):2708–2719.
23 Whalen, K.A., White, B.H., Quinn, J.M., et al. Targeting the somatostatin receptor 2 with the miniaturized drug conjugate, PEN-221: a potent and novel therapeutic for the treatment of small cell lung cancer. *Mol Cancer Ther.* 2019;18(11):1926–1936.
24 Lahdenranta, J., Bennett, G., Huxley, P., et al. Abstract 1319: Rapid accumulation of cytotoxic payload in tumor tissue drives BT5528 activity in tumor models. *Cancer Res.* 2021;81(Suppl 13):1319.
25 Bennett, G., Brown, A., Mudd, G., et al. MMAE delivery using the bicycle toxin conjugate BT5528. *Mol Cancer Ther.* 2020;19(7):1385–1394.
26 Mahalingam, D., Wilding, G., Denmeade, S., et al. Mipsagargin, a novel thapsigargin-based PSMA-activated prodrug: results of a first-in-man Phase I clinical trial in patients with refractory, advanced or metastatic solid tumours. *Br J Cancer.* 2016;114(9):986–994.
27 Andersen, T., López, C., Manczak, T., Martinez, K., and Simonsen, H. Thapsigargin—from Thapsia L. to mipsagargin. *Molecules.* 2015;20(4):6113–6127.
28 Kurzrock, R., Gabrail, N., Chandhasin, C., et al. Safety, pharmacokinetics, and activity of GRN1005, a novel conjugate of angiopep-2, a peptide facilitating brain penetration, and paclitaxel, in patients with advanced solid tumors. *Mol Cancer Ther.* 2012;11(2):308–316.
29 Drappatz, J., Brenner, A., Wong, E.T., et al. Phase I study of GRN1005 in recurrent malignant glioma. *Clin Cancer Res.* 2013;19(6):1567–1576.
30 Demeule, M., Charfi, C., Currie, J., et al. TH1902, a new docetaxel-peptide conjugate for the treatment of sortilin-positive triple-negative breast cancer. *Cancer Sci.* 2021;112(10):4317–4334.
31 Finan, B., Yang, B., Ottaway, N., et al. Targeted estrogen delivery reverses the metabolic syndrome. *Nat Med.* 2012;18(12):1847–1856.
32 Quarta, C., Stemmer, K., Novikoff, A., et al. GLP-1-mediated delivery of tesaglitazar improves obesity and glucose metabolism in male mice. *Nat Metab.* 2022;4(8):1071–1083.
33 Finan, B., Clemmensen, C., Zhu, Z., et al. Chemical hybridization of glucagon and thyroid hormone optimizes therapeutic impact for metabolic disease. *Cell.* 2016;167(3):843–857.
34 Wittrisch, S., Klöting, N., Mörl, K., Chakaroun, R., Blüher, M., and Beck-Sickinger, A.G. NPY1R-targeted peptide-mediated delivery of a dual

PPARα/γ agonist to adipocytes enhances adipogenesis and prevents diabetes progression. *Mol Metab*. 2019;31:163–180.

35 Weckbecker, G., Lewis, I., Albert, R., Schmid, H.A., Hoyer, D., and Bruns C. Opportunities in somatostatin research: biological, chemical and therapeutic aspects. *Nat Rev Drug Discov*. 2003;2(12):999–1017.

36 Ibáñez-Costa, A., López-Sánchez, L.M., Gahete, M.D., et al. BIM-23A760 influences key functional endpoints in pituitary adenomas and normal pituitaries: molecular mechanisms underlying the differential response in adenomas. *Sci Rep*. 2017;7(1):42002.

37 Watanabe, H., Fujishima, F., Unno, M., Sasano, H., and Suzuki, T. Somatostatin receptor 2 in 10 different types of human non-neoplastic gastrointestinal neuroendocrine cells. *Pathol Res Pr*. 2023;244:154418.

38 Vrettos, E.I., Mező, G., and Tzakos, A.G. On the design principles of peptide–drug conjugates for targeted drug delivery to the malignant tumor site. *Beilstein J Org Chem*. 2018;14(1):930–954.

39 Halperin, D.M., Johnson, M.L., Chan, J.A., et al. The safety and efficacy of PEN-221 somatostatin analog (SSA)-DM1 conjugate in patients (PTS) with advanced GI mid-gut neuroendocrine tumor (NET): phase 2 results. *J Clin Oncol*. 2021;39(Suppl 15):4110.

40 Ireton, R. and Chen, J. EphA2 receptor tyrosine kinase as a promising target for cancer therapeutics. *Curr Cancer Drug Targets*. 2005;5(3):149–157.

41 Zhao, P., Jiang, D., Huang, Y., and Chen, C. EphA2: a promising therapeutic target in breast cancer. *J Genet Genom*. 2021;48(4):261–7.

42 Wilson, K., Shiuan, E., and Brantley-Sieders, D.M. Oncogenic functions and therapeutic targeting of EphA2 in cancer. *Oncogene*. 2021;40(14):2483–2495.

43 Xiao, T., Xiao, Y., Wang, W., Tang, Y.Y., Xiao, Z., and Su, M. Targeting EphA2 in cancer. *J Hematol Oncol*. 2020;13(1):114.

44 Therapeutics B. Bicycle therapeutics announces interim BT5528 Phase I clinical trial results and preliminary results from ongoing BT8009 Phase I clinical trial [Internet]. [cited 2023 Sep 17]. Available from: https://investors.bicycletherapeutics.com/news-releases/news-release-details/bicycle-therapeutics-announces-interim-bt5528-phase-i-clinical#:~:text=A%20total%20of%20eight%20BT5528, under%20RECIST%20version%201.1%20criteria.

45 Silver, D.A., Pellicer, I., Fair, W.R., Heston, W.D., and Cordon-Cardo, C. Prostate-specific membrane antigen expression in normal and malignant human tissues. *Clin Cancer Res*. 1997;3:81–85.

46 Kinoshita, Y., Kuratsukuri, K., Landas, S., et al. Expression of prostate-specific membrane antigen in normal and malignant human tissues. *World J Surg*. 2006;30(4):628–636.

47 Denmeade, S.R. and Isaacs, J.T. Engineering enzymatically activated “molecular grenades” for cancer. *Oncotarget*. 2012;3(7):666–667.

48 Denmeade, S.R., Mhaka, A.M., Rosen, D.M., et al. Engineering a prostate-specific membrane antigen–activated tumor endothelial cell prodrug for cancer therapy. *Sci Transl Med*. 2012;4(140):140ra86.

49 Isaacs, J.T., Brennen, W.N., Christensen, S.B., and Denmeade, S.R. Mipsagargin: the beginning—not the end—of thapsigargin prodrug-based cancer therapeutics. *Molecules*. 2021;26(24):7469.

50 Hennenfent, K.L. and Govindan, R. Novel formulations of taxanes: a review. Old wine in a new bottle? *Ann Oncol*. 2006;17(5):735–749.

51 Spencer, C.M. and Faulds, D. Paclitaxel. A review of its pharmacodynamic and pharmacokinetic properties and therapeutic potential in the treatment of cancer. *Drugs*. 1994;48:794–847.

52 Demeule, M., Régina, A., Ché, C., et al. Identification and design of peptides as a new drug delivery system for the brain. *J Pharmacol Exp Ther*. 2008;324(3):1064–1072.

53 Shao, K., Huang, R., Li, J., et al. Angiopep-2 modified PE-PEG based polymeric micelles for amphotericin B delivery targeted to the brain. *J Control Release*. 2010;147(1):118–126.

54 Régina, A., Demeule, M., Ché, C., et al. Antitumour activity of ANG1005, a conjugate between paclitaxel and the new brain delivery vector Angiopep-2. *Br J Pharmacol*. 2008;155(2):185–197.

55 Li, F. and Tang, S.C. Targeting metastatic breast cancer with ANG1005, a novel peptide-paclitaxel conjugate that crosses the blood-brain-barrier (BBB). *Genes Dis*. 2017;4(1):1–3.

56 Kumthekar, P., Tang, S.C., Brenner, A.J., et al. ANG1005, a brain-penetrating peptide–drug conjugate, shows activity in patients with breast cancer with leptomeningeal carcinomatosis and recurrent brain metastases. *Clin Cancer Res*. 2020;26(12):2789–2799.

57 ClinicalTrials.Gov. ANG1005 in leptomeningeal disease from breast cancer (ANGLeD) [Internet]. [cited 2023 Sep 17]. Available from: https://clinicaltrials.gov/study/NCT03613181?a=5.

58 Ghaemimanesh, F., Mehravar, M., Milani, S., Poursani, E.M., and Saliminejad, K. The multifaceted role of sortilin/neurotensin receptor 3 in human cancer development. *J Cell Physiol*. 2021;236(9):6271–6281.

59 Mazella, J., Zsürger, N., Navarro, V., et al. The 100-kDa neurotensin receptor is gp95/sortilin, a non-G-protein-coupled receptor*. *J Biol Chem*. 1998;273(41):26273–26276.

60 Hu, F., Padukkavidana, T., Vægter, C.B., et al. Sortilin-mediated endocytosis determines levels of the frontotemporal dementia protein, progranulin. *Neuron*. 2010;68(4):654–667.

61 Carlo, A.S., Gustafsen, C., Mastrobuoni, G., et al. The pro-neurotrophin receptor sortilin is a major neuronal apolipoprotein E receptor for catabolism of amyloid-β peptide in the brain. *J Neurosci*. 2013;33(1):358–370.

62 Currie, J.C., Demeule, M., Charfi, C., et al. The peptide-drug conjugate TH1902: a new sortilin receptor-mediated cancer therapeutic against ovarian and endometrial cancers. *Cancers*. 2022;14(8):1877.

63 Theratechnologies. Theratechnologies to present preliminary safety and efficacy data from Phase 1 trial of sudocetaxel zendusortide in heavily pretreated cancer patients at ASCO 2023 [Internet]. [cited 2023 Sep 19]. Available from: https://www.theratech.com/news-releases/news-release-details/theratechnologies-present-preliminary-safety-and-efficacy-data.

64 Vogel, H., Wolf, S., Rabasa, C., et al. GLP-1 and estrogen conjugate acts in the supramammillary nucleus to reduce food-reward and body weight. *Neuropharmacology*. 2016;110(Pt A):396–406.

65 Gupte, A.A., Pownall, H.J., and Hamilton, D.J. Estrogen: an emerging regulator of insulin action and mitochondrial function. *J Diabetes Res*. 2015; 2015:916585.

66 Mauvais-Jarvis, F. Estrogen and androgen receptors: regulators of fuel homeostasis and emerging targets for diabetes and obesity. *Trends Endocrinol Metab*. 2011;22(1):24–33.

67 Pereira, R.I., Casey, B.A., Swibas, T.A., Erickson, C.B., Wolfe, P., and Pelt, R.E.V. Timing of estradiol treatment after menopause may determine benefit or harm to insulin action. *J Clin Endocrinol Metab*. 2015;100(12):4456–4462.

68 Bonds, D.E., Lasser, N., Qi, L., et al. The effect of conjugated equine oestrogen on diabetes incidence: the women's health initiative randomised trial. *Diabetologia*. 2006;49(3):459–468.

69 Rossouw, J.E., Anderson, G.L., Prentice, R.L., et al. Risks and benefits of estrogen plus progestin in healthy postmenopausal women: principal results from the women's health initiative randomized controlled trial. *JAMA*. 2002;288:321–333.

70 Kim, J.H., Meyers, M.S., Khuder, S.S., et al. Tissue-selective estrogen complexes with bazedoxifene prevent metabolic dysfunction in female mice. *Mol Metab*. 2014;3(2):177–190.

71 Barrera, J., Chambliss, K.L., Ahmed, M., et al. Bazedoxifene and conjugated estrogen prevent diet-induced obesity, hepatic steatosis, and type 2 diabetes in mice without impacting the reproductive tract. *Am J Physiol-Endocrinol Metab*. 2014;307(3):E345–E354.

72 Cao, X., Xu, P., Oyola, M.G., et al. Estrogens stimulate serotonin neurons to inhibit binge-like eating in mice. *J Clin Investig*. 2014;124(10):4351–4362.

73 Tiano, J.P., Tate, C.R., Yang, B.S., DiMarchi, R., and Mauvais-Jarvis, F. Effect of targeted estrogen delivery using glucagon-like peptide-1 on insulin secretion, insulin sensitivity and glucose homeostasis. *Sci Rep*. 2015;5(1):10211.

74 Schwenk, R.W., Baumeier, C., Finan, B., et al. GLP-1–oestrogen attenuates hyperphagia and protects from beta cell failure in diabetes-prone New Zealand obese (NZO) mice. *Diabetologia*. 2015;58(3):604–614.

75 Sachs, S., Bastidas-Ponce, A., Tritschler, S., et al. Targeted pharmacological therapy restores β-cell function for diabetes remission. *Nat Metab.* 2020;2(2):192–209.
76 Fiévet, C., Fruchart, J.C., and Staels, B. PPARα and PPARγ dual agonists for the treatment of type 2 diabetes and the metabolic syndrome. *Curr Opin Pharmacol.* 2006;6(6):606–614.
77 Takada, I. and Makishima, M. Peroxisome proliferator-activated receptor agonists and antagonists: a patent review (2014–present). *Expert Opin Ther Pat.* 2020;30(1):1–13.
78 Bays, H., McElhattan, J., Bryzinski, B.S., and Group G6S. A double-blind, randomised trial of tesaglitazar versus pioglitazone in patients with type 2 diabetes mellitus. *Diabetes Vasc Dis Res.* 2007;4(3):181–193.
79 Hong, F., Xu, P., and Zhai, Y. The opportunities and challenges of peroxisome proliferator-activated receptors ligands in clinical drug discovery and development. *Int J Mol Sci.* 2018;19(8):2189.
80 Müller, T.D., Finan, B., Clemmensen, C., DiMarchi, R.D., and Tschöp, M.H. The new biology and pharmacology of glucagon. *Physiol Rev.* 2017;97(2):721–766.
81 Grozinsky-Glasberg, S., Fraser, A., Nahshoni, E., Weizman, A., and Leibovici, L. Thyroxine-triiodothyronine combination therapy versus thyroxine monotherapy for clinical hypothyroidism: meta-analysis of randomized controlled trials. *J Clin Endocrinol Metab.* 2006;91(7):2592–2599.
82 Mullur, R., Liu, Y.Y., and Brent, G.A. Thyroid hormone regulation of metabolism. *Physiol Rev.* 2014;94(2):355–382.
83 Sinha, R.A., Singh, B.K., and Yen, P.M. Thyroid hormone regulation of hepatic lipid and carbohydrate metabolism. *Trends Endocrinol Metab.* 2014;25(10):538–545.
84 Ochs, N., Auer, R., Bauer, D.C., et al. Meta-analysis: subclinical thyroid dysfunction and the risk for coronary heart disease and mortality. *Ann Intern Med.* 2008;148:832–845.
85 Baxter, J.D. and Webb, P. Thyroid hormone mimetics: potential applications in atherosclerosis, obesity and type 2 diabetes. *Nat Rev Drug Discov.* 2009;8(4):308–320.
86 Klöting, N. and Blüher, M. Adipocyte dysfunction, inflammation and metabolic syndrome. *Rev Endocr Metab Disord.* 2014;15(4):277–287.
87 Horwitz, A. and Birk, R. Adipose tissue hyperplasia and hypertrophy in common and syndromic obesity—the case of BBS obesity. *Nutrients.* 2023;15(15):3445.
88 Ahmadian, M., Suh, J.M., Hah, N., et al. PPARγ signaling and metabolism: the good, the bad and the future. *Nat Med.* 2013;19(5):557–566.
89 Berger, J. and Moller, D.E. The mechanisms of action of PPARS. *Annu Rev Med.* 2002;53(1):409–435.

90 Park, C., Kim, J., Ko, S.B., et al. Author correction: structural basis of neuropeptide Y signaling through Y1 receptor. *Nat Commun.* 2022;13(1):1126.
91 Zhang, P., Cheetham, A.G., Lock, L.L., and Cui, H. Cellular uptake and cytotoxicity of drug–peptide conjugates regulated by conjugation site. *Bioconjugate Chem.* 2013;24(4):604–613.
92 Mohan, G., Lay, E.Y.A., Berka, H., et al. A novel hybrid compound LLP2A-Ale both prevented and rescued the osteoporotic phenotype in a mouse model of glucocorticoid-induced osteoporosis. *Calcif Tissue Int.* 2017;100(1):67–79.
93 Wang, Z., Zhao, Y., Zhang, D., et al. A novel hybrid compound LLP2A-alendronate accelerates open fracture healing in a rabbit model. *Drug Des Dev Ther.* 2019;13:1077–1086.
94 Sangar, M.L.C., Girard, E.J., Hopping, G., et al. A potent peptide-steroid conjugate accumulates in cartilage and reverses arthritis without evidence of systemic corticosteroid exposure. *Sci Transl Med.* 2020;12(533):eaay1041.
95 Zhang, T., Ouyang, X., Gou, S., et al. Novel synovial targeting peptide-sinomenine conjugates as a potential strategy for the treatment of rheumatoid arthritis. *Int J Pharm.* 2022;617:121628.

13

Emerging Landscape of Therapeutic Oligonucleotide Delivery by Cell-penetrating Peptides

Xiang Li, Xiulong Shen, Ashweta Sahni and Ziqing Qian

Entrada Therapeutics, Inc., Boston, MA, USA

13.1 Introduction

Cell-penetrating peptides (CPPs), also referred to as protein transduction domains, are a class of short peptides (typically 5–30 residues in length) that possess the ability to internalize into cells, thereby facilitating the intracellular delivery of various cargos [1]. First discovered in 1988, this property has attracted increased interest in employing CPPs as delivery tools for a diverse range of therapeutic agents that encounter challenges in penetrating cellular barriers, including therapeutic oligonucleotides.

Therapeutic oligonucleotides are short synthetic nucleic acids that can be used to treat diseases by binding to their RNA targets, thereby modulating corresponding RNA and respective downstream protein levels [2,3]. This distinct mechanism offers the potential to address therapeutic targets previously considered "undruggable" for small molecule and antibody modalities [2]. The development of *N*-acetylgalactosamine (GalNAc) and lipid nanoparticle-based delivery systems enabled the delivery of oligonucleotides to the liver with clinical success [3–5]. However, significant challenges persist in achieving effective delivery of oligonucleotides to extrahepatic tissues due to poor biodistribution, limited cellular uptake, and low efficiency of endosomal escape [2,5,6]. To overcome these limitations, various CPP-based platforms have been developed for extrahepatic therapeutic oligonucleotide delivery [2], with multiple agents being

Peptide Science: Chemical Ligation, Lead Generation, and Therapeutic Advances, First Edition. Fa Liu.

evaluated in clinical studies [1,7]. In this review, we present the introduction of CPPs and summarize their use as a delivery system for therapeutic oligonucleotides. Additionally, we highlight the challenges faced by this field and offer insights into future perspectives.

13.2 Cell-penetrating Peptides

13.2.1 Discovery of CPPs

Since the turn of the century, CPPs derived from natural sources have been discovered (e.g. viruses, venoms, and toxins). Additionally, an increasing number of CPPs have also been rationally designed [1]. In 1988, research on human immunodeficiency virus (HIV) identified that the 86-amino acid trans-activator of transcription domain could rapidly enter mammalian cells in culture and translocate across the nuclear membrane, even at low extracellular concentrations [8,9]. Subsequent research showed that the translocation of the protein was due to a small stretch of cationic amino acids, thereafter, named Tat peptide (GRKKRRQRRR, Table 13.1) [10]. In 1991, the homeodomain of Antennapedia from *Drosophila* was reported to enter neuronal cells in culture and accumulate in nuclei [11]. Later, researchers found that a 16-amino

Table 13.1 Representative cell-penetrating peptides.

Class	Peptide	Sequence	Origin
Cationic	Tat [14]	GRKKRRQRRR	HIV trans-activator of transcription
	Penetratin [13]	RQIKIWFQNRRMKWKK	Antennapedia homeodomain
	R9 [15]	RRRRRRRRR	Rational design
Amphipathic	Pep-1 [16]	KETWWETWWTEWSQPKKKRKV	SV40 NLS
	Transportan [14,17]	GWTLNSAGYLLGKINLKALAALAKKIL	Galanin and mastoparan
Hydrophobic	Pep-7 [18]	SDLWEMMMVSLACQY	Phage display
Cyclic	cTat [19]	*cyclo*(CYGRKKRRQRRRC)	Rational design
	CPP12 [20]	*cyclo*(FfΦRrRrQ)	

Φ, L-2-naphthylalanine; CPP, cell-penetrating peptide; f, D-phenylalanine; HIV, human immunodeficiency virus; NLS, nuclear localization sequence; r, D-arginine; SV40, simian virus 40.

acid stretch derived from the third helix of the homeodomain of Antennapedia, named penetratin (RQIKIWFQNRRMKWKK, Table 13.1), was able to cross the plasma membrane [12,13].

CPPs rationally designed to deliver therapeutic cargos soon followed. In 2001, Pep-1 (KETWWETWWTEWSQPKKKRKV, Table 13.1) was introduced as a vector for non-covalent intracellular delivery of peptides and proteins [16]. *In vivo* proof of concept was first reported in 1998, when Pooga et al. showed that peptide nucleic acids (PNAs) conjugated with the peptide penetratin could be delivered to neurons in rats after intrathecal injection [21]. A year later, it was reported that a protein genetically fused to Tat could be delivered to all tested tissues in mice after intraperitoneal injection, including the brain [22]. Subsequently, sequences containing structural motifs (e.g. cyclization, cystine knots, β-turns and α-helices), synthetic amino acids, non-peptidic motifs, and novel linkers have been incorporated into CPP design to enhance their ability to deliver cargo [14,23,24]. As of 2016, ~1850 CPPs were reported [25], and interest in designing CPPs for functional cargo delivery has continued to grow exponentially.

13.2.2 Structural Features of CPPs

CPPs can be categorized as cationic, amphipathic, or hydrophobic depending on their sequences [14,26,27]. Table 13.1 shows some representative sequences of CPPs from these classes. CPPs that have a high abundance of positively charged amino acids (e.g. arginine or lysine) are classified as cationic CPPs, examples being Tat and penetratin [14]. Deletion or substitution of positively charged residues led to a decrease in cell penetration of Tat and penetratin [10,28,29], demonstrating that these residues are essential for uptake.

The positive charge from lysine and arginine residues has long been thought to facilitate interactions with the plasma membrane [26,30]. Poly-arginine stretches are uniquely effective at cellular internalization. Studies have identified that the optimum number of arginines needed for translocation is six to nine consecutive residues [15,28]. For instance, R9 (RRRRRRRRR, Table 13.1) was reported to have approximately two- to threefold higher uptake efficiency than Tat [20,28]. In further work, it was demonstrated that polypeptides consisting of at least six arginine residues entered cells more easily than those consisting of ornithine or lysine [31], suggesting that arginine is more effective at promoting internalization than other positively charged amino acids. This effect has been attributed to bidentate interactions formed between the guanidinium group in arginine and negatively charged cell surface groups (such as phosphates in proteoglycans and membrane lipids, heparan sulfates, and carboxylates), which play a key role in enhancing uptake [32]. Therefore, stretches of arginine residues are posited to promote extensive networks of charge–charge interactions and hydrogen bonds between the CPP and plasma membrane. It should be noted that despite the promise of traditional arginine-rich cationic CPPs for intracellular delivery of therapeutic cargos [26], their utility is tempered by their liability for toxicity, which increases with arginine content [33–35].

Amphipathic CPPs such as Pep-1 [16] and transportan [14] (Table 13.1) contain both polar (hydrophilic) and nonpolar (hydrophobic) residues [36]. Primary amphipathic peptides contain sequential hydrophobic and hydrophilic domains [14,26]. Secondary amphipathic CPPs, typically smaller than primary amphipathic peptides, display amphipathic properties by adopting an α-helical or β-sheet structure when binding to the plasma membrane [14,26]. It is thought that nonpolar residues (e.g. tryptophan and phenylalanine) facilitate the insertion of the CPP into the hydrophobic core of the phospholipid bilayer [36,37]. In fact, CPPs with both hydrophobic residues and multiple arginines have been reported to have very high cytosolic delivery efficiencies due to the combined effect of membrane insertion and bidentate membrane interactions, respectively [23,38]. Many amphipathic CPPs have been derived from natural proteins, but a number have been rationally designed by attaching a membrane-binding hydrophobic domain to a cationic nuclear localization sequence (NLS) [14]. For example, Pep-1 (Table 13.1) contains a cationic domain (PKKKRKV) derived from the NLS of the simian virus 40 protein [16]. This domain is connected via a short linker peptide to a tryptophan-rich cluster (KETWWETWWTEW) [16].

Although most CPPs can be classified as cationic or amphipathic, hydrophobic CPPs, such as Pep-7 (SDLWEMMMVSLACQY, Table 13.1) have also been reported [14,18]. However, the hydrophobicity of such sequences poses additional challenges, including aggregation [38], that limit their potential to deliver therapeutic cargos.

Structurally, CPPs can also be classified as linear or cyclic [27]. It is well established that cyclization of a CPP sequence increases its cellular uptake efficiency, endosomal escape, and metabolic stability [23,39]. The rigidity of the peptide backbone in a cyclic peptide is thought to reduce its conformational freedom, lowering the entropic cost of binding to the plasma membrane [23]. Some examples of cyclic CPPs that have been used for cargo delivery include CPP12 (*cyclo*-[FfΦRrRrQ], where Φ is L-2-naphthylalanine) and cTat (*cyclo*[CYGRKKRRQRRRC], where the two cysteine residues form a disulfide bond) [20,40] (Table 13.1). CPP12 has been used to deliver a wide range of cargos, such as peptides, proteins, and nucleic acids [41–44]. cTat has been demonstrated to deliver green fluorescent protein (GFP) to mammalian cells in culture [19].

13.2.3 Mechanisms of Internalization

With the rising popularity of CPPs for therapeutic payload delivery, understanding their mechanism of internalization is crucial for the design of CPPs with high cytosolic delivery potential. For most cationic and amphipathic CPPs, two distinct mechanisms have been proposed [23,26]: (i) direct translocation across the plasma membrane and (ii) endocytosis followed by endosomal escape [45] (Figure 13.1). The two mechanisms are not mutually exclusive, and CPPs can utilize both to enter the cell [23,45]. Importantly, the experimental conditions and properties of the cargo molecules can affect the uptake mechanism even for the same CPP [46–48]. For example, high concentrations of R9 ($\geq$10 μM) trigger direct translocation, while low concentrations typically lead to internalization through endocytic pathways [45]. Interestingly, it is reported that not all of these pathways lead to "productive" uptake, i.e. release into the cytosol. Regardless of the pathway taken, the CPP must ultimately traverse across a phospholipid bilayer membrane, be it the plasma or endosome membrane, in order to enter the cytosol of the cell. This biophysical process is not fully understood in spite of decades of research, but several hypotheses, such as pore formation, vesicle budding and collapse, and membrane destabilization, have been proposed [26,49,50].

A small subset of hydrophobic CPPs can passively diffuse across the plasma membrane (along the concentration gradient), but these peptides are typically

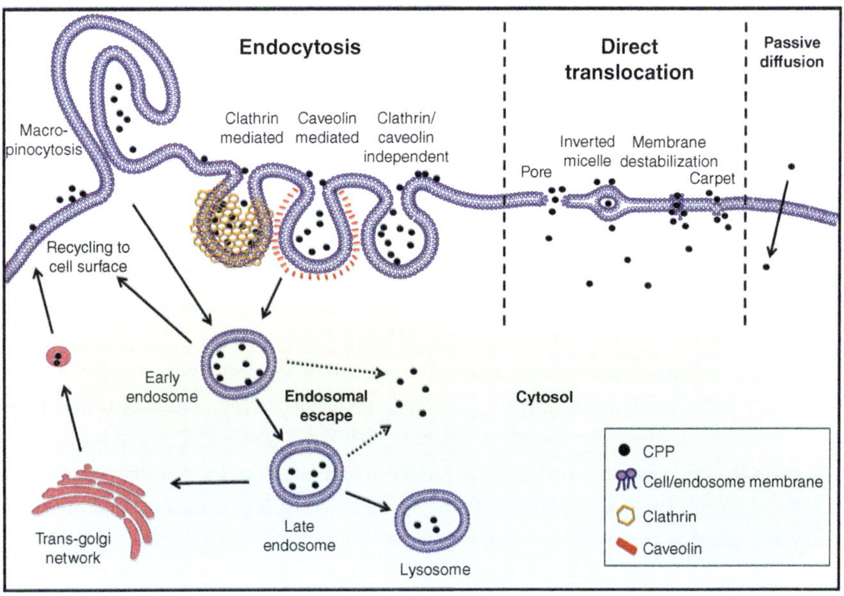

Figure 13.1 Overview of mechanisms of cellular uptake of CPPs. Most cationic and amphipathic CPPs enter the cell via endocytosis followed by endosomal escape (left). Some CPPs, under certain conditions, can directly translocate across the cell membrane and gain direct access to the cytosol (middle). A subset of hydrophobic CPPs can passively diffuse across the cell membrane (right). CPP, cell-penetrating peptide.

hydrophobic and may not be suitable for delivering large and/or hydrophilic cargos as they cannot diffuse through the hydrophobic interior of the lipid bilayer membrane [51].

13.2.3.1 Direct Translocation

Direct translocation of CPPs refers to the mechanism of uptake wherein cationic or amphipathic CPPs directly enter the cell cytosol by crossing the plasma membrane in an energy-independent process [23,26,45]. This process bypasses the challenge of endosomal escape and results in the release of the CPP into the cytosol in a much shorter timescale (a few minutes) versus endocytosis (a few hours). However, reports of cargo delivery have been largely skewed toward small molecule delivery, e.g. fluorophores [23,45,52], with only a few instances of protein delivery *in vitro* [15,40,47,53]. Recently, it was reported that the direct translocation of fluorophore-labeled CPPs is exclusively responsible for the productive delivery of CPPs into the cytosol and that these CPPs are incapable of escape from endosomes [52]. However, it is unclear whether this holds true for large, hydrophilic cargos, such as proteins and oligonucleotides.

Based on proposed direct translocation mechanisms, the interactions between CPPs and the plasma membrane are foundational to their internalization. These interactions reportedly cause transient, local reorganization of the phospholipid bilayer [23,45,54,55]. Other models have also been proposed, such as pore formation [56–59], inverted micelle formation [60], carpet model [61], and osmotic pressure-mediated local membrane phase transition [62] (Figure 13.1).

13.2.3.2 Endocytosis and Endosomal Escape

Most CPPs gain intracellular access in an energy-dependent manner via endocytosis and subsequent endosomal escape [26,63]. The initial step of endocytic uptake is either receptor-mediated or receptor-independent (e.g. through interactions with cell surface proteoglycans, phosphate head groups in the outer membrane) [23,64–68]. The subsequent endocytic pathways are dependent on the sequence and physicochemical properties of the CPP and can be categorized as clathrin-mediated endocytosis, macropinocytosis, caveolin/lipid raft-mediated endocytosis, and clathrin/caveolin-independent endocytosis [23,69–72] (Figure 13.1). CPPs can enter the cell through multiple simultaneous endocytic pathways, and the mechanism(s) utilized are affected by the cargo and experimental conditions (e.g. concentration of CPP, temperature, cell, or tissue type) [46,47,73,74].

After internalization into endosomes, a much bigger hurdle is faced – escape from the endosome – which is often inefficient for most CPPs [49]. Upon entering endocytic pathways, CPPs are trafficked along a highly regulated endosome maturation pathway – from the early endosome to the late endosome and finally to the lysosome [49]. If the CPP is unable to escape from the endosome during this process, it may be degraded in the lysosome, undergo retrograde transport to the endoplasmic reticulum/Golgi apparatus, or get recycled back to the cell surface [49,75], all of which are likely to be nonproductive. Therefore, endosomal escape is one of the most significant bottlenecks for the delivery of intracellular therapeutics and ultimately governs the kinetics and extent of active payload delivered to the cytosol [49,75].

Significant efforts have been undertaken to understand the mechanisms of endosomal escape to engineer more effective CPPs. Various models have been proposed, including vesicle budding and collapse [76], the proton sponge effect [77], pore formation [30], and leakage from late endosomes [49,78,79]. Many of these models are contentious due to contradictory results, yet some have been supported with data from live cell models [76]. For instance, it was demonstrated that cyclic CPPs (e.g. CPP12, Table 13.1) among others, promote endosomal escape by facilitating the budding of CPP-enriched vesicles from the endosome. These budded vesicles ultimately collapse into lipid-peptide aggregates that rapidly disperse into the cytoplasm, thereby resulting in the release of the CPP [20,76]. However, few mechanistic studies have been performed for CPP-oligonucleotide conjugates to gain an understanding that could be used to optimize the delivery platform.

13.2.4 Overview of CPPs for the Delivery of Therapeutic Cargos

Peptides, proteins, nucleic acids, and other biomolecules are all capable of influencing cellular function and serving as therapeutic agents against intracellular targets. However, many of these molecules display little or no biological activity due to their limited membrane permeability [37]. CPPs have been explored as a means to facilitate cellular uptake of a wide range of biomolecules, including peptides [80,81], proteins [15,82–85], nucleic acids [86–89], and nanoparticles [90–92]. For instance, Tat was reported to improve the uptake and activity of a tumor suppressor p53-derived peptide to restore p53 functionality in cancer cells [80]. Additionally, genetically engineered Tat-fusion proteins have been employed to deliver antibodies into cells [93]. CPPs have also been used to deliver a variety of nucleic acid modalities, such as plasmid DNA [94,95], antisense oligonucleotides (ASOs) [96], small interfering RNAs (siRNAs) [87,97], and others.

Over the past few decades, there has been significant advancement in the field of delivering biomacromolecules, including peptides, via conjugated CPPs, with many now being evaluated in clinical trials, as summarized in Table 13.2. For example, PEP-010 was developed by conjugating a CPP with a peptide inhibiting caspase-9 and protein phosphatase 2A interactions [98], and it is being evaluated in the treatment of metastatic solid cancers [99]. Oligonucleotide therapeutics have also recently undergone improvements via conjugation to CPPs. Impressively, in the past 5 years, there has been a surge in clinical studies of CPP-conjugated oligonucleotide therapeutics, particularly in the treatment of neuromuscular disorders (Table 13.2).

Table 13.2 Summary of ongoing clinical trials with CPP-conjugated therapeutics.

ID	Company	Cargo molecules	Indication	Status
ATX-101 [100]	Therapim Pty	A PCNA-interacting peptide	Advanced solid tumor	NCT04814875 Phase 1/2
PEP-010 [98]	PEP-Therapy	A peptidyl inhibitor interfering with the interaction between caspase-9 and PP2A	Metastatic solid tumor	NCT04733027 Phase 1
ATP128 [101]	AMAL Therapeutics	A chimeric sequence composed of a TLR agonist peptide and a multi-antigenic domain	Colorectal cancer	NCT04046445 Phase 1b
PGN-EDO51	PepGen	A *DMD* exon 51-skipping PMO	DMD	NCT06079736 Phase 2
PGN-EDODM1	PepGen	A CUG-repeat blocking PMO	DM1	NCT06204809 Phase 1
ENTR-601-44 [102]	Entrada Therapeutics	A *DMD* exon 44-skipping PMO	DMD	Phase 1
VX-670	Vertex Pharmaceuticals/ Entrada Therapeutics	A CUG-repeat blocking PMO	DM1	NCT06185764 Phase 1/2
VP-001	PYC Therapeutics	An exon skipping PMO for *CNOT3*	RP11	NCT05902962 Phase 1

CNOT3, CCR4-NOT transcription complex subunit 3; CPP, cell-penetrating peptide; DM1, myotonic dystrophy Type 1; DMD, Duchenne muscular dystrophy; PCNA, proliferating cell nuclear antigen; PMO, phosphorodiamidate morpholino oligomer; PP2A, protein phosphatase 2A; RP11, retinitis pigmentosa Type 11; TLR, toll-like receptor.

13.3 Applications of CPPs for the Delivery of Therapeutic Oligonucleotides

13.3.1 Overview of Therapeutic Oligonucleotides

Therapeutic oligonucleotides are short synthetic nucleic acids, typically consisting of 15–30 nucleotides, that modulate transcript levels by targeting RNA transcripts, mostly through Watson–Crick–Franklin base pairing [2,3]. Distinguished by their mechanisms of action, there are primarily three types of oligonucleotide therapeutics: siRNAs, ASOs, and aptamers [3]. siRNAs, consisting of passenger and guide strands, function through the RNA interference pathway. In this process, the guide strand of siRNA is loaded into the RNA-induced silencing complex, subsequently binding to the target RNA transcript. The passenger strand serves to protect the guide strand from being degraded [4].

In contrast, ASOs are single-stranded and can function as either catalytic ASO gapmers or splice-switching oligonucleotides (SSOs). Gapmers contain a central DNA "gap" flanked by modified RNA bases. Upon binding to the RNA targets, the DNA gap acts to recruit RNase H, thereby cleaving the target RNA transcript. SSOs, on the other hand, bind to the target RNA without inducing RNase H-dependent cleavage and obstruct the interaction between the target RNA and its RNA/protein partners [4]. In certain cases, SSOs lead to splicing modulation of pre-messenger RNA (pre-mRNA) transcripts [2,4]. In addition to binding to RNA targets, oligonucleotides can be designed to function as aptamers, which bind to protein targets via their secondary structure [4]. They can be selected through a PCR-based process named Systematic Evolution of Ligands by EXponential Enrichment (SELEX) [5].

As of December 2023, there are 16 approved oligonucleotide therapeutics by the Food and Drug Administration (FDA) [2], as outlined in Table 13.3.

As most therapeutic oligonucleotides engage with targets by binding to intracellular RNA, robust efficacy requires efficient delivery of them into target cells and then into the cytoplasm and/or nucleus. As a result, initial clinical success of approved oligonucleotide drugs focused on their delivery via direct, local administration in the treatment of ocular and central nervous system (CNS) disorders: intrathecal injection in neurological disorders, e.g. nusinersen, or intravitreal injection in ocular disorders, e.g. fomivirsen [2,3]. In recent years, significant advancements in hepatic delivery platforms such as GalNAc- or lipid nanoparticle-based systems [3–5] have been made. These advancements enabled multiple approved treatments by addressing hepatic targets via systemic administration. To overcome the limitations of poor target tissue distribution, limited cellular uptake, and low endosomal escape, CPPs were subsequently investigated to enhance the delivery of oligonucleotide cargos [120]. CPPs may

Table 13.3 FDA-approved oligonucleotide therapeutics.

Mechanism of action	Oligo	Name	Indications
RNA interference	siRNA	Patisiran [103]	Hereditary transthyretin-mediated amyloidosis
		Givosiran [104]	Acute hepatic porphyria
		Lumasiran [105]	Primary hyperoxaluria Type 1
		Inclisiran [106]	Familial hypercholesterolemia
		Vutrisiran [107]	Hereditary transthyretin-mediated amyloidosis
RNase H	ASO	Tofersen [108]	*SOD1* ALS
		Inotersen[a] [109]	Hereditary transthyretin-mediated amyloidosis
		Fomivirsen[a] [110]	Cytomegalovirus retinitis
		Mipomersen[a] [111]	Familial hypercholesterolemia
		Eplontersen [112]	Hereditary amyloid transthyretin amyloidosis
Splice modulation	ASO	Nusinersen [113]	Spinal muscular dystrophy
		Eteplirsen [114]	DMD exon 51 skipping
		Golodirsen [115]	DMD exon 53 skipping
		Viltolarsen [116]	DMD exon 53 skipping
		Casimersen [117]	DMD exon 45 skipping
Protein binding	Aptamer	Pegaptanib[a] [118]	Age-related macular degeneration
		Avacincaptad pegol [119]	Geographic atrophy

ASO, antisense oligonucleotides; DMD, Duchenne muscular dystrophy; FDA, Food and Drug Administration; SOD1 ALS, superoxide dismutase 1 gene mutated amyloid lateral sclerosis; siRNA, small interfering RNA.

[a]Drug discontinued.

Source: Adapted from *Cell Chemical Biology*, Volume 31/Issue 1, Hofman et al., Targeting RNA with synthetic oligonucleotides: clinical success invites new challenges. Ref. [2]/with permission of Elsevier.

be connected with oligonucleotides either through non-covalent complexation or covalent conjugation. In the case of negatively charged oligonucleotide cargos (e.g. siRNAs and gapmers), their negatively charged backbones (e.g. phosphodiester or phosphorothioate) may non-covalently bond to positively charged cationic CPPs through charge–charge interactions, leading to complexation into particles that have translocation potential across the cell membrane [121]. On the other hand, neutrally charged backbones, such as phosphorodiamidate bonds in phosphorodiamidate morpholino oligomers (PMOs) and amide bonds in PNAs, are suited for covalent conjugation with CPPs [121]. Though PMOs and PNAs have high biochemical affinity and stability, these neutrally charged molecules generally have limited cellular uptake and endosomal escape, thus having low efficacy *in vivo* [44,121–123]. For the rest of the chapter, we will review CPP-mediated delivery of therapeutic PMO and PNA oligonucleotides through covalent conjugation.

13.3.2 Development of CPPs for PMO Delivery

PMOs are artificial, neutrally charged DNA analogs, in which a morpholine ring replaces ribose and phosphorodiamidite replaces phosphodiester [123]. PMOs have been studied widely and show great promise in the regulation of gene expression through steric blockage of complementary RNA sequences [123,124]. PMOs include a number of clinically approved oligonucleotides (e.g. eteplirsen, golodirsen, viltolarsen, and casimersen, Table 13.3) that act via the mechanism of SSO in the treatment of Duchenne muscular dystrophy (DMD). However, despite these advancements, PMOs do not efficiently enter cells unaided as they have limited protein and membrane interactions [125]. Conjugation of PMOs with CPPs has been employed as an approach to tackle this challenge.

Many CPPs have been evaluated in both preclinical and clinical studies (Tables 13.2 and 13.4), with most sequences containing a high percentage of cationic and hydrophobic residues [126–129]. While some of these CPPs are naturally derived, most are rationally designed with some of which having a cyclic structure [44,130,131]. Seminal work reported that a poly-lysine peptide delivering PMO restored enhanced GFP (EGFP) protein expression in cells carrying EGFP interrupted by the human β-globin IVS2-654 mutation [132,133]. Follow-up studies pioneered applying polyarginine-derived CPPs, such as R9F2 (RRRRRRRRRFF, Table 13.4), to deliver PMO in multiple models [126–129]. In another study, peptides composed of six to eight arginines spaced with 6-aminohexanoic acid (X) and β-alanine (B) residues have shown high delivery efficacy [134]. Among these arginine-rich CPP sequences, PepB peptide (RXRRBRRXRRBRXB, Table 13.4) was widely studied due to its high delivery efficacy [134–139].

The development of CPP-PMO conjugates was further advanced by Gait and Wood groups with the introduction of PNA-internalization peptides (Pips). Although initially designed for PNA delivery, Pips have been extensively applied to deliver PMOs. Pip1 (RXRRXRRXRIKILFQNRRMKWKK, Table 13.4) was developed from penetratin fused with hexa-arginine subsequently replaced by a R-X-R motif and the hydrophobic region shortened to IKILFQN [146]. Since the initial proof-of-concept study on Pip1, further optimizations were performed to screen for hydrophobic region sequences and arginine-rich domains. This eventually led

Table 13.4 Representative CPPs used for the delivery of therapeutic oligonucleotides.

Peptides	Sequences	Applications
PepB	RXRRBRRXRRBRXB	DMD [135,136,140] Ataxia-telangiectasia [138] SMA [141] Pompe disease [142] Antiviral [137]
PepB-MSP	RXRRBRRXRRBRXB-ASSLNIAX	DMD [143,144]
R6G	RRRRRRG	HGPS [145]
Pip1	RXRRXRRXRIKILFQNRRMKWKK	DMD [146]
Pip5	A family of sequences with Pip5e: RXRRBRRXRILFQYRXRBRXRB	DMD [147]
Pip6	A family of sequences with Pip6a: RXRRBRRXRYQFLIRXRBRXRB Pip6aKC: KXKKBKKXKYQFLIKXKBKXKB	DMD [148–153] DM1 [154,155] miRNA regulation [156] SMA [157]
Pip9b2	Ac-RXRRBRRFQILYRBRXRB	DM1 [155] DMD [158]
EEV2	Ac-PKKKRKV-K(*cyclo*[FfΦRrRrQ])	DMD [44] FECD [131] FSHD [44]
Br-ApoE	(Ac-LRALRARLLRGG)$_2$KX-Bpg-G	SMA [159]
DG9	YArVRRrGPRGYArVRRrGPRr	SMA [160]
R9F2	RRRRRRRRRFF	Antiviral [127–129,161–163]
(RXR)$_4$	RXRRXRRXRRXR	Antiviral and antibacterial [164–166]
Tat	GRKKRRQRRRPPG	HBV [167]
R8	RRRRRRRR	miRNA regulation [156,168]

Φ, L-2-naphthylalanine (Nal); Ac, acetyl; B, β-alanine; Bpg, bishomopropargylglycine; CPP, cell-penetrating peptide; DM1, myotonic dystrophy type 1; DMD, Duchenne muscular dystrophy; f, D-phenylalanine; FECD, Fuchs endothelial corneal dystrophy; FSHD, facioscapulohumeral muscular dystrophy; HBV, hepatitis B virus; HGPS, Hutchinson–Gilford progeria syndrome; miRNA, microRNA; MSP, muscle-targeting heptapeptide; pip, PNA-internalization peptide; PNA, peptide nucleic acids; r, D-arginine; SMA, spinal muscular atrophy; X, 6-aminohexanoic acid.

to the generation of Pip5 [147] and Pip6 [148] series, which delivered PMOs with high efficiency in *mdx* mice, a mouse model for DMD. Further optimizations were carried out to reduce the number of arginine residues, resulting in the Pip9 [169] series (Table 13.4).

The synergistic effects of chimeric CPPs, which involve the fusion of multiple CPPs into one single sequence, were recently reported [170]. Several chimeric sequences were evaluated in the HeLa cell line harboring the EGFP-654 reporter gene discussed earlier. For example, penetratin-Bpep (RQIKIWFQNRRJKWKK-RXRRBRRXRRBR) achieved an almost 70-fold increase in EGFP restoration compared with PMO alone and outperformed individual CPPs by ~1.5-fold [170]. The same group also utilized a machine-learning approach to predict the effectiveness of CPP sequences for PMO delivery [171]. By synthesizing and evaluating a library of 64 CPP-PMO conjugates, a computational model was built. They synthesized and tested seven novel sequences generated from this predictive model and confirmed that all computationally predicted sequences were effective in the EGFP-654 reporter cell line. These findings underscore the potential of *in silico* tools in optimizing CPP sequences for oligonucleotide delivery.

As discussed in Section 13.2.2, cyclic CPPs have shown improved proteolytic stability, enhanced cellular uptake, and endosomal escape when conjugated to peptides and other biomolecules [20,172,173]. To overcome current limitations of oligonucleotide therapeutic delivery, cyclic CPPs were leveraged and further developed as the Endosomal Escape Vehicle (EEV™) (Entrada Therapeutics, Boston, MA, USA) technology [44,131]. Recently, a novel CPP was developed by combining the CPP12 sequence (Table 13.1) with an NLS (PKKKRKV), and the resulting peptide, EEV2 (Ac-PKKKRKV-K[*cyclo*(FfΦRrRrQ)], Table 13.4), was shown to efficiently deliver oligonucleotides in several disease models [44,131]. Representative CPPs and their applications for the delivery of therapeutic oligonucleotides are summarized in Table 13.4.

13.3.3 Applications of CPP-PMOs to Modulate Pre-mRNA Splicing

CPP-PMOs have been widely studied for the regulation of pre-mRNA splicing. This approach demonstrated its potential in achieving both exon skipping and exon inclusion, showing promise as a therapeutic strategy for addressing various devastating diseases, including DMD and spinal muscular atrophy (SMA). Initial proof-of-concept studies to demonstrate the efficacy of CPP-PMOs were performed in HeLa cell lines harboring cryptic splicing sites in reporter genes (EGFP or luciferase) [126,132,170,174,175]. For instance, Moulton and colleagues examined a series of arginine-rich CPPs and observed enhanced splicing activity upon peptide conjugation [126]. Among the CPPs studied, the most effective CPP, R9F2, was shown to correct mis-splicing of luciferase pre-mRNA at low micromolar concentrations. Notably, the R9F2 peptide outperformed other CPPs, including Tat and penetratin peptides, underlining the additive effects of arginine and hydrophobic residues for effective intracellular delivery [126]. Additional cationic CPPs were evaluated for their efficacy to deliver PMO in the EGFP-654 transgenic mice that harbors an aberrant splicing of *EGFP* transcript [134]. Of all tested CPP-PMOs,

PepB (Table 13.4) emerged as the most encouraging candidate. In EGFP-654 mice, intravenous administration of PepB-PMO at 12 mg/kg daily for 4 days led to 90% splice correction in the heart 1 day after the final injection. One hundred percent splice correction could be detected in diaphragm and quadriceps tissue for up to 3 and 12 weeks, respectively [134].

13.3.3.1 Duchenne Muscular Dystrophy

DMD is a genetic disorder marked by progressive muscle degeneration due to dystrophin gene mutations and subsequent loss of functional dystrophin protein. To develop therapeutics for DMD, PMOs have been designed to bind dystrophin pre-mRNA and restore the reading frame [176] by exon skipping, ultimately leading to the production of an internally truncated yet functional dystrophin protein [114–117]. Although several unconjugated PMO-based therapies for DMD are already approved [114–117] (Table 13.3), they are limited by their efficacy, exemplified by the poor restoration of dystrophin protein levels [122]. To address this challenge, studies have been conducted to improve intracellular delivery of PMO by conjugation with CPPs.

As mentioned earlier, PepB has been the subject of intense investigations for PMO delivery. Wu et al. reported the conjugation of PepB with PMO-M23D, a PMO designed to skip mouse *Dmd* exon 23 that contains a premature stop codon and is used in the *mdx* mouse model of DMD to evaluate exon skipping potential [135]. A single intravenous injection of PepB-PMO-M23D at a dose of 30 mg/kg into adult *mdx* mice resulted in robust dystrophin expression in 100% of fibers across all studied skeletal muscles, a significant improvement compared to the 5% or less observed after the administration of naked PMO-M23D. Efficient dystrophin restoration was also observed with Western blot analysis. A biweekly intravenous injection for 3 months led to near-normal dystrophin expression in skeletal muscles, significant dystrophin expression levels in cardiac muscle, and improved muscle pathology. No pathological changes were detected in the liver or kidneys 2 weeks after the last injection [135].

The same group carried out a 12-month study to assess the therapeutic potential of PepB-PMO-M23D [139]. Biweekly injections of 6 mg/kg PepB-PMO-M23D for 12 months resulted in over 20% dystrophin expression across all skeletal muscles. Monthly 30 mg/kg injections of PepB-PMO-M23D restored dystrophin to over 50% of normal levels in skeletal muscle and 15% in cardiac muscle. In both cases, improvements in muscle function and pathology, as well as a decrease in serum creatine kinase, a biomarker for muscle damage, were observed. The acute toxicity of PepB-PMO-M23D was also evaluated. All *mdx* mice survived after a dose of 30 mg/kg, with mouse mortality only observed in the 60 mg/kg group.

In addition to PepB, Yin et al. further designed a fusion peptide combining PepB with a muscle-targeting sequence identified by phage display and tested the resulting PepB-muscle-targeting heptapeptide (MSP) (RXRRBRRXRRBRXB-ASSLNIAX, Table 13.4) for its delivery of PMO-M23D in *mdx* mice [143]. PepB-MSP-PMO-M23D exhibited significantly higher efficacy than PepB-PMO-M23D at low doses. After six weekly intravenous injections at 3 mg/kg, PepB-MSP-PMO-M23D resulted in ~5% dystrophin restoration of normal level in skeletal muscles versus only about ~1% observed in the PepB-PMO-M23D group. After three weekly intravenous injections at 6 mg/kg, *mdx* mice treated with PepB-MSP-PMO-M23D were associated with up to 25% restoration of the normal dystrophin protein levels in skeletal muscles, compared with ~10% achieved with PepB-PMO-M23D. These encouraging results prompted additional characterization of this chimeric peptide using the murine *mdx* model [144,177].

Based on the hexa-arginine-penetratin fusion peptide, groups led by Gait and Wood employed the highly efficacious arginine-rich R-X-R motif and optimized the hydrophobic region of penetratin and developed Pips, such as Pip5 series [147] and Pip6 series [148] (Table 13.4). The Pip5 series contains a central ILFQY hydrophobic region flanked by domains consisting of R, X, and B residues, including Pip5e (RXRRBRRXRILFQYRXRBRXRB, Table 13.4) [147]. When Pip5 were used to deliver PMO-M23D in the *mdx* model, high exon-skipping efficacy and protein restoration were achieved. Single intravenous administration of Pip5e-PMO-M23D conjugate at 25 mg/kg was associated with significant restoration of skeletal and cardiac dystrophin expression in *mdx* mice. Additional improvements were made to its hydrophobic residues and generated the Pip6 series (Table 13.4) [148]. When Pip6 were tested for their delivery of PMO-M23D, it was discovered that the hydrophobic core region is vital for dystrophin production in cardiac tissue. Of the various Pip6 analogs, Pip6a (RXRRBRRXRYQFLIRXRBRXRB, Table 13.4) conjugated PMO showed greater dystrophin production than Pip5e-PMO [148]. In attempts to further reduce the toxicity of Pip-series, further optimizations continued to reduce the number of arginine residues, resulting in several improvements, such as Pip9b2 (RXRRBRRFQILYRBRXRB, Table 13.4) [169]. Additionally, Pips for the functional delivery of PMO in various DMD models have been extensively evaluated [149–155,157,158,178,179]. Further developments of Pip peptide-based PMO conjugates led to multiple clinical programs (Table 13.3).

In a recent study, RC-1001, a peptide-PMO conjugate composed of a proprietary CPP and a PMO targeting *Dmd* mutation in *mdx* mice, was reported [180]. Single doses of RC-1001 at 80 mg/kg led to 94% exon skipping and 27% wild-type dystrophin protein expression in biceps tissue. Dystrophin protein levels were sustained for 60 days. Dose-dependent improvements in muscle function were also observed. Furthermore, three monthly doses resulted in exon skipping of 99% and dystrophin protein restoration of 39% in quadriceps tissue. Given the promising results supporting monthly dosing, the same peptide was applied to develop vesleteplirsen [180], a CPP-PMO designed for treating exon 51 skipping amenable

DMD patients, which is currently being evaluated in patients with DMD [180–182] (Table 13.3). Preliminary clinical trial results showed greater exon 51 skipping and dystrophin production with the CPP-PMO compared to an unconjugated exon 51 skipping PMO [183]. However, further development of vesleteplirsen (SRP-5051) is discontinued due to long-term safety and tolerability issues [184].

In addition to the aforementioned studies employing canonical linear CPPs, EEV platform has been applied to enhance intracellular delivery of PMO for DMD [44]. Leveraging the advantages of cyclic CPPs, the optimized peptide, EEV2 (Table 13.4) was conjugated with PMO-M23D. The resulting EEV2-PMO-M23D was associated with >70% exon skipping across skeletal muscles and, notably, 61% in cardiac muscle 1 week after a single 20 mg/kg intravenous injection in *mdx* mice. A regimen of four weekly intravenous injections at 10 mg/kg also resulted in significant exon skipping (>30% in the heart and >50% in skeletal muscles) and restoration of dystrophin protein (>30% in all evaluated tissues) 1 week after the last injection, while PMO-M23D alone resulted in negligible effects. Moreover, a marked reduction in serum creatine kinase levels was observed, indicating restored muscle integrity [44]. These findings underscore the therapeutic potential of the EEV platform in a broad range of neuromuscular applications and led to several ongoing clinical programs, including for ENTR-601-44 for exon 44 skip-amenable DMD, which has since reported positive data from the Phase 1 ENTR-601-44-101 trial [185] (Table 13.2).

13.3.3.2 Spinal Muscular Atrophy

SMA is a severe neurodegenerative disease caused by mutations in the survival motor neuron 1 (*SMN1*) gene. This leads to a deficiency of the SMN protein, which is crucial for the functioning and survival of motor neurons [186]. One effective strategy is through the use of nusinersen, an unconjugated ASO approved for the treatment of SMA in pediatric and adult patients [113]. This compound enhances exon 7 inclusion of *SMN2* gene, the second paralog gene of *SMN*, and restores production of the full-length SMN protein. Therapeutically, ASOs typically do not cross the blood–brain barrier, and nusinersen is administered intrathecally [113,187]. The same mechanism of action may also be exploited by a CPP-PMO modality, and the studies discussed below examined whether this approach could increase the potency of treatment or leverage less invasive administration routes [141,157,159,160,188,189].

Four cationic CPPs (Tat, R6, r6, and PepB) were conjugated with PMOs targeting the ISS-N1 element of *SMN2* intron 7 and screened for SMN protein restoration after neonatal injections in SMNΔ7 mice, a murine model of SMA [141]. PepB- and r6-PMO conjugates were selected for further characterization due to their high efficacy. After intraperitoneal injection in a disease model at postnatal day 5, both PepB-PMO and r6-PMO significantly enhanced SMN protein expression in the CNS as well as significantly enhanced the survival and functional status of symptomatic SMA mice. In comparison with unconjugated PMO, median survival increased from 17 days to 41.4 days in the r6-PMO group and to 23 days in the PepB-PMO group. Additionally, improved numbers of innervated neuromuscular junctions and motor neurons were observed at P10 and were sustained up to P30 after the single dose. However, when the same treatment was administered at P7 and P10, no improvement in survival was observed compared with mice treated with the naked PMO.

Pip6a-conjugated PMO (Pip6a-PMO), targeting the ISS-N1 element of *SMN2* intron 7, was shown to be efficiently delivered following systemic administration [157]. Pip6a-PMO was first assessed in Taiwanese severe SMA mice. A single facial vein dose of Pip6a-PMO was associated with greater self-righting behavior and strength, and significantly enhanced full-length *SMN* mRNA and SMN protein expression in all tissues 7 days after neonatal injection. Single doses of Pip6a-PMO dose-dependently increased median survival. Two doses of 5 and 10 μg/g of Pip6a-PMO extended median survival to 283 days and 457 days, respectively. In an adult SMA murine model carrying the human *SMN2* transgene, two intravenous doses of Pip6a-PMO at 18 μg/g significantly increased full-length *SMN2* transcript levels in all studied parts of the brain, spinal cord, skeletal muscle, and liver. Increased SMN protein levels were detected only in skeletal muscle and the liver, potentially due to limited delivery to CNS tissues [157].

The same research team continued to investigate other CPPs to enhance the CNS delivery of PMO [159]. For instance, a branched derivative of apolipoprotein E sequence, Br-ApoE ($[\text{Ac-LRALRARLLRGG}]_2$KX-Bpg-G, where Bpg refers to bishomopropargylglycine, Table 13.4), was developed and conjugated with a PMO targeting *SMN2*. Br-ApoE-PMO was able to increase levels of full-length *SMN2* transcript in the brain, kidney, and quadriceps after two systemic doses at 10 mg/kg in a neonatal model of SMA. Treatment also significantly increased the median lifespan from 10.5 to 78 days. Weight gain, righting reflex, and strength measures were also improved. In an adult mouse model, six weekly administrations of the conjugate at 10 mg/kg resulted in a significant increase in the level of *SMN2* transcript in the CNS and peripheral tissues, but significant SMN protein restoration was only observed in the kidney.

In a recent study from Yokota's group, a novel peptide derived from human polyhomeotic 1 homolog transcription factor, DG9 peptide (YArVRRrGPRGYArVRRrGPRr, Table 13.4) was conjugated with a PMO derived from a sequence identical to nusinersen and used to examine restoration of SMN protein expression in Taiwanese severe SMA mice [160]. After neonatal administration of one subcutaneous dose at 40 mg/kg, median survival increased from 12 days for the unconjugated PMO group to 58 days for DG9-PMO treated group. The median survival increased from 57 days for unconjugated PMO to 121 days for DG9-PMO after a single subcutaneous dose at 80 mg/kg. No toxicity or immune dysfunction was observed. Furthermore, the treated mice exhibited mitigated disease phenotypes, including improved muscle strength, pathology and respiratory parameters [160].

Another study from Hammond et al. explored the delivery of the PMO cargos across the blood–brain barrier using an antibody targeting the murine CD71, which further attached to the PMO via a cationic CPP linker [190]. When administered systemically in an adult SMA murine model carrying the human *SMN2* transgene, this conjugate demonstrated significantly increased bioavailability and increased SMN protein expression in the brain. In further investigations, extended median survival was demonstrated in a neonatal mouse model of severe SMA, compared with vehicle alone or the same CPP-PMO conjugated with an isotype control antibody [190].

13.3.3.3 Other Indications

Besides applications in DMD and SMA, CPP-PMOs have been explored to modulate aberrant splicing and thereby treat other diseases, including improving the splicing correction efficiency of the *ATM* gene as a potential treatment of ataxia-telangiectasia [138], inhibition of p53 via alternative splicing in human tumor cells [191], and correction of splicing defects in cystic fibrosis transmembrane conductance regulator in the potential treatment of cystic fibrosis caused by 3849 + 10 kb C → T mutation [192].

One extensive case study used R6G (RRRRRRG, Table 13.4) peptide and a peptide-PMO conjugate, SRP-2001, was developed for treating Hutchinson–Gilford progeria syndrome (HGPS) [145]. HGPS is a rare genetic disorder marked by accelerated aging in children and characterized by premature death from myocardial infarction or stroke [193]. Approximately 90% of HGPS cases are caused by a rare single-nucleotide mutation in the *LMNA* gene that encodes the lamin A protein [194]. This results in a cryptic splice donor within exon 11 of *LMNA* and the production of a toxic form of lamin A, termed progerin [193,194]. In this recent study, researchers developed an antisense PMO blocking the cryptic splicing site and tested SRP-2001, which comprised a CPP and the functional PMO targeting the cryptic splicing site [145]. The treatment of human HGPS fibroblasts with SRP-2001 for 2 weeks reduced progerin transcript levels by 92%, decreased progerin protein expression, and restored cellular proliferation. A longevity study was subsequently carried out in the HGPS disease model ($LMNA^{G/G}$ mice). Long-term intravenous administration of SRP-2001 at 60 mg/kg led to an over sevenfold increase in vascular smooth muscle cell density, a 53% reduction in adventitial area, and a 62% increase in lifespan. These results indicated that SRP-2001 can successfully rescue vascular smooth muscle cell loss and reduce vessel damage after long-term systematic administration, providing promising insights into the disease biology and a basis for potential evaluation in the clinic [145].

Another extensive study used CPP-PMO conjugate as a potential treatment for glycogen storage disease type II (Pompe disease) [142]. Pompe disease is an autosomal recessive disorder caused by a deficiency in acid alpha-glucosidase (GAA), a lysosomal enzyme that catalyzes the breakdown of glycogen to glucose, ultimately resulting in glycogen accumulation in muscles and impairing their function [195]. Clayton and colleagues developed a CPP-PMO as a substrate reduction therapy for Pompe disease by targeting the glycogen synthase (Gys1), which is critical for glycogen synthesis [142]. The PepB-conjugated PMO was designed to skip exon 6 of the *Gys1* transcript, create a premature stop codon, and effectively downregulate Gys1 levels. Biweekly administration of CPP-PMOs intravenously to GAA-deficient mice ($Gaa^{-/-}$ mice) at 15 or 30 mg/kg for 12 weeks dose-dependently reduced glycogen synthase transcripts and glycogen accumulation in both the quadriceps and the diaphragm [142].

13.3.4 Applications of CPP-PMOs to Target mRNA with Aberrant Repeat Expansions

Short tandem repeats (STRs) of DNA motifs (usually 1–6 bp) are abundant throughout the genomes of many species, including humans [196,197]. STRs are a large source of genetic variation. The lengths of STRs frequently fluctuate due to contractions and expansions during cell division [197]. Coding and noncoding STRs have been found to play crucial roles in cellular function, including the regulation of gene expression. However, the high mutation rate inherent in STRs makes those repeats beyond a certain length highly unstable, and the resultant large repeat expansions may cause a range of diseases, including degenerative CNS disorders [197–199], such as Huntington's disease. A promising approach for addressing such diseases involves using ASOs to target mRNA transcripts containing repeat expansions, consequently disrupting these transcripts and restoring normal cellular functions [131,154,200,201].

Myotonic dystrophy Type 1 (DM1) is one of the most prevalent neuromuscular disorders in adults and is caused by pathological expansions of CTG repeats in the 3′-untranslated region (3′ UTR) of the dystrophia myotonica protein kinase transcript (*DMPK*), over 50 of which lead to disease symptoms [202]. The resultant mutant CUG repeat-containing *DMPK* transcripts can sequester muscleblind-like (MBNL) proteins, such as MBNL1 and MBNL2 proteins and other splicing factors, form nuclear foci, and result in mis-splicing in dozens of downstream transcripts [202]. Klein et al. first demonstrated that pathologic *DMPK* transcripts can be targeted using a $(CAG)_7$ PMO (PMO-CAG) conjugated with Pip6a in an allele-specific manner [154]. In a DM1 mouse model (HSA^{LR} mice), a single intravenous injection of 12.5 mg/kg Pip6a-PMO-CAG led to significant correction of MBNL1-dependent splicing defects in skeletal muscle at 2 weeks post injection.

In contrast, no discernible effect was observed with high doses of unconjugated PMO. An analysis of the global transcriptome profile also revealed a correction in wild-type mice after Pip6a-PMO-CAG treatment, with an average of 76% for the splicing correction index. Furthermore, a significant portion of these splicing corrections remained evident 6 months after treatment, indicating sustained activity for the CPP-PMO. These results indicated that Pip6a-PMO-CAG was able to target the disease-causing transcript, release MBNL1 from nuclear foci, and restore its subsequent mis-splicing [154]. Another CPP-PMO, VX-670, is currently under investigation for the treatment of DM1. Like Pip6a-PMO-CAG, VX-670 targets *DMPK* via a CUG-repeat steric blocking approach and has demonstrated splicing correction in animal models [203]. A Phase 1/2 clinical trial for VX-670 in patients with DM1 is ongoing [204].

A similar approach has been developed for Fuchs endothelial corneal dystrophy (FECD), which is a leading cause of vision loss [205]. FECD is caused by intronic CTG repeat expansion in the transcription factor 4 (*TCF4*) gene [206]. In an effort to treat FECD, a CPP-PMO conjugate was developed based on the EEV platform, wherein EEV2 (Table 13.4) was conjugated with a PMO-CAG designed to target CUG-repeat expansion and block the formation of nuclear foci [131]. The cellular uptake of EEV2-PMO-CAG was assessed in a patient-derived corneal endothelial cell line containing ~1500 CUG repeats. EEV2-PMO-CAG demonstrated high delivery efficiency and significantly reduced the number of cells with nuclear foci and the number of foci per cell [131].

13.3.5 Applications of CPP-PMOs to Inhibit mRNA Translation

The concept of using ASOs to bind with target mRNA and block protein translation was initially demonstrated in 1978. In a pioneering report, Zamecnik and Stephenson demonstrated that an artificial tridecamer DNA sequence could hybridize with complementary viral RNA, and the resulting hybrid could block viral gene expression, thereby preventing viral replication [207]. CPP-PMOs have been explored in the inhibition of mRNA translation, especially in antiviral and antibacterial applications [127–129,137,161–166,208–226]. In these studies, PMOs were designed and screened by targeting different mRNA regions to inhibit the translation process.

A wide variety of CPPs have been investigated, with those incorporating peptide $(RXR)_4$ (RXRRXRRXRRXR) [164,165,208–212,215–225] and peptide R9F2 [163,211,218] (Table 13.4) being among the most characterized. For instance, the CPP-PMO approach was employed to target coronavirus replication, a goal that had previously proved elusive [127]. In this *in vitro* study, PMO sequences were screened and optimized for antiviral activity in cell-free cultures of mouse hepatitis virus (a member of the coronavirus family). A PMO targeting the start codon of a viral polyprotein gene was conjugated with R9F2 (R9F2-PMO). Under serum-free cell culture conditions, R9F2-PMO was shown to be delivered to >80% of transfected murine cells at 10 μM. The conjugate was also shown to exhibit a sequence-specific, dose-responsive inhibition of virus titers, resulting in reductions up to 100-fold. This level of viral inhibition was on par with that of the aminoglycoside hygromycin B, the only previously known inhibitor of coronavirus infection. However, R9F2-PMO was ineffective when tested in the presence of culture medium containing 10% fetal bovine serum, suggesting that this conjugate is likely to have little activity for *in vivo* applications [127].

In a recent study, CPP-PMOs were developed with the goal of inhibiting the growth of severe acute respiratory syndrome coronavirus 2 (SARS-CoV-2) [166]. Here, the $(RXR)_4$ peptide was coupled with five PMOs designed to target either the 5′ UTR or the leader transcription regulatory sequence region of SARS-CoV-2 positive-sense genomic RNA. Four out of the five CPP-PMOs demonstrated high efficacy in monkey renal cell cultures, reducing viral titers by 4–6 $\log_{10}$ 48–72 hours post infection in a nontoxic and dose-responsive manner [166]. Both studies suggest that CPP-PMO is a promising approach to treating patients with viral infections.

13.3.6 Applications of CPP-PMOs to Modulate RNA Levels

Several research groups have reported applying CPP-PMOs to modulate RNA levels by targeting the UTR of mRNA transcripts, such as the polyadenylation (polyA) signal [44] or inhibiting microRNA (miRNA) and UTR interactions [179]. Facioscapulohumeral muscular dystrophy (FSHD) is an autosomal dominant muscular dystrophy caused by genetic mutations that lead to the misexpression of the distal double homeobox protein 4 (DUX4) in muscle [227]. Suppressing DUX4 expression is a potential approach to treating patients with FSHD, and PMOs have been developed to block mRNA polyadenylation sites, decrease mRNA stability, and downregulate DUX4 levels [228–230]. Chen et al. designed several PMOs targeting the *DUX4* transcript, the most efficacious sequence of which, PMO-FM10, targets the polyadenylation signal of *DUX4* [231]. After a 4-day treatment of FSHD myotubes with 10 μM FM10 by free uptake, a notable reduction was observed in both DUX4 protein and downstream gene expression [231]. Subsequent PMO optimization led to the development of PMO-PACS4. This was conjugated to an octa-guanidine dendrimer to enhance delivery efficiency, and the resulting vivo-PMO-PACS4 reduced *DUX4* and downstream mRNA levels by 50% after four weekly intraperitoneal treatments in a murine model of FSHD [229].

In a recent study, the *DUX4*-targeting PMO-FM10 was conjugated with EEV2 peptide (Table 13.4), and the EEV2-PMO-FM10 conjugate was evaluated in two FSHD patient-derived cell lines. A dose-dependent reduction in *DUX4* level and its downstream gene expressions were observed after 24 hours of treatment at low micromolar concentrations [44]. These examples confirmed the potential to downregulate a target gene by targeting the polyA site with CPP-PMO strategy.

In order to upregulate a target transcript, PMOs have been developed to target miRNA or its binding sites on the mRNA [232]. miRNAs are a class of small noncoding RNA molecules that are able to interact with the target transcript and result in mRNA degradation [233]. Several CPP-PMOs have been reported to block miRNA and mRNA interactions [155,179,234–240]. In one such study, Pips (e.g. Pip9b2, Table 13.3) were conjugated with anti-miRNA PMO sequences (antimiR-23b or antimiR-218). Both miR-23b and miR-218 were upregulated in DM1 and implicated in the degradation of *MBNL* transcripts [155]. These CPP-PMOs were found to elevate MBNL protein expression following free uptake in cellular models of DM1 [155]. They were subsequently shown to enhance molecular, histopathological, and functional characteristics of skeletal muscle without evident toxicity after a single intravenous administration at 9.2 mg/kg in an HSA^{LR} murine model of DM1. All conjugates significantly decreased levels of miR-23b and miR-218, significantly increased both *Mbnl1* transcript and MBNL1 protein levels in gastrocnemius and quadriceps, and decreased the pathologic presence of nuclear foci [155].

One potential challenge with targeting miRNAs is that specificity could be limited as a single miRNA can act on a broad range of transcripts [232]. To circumvent this, a PMO was designed to bind with the miR-23b's binding site on the 3′ UTR regions of *Mbnl1* and *Mbnl2* transcripts. The resulting PMO was conjugated with Pip9b2 (Table 13.4) and tested in DM1 models [179]. In DM1 patient-derived myotubes, and following free uptake of the conjugate, cells exhibited an increase in MBNL1 protein immunofluorescence and significant correction of mis-splicing. This Pip9b2-PMO was then characterized in HSA^{LR} DM1 mice. Four days after the single intravenous dose at 10 mg/kg, mice treated with Pip9b2-PMO exhibited significant correction of abnormal splicing in downstream *Mbnl* transcripts, increased MBNL1 protein levels, reduced nuclear foci in quadriceps muscle fibers, and significantly improved grip strength [179].

13.3.7 Applications of CPP for PNA Delivery

In addition to PMO cargos, PNAs have been conjugated with CPPs for various *in vitro* and *in vivo* applications. PNAs are synthetic, neutrally charged nucleic acid oligomers in which the sugar-phosphate backbone is replaced by a peptide structure [241,242]. As PNAs have limited cellular uptake, studies have examined conjugation with CPPs to improve their intracellular delivery [242].

Some early studies investigated the functional delivery of CPP-conjugated PNAs by measuring the extent of splicing correction and subsequent restoration of luciferase expression in a cell line with an interrupted luciferase gene [146,243–247].

Several CPPs, including transportan, were linked to PNAs with varying conjugation chemistries (continuous synthesis, maleimide coupling, and ester or disulfide linkage) [243]. It was found that the conjugation chemistry of the CPP-PNA markedly affected delivery efficiency, with disulfide linkage conferring the highest efficiency. Interestingly, the presence of serum markedly inhibited the delivery efficiency of CPP-PNAs. Pips have also been tested for PNA delivery and evaluated in the modulation of splicing in the *mdx* murine model [146]. Pip-PNA conjugates were designed to skip exon 23 in the mouse dystrophin gene, and local injection of these conjugates into the tibialis anterior muscles in *mdx* mice resulted in approximately threefold higher numbers of dystrophin-positive fibers than the PNA-alone treatment group [146].

In addition to splicing modulation, CPP-PNAs have also been investigated as a method of targeting promoter regions of target genes [248], inhibiting miRNA [41,156,168,249], and inhibiting bacterial and viral RNA [167,250–254]. In one study, anti-gene PNAs were shown to inhibit progesterone receptor expression in a breast cancer cell line [248]. Inhibition of gene expression was observed with cationic CPP-conjugated PNAs, with conjugates containing more than eight arginine or lysine residues showing the highest potency. CPP-PNAs have been developed to target miRNA-155 with either linear or cyclic CPPs for their anti-miRNA activity [41,156]. In one case, the PNA sequence was designed to inhibit miRNA-155, an immune-associated noncoding RNA regulating inflammatory activity [255,256], and the PNA was conjugated to different CPPs to enhance uptake [156]. R8 (RRRRRRRR, Table 13.4) peptide was able to deliver the PNA cargo to ~100% cells at 200 nM and achieved global de-repression of the miRNA-155 regulome as evaluated by RNA sequencing [156].

Several *in vivo* studies have been performed to evaluate the delivery of CPP-PNAs [21,146,167,257,258]. One example evaluated the delivery of an antisense PNA targeting direct repeat of the hepatitis B virus (HBV) gene by Tat sequence (GRKKRRQRRRPPG, Table 13.4). This Tat-PNA conjugate inhibited the expression of HBV proteins and the synthesis of both HBV DNA and RNA in *in vitro* and *in vivo* models. Intravenous administration of Tat-PNA in an acute hepatitis B mouse model reduced serum levels of HBeAg, indicating significant viral clearance [167].

13.4 Future Perspectives and Summary

Over the past few decades, there has been significant growth in the field of delivering therapeutic oligonucleotides via conjugation to CPPs. This growth in research has affirmed the advantages of CPPs as an emerging approach for delivering the oligonucleotide modality and spurred multiple clinical studies in a broader spectrum of indications (Table 13.2). Nevertheless, several significant challenges remain under evaluation, presenting obstacles to the wider application of this strategy. (i) Although CPP conjugation significantly enhances delivery efficiency, there is concern over reports of dose-dependent toxicity [139,259]. Early medicinal chemistry of CPPs focused primarily on improving cellular efficacy of conjugated oligonucleotides. Yet, systematic characterization of *in vivo* safety and therapeutic index profiling with a larger set of compounds has been lacking. As more experimental data become available, the construction of machine-learning models to predict the efficacy and toxicity of CPPs could lead to the design of new CPPs and further enhance the therapeutic potential of CPP-oligonucleotide conjugates. (ii) Another important aspect of enhancing the functional delivery of CPP-based oligonucleotides is focusing on improving tissue targeting. This can be achieved by either optimizing the cell/tissue specificity of CPPs or by combining them with targeting ligands. Such advancements in CPP-oligonucleotide conjugates may significantly boost their therapeutic efficacy while simultaneously reducing the risk of toxicity that arises from the unintended deposition in nontarget tissues [143,144]. (iii) As discussed in Section 13.2.3, the internalization mechanisms of CPPs are yet to be fully elucidated. Albeit numerous examples, as illustrated in this review, there are very few studies dedicated to understanding the internalization mechanism for CPP-oligonucleotide therapeutics. A comprehensive understanding of the mechanism could be the cornerstone to designing more efficient CPPs for functional delivery. (iv) Current applications of CPP-based delivery approach by covalent conjugation are predominantly limited to oligonucleotides with neutrally charged backbones, such as PNA and PMO [260]. Expanding the cargo space to negatively charged gapmers and siRNAs could markedly widen the spectrum of oligonucleotides able to be delivered via CPPs. Efforts have been made to develop CPPs that are compatible with siRNAs and gapmers by engineering new linker and bioconjugation strategies [261,262]. One more interesting approach reported a prodrug strategy by shielding the negative charges of siRNA, which has the potential to be compatible with CPP-based delivery methods [97]. In conclusion, CPPs have demonstrated their value and potential as a promising delivery platform for therapeutic oligonucleotides. These developments provide a valuable approach and the potential for treating patients with many devastating diseases with high unmet medical needs.

References

1 Zhang, H., Zhang, Y., Zhang C., et al. Recent advances of cell-penetrating peptides and their application as vectors for delivery of peptide and protein-based cargo molecules. *Pharmaceutics*. 2023;15(8):2093.

2 Hofman, C.R and Corey, D.R. Targeting RNA with synthetic oligonucleotides: clinical success invites new challenges. *Cell Chem Biol*. 2024;31(1):125–138.

3 Jo, S.J., Chae, S.U., Lee, C.B., and Bae, S.K. Clinical pharmacokinetics of approved RNA therapeutics. *Int J Mol Sci*. 2023;24(1):746.

4 Shen, X. and Corey, D.R. Chemistry, mechanism and clinical status of antisense oligonucleotides and duplex RNAs. *Nucleic Acids Res*. 2018;46(4):1584–1600.

5 Egli, M. and Manoharan, M. Chemistry, structure and function of approved oligonucleotide therapeutics. *Nucleic Acids Res*. 2023;51(6):2529–2573.

6 Takakusa, H., Iwazaki, N., Nishikawa, M., Yoshida, T., Obika, S., and Inoue, T. Drug metabolism and pharmacokinetics of antisense oligonucleotide therapeutics: typical profiles, evaluation approaches, and points to consider compared with small molecule drugs. *Nucleic Acid Ther*. 2023;33(2):83–94.

7 Timotievich, E.D., Shilovskiy, I.P., and Khaitov, M.R. Cell-penetrating peptides as vehicles for delivery of therapeutic nucleic acids. Mechanisms and application in medicine. *Biochemistry (Mosc)*. 2023;88(11):1800–1817.

8 Frankel, A.D. and Pabo, C.O. Cellular uptake of the tat protein from human immunodeficiency virus. *Cell*. 1988;55(6):1189–1193.

9 Green, M. and Loewenstein, P.M. Autonomous functional domains of chemically synthesized human immunodeficiency virus tat trans-activator protein. *Cell*. 1988;55(6):1179–1188.

10 Vivès, E., Brodin, P., and Lebleu, B. A truncated HIV-1 Tat protein basic domain rapidly translocates through the plasma membrane and accumulates in the cell nucleus. *J Biol Chem*. 1997;272(25):16010–16017.

11 Joliot, A., Pernelle, C., Deagostini-Bazin, H., and Prochiantz, A. Antennapedia homeobox peptide regulates neural morphogenesis. *Proc Natl Acad Sci U S A*. 1991;88(5):1864–1868.

12 Derossi, D., Joliot, A.H., Chassaing, G., and Prochiantz, A. The third helix of the Antennapedia homeodomain translocates through biological membranes. *J Biol Chem*. 1994;269(14):10444–10450.

13 Derossi, D., Chassaing, G., and Prochiantz, A. Trojan peptides: the penetratin system for intracellular delivery. *Trends Cell Biol*. 1998;8(2):84–87.

14 Milletti, F. Cell-penetrating peptides: classes, origin, and current landscape. *Drug Discov Today*. 2012;17(15–16):850–860.

15 Futaki, S., Suzuki, T., Ohashi, W., et al. Arginine-rich peptides. An abundant source of membrane-permeable peptides having potential as carriers for intracellular protein delivery. *J Biol Chem*. 2001;276(8):5836–5840.

16 Morris, M.C., Depollier, J., Mery, J., Heitz, F., and Divita, G. A peptide carrier for the delivery of biologically active proteins into mammalian cells. *Nat Biotechnol*. 2001;19(12):1173–1176.

17 Pooga, M, Hällbrink, M., Zorko, M., and Langel, U. Cell penetration by transportan. *FASEB J*. 1998;12(1):67–77.

18 Gao, C., Mao, S., Ditzel, H.J., et al. A cell-penetrating peptide from a novel pVII-pIX phage-displayed random peptide library. *Bioorg Med Chem.* 2002;10(12):4057–4065.

19 Patel, S.G., Sayers, E.J., He, L., et al. Cell-penetrating peptide sequence and modification dependent uptake and subcellular distribution of green florescent protein in different cell lines. *Sci Rep.* 2019;9(1):6298.

20 Qian, Z., Martyna, A., Hard, R.L., et al. Discovery and mechanism of highly efficient cyclic cell-penetrating peptides. *Biochemistry.* 2016;55(18):2601–2612.

21 Pooga, M., Soomets, U., Hällbrink, M., et al. Cell penetrating PNA constructs regulate galanin receptor levels and modify pain transmission in vivo. *Nat Biotechnol.* 1998;16(9):857–861.

22 Schwarze, S.R., Ho, A., Vocero-Akbani, A., and Dowdy, S.F. In vivo protein transduction: delivery of a biologically active protein into the mouse. *Science.* 1999;285(5433):1569–1572.

23 Dougherty, P.G., Sahni, A., and Pei, D. Understanding cell penetration of cyclic peptides. *Chem Rev.* 2019;119(17):10241–10287.

24 Wang, L., Wang, N., Zhang, W., et al. Therapeutic peptides: current applications and future directions. *Signal Transduct Target Ther.* 2022;7(1):48.

25 Agrawal, P., Bhalla, S., Usmani, S.S., et al. CPPsite 2.0: a repository of experimentally validated cell-penetrating peptides. *Nucleic Acids Res.* 2016;44(D1):D1098–D1103.

26 Madani, F., Lindberg, S., Langel, U., Futaki, S., and Gräslund, A. Mechanisms of cellular uptake of cell-penetrating peptides. *J Biophys.* 2011;2011414729.

27 Zorko, M., Jones, S., and Langel, Ü. Cell-penetrating peptides in protein mimicry and cancer therapeutics. *Adv Drug Deliv Rev.* 2022;180:114044.

28 Wender, P.A, Mitchell, D.J, Pattabiraman, K., Pelkey, E.T., Steinman, L., and Rothbard, J.B. The design, synthesis, and evaluation of molecules that enable or enhance cellular uptake: peptoid molecular transporters. *Proc Natl Acad Sci U S A.* 2000;97(24):13003–13008.

29 Fischer P.M, Zhelev N.Z, Wang S, Melville J.E, Fahraeus R, and Lane D.P. Structure-activity relationship of truncated and substituted analogues of the intracellular delivery vector penetratin. *J Pept Res.* 2000;55(2):163–172.

30 Herce, H.D., Garcia, A.E., Litt, J., et al. Arginine-rich peptides destabilize the plasma membrane, consistent with a pore formation translocation mechanism of cell-penetrating peptides. *Biophys J.* 2009;97(7):1917–1925.

31 Mitchell, D.J., Kim, D.T., Steinman, L., Fathman, C.G., and Rothbard, J.B. Polyarginine enters cells more efficiently than other polycationic homopolymers. *J Pept Res.* 2000;56(5):318–325.

32 Futaki, S. and Nakase, I. Cell-surface interactions on arginine-rich cell-penetrating peptides allow for multiplex modes of internalization. *Acc Chem Res.* 2017;50(10):2449–2456.

33 Basheer, M., Schwalb, H., Shefler, I., Levdansky, L., Mekori, Y.A., and Gorodetsky, R. Blood pressure modulation following activation of mast cells by cationic cell penetrating peptides. *Peptides*. 2011;32(12):2444–2451.

34 Li, Q., Xu, M., Cui, Y., Huang, C., and Sun, M. Arginine-rich membrane-permeable peptides are seriously toxic. *Pharmacol Res Perspect*. 2017;5(5):e00334.

35 Lafarga, V., Sirozh, O., Díaz-López, I., et al. Widespread displacement of DNA- and RNA-binding factors underlies toxicity of arginine-rich cell-penetrating peptides. *EMBO J*. 2021;40(13):e103311.

36 Bottens, R.A. and Yamada, T. Cell-penetrating peptides (CPPs) as therapeutic and diagnostic agents for cancer. *Cancers (Basel)*. 2022;14(22):5546.

37 Xie, J., Bi, Y., Zhang, H., et al. Cell-penetrating peptides in diagnosis and treatment of human diseases: from preclinical research to clinical application. *Front Pharmacol*. 2020;11:697.

38 Allen, J. and Pellois, J.P. Hydrophobicity is a key determinant in the activity of arginine-rich cell penetrating peptides. *Sci Rep*. 2022;12(1):15981.

39 Lättig-Tünnemann, G., Prinz, M., Hoffmann, D., et al. Backbone rigidity and static presentation of guanidinium groups increases cellular uptake of arginine-rich cell-penetrating peptides. *Nat Commun*. 2011;2:453.

40 Herce, H.D., Schumacher, D., Schneider, A.F.L., et al. Cell-permeable nanobodies for targeted immunolabelling and antigen manipulation in living cells. *Nat Chem*. 2017;9(8):762–771.

41 Soudah, T., Khawaled, S., Aqeilan, R.I., and Yavin, E. AntimiR-155 cyclic peptide-PNA conjugate: synthesis, cellular uptake, and biological activity. *ACS Omega*. 2019;4(9):13954–13961.

42 Zhang, W., Lin, M., Yan, Q., et al. An intracellular nanobody targeting T4SS effector inhibits Ehrlichia infection. *Proc Natl Acad Sci U S A*. 2021;118(18):e2024102118.

43 Buyanova, M., Sahni, A., Yang, R., Sarkar, A., Salim, H., and Pei, D. Discovery of a cyclic cell-penetrating peptide with improved endosomal escape and cytosolic delivery efficiency. *Mol Pharm*. 2022;19(5):1378–1388.

44 Li, X., Kheirabadi, M., Dougherty, P.G., et al. The endosomal escape vehicle platform enhances delivery of oligonucleotides in preclinical models of neuromuscular disorders. *Mol Ther Nucleic Acids*. 2023;33:273–285.

45 Duchardt, F., Fotin-Mleczek, M., Schwarz, H., Fischer, R., and Brock, R. A comprehensive model for the cellular uptake of cationic cell-penetrating peptides. *Traffic*. 2007;8(7):848–866.

46 Maiolo, J.R., Ferrer, M., and Ottinger, E.A. Effects of cargo molecules on the cellular uptake of arginine-rich cell-penetrating peptides. *Biochim Biophys Acta*. 2005;1712(2):161–172.

47 Tünnemann, G., Martin, R.M., Haupt, S., Patsch, C., Edenhofer, F., and Cardoso, M.C. Cargo-dependent mode of uptake and bioavailability of TAT-containing proteins and peptides in living cells. *FASEB J.* 2006;20(11):1775–1784.

48 Szabó, I., Yousef, M., Soltész, D., Bató, C., Mező, G., and Bánóczi, Z. Redesigning of cell-penetrating peptides to improve their efficacy as a drug delivery system. *Pharmaceutics.* 2022;14(5):907.

49 Pei, D. and Buyanova, M. Overcoming endosomal entrapment in drug delivery. *Bioconjug Chem.* 2019;30(2):273–283.

50 Pei, D. How do biomolecules cross the cell membrane? *Acc Chem Res.* 2022;55(3):309–318.

51 Yang, N.J. and Hinner, M.J. Getting across the cell membrane: an overview for small molecules, peptides, and proteins. *Methods Mol Biol.* 2015; 1266:29–53.

52 Serulla, M., Anees, P., Hallaj, A., et al. Plasma membrane depolarization reveals endosomal escape incapacity of cell-penetrating peptides. *Eur J Pharm Biopharm.* 2023;184:116–124.

53 Schneider, A.F.L., Kithil, M., Cardoso, M.C., Lehmann, M., and Hackenberger, C.P.R. Cellular uptake of large biomolecules enabled by cell-surface-reactive cell-penetrating peptide additives. *Nat Chem.* 2021;13(6):530–539.

54 Hirose, H., Takeuchi, T., Osakada, H., et al. Transient focal membrane deformation induced by arginine-rich peptides leads to their direct penetration into cells. *Mol Ther.* 2012;20(5):984–993.

55 Ida, H., Takahashi, Y., Kumatani, A., et al. Nanoscale visualization of morphological alteration of live-cell membranes by the interaction with oligoarginine cell-penetrating peptides. *Anal Chem.* 2021;93(13):5383–5393.

56 Yang, L., Harroun, T.A., Weiss, T.M., Ding, L., and Huang, H.W. Barrel-stave model or toroidal model? A case study on melittin pores. *Biophys J.* 2001;81(3):1475–1485.

57 Herce, H.D. and Garcia, A.E. Molecular dynamics simulations suggest a mechanism for translocation of the HIV-1 TAT peptide across lipid membranes. *Proc Natl Acad Sci U S A.* 2007;104(52):20805–20810.

58 Allolio, C., Magarkar, A., Jurkiewicz, P., et al. Arginine-rich cell-penetrating peptides induce membrane multilamellarity and subsequently enter via formation of a fusion pore. *Proc Natl Acad Sci U S A.* 2018;115(47):11923–11928.

59 Trofimenko, E., Grasso, G., Heulot, M., et al. Genetic, cellular, and structural characterization of the membrane potential-dependent cell-penetrating peptide translocation pore. *Elife.* 2021;10:e69832.

60 Kawamoto, S., Takasu, M., Miyakawa, T., et al. Inverted micelle formation of cell-penetrating peptide studied by coarse-grained simulation: importance of attractive force between cell-penetrating peptides and lipid head group. *J Chem Phys.* 2011;134(9):095103.

61 Alves, I.D., Goasdoué, N., Correia, I., et al. Membrane interaction and perturbation mechanisms induced by two cationic cell penetrating peptides with distinct charge distribution. *Biochim Biophys Acta*. 2008;1780(7–8): 948–959.
62 Sakamoto, K., Morishita, T., Aburai, K., et al. Direct entry of cell-penetrating peptide can be controlled by maneuvering the membrane curvature. *Sci Rep*. 2021;11(1):31.
63 Ruseska, I. and Zimmer, A. Internalization mechanisms of cell-penetrating peptides. *Beilstein J Nanotechnol*. 2020;11:101–123.
64 Gonçalves, E., Kitas, E., and Seelig, J. Binding of oligoarginine to membrane lipids and heparan sulfate: structural and thermodynamic characterization of a cell-penetrating peptide. *Biochemistry*. 2005;44(7):2692–2702.
65 Nakase, I., Tadokoro, A., Kawabata, N., et al. Interaction of arginine-rich peptides with membrane-associated proteoglycans is crucial for induction of actin organization and macropinocytosis. *Biochemistry*. 2007;46(2):492–501.
66 Ezzat, K., Helmfors, H., Tudoran, O., et al. Scavenger receptor-mediated uptake of cell-penetrating peptide nanocomplexes with oligonucleotides. *FASEB J*. 2012;26(3):1172–1180.
67 Tanaka, G., Nakase, I., Fukuda, Y., et al. CXCR4 stimulates macropinocytosis: implications for cellular uptake of arginine-rich cell-penetrating peptides and HIV. *Chem Biol*. 2012;19(11):1437–1446.
68 Christianson, H.C. and Belting, M. Heparan sulfate proteoglycan as a cell-surface endocytosis receptor. *Matrix Biol*. 2014;35:51–55.
69 Conner, S.D. and Schmid, S.L. Regulated portals of entry into the cell. *Nature*. 2003;422(6927):37–44.
70 Ferrari, A., Pellegrini, V., Arcangeli, C., Fittipaldi, A., Giacca, M., and Beltram F. Caveolae-mediated internalization of extracellular HIV-1 tat fusion proteins visualized in real time. *Mol Ther*. 2003;8(2):284–294.
71 Nakase, I., Niwa, M., Takeuchi, T., et al. Cellular uptake of arginine-rich peptides: roles for macropinocytosis and actin rearrangement. *Mol Ther*. 2004;10(6):1011–1022.
72 Kawaguchi, Y., Takeuchi, T., Kuwata, K., et al. Syndecan-4 is a receptor for clathrin-mediated endocytosis of arginine-rich cell-penetrating peptides. *Bioconjug Chem*. 2016;27(4):1119–1130.
73 Richard, J.P., Melikov, K., Brooks, H., Prevot, P., Lebleu, B., and Chernomordik, L.V. Cellular uptake of unconjugated TAT peptide involves clathrin-dependent endocytosis and heparan sulfate receptors. *J Biol Chem*. 2005;280(15):15300–15306.
74 Arukuusk, P., Pärnaste, L., Margus, H., et al. Differential endosomal pathways for radically modified peptide vectors. *Bioconjug Chem*. 2013;24(10):1721–1732.

75 LeCher, J.C., Nowak, S.J., and McMurry, J.L. Breaking in and busting out: cell-penetrating peptides and the endosomal escape problem. *Biomol Concepts.* 2017;8(3–4):131–141.

76 Sahni, A., Qian, Z., and Pei, D. Cell-penetrating peptides escape the endosome by inducing vesicle budding and collapse. *ACS Chem Biol.* 2020;15(9):2485–2492.

77 Behr, J.-P. The proton sponge: a trick to enter cells the viruses did not exploit. *Chimia.* 1997;51(1–2):34–36.

78 Erazo-Oliveras, A., Muthukrishnan, N., Baker, R., Wang, T.Y., and Pellois, J.P. Improving the endosomal escape of cell-penetrating peptides and their cargos: strategies and challenges. *Pharmaceuticals (Basel).* 2012;5(11):1177–1209.

79 Brock, D.J., Kustigian, L., Jiang, M., et al. Efficient cell delivery mediated by lipid-specific endosomal escape of supercharged branched peptides. *Traffic.* 2018;19(6):421–435.

80 Snyder, E.L., Meade, B.R., Saenz, C.C., and Dowdy, S.F. Treatment of terminal peritoneal carcinomatosis by a transducible p53-activating peptide. *PLoS Biol.* 2004;2(2):E36.

81 Kristensen, M., Birch, D., and Morck Nielsen, H. Applications and challenges for use of cell-penetrating peptides as delivery vectors for peptide and protein cargos. *Int J Mol Sci.* 2016;17(2):185.

82 Schneider, A.F.L., Wallabregue, A.L.D., Franz, L., and Hackenberger, C.P.R. Targeted subcellular protein delivery using cleavable cyclic cell-penetrating peptides. *Bioconjug Chem.* 2019;30(2):400–404.

83 Fawell, S., Seery, J., Daikh, Y., et al. Tat-mediated delivery of heterologous proteins into cells. *Proc Natl Acad Sci U S A.* 1994;91(2):664–668.

84 Wadia, J.S. and Dowdy, S.F. Transmembrane delivery of protein and peptide drugs by TAT-mediated transduction in the treatment of cancer. *Adv Drug Deliv Rev.* 2005;57(4):579–596.

85 Dinca, A., Chien, W.M., and Chin, M.T. Intracellular delivery of proteins with cell-penetrating peptides for therapeutic uses in human disease. *Int J Mol Sci.* 2016;17(2):263.

86 Hoyer, J. and Neundorf, I. Peptide vectors for the nonviral delivery of nucleic acids. *Acc Chem Res.* 2012;45(7):1048–1056.

87 Muratovska, A. and Eccles, M.R. Conjugate for efficient delivery of short interfering RNA (siRNA) into mammalian cells. *FEBS Lett.* 2004;558(1–3):63–68.

88 El Andaloussi, S., Lehto, T., Mäger, I., et al. Design of a peptide-based vector, PepFect6, for efficient delivery of siRNA in cell culture and systemically in vivo. *Nucleic Acids Res.* 2011;39(9):3972–3987.

89 Nakase, I., Akita, H., Kogure, K., et al. Efficient intracellular delivery of nucleic acid pharmaceuticals using cell-penetrating peptides. *Acc Chem Res.* 2012;45(7):1132–1139.

90 Kapur, A., Medina, S.H., Wang, W., Palui, G., Schneider, J.P., and Mattoussi, H. Intracellular delivery of gold nanocolloids promoted by a chemically conjugated anticancer peptide. *ACS Omega*. 2018;3(10):12754–12762.

91 Dos Santos Rodrigues, B., Lakkadwala, S., Kanekiyo, T., and Singh, J. Development and screening of brain-targeted lipid-based nanoparticles with enhanced cell penetration and gene delivery properties. *Int J Nanomedicine*. 2019;14:6497–6517.

92 Silva, S., Almeida, A.J., and Vale, N. Combination of cell-penetrating peptides with manoparticles for therapeutic application: a review. *Biomolecules*. 2019;9(1):22.

93 Niesner, U., Halin, C., Lozzi, L., et al. Quantitation of the tumor-targeting properties of antibody fragments conjugated to cell-permeating HIV-1 TAT peptides. *Bioconjug Chem*. 2002;13(4):729–736.

94 Rudolph, C., Plank, C., Lausier, J., Schillinger, U., Müller, R.H., and Rosenecker, J. Oligomers of the arginine-rich motif of the HIV-1 TAT protein are capable of transferring plasmid DNA into cells. *J Biol Chem*. 2003;278(13):11411–11418.

95 Liu, Z., Li, M., Cui, D., and Fei, J. Macro-branched cell-penetrating peptide design for gene delivery. *J Control Release*. 2005;102(3):699–710.

96 Astriab-Fisher, A., Sergueev, D., Fisher, M., Shaw, B.R., and Juliano, R.L. Conjugates of antisense oligonucleotides with the Tat and antennapedia cell-penetrating peptides: effects on cellular uptake, binding to target sequences, and biologic actions. *Pharm Res*. 2002;19(6):744–754.

97 Meade, B.R., Gogoi, K., Hamil, A.S., et al. Efficient delivery of RNAi prodrugs containing reversible charge-neutralizing phosphotriester backbone modifications. *Nat Biotechnol*. 2014;32(12):1256–1261.

98 Lebel-Binay, S., Nemati, F., Dominguez-Berrocal, L., et al. PEP-010, a cell penetrating & interfering peptide as a new therapeutic approach in breast cancer. *Cancer Res*. 2018; 78(Suppl 13):3904

99 First-in-human phase I to evaluate PEP-010 as single agent and in combination with paclitaxel or with gemcitabine (CleverPeptide). Clinicaltrials.gov identifier: NCT04733027. Updated December 21, 2023. [cite 2024 Feb 9]. Accessed from: https://clinicaltrials.gov/study/NCT04733027.

100 Muller, R., Misund, K., Holien, T., et al. Targeting proliferating cell nuclear antigen and its protein interactions induces apoptosis in multiple myeloma cells. *PLoS One*. 2013;8(7):e70430.

101 Pascolutti, R., Yeturu, L., Philippin, G., et al. ATP128 clinical therapeutic cancer vaccine activates NF-κB and IRF3 pathways through TLR4 and TLR2 in human monocytes and dendritic cells. *Cancers (Basel)*. 2022;14(20):5134.

102 Tang, A. and Yokota, T. Duchenne muscular dystrophy: promising early-stage clinical trials to watch. *Expert Opin Investig Drugs*. 2024;33:1–17.

103 Onpattro (patisiran). https://www.alnylam.com/sites/default/files/pdfs/ONPATTRO-Prescribing-Information.pdf. Cambridge, MA, USA: Alnylam Pharmaceuticals, Inc.; 2023. Accessed May 6, 2025.

104 Givlaari (givosiran). https://www.alnylam.com/sites/default/files/pdfs/GIVLAARI-Prescribing-Information.pdf. Cambridge, MA, USA: Alnylam Pharmaceuticals, Inc.; 2024. Accessed May 6, 2025.

105 Oxlumo (lumasiran). https://www.alnylam.com/sites/default/files/pdfs/OXLUMO-Prescribing-Information.pdf. Cambridge, MA, USA: Alnylam Pharmaceuticals, Inc.; 2025. Accessed May 6, 2025.

106 Leqvio (inclisiran). https://www.novartis.com/us-en/sites/novartis_us/files/leqvio.pdf. East Hanover, NJ, USA: Novartis Pharmaceuticals Corporation; 2024. Accessed May 6, 2025.

107 Amvuttra (vutrisiran). https://www.alnylam.com/sites/default/files/pdfs/amvuttra-us-prescribing-information.pdf. Cambridge, MA, USA: Alnylam Pharmaceuticals, Inc.; 2025. Accessed May 6, 2025.

108 Qalsody (tofersen). https://www.biogencdn.com/us/pdfs/qalsody-prescribing-information.pdf. Cambridge, MA, USA: Biogen MA Inc.; 2023. Accessed May 6, 2025.

109 Keam SJ. Inotersen: First Global Approval. Drugs. 2018 Sep;78(13):1371-1376. https://doi.org/10.1007/s40265-018-0968-5.

110 Piascik, P. Fomiversen sodium approved to treat CMV retinitis. *J Am Pharm Assoc (Wash)*. 1999;39(1):84–85.

111 Hair, P., Cameron, F., and McKeage, K. Mipomersen sodium: first global approval. *Drugs*. 2013;73(5):487–493.

112 Wainua (eplontersen). https://www.accessdata.fda.gov/drugsatfda_docs/label/2025/217388Orig1s004lbl.pdf. Wilmington, DE, USA: Astrazeneca Pharmaceuticals LP; 2025. Accessed May 6, 2025.

113 Spinraza (nusinersen). https://www.spinraza.com/content/dam/commercial/spinraza/caregiver/en_us/pdf/spinraza-prescribing-information.pdf. Cambridge, MA, USA: Biogen; 2024. Accessed May 6, 2025.

114 Exondys 51 (eteplirsen). https://www.exondys51.com/pi. Cambridge, MA, USA: Sarepta Therapeutics, Inc.; 2024. Accessed May 6, 2025.

115 Vyondys 53 (golodirsen). https://www.vyondys53.com/PI. Cambridge, MA, USA: Sarepta Therapeutics, Inc.; 2024. Accessed May 6, 2025.

116 Viltepso (vitolarsen). https://www.viltepso.com/prescribing-information?utm_medium_p=gcpc. Paramus, NJ, USA: NS Pharma, Inc.; 2021. Accessed May 6, 2025.

117 Amondys 45 (casimersen). https://www.amondys45.com/pi. Cambridge, MA, USA: Sarepta Therapeutics, Inc.; 2024. Accessed May 6, 2025.

118 Doggrell, S.A. Pegaptanib: the first antiangiogenic agent approved for neovascular macular degeneration. *Expert Opin Pharmacother*. 2005;6(8):1421–1423.

119 Izervay (avacincaptad pegol). https://www.astellas.com/us/system/files/izervay_pi.pdf. Parsippany, NJ, USA: IVERIC bio, Inc.; 2025. Accessed May 6, 2025.

120 Goto, A., Yamamoto, S., and Iwasaki, S. Biodistribution and delivery of oligonucleotide therapeutics to the central nervous system: advances, challenges, and future perspectives. *Biopharm Drug Dispos.* 2023;44(1):26–47.

121 McClorey, G. and Banerjee, S. Cell-penetrating peptides to enhance delivery of oligonucleotide-based therapeutics. *Biomedicines.* 2018;6(2):51.

122 Aartsma-Rus, A. Good news for the mdx mouse community: improved dystrophin restoration after skipping mouse dystrophin exon 23. *Mol Ther Nucleic Acids.* 2022;30:355–356.

123 Wickstrom, E. DNA and RNA derivatives to optimize distribution and delivery. *Adv Drug Deliv Rev.* 2015;87:25–34.

124 Flynt, A.S. Blocking zebrafish microRNAs with morpholinos. In: *Morpholino Oligomers: Methods and Protocols* (ed. H.M. Moulton and J.D. Moulton). New York, NY, United States: Humana Press; 2017; pp. 59–78.

125 Moulton, J.D. and Yan, Y.L. Using morpholinos to control gene expression. *Curr Protoc Mol Biol.* 2008;83(1):26.28.21–26.28.29.

126 Moulton, H.M., Nelson, M.H., Hatlevig, S.A., Reddy, M.T., and Iversen, P.L. Cellular uptake of antisense morpholino oligomers conjugated to arginine-rich peptides. *Bioconjug Chem.* 2004;15(2):290–299.

127 Neuman, B.W., Stein, D.A., Kroeker, A.D., et al. Antisense morpholino-oligomers directed against the 5′ end of the genome inhibit coronavirus proliferation and growth. *J Virol.* 2004;78(11):5891–5899.

128 Neuman, B.W., Stein, D.A., Kroeker, A.D., et al. Inhibition, escape, and attenuated growth of severe acute respiratory syndrome coronavirus treated with antisense morpholino oligomers. *J Virol.* 2005;79(15):9665–9676.

129 Neuman, B.W., Stein, D.A., Kroeker, A.D., et al. Inhibition and escape of SARS-CoV treated with antisense morpholino oligomers. *Adv Exp Med Biol.* 2006;581:567–571.

130 Wolfe, J.M., Fadzen, C.M., Holden, R.L., Yao, M., Hanson, G.J., and Pentelute, B.L. Perfluoroaryl bicyclic cell-penetrating peptides for delivery of antisense oligonucleotides. *Angew Chem Int Ed Engl.* 2018;57(17):4756–4759.

131 Hu, J., Shen, X., Kheirabadi, M., et al. Targeting the expanded TCF4/Fuchs' endothelial corneal dystrophy CUG repeat with morpholino peptide conjugates. *ACS Omega.* 2023;8(45):42797–42802.

132 Sazani, P., Kang, S.H., Maier, M.A., et al. Nuclear antisense effects of neutral, anionic and cationic oligonucleotide analogs. *Nucleic Acids Res.* 2001;29(19):3965–3974.

133 Sazani, P., Gemignani, F., Kang, S.H., et al. Systemically delivered antisense oligomers upregulate gene expression in mouse tissues. *Nat Biotechnol.* 2002;20(12):1228–1233.

134 Jearawiriyapaisarn, N., Moulton, H.M., Buckley, B., et al. Sustained dystrophin expression induced by peptide-conjugated morpholino oligomers in the muscles of mdx mice. *Mol Ther*. 2008;16(9):1624–1629.
135 Wu, B., Moulton, H.M., Iversen, P.L., et al. Effective rescue of dystrophin improves cardiac function in dystrophin-deficient mice by a modified morpholino oligomer. *Proc Natl Acad Sci U S A*. 2008;105(39):14814–14819.
136 Yin, H., Moulton, H.M., Seow, Y., et al. Cell-penetrating peptide-conjugated antisense oligonucleotides restore systemic muscle and cardiac dystrophin expression and function. *Hum Mol Genet*. 2008;17(24):3909–3918.
137 Böttcher-Friebertshäuser, E., Stein, D.A., Klenk, H.D., and Garten, W. Inhibition of influenza virus infection in human airway cell cultures by an antisense peptide-conjugated morpholino oligomer targeting the hemagglutinin-activating protease TMPRSS2. *J Virol*. 2011;85(4):1554–1562.
138 Du, L., Kayali, R., Bertoni, C., et al. Arginine-rich cell-penetrating peptide dramatically enhances AMO-mediated ATM aberrant splicing correction and enables delivery to brain and cerebellum. *Hum Mol Genet*. 2011;20(16):3151–3160.
139 Wu, B., Lu, P., Cloer, C., et al. Long-term rescue of dystrophin expression and improvement in muscle pathology and function in dystrophic mdx mice by peptide-conjugated morpholino. *Am J Pathol*. 2012;181(2):392–400.
140 Echigoya, Y., Nakamura, A., Nagata, T., et al. Effects of systemic multiexon skipping with peptide-conjugated morpholinos in the heart of a dog model of Duchenne muscular dystrophy. *Proc Natl Acad Sci U S A*. 2017;114(16):4213–4218.
141 Bersani, M., Rizzuti, M., Pagliari, E., et al. Cell-penetrating peptide-conjugated morpholino rescues SMA in a symptomatic preclinical model. *Mol Ther*. 2022;30(3):1288–1299.
142 Clayton, N.P., Nelson, C.A., Weeden, T., et al. Antisense oligonucleotide-mediated suppression of muscle glycogen synthase 1 synthesis as an approach for substrate reduction therapy of pompe disease. *Mol Ther Nucleic Acids*. 2014;3(10):e206.
143 Yin, H., Moulton, H.M., Betts, C., et al. A fusion peptide directs enhanced systemic dystrophin exon skipping and functional restoration in dystrophin-deficient mdx mice. *Hum Mol Genet*. 2009;18(22):4405–4414.
144 Yin, H., Boisguerin, P., Moulton, H.M., et al. Context dependent effects of chimeric peptide morpholino conjugates contribute to dystrophin exon-skipping efficiency. *Mol Ther Nucleic Acids*. 2013;2(9):e124.
145 Erdos, M.R., Cabral, W.A., Tavarez, U.L., et al. A targeted antisense therapeutic approach for Hutchinson–Gilford progeria syndrome. *Nat Med*. 2021;27(3):536–545.

146 Ivanova, G.D., Arzumanov, A., Abes, R., et al. Improved cell-penetrating peptide-PNA conjugates for splicing redirection in HeLa cells and exon skipping in mdx mouse muscle. *Nucleic Acids Res.* 2008;36(20):6418–6428.

147 Yin, H., Saleh, A.F., Betts, C., et al. Pip5 transduction peptides direct high efficiency oligonucleotide-mediated dystrophin exon skipping in heart and phenotypic correction in mdx mice. *Mol Ther.* 2011;19(7):1295–1303.

148 Betts, C., Saleh, A.F., Arzumanov, A.A., et al. Pip6-PMO, a new generation of peptide-oligonucleotide conjugates with improved cardiac exon skipping activity for DMD treatment. *Mol Ther Nucleic Acids.* 2012;1(8):e38.

149 Betts, C.A., Saleh, A.F., Carr, C.A., et al. Prevention of exercised induced cardiomyopathy following Pip-PMO treatment in dystrophic mdx mice. *Sci Rep.* 2015;5:8986.

150 Godfrey, C., Muses, S., McClorey, G., et al. How much dystrophin is enough: the physiological consequences of different levels of dystrophin in the mdx mouse. *Hum Mol Genet.* 2015;24(15):4225–4237.

151 Shabanpoor, F., McClorey, G., Saleh, A.F., Järver, P., Wood, M.J., and Gait, M.J. Bi-specific splice-switching PMO oligonucleotides conjugated via a single peptide active in a mouse model of Duchenne muscular dystrophy. *Nucleic Acids Res.* 2015;43(1):29–39.

152 Blain, A.M., Greally, E., McClorey, G., et al. Peptide-conjugated phosphodiamidate oligomer-mediated exon skipping has benefits for cardiac function in mdx and Cmah-/-mdx mouse models of Duchenne muscular dystrophy. *PLoS One.* 2018;13(6):e0198897.

153 Betts, C.A., McClorey, G., Healicon, R., et al. Cmah-dystrophin deficient mdx mice display an accelerated cardiac phenotype that is improved following peptide-PMO exon skipping treatment. *Hum Mol Genet.* 2019;28(3):396–406.

154 Klein, A.F., Varela, M.A., Arandel, L., et al. Peptide-conjugated oligonucleotides evoke long-lasting myotonic dystrophy correction in patient-derived cells and mice. *J Clin Invest.* 2019;129(11):4739–4744.

155 González-Martínez, I., Cerro-Herreros, E., Moreno, N., et al. Peptide-conjugated antimiRs improve myotonic dystrophy type 1 phenotypes by promoting endogenous MBNL1 expression. *Mol Ther Nucleic Acids.* 2023;34:102024.

156 Linden, G., Janga, H., Franz, M., et al. Efficient antisense inhibition reveals microRNA-155 to restrain a late-myeloid inflammatory programme in primary human phagocytes. *RNA Biol.* 2021;18(5):604–618.

157 Hammond, S.M., Hazell, G., Shabanpoor, F., et al. Systemic peptide-mediated oligonucleotide therapy improves long-term survival in spinal muscular atrophy. *Proc Natl Acad Sci U S A.* 2016;113(39):10962–10967.

158 Chwalenia, K., Oieni, J., Zemła, J., et al. Exon skipping induces uniform dystrophin rescue with dose-dependent restoration of serum miRNA

biomarkers and muscle biophysical properties. *Mol Ther Nucleic Acids.* 2022;29:955–968.

159 Shabanpoor, F., Hammond, S.M., Abendroth, F., Hazell, G., Wood, M.J.A., and Gait, M.J. Identification of a peptide for systemic brain delivery of a morpholino oligonucleotide in mouse models of spinal muscular atrophy. *Nucleic Acid Ther.* 2017;27(3):130–143.

160 Aslesh, T., Erkut, E., Ren, J., et al. DG9-conjugated morpholino rescues phenotype in SMA mice by reaching the CNS via a subcutaneous administration. *JCI Insight.* 2023;8(5):e160516.

161 Holden, K.L., Stein, D.A., Pierson, T.C., et al. Inhibition of dengue virus translation and RNA synthesis by a morpholino oligomer targeted to the top of the terminal 3′ stem-loop structure. *Virology.* 2006;344(2):439–452.

162 Vagnozzi, A., Stein, D.A., Iversen, P.L., and Rieder, E. Inhibition of foot-and-mouth disease virus infections in cell cultures with antisense morpholino oligomers. *J Virol.* 2007;81(21):11669–11680.

163 Stone, J.K., Rijnbrand, R., Stein, D.A., et al. A morpholino oligomer targeting highly conserved internal ribosome entry site sequence is able to inhibit multiple species of picornavirus. *Antimicrob Agents Chemother.* 2008;52(6):1970–1981.

164 Eide, K., Moerdyk-Schauwecker, M., Stein, D.A., Bildfell, R., Koelle, D.M., and Jin, L. Reduction of herpes simplex virus type-2 replication in cell cultures and in rodent models with peptide-conjugated morpholino oligomers. *Antivir Ther.* 2010;15(8):1141–1149.

165 Geller, B.L., Marshall-Batty, K., Schnell, F.J., McKnight, M.M., Iversen, P.L., and Greenberg, D.E. Gene-silencing antisense oligomers inhibit acinetobacter growth in vitro and in vivo. *J Infect Dis.* 2013;208(10):1553–1560.

166 Rosenke, K., Leventhal, S., Moulton, H.M., et al. Inhibition of SARS-CoV-2 in vero cell cultures by peptide-conjugated morpholino oligomers. *J Antimicrob Chemother.* 2021;76(2):413–417.

167 Zeng, Z., Han, S., Hong, W., et al. A Tat-conjugated peptide nucleic acid Tat-PNA-DR inhibits hepatitis B virus replication in vitro and in vivo by targeting LTR direct repeats of HBV RNA. *Mol Ther Nucleic Acids.* 2016;5(3):e295.

168 Brognara, E., Fabbri, E., Bazzoli, E., et al. Uptake by human glioma cell lines and biological effects of a peptide-nucleic acids targeting miR-221. *J Neurooncol.* 2014;118(1):19–28.

169 Gait, M.J., Arzumanov, A.A., McClorey, G., et al. Cell-penetrating peptide conjugates of steric blocking oligonucleotides as therapeutics for neuromuscular diseases from a historical perspective to current prospects of treatment. *Nucleic Acid Ther.* 2019;29(1):1–12.

170 Fadzen, C.M., Holden, R.L., Wolfe, J.M., et al. Chimeras of cell-penetrating peptides demonstrate synergistic improvement in antisense efficacy. *Biochemistry*. 2019;58(38):3980–3989.

171 Wolfe, J.M., Fadzen, C.M., Choo, Z.N., et al. Machine learning to predict cell-penetrating peptides for antisense delivery. *ACS Cent Sci*. 2018;4(4):512–520.

172 Qian, Z., Liu, T., Liu, Y.Y., et al. Efficient delivery of cyclic peptides into mammalian cells with short sequence motifs. *ACS Chem Biol*. 2013;8(2):423–431.

173 Qian, Z., LaRochelle, J.R., Jiang, B., et al. Early endosomal escape of a cyclic cell-penetrating peptide allows effective cytosolic cargo delivery. *Biochemistry*. 2014;53(24):4034–4046.

174 Abes, S., Moulton, H.M., Clair, P., et al. Vectorization of morpholino oligomers by the (R-Ahx-R)4 peptide allows efficient splicing correction in the absence of endosomolytic agents. *J Control Release*. 2006;116(3):304–313.

175 Abes, R., Moulton, H.M., Clair, P., et al. Delivery of steric block morpholino oligomers by (R-X-R)4 peptides: structure-activity studies. *Nucleic Acids Res*. 2008;36(20):6343–6354.

176 Elangkovan, N. and Dickson, G. Gene therapy for Duchenne muscular dystrophy. *J Neuromuscul Dis*. 20218(s2):S303–S316.

177 Yin, H., Moulton, H.M., Betts, C., et al. Functional rescue of dystrophin-deficient mdx mice by a chimeric peptide-PMO. *Mol Ther*. 2010;18(10):1822–1829.

178 Lehto, T., Castillo Alvarez, A., Gauck, S., et al. Cellular trafficking determines the exon skipping activity of Pip6a-PMO in mdx skeletal and cardiac muscle cells. *Nucleic Acids Res*. 2014;42(5):3207–3217.

179 Overby, S.J., Cerro-Herreros, E., González-Martínez, I., et al. Proof of concept of peptide-linked blockmiR-induced MBNL functional rescue in myotonic dystrophy type 1 mouse model. *Mol Ther Nucleic Acids*. 2022;27:1146–1155.

180 Gan, L., Wu, L.C.L, Wood, J.A., et al. A cell-penetrating peptide enhances delivery and efficacy of phosphorodiamidate morpholino oligomers in mdx mice. *Mol Ther Nucleic Acids*. 2022;30:17–27.

181 Two-part study for dose determination of vesleteplirsen (SRP-5051) (part A), then dose efficacy (part B) in participants with Duchenne muscular dystrophy amenable to exon 51-skipping treatment (MOMENTUM). Clinicaltrials.gov identifier: NCT04004065. Updated December 1, 2023 [cited 2024 Feb 9]. Accessed from: https://clinicaltrials.gov/study/NCT04004065.

182 A study to evaluate the safety, tolerability, and pharmacokinetics of a single dose of SRP-5051 (vesleteplirsen) in patients with Duchenne muscular dystrophy (DMD). Clinicaltrials.gov identifier: NCT03375255. Updated July 6, 2022 [cited 2024 March 11]. Accessed from: https://www.clinicaltrials.gov/study/NCT03375255.

183 Campbell, C., Mathews, K., van de Rijn, M., et al. Phase 2 multiple-ascending-dose study of SRP-5051 PPMO in patients with DMD amenable to exon 51 skipping: part A results. In: *Poster Session Presented at MDA 2022: Annual Clinical and Scientific Conference of the Muscular Dystrophy Association*. Nashville, TN; 2022 March 13–16.

184 Sarepta Therapeutics. (2024, October 23). *Sarepta Therapeutics announces third quarter 2024 financial results*. Retrieved May 5, 2025, from https://investorrelations.sarepta.com/news-releases/news-release-details/sarepta-therapeutics-announces-third-quarter-2024-financial.

185 Entrada Therapeutics. (2024, October 9). *Entrada Therapeutics presents new data supporting its expanding Duchenne franchise at the 29th Annual Congress of the World Muscle Society*. Retrieved May 5, 2025, from https://ir.entradatx.com/news-releases/news-release-details/entrada-therapeutics-presents-new-data-supporting-its-expanding.

186 Fay, A. Spinal muscular atrophy: a (now) treatable neurodegenerative disease. *Pediatr Clin North Am*. 2023;70(5):963–977.

187 Geary, R.S., Norris, D., Yu, R., and Bennett, C.F. Pharmacokinetics, biodistribution and cell uptake of antisense oligonucleotides. *Adv Drug Deliv Rev*. 2015;87:46–51.

188 Tajik-Ahmadabad, B., Polyzos, A., Separovic, F., and Shabanpoor, F. Amphiphilic lipopeptide significantly enhances uptake of charge-neutral splice switching morpholino oligonucleotide in spinal muscular atrophy patient-derived fibroblasts. *Int J Pharm*. 2017;532(1):21–28.

189 Dastpeyman, M., Sharifi, R., Amin, A., et al. Endosomal escape cell-penetrating peptides significantly enhance pharmacological effectiveness and CNS activity of systemically administered antisense oligonucleotides. *Int J Pharm*. 2021;599:120398.

190 Hammond, S.M., Abendroth, F., Goli, L., et al. Antibody-oligonucleotide conjugate achieves CNS delivery in animal models for spinal muscular atrophy. *JCI Insight*. 2022;7(24):e154142.

191 Ma, W., Lin, Y., Xuan. W., Iversen, P.L., Smith, L.J., and Benchimol, S. Inhibition of p53 expression by peptide-conjugated phosphorodiamidate morpholino oligomers sensitizes human cancer cells to chemotherapeutic drugs. *Oncogene*. 2012;31(8):1024–1033.

192 Dang, Y., van Heusden, C., Nickerson, V., et al. Enhanced delivery of peptide-morpholino oligonucleotides with a small molecule to correct splicing defects in the lung. *Nucleic Acids Res*. 2021;49(11):6100–6113.

193 Kim, T.J., Cho, K.S., and Koo, K.C. Current status and future perspectives of immunotherapy for locally advanced or metastatic urothelial carcinoma: a comprehensive review. *Cancers (Basel)*. 2020;12(1):192.

194 Eriksson, M., Brown, W.T., Gordon, L.B., et al. Recurrent de novo point mutations in lamin A cause Hutchinson–Gilford progeria syndrome. *Nature*. 2003;423(6937):293–298.

195 Kohler, L., Puertollano, R., and Raben, N. Pompe disease: from basic science to therapy. *Neurotherapeutics*. 2018;15(4):928–942.

196 Fotsing, S.F., Margoliash, J., Wang, C., et al. The impact of short tandem repeat variation on gene expression. *Nat Genet*. 2019;51(11):1652–1659.

197 Verbiest, M., Maksimov, M., Jin, Y., Anisimova, M., Gymrek, M., and Bilgin Sonay, T. Mutation and selection processes regulating short tandem repeats give rise to genetic and phenotypic diversity across species. *J Evol Biol*. 2023;36(2):321–336.

198 Flower, M., Lomeikaite, V., Ciosi, M., et al. MSH3 modifies somatic instability and disease severity in Huntington's and myotonic dystrophy type 1. *Brain*. 2019;142(7):1876–1886.

199 Wieben, E.D., Aleff, R.A., Tang, X., et al. Trinucleotide repeat expansion in the transcription factor 4 (TCF4) gene leads to widespread mRNA splicing changes in Fuchs' endothelial corneal dystrophy. *Invest Ophthalmol Vis Sci*. 2017;58(1):343–352.

200 Hu, J., Matsui, M., Gagnon, K.T., et al. Allele-specific silencing of mutant huntingtin and ataxin-3 genes by targeting expanded CAG repeats in mRNAs. *Nat Biotechnol*. 2009;27(5):478–484.

201 Leger, A.J., Mosquea, L.M., Clayton, N.P., et al. Systemic delivery of a peptide-linked morpholino oligonucleotide neutralizes mutant RNA toxicity in a mouse model of myotonic dystrophy. *Nucleic Acid Ther*. 2013;23(2):109–117.

202 Stoodley, J., Vallejo-Bedia, F., Seone-Miraz, D., Debasa-Mouce, M., Wood, M.J.A., and Varela, M.A. Application of antisense conjugates for the treatment of myotonic dystrophy type 1. *Int J Mol Sci*. 2023;24(3):2697.

203 Kreher, N.C., Estrella, N.L., Hicks, A.N., et al. A novel EEV-conjugated PMO, ENTR-701, reduces nuclear foci and corrects aberrant splicing in myotonic dystrophy type 1 preclinical models. In: *Poster Session Presented at MDF 2022: Annual Conference of the Myotonic Dystrophy Foundation*. San Diego, CA; 2022 September 9–10.

204 A phase 1/2 study of VX-670 in adult participants with myotonic dystrophy 1 (DM1).(Galileo). Clinicaltrials.gov identifier: NCT06185764. Updated March 1, 2024. [cited 2024 March 11]. Available from: https://www.clinicaltrials.gov/study/NCT06185764.

205 Fautsch, M.P., Wieben, E.D., Baratz, K.H., et al. TCF4-mediated Fuchs endothelial corneal dystrophy: Insights into a common trinucleotide repeat-associated disease. *Prog Retin Eye Res*. 2021;81:100883.

206 Mootha, V.V., Gong, X., Ku, H.C., and Xing, C. Association and familial segregation of CTG18.1 trinucleotide repeat expansion of TCF4 gene in Fuchs' endothelial corneal dystrophy. *Invest Ophthalmol Vis Sci*. 2014;55(1):33–42.

207 Zamecnik, P.C. and Stephenson, M.L. Inhibition of Rous sarcoma virus replication and cell transformation by a specific oligodeoxynucleotide. *Proc Natl Acad Sci U S A*. 1978;75(1):280–284.

208 Deas, T.S., Binduga-Gajewska, I., Tilgner, M., et al. Inhibition of flavivirus infections by antisense oligomers specifically suppressing viral translation and RNA replication. *J Virol*. 2005;79(8):4599–4609.

209 Ge, Q., Pastey, M., Kobasa, D., et al. Inhibition of multiple subtypes of influenza A virus in cell cultures with morpholino oligomers. *Antimicrob Agents Chemother*. 2006;50(11):3724–3733.

210 Yuan, J., Stein, D.A., Lim, T., et al. Inhibition of coxsackievirus B3 in cell cultures and in mice by peptide-conjugated morpholino oligomers targeting the internal ribosome entry site. *J Virol*. 2006;80(23):11510–11519.

211 Burrer, R., Neuman, B.W., Ting, J.P., et al. Antiviral effects of antisense morpholino oligomers in murine coronavirus infection models. *J Virol*. 2007;81(11):5637–5648.

212 Deas, T.S., Bennett, C.J., Jones, S.A., et al. In vitro resistance selection and in vivo efficacy of morpholino oligomers against West Nile virus. *Antimicrob Agents Chemother*. 2007;51(7):2470–2482.

213 Tilley, L.D., Mellbye, B.L., Puckett, S.E., Iversen, P.L., and Geller, B.L. Antisense peptide-phosphorodiamidate morpholino oligomer conjugate: dose-response in mice infected with Escherichia coli. *J Antimicrob Chemother*. 2007;59(1):66–73.

214 Bok, K., Cavanaugh, V.J., Matson, D.O., et al. Inhibition of norovirus replication by morpholino oligomers targeting the 5′-end of the genome. *Virology*. 2008;380(2):328–337.

215 Gabriel, G., Nordmann, A., Stein, D.A., Iversen, P.L., and Klenk, H.D. Morpholino oligomers targeting the PB1 and NP genes enhance the survival of mice infected with highly pathogenic influenza A H7N7 virus. *J Gen Virol*. 2008;89(Pt 4):939–948.

216 Lai, S.H., Stein, D.A., and Guerrero-Plata, A., et al. Inhibition of respiratory syncytial virus infections with morpholino oligomers in cell cultures and in mice. *Mol Ther*. 2008;16(6):1120–1128.

217 Lupfer, C., Stein, D.A., Mourich, D.V., Tepper, S.E., Iversen, P.L., and Pastey, M. Inhibition of influenza A H3N8 virus infections in mice by morpholino oligomers. Arch Virol. 2008;153(5):929–937.

218 Paessler, S., Rijnbrand, R., Stein, D.A., et al. Inhibition of alphavirus infection in cell culture and in mice with antisense morpholino oligomers. *Virology*. 2008;376(2):357–370.

219 Stein, D.A., Huang, C.Y., Silengo, S., et al. Treatment of AG129 mice with antisense morpholino oligomers increases survival time following challenge with dengue 2 virus. *J Antimicrob Chemother*. 2008;62(3):555–565.

220 Mellbye, B.L., Puckett, S.E., Tilley, L.D., Iversen, P.L., and Geller, B.L. Variations in amino acid composition of antisense peptide-phosphorodiamidate morpholino oligomer affect potency against Escherichia coli in vitro and in vivo. *Antimicrob Agents Chemother*. 2009;53(2):525–530.

221 Sleeman, K., Stein, D.A., Tamin, A., Reddish, M., Iversen, P.L., and Rota, P.A. Inhibition of measles virus infections in cell cultures by peptide-conjugated morpholino oligomers. *Virus Res*. 2009;140(1–2):49–56.

222 Anantpadma, M., Stein, D.A., and Vrati, S. Inhibition of Japanese encephalitis virus replication in cultured cells and mice by a peptide-conjugated morpholino oligomer. *J Antimicrob Chemother*. 2010;65(5):953–961.

223 Mellbye, B.L., Weller, D.D., Hassinger, J.N., et al. Cationic phosphorodiamidate morpholino oligomers efficiently prevent growth of Escherichia coli in vitro and in vivo. *J Antimicrob Chemother*. 2010;65(1):98–106.

224 Zhang, J., Stein, D.A., Timoney, P.J., and Balasuriya, U.B. Curing of HeLa cells persistently infected with equine arteritis virus by a peptide-conjugated morpholino oligomer. *Virus Res*. 2010;150(1–2):138–142.

225 Neuman, B.W., Bederka, L.H., Stein, D.A., Ting, J.P., Moulton, H.M., and Buchmeier, M.J. Development of peptide-conjugated morpholino oligomers as pan-arenavirus inhibitors. *Antimicrob Agents Chemother*. 2011;55(10):4631–4638.

226 Opriessnig, T., Patel, D., Wang, R., et al. Inhibition of porcine reproductive and respiratory syndrome virus infection in piglets by a peptide-conjugated morpholino oligomer. *Antiviral Res*. 2011;91(1):36–42.

227 Tihaya, M.S., Mul, K., Balog, J., et al. Facioscapulohumeral muscular dystrophy: the road to targeted therapies. *Nat Rev Neurol*. 2023;19(2):91–108.

228 Ansseau, E., Vanderplanck, C., Wauters, A., Harper, S.Q., Coppée, F., and Belayew, A. Antisense oligonucleotides used to target the DUX4 mRNA as therapeutic approaches in faciosscapulohumeral muscular dystrophy (FSHD). *Genes (Basel)*. 2017;8(3):93.

229 Lu-Nguyen, N., Malerba, A., Herath, S., Dickson, G., and Popplewell, L. Systemic antisense therapeutics inhibiting DUX4 expression ameliorates FSHD-like pathology in an FSHD mouse model. *Hum Mol Genet*. 2021;30(15):1398–1412.

230 Lu-Nguyen, N., Dickson, G., Malerba, A., and Popplewell, L. Long-term systemic treatment of a mouse model displaying chronic FSHD-like pathology with antisense therapeutics that inhibit DUX4 expression. *Biomedicines*. 2022;10(7):1623.

231 Chen, J.C., King, O.D., Zhang, Y., et al. Morpholino-mediated knockdown of DUX4 toward facioscapulohumeral muscular dystrophy therapeutics. *Mol Ther.* 2016;24(8):1405–1411.

232 Diener, C., Keller, A., and Meese, E. Emerging concepts of miRNA therapeutics: from cells to clinic. *Trends Genet.* 2022;38(6):613–626.

233 Cai, Y., Yu, X., Hu, S., and Yu, J. A brief review on the mechanisms of miRNA regulation. *Genom Proteom Bioinf.* 2009;7(4):147–154.

234 Carraro, G., Shrestha, A., Rostkovius, J., et al. miR-142-3p balances proliferation and differentiation of mesenchymal cells during lung development. *Development.* 2014;141(6):1272–1281.

235 Glaser, S., Meng, F., Han, Y., et al. Secretin stimulates biliary cell proliferation by regulating expression of microRNA 125b and microRNA let7a in mice. *Gastroenterology.* 2014;146(7):1795–1808.

236 Chang, R.M., Xiao, S., Lei, X., Yang, H., Fang, F., and Yang, L.Y. miRNA-487a promotes proliferation and metastasis in hepatocellular carcinoma. *Clin Cancer Res.* 2017;23(10):2593–2604.

237 Hall, C., Ehrlich, L., Meng, F., et al. Inhibition of microRNA-24 increases liver fibrosis by enhanced menin expression in Mdr2(−/−) mice. *J Surg Res.* 2017;217:160–169.

238 Hazra, B., Kumawat, K.L., and Basu, A. The host microRNA miR-301a blocks the IRF1-mediated neuronal innate immune response to Japanese encephalitis virus infection. *Sci Signal.* 2017;10(466):eaaf5185.

239 Yin, Y., Xu, L., Chang, Y., et al. N-myc promotes therapeutic resistance development of neuroendocrine prostate cancer by differentially regulating miR-421/ATM pathway. *Mol Cancer.* 2019;18(1):11.

240 Al Qaryoute, A., Fallatah, W., Dhinoja, S., Raman, R., and Jagadeeswaran, P. Role of microRNAs and their downstream target transcription factors in zebrafish thrombopoiesis. *Sci Rep.* 2023;13(1):16066.

241 Uhlmann, E. Peptide nucleic acids (PNA) and PNA-DNA chimeras: from high binding affinity towards biological function. *Biol Chem.* 1998;379(8–9):1045–1052.

242 Ndeboko, B., Hantz, O., Lemamy, G.J., and Cova, L. Developments in cell-penetrating peptides as antiviral agents and as vehicles for delivery of peptide nucleic acid targeting hepadnaviral replication pathway. *Biomolecules.* 2018;8(3):55.

243 Bendifallah, N., Rasmussen, F.W., Zachar, V., Ebbesen, P., Nielsen, P.E., and Koppelhus, U. Evaluation of cell-penetrating peptides (CPPs) as vehicles for intracellular delivery of antisense peptide nucleic acid (PNA). *Bioconjug Chem.* 2006;17(3):750–758.

244 Abes, S., Turner, J.J., Ivanova, G.D., et al. Efficient splicing correction by PNA conjugation to an R6-Penetratin delivery peptide. *Nucleic Acids Res.* 2007;35(13):4495–4502.

245 Wright, D.G., Zhang, Y., and Murphy, J.R. Effective delivery of antisense peptide nucleic acid oligomers into cells by anthrax protective antigen. *Biochem Biophys Res Commun*. 2008;376(1):200–205.
246 Deuss, P.J., Arzumanov, A.A., Williams, D.L., and Gait, M.J. Parallel synthesis and splicing redirection activity of cell-penetrating peptide conjugate libraries of a PNA cargo. *Org Biomol Chem*. 2013;11(43):7621–7630.
247 Ghavami, M., Shiraishi, T., and Nielsen, P.E. Cooperative cellular uptake and activity of octaarginine antisense peptide nucleic acid (PNA) conjugates. *Biomolecules*. 2019;9(10):554.
248 Hu, J. and Corey, D.R. Inhibiting gene expression with peptide nucleic acid (PNA)--peptide conjugates that target chromosomal DNA. *Biochemistry*. 2007;46(25):7581–7589.
249 Fabani, M.M. and Gait, M.J. miR-122 targeting with LNA/2′-O-methyl oligonucleotide mixmers, peptide nucleic acids (PNA), and PNA-peptide conjugates. *Rna*. 2008;14(2):336–346.
250 Good, L., Awasthi, S.K., Dryselius, R., Larsson, O., and Nielsen, P.E. Bactericidal antisense effects of peptide-PNA conjugates. *Nat Biotechnol*. 2001;19(4):360–364.
251 Bai, H., You, Y., Yan, H., et al. Antisense inhibition of gene expression and growth in gram-negative bacteria by cell-penetrating peptide conjugates of peptide nucleic acids targeted to rpoD gene. *Biomaterials*. 2012;33(2):659–667.
252 Ma, S., Schroeder, B., Sun, C., et al. Electroporation-based delivery of cell-penetrating peptide conjugates of peptide nucleic acids for antisense inhibition of intracellular bacteria. *Integr Biol (Camb)*. 2014;6(10):973–978.
253 Frimodt-Møller, J., Koulouktsis, A., Charbon, G., Otterlei, M., Nielsen, P.E., and Løbner-Olesen, A. Activating the Cpx response induces tolerance to antisense PNA delivered by an arginine-rich peptide in Escherichia coli. *Mol Ther Nucleic Acids*. 2021;25:444–454.
254 Yokoi, Y., Kawabuchi, Y., Zulmajdi, A.A., et al. Cell-penetrating peptide-peptide nucleic acid conjugates as a tool for protein functional elucidation in the native bacterium. *Molecules*. 2022;27(24):8944.
255 Schulte, L.N., Westermann, A.J., and Vogel, J. Differential activation and functional specialization of miR-146 and miR-155 in innate immune sensing. *Nucleic Acids Res*. 2013;41(1):542–553.
256 Janga, H., Aznaourova, M., Boldt, F., Damm, K., Grünweller, A., and Schulte, L.N. Cas9-mediated excision of proximal DNaseI/H3K4me3 signatures confers robust silencing of microRNA and long non-coding RNA genes. *PLoS One*. 2018;13(2):e0193066.
257 Tan, X.X., Actor, J.K., and Chen, Y. Peptide nucleic acid antisense oligomer as a therapeutic strategy against bacterial infection: proof of principle using mouse intraperitoneal infection. *Antimicrob Agents Chemother*. 2005;49(8):3203–3207.

258 Wancewicz, E.V., Maier, M.A., Siwkowski, A.M., et al. Peptide nucleic acids conjugated to short basic peptides show improved pharmacokinetics and antisense activity in adipose tissue. *J Med Chem*. 2010;53(10):3919–3926.

259 Amantana, A., Moulton, H.M., Cate, M.L., et al. Pharmacokinetics, biodistribution, stability and toxicity of a cell-penetrating peptide-morpholino oligomer conjugate. *Bioconjug Chem*. 2007;18(4):1325–1331.

260 Boisguerin, P., Deshayes, S., Gait, M.J., et al. Delivery of therapeutic oligonucleotides with cell penetrating peptides. *Adv Drug Deliv Rev*. 2015;87:52–67.

261 Ye, J., Liu, E., Gong, J., et al. High-yield synthesis of monomeric LMWP(CPP)-siRNA covalent conjugate for effective cytosolic delivery of siRNA. *Theranostics*. 2017;7(9):2495–2508.

262 Ye, J., Pei, X., Cui, H., et al. Cellular uptake mechanism and comparative in vitro cytotoxicity studies of monomeric LMWP-siRNA conjugate. *J Ind Eng Chem*. 2018;63:103–111.

Section V

Peptide CMC and Regulatory

14

Peptides Chemistry, Manufacturing, and Controls

Jack Xu

Elixtide LLC, San Marcos, CA, USA

14.1 Introduction

Peptides are a promising class of therapeutic modalities and have received more attention as a significant portion of the drugs approved by the Food and Drug Administration (FDA). Peptides are considered to fall between small molecules and large biological molecules (e.g. proteins or antibodies) and exhibit the combined advantages of both. To date, over 80 peptides have been commercialized in the global market, over 150 peptides are in active clinical development, and several hundred are in preclinical studies [1,2].

The FDA considers any alpha amino acid polymer composed of 40 or fewer amino acids to be a peptide, not a protein. Peptides are regulated as a drug under the FD&C Act unless the peptide otherwise meets the statutory definition of a "biological product" (e.g. a peptide vaccine) [3]. Regulatory challenges persist relating to establishing quality standards for synthetic peptides; no official industry guidance has been provided by the FDA or the International Conference on Harmonisation (ICH), though the European Medicines Agency has recently drafted a guideline on the development and manufacture of synthetic peptides for public consultation [4–6].

When a new chemical entity (NCE) peptide is nominated for therapeutic development after discovery and lead optimization, the chemistry, manufacturing, and controls (CMC) development starts and continues through all stages of the product life cycle. Sufficient CMC information is required to be submitted to assure the proper identification, quality, purity, and strength of the investigational drug, while the amount of information needed to make that assurance will vary with the stage of investigation. CMC is a crucial element of drug development,

Peptide Science: Chemical Ligation, Lead Generation, and Therapeutic Advances, First Edition. Fa Liu.

and adequate CMC information must be provided to the regulatory authorities for review. In the common technical document (CTD) format developed by the ICH, the CMC section is presented as Module 3 (Quality) and consists of a drug substance (DS) or active pharmaceutical ingredient (API) part (3.2.S) and a drug product (DP) part (3.2.P) for an NCE therapeutic [7,8]. Both DS and DP parts contain technical information, processes, and key parameters, and various justifications supported by qualification and/or validation studies. In the CTD format, Arabic numbers are assigned to specific sections. For example, the designation 3.2 before S and P indicates Module 3, Body of Data Section 2.

This chapter begins with an overview of the CMC content and CTD format generally recommended for an investigational drug application (IND). The latter sections mainly discuss the peptide CMC development from the perspectives of manufacturing process development, phase-appropriate considerations, and API control strategies.

14.2 CMC Information and Format for DS

The DS part describes the structure, manufacture, and control of the DS and consists of seven sections (3.2.S.1–3.2.S.7). For the purpose of brevity, the initial designation 3.2 is not repeated throughout the rest of the document (e.g. 3.2.S.1–3.2.S.7 reads S.1–S.7, with the same applicable to 3.2.P.1–3.2.P.8 in the Section 14.3 description for DP).

S.1 General Information

In this section, nomenclature of the peptide (IUPAC name, other applicable identifiers, and names) should be provided. However, in Phase 1 INDs, the international nonproprietary names (INN) and the chemical abstracts service (CAS) registry number may not be available and are not required. Peptide structural information, such as the sequence formula (three-letter amino acid code, nonnatural amino acid), molecular formula, molecular weight, and chemical structure, should be included.

The content of this section should also include the general properties of the peptide, such as solubility, optical rotation, hygroscopicity, partition coefficient, isoelectric point (pI), particle size, polymorphs (if applicable), and biological activity. Generally, a brief description of physical, chemical, and biological properties is sufficient for early-phase INDs.

S.2 Manufacture

In this section, information about the manufacturer and the manufacturing method, controls of materials, critical steps and intermediates, process validation and/or evaluation, and process development, should be provided.

S.2.1 Manufacturer(s)

The manufacturer of the peptide should be described, including descriptions of the name, address of the manufacturer and manufacturing and testing facility, any regulatory identification (such as facility establishment identification (FEI#), data universal numbering system (DUNS#), and additional manufacturers when added at later stages.

S.2.2 Description of Manufacturing Process and Process Controls

This section provides a narrative description of the peptide manufacturing process with key unit operations and process controls. The description should include but not be limited to batch size, material usage (amino acid derivatives, intermediates, reagents, solvents, etc.), reaction times and temperatures for coupling, deprotection, cyclization, cleavage, etc., purification condition, including lyophilization, parameters, and associated in-process controls. Furthermore, a flow diagram of the peptide synthetic process should be provided schematically, depicting starting materials, reactions, intermediates, and products within the process. As the development phase progresses, more information will need to be added. Any changes to process and in-process controls, identification of critical process parameters (CPPs), etc., should be included in the late-stage CMC dossier along with the final peptide manufacturing process with all in-process controls, the batch size range, the ratio of reagents, operating conditions, etc.

S.2.3 Control of Materials

Information on all materials used in the peptide manufacturing process, such as amino acid building blocks, other materials (coupling reagents, catalysts, etc.), and organic solvents, should be provided. Such information includes the manufacturer(s), basis for selecting starting materials, and specifications. In solid-phase peptide synthesis (SPPS), typically preloaded resin with the first amino acid (through a linker), protected amino acid derivatives are the starting materials, and the resin substitution/loading and the chiral purity of amino acids are important specification parameters. Chromatographic materials used in the peptide downstream process should also be provided. More information about starting materials will be required in the late-stage CMC dossiers.

S.2.4 Controls of Critical Steps and Intermediates

Critical steps and ranges of process parameters and control of specifications of intermediates should be provided. Controls for critical steps in the SPPS, such as N^{α}-deprotection, amino acid coupling, acetylation, global deprotection and cleavage, crude isolation, and drying, should be adequately described. For the downstream process, acceptable criteria for fractions and pooling in each purification stage should be defined and provided. These criteria typically include purity of product, individual impurities, and unspecified impurities. Control for microfiltration and lyophilization steps should be provided. The levels of detail increase from the early to late-stage CMCs.

S.2.5 Process Validation and/or Evaluation

Process validation is not required in Phase 1 and Phase 2 clinical stages but must be conducted and completed before commercial distribution and provided in this subsection. Generally, the process validation and/or evaluation is carried out in the Phase 3 stage.

S.2.6 Manufacturing Process Development

A description and discussion of peptide synthesis route selection/evaluation. Significant changes to the manufacturing process and/or manufacturing site, etc., should be included here.

S.3 Characterization

Peptide characterization includes the elucidation of structure and other characteristics (such as solid-state characteristics if applicable) and impurity characterization.

S.3.1 Elucidation of Structure and Other Characteristics

The structure of the peptide, including the primary, secondary, tertiary, and/or quaternary structure, should be confirmed by analytical data where relevant. Confirmation of the peptide structure based on the synthesis route and relevant spectral analyses should be provided. Information such as the potential for isomerism and the identification of stereochemistry should also be included. The peptide sequence (primary structure) should be confirmed by appropriate analytical technologies, such as mass spectrometry (MS), nuclear magnetic resonance spectrometry (NMR), amino acid analysis (AAA), etc. NMR, circular dichroism, ultraviolet (UV), and infrared (FT-IR) spectroscopy can be used to confirm the secondary structure. Physicochemical characterization could include solubility and hygroscopicity studies, determination of the pI, and thermogravimetric studies.

S.3.2 Impurities

For synthetic peptides, purity is one of the most important critical quality attributes (CQAs). Their impurities are typically classified into peptide-related impurities (related substances) and non-peptidic impurities. Peptide-related impurities involve structural elements of the synthetic peptide sequence and can be deletions, insertions, degraded peptides, diastereoisomers, modified peptides and aggregates, etc. Non-peptidic impurities include process reagents and their byproducts, residual solvents, and elemental impurities, which may be potentially genotoxic.

S.4 Control of the DS

This section contains peptide API specification and justification, analytical procedures and method validation information, a description of the batches, and their results. The specification provided should cover the CQAs, consisting of a list of tests, references to analytical procedures, and appropriate acceptance criteria or ranges.

S.4.1 Specification

A typical specification for a peptide DS includes appearance, identity, purity, related substances (single impurities, total impurities, unspecified/unidentified impurities), peptide content (assay), aggregate (if applicable), counterion, residual trifluoroacetic acid (TFA) content (if applicable), moisture content, residual solvents, elemental impurities, endotoxins, and bioburden. Other attributes may also be required depending upon the use of the peptide.

S.4.2 Analytical Procedures

The analytical procedures used for testing the DS should be provided. Analytical methods routinely used for the release of a synthetic peptide are high-performance liquid chromatography (HPLC), ultra-high-performance liquid chromatography (UHPLC), liquid chromatography-mass spectrometry (LC-MS), mass spectrometry (MS), and AAA. Chromatographic methods are most useful in determining product purity and related substances due to their high resolving power. For counterions, water, residual solvents, elemental impurities, specific optical rotation, bacterial endotoxins, and microbiology, compendial methods or suitable in-house methods can be employed.

S.4.3 Validation of Analytical Procedures

Analytical method validation information, including experimental data for the analytical procedures used for testing the peptide API, should be provided when appropriate (full validation is typically conducted during late Phase 2 or Phase 3).

S.4.4 Batch Analyses

A description of batches and results of analyses should be provided in this section. Batches and analyses, including the lots used in safety, efficacy, bioavailability/bioequivalence, and primary stability studies, should be provided. A comparison of results from the batches used in toxicology studies versus those used for clinical studies should be conducted. The levels of each impurity type in the clinical batches typically should not exceed those of the toxicology batches.

S.4.5 Justification of Specification

Justification for the proposed specification should be provided. A summary of data from other sections with a cross-reference to the detailed information can be provided to support the justification. The justification should be based on relevant development data, pharmacopeial standards, test data for DSs and DPs used in toxicology and clinical studies, and stability results as appropriate. A reasonable range of expected analytical and manufacturing variability should be considered.

S.5 Reference Standards and Materials Information

A reference standard, or reference material, is a substance prepared for use as the standard in an assay, identification, or purity test. Reference substances must be characterized by multiple analytical techniques and qualified for their intended uses. All analytical results of reference standards (and/or reference materials used as reference substances) should be provided.

S.6 Container Closure System(s)

A description of the container closure system(s) should be provided, including the identity of materials of construction of each packaging component (primary and secondary), their specifications, and the basis and support of the selection (for example, choice of materials, protection from moisture and light, compatibility).

S.7 Stability

Information provided should include a stability summary and conclusions, a post-approval stability protocol and stability commitment when applicable, and stability data.

S.7.1 Stability Summary and Conclusions

The types of studies conducted, protocols used, and the results of the studies of all stability batches should be summarized. The summary should include results, for example, from forced degradation studies and stress conditions, as well as conclusions regarding storage conditions and retest date or shelf life. The degradation pathway for the DS and any critical intermediates should be elucidated as appropriate. Conclusions with respect to storage conditions and retest date or shelf life should be provided as appropriate.

S.7.2 Post-approval Stability Protocol and Stability Commitment

The post-approval stability protocol and stability commitment should be provided if a new drug application (NDA) submission does not include long-term stability data on three production batches. This is typically a commitment to initiate or continue studies and provide the data to the FDA.

S.7.3 Stability Data

Results of the stability studies, including stress conditions, should be presented in an appropriate format, such as tabular, graphical, or narrative. Information on the analytical procedures used and, if applicable, validation of these procedures should be included or referenced.

14.3 CMC Information and Format for DP

The DP part (3.2.P) describes the composition, manufacture, and control of the DP. Drug product CMC information is organized into eight sections (P.1–P.8).

P.1 Description and Composition

A description of the DP and its composition should be provided. All components of DPs, their amounts on a per-unit basis, functions, and quality (e.g. manufacturer's specifications or compendial monographs) should be included. Information about the type of container and closure for the DP should also be provided.

P.2 Pharmaceutical Development

Development studies and results should be provided to demonstrate that the dosage form, the formulation, the manufacturing process, the container closure system, the microbiological attributes, and the usage instructions are appropriate for the purpose specified in the application.

P.2.1 Components of the DP

This section includes the two categories of components: peptide API and excipients (inactive ingredients). The properties of the peptide API that can influence the performance of the DP should be discussed. The compatibility of the peptide with excipients and with other DSs (if applicable) should be discussed. The identity and concentration of excipients in the DP as well as characteristics that can influence DP performance should be discussed relative to their respective functions. If novel excipients are used, manufacturing information needs to be provided.

P.2.2 Drug Product

This section provides information of DP on formulation development, overages, and physicochemical and biological properties.

A description of the formulation development for the proposed route of administration and usage should be provided, which should include a brief history of how the final formulation was developed (e.g. the formulations evaluated and the formulation selected for further development). The differences in formulation between clinical formulations and the formulation described in 3.2.P.1 should be discussed.

Any overages in the formulation(s) described in 3.2.P.1 should be justified. Information should be provided on the amount of overage, the reason for the overage, and the justification for the amount of overage.

The physicochemical and biological properties relevant to the safety, performance, or manufacturability of the DP should be discussed. This includes the physiological implications of API and formulation attributes. Parameters should be addressed including pH, ionic strength, dissolution, redispersion, reconstitution, particle size distribution, particle size of the lend/granules flow, properties that might affect capsule filling or tableting, aggregation, polymorphism, rheological properties, biological activity or potency, and/or immunological activity.

P.2.3 Manufacturing Process Development

The selection, control, and any improvement of the manufacturing process described in 3.2.P.3.3 should be explained. Process development studies should provide the basis for process improvement, process validation, continuous process verification (where applicable), and any process control requirements. Appropriateness of the equipment used for the intended products should be discussed.

Manufacturing process development should identify any CPPs that should be monitored or controlled to ensure that the product is of the desired quality. The discussion should include an emphasis on the reproducibility of the CQAs of the DP. Changes to the method of manufacture should be detailed as the process moves from initial phase 1 studies through to the final commercial process. Significant differences between the manufacturing processes used to produce batches for pivotal clinical trials (safety, efficacy, bioavailability, and bioequivalence) or primary stability studies and the process described in 3.2.P.3.3 should be discussed. The discussion should summarize the influence of the differences on the performance, manufacturability, and quality of the product.

For sterile products, an appropriate method of sterilization for the DP and primary packaging material should be chosen and justified.

P.2.4 Container Closure System

The container closure system used for the storage, transportation (shipping), and use of the DP should be discussed concerning the selection of materials, protection from moisture and light, compatibility and safety of the materials of construction, and performance. The selection of primary packaging should be based on the knowledge of the physical and chemical behavior of the DP in pre-formulation and subsequent stability studies, and it should be adequately justified.

P.2.5 Microbiological Attributes

The microbiological attributes of the dosage form should be discussed. For non-sterile products, the rationale for not performing microbial limits testing and the selection and effectiveness of preservative systems in products containing antimicrobial preservatives should be provided and discussed. For sterile products, quality expectations are clearly delineated in the United States and European Union to meet compendial requirements. A strategy to control endotoxins in excipients must be developed, corresponding to the route of administration and dosing regimen of the sterile product. The integrity of the container closure system to prevent microbial contamination should also be addressed.

P.2.6 Compatibility

Where relevant, compatibility of the DP with reconstitution diluents or dosage devices (e.g. precipitation of API in solution, sorption on injection vessels, stability) should be addressed to provide appropriate and supportive information for the labeling.

P.3 Manufacture

This section provides information regarding the manufacturer, batch formula, manufacturing process and process control, controls of critical steps and intermediates, and process validation and/or evaluation.

P.3.1 Manufacturer(s)

The manufacturer information includes the name, address, FEI number if applicable, and responsibility of each DP manufacturer, including manufacturing facility, testing facility, labeling site, and contractors if relevant activities are outsourced.

P.3.2 Batch Formula

A batch formula should be provided that includes a list of all components of the dosage form to be used in the manufacturing process, their amounts on a per-batch basis, including overages, and a reference to their quality standards.

P.3.3 Description of Manufacturing Process and Process Controls

A process flow diagram should be provided showing the manufacturing process, in-process controls, and steps where materials enter the process, eventually leading to the final DP. A narrative description of the manufacturing process should be provided, including step-by-step unit operations and packaging that represent the sequence of steps undertaken and the scale of production. Steps in the process should have the appropriate process parameters identified, such as time, temperature, or pH. The critical steps and points at which process controls, intermediate tests, or final product controls are conducted should be identified. Equipment utilized should be identified by type and capacity.

P.3.4 Control of Critical Steps and Intermediates

Tests and acceptance criteria should be provided and justified for critical in-process controls and/or parameters identified in the manufacture of the DP to demonstrate that the process is adequately controlled. If there are any intermediate products isolated in the process, information on the quality and control of intermediates should also be provided.

P.3.5 Process Validation/or Evaluation

Information on the status of process validation should be provided. When appropriate, descriptions, documentation, and results of the validation and/or evaluation studies should be provided for critical steps or assays used in the manufacturing process (e.g. validation of the sterilization process or aseptic processing or filling if applicable).

P.4 Control of Excipients

Information to be provided in this section includes excipients used, their specifications and justifications, and analytical procedures used for testing and method validation.

P.4.1 Specification(s)

All excipients used and their information on grade and standard should be provided. For pharmacopoeial excipients, generally, it is sufficient to state that the excipient is tested according to the requirements of that standard, and no specification is required. For non-pharmacopoeial excipients, an appropriate in-house specification should be established based on physical, identification, purity, assay, and other relevant tests.

P.4.2 Analytical Procedures

Generally, the analytical procedures used for testing the excipients should be provided when no references to a pharmacopoeial monograph can be made.

P.4.3 Validation of Analytical Procedures

When appropriate, analytical validation information, including experimental data, for the non-pharmacopoeia analytical procedures used for testing the excipients should be provided.

P.4.4 Justification of Specifications

Justification for the proposed excipient specifications should be provided where appropriate.

P.4.5 Excipients of Human or Animal Origin

For excipients of human or animal origin, information regarding adventitious agents should be provided (e.g. sources, specifications, description of the testing performed, and viral safety data). Adequate measures should be taken to ensure that any animal-derived ingredients used in the product are free from transmissible spongiform encephalopathy (TSE) contamination. For excipients manufactured from raw material obtained from sources that have the potential to transmitting bovine spongiform encephalopathy (BSE) and/or TSE agents, a letter of attestation should be provided confirming that the excipients used to manufacture the DP are without risk of transmitting agents of BSE and/or TSE.

P.4.6 Novel Excipients

For novel excipients used for the first time in a DP or by a new route of administration, full details of manufacture, characterization, and controls, with cross-references to supporting safety data (nonclinical and/or clinical), should be provided (the requirements similar to those for an NCE DS).

P.5 Control of DP

Information provided in this section includes DP specifications and justification, analytical procedures used and validation of the procedures, batch analyses, and characterization of impurities (e.g. degradation products).

P.5.1 Specifications

The specification(s) for the DP should be provided. The specification(s) should preferably be tabulated, indicating the methods used for each test (including references to international quality standards where applicable) and the acceptance criteria. Specifications applicable for release and shelf life should be clearly identified.

P.5.2 Analytical Procedures

The analytical procedures used for testing the DP should be provided. Each analytical procedure should be identified as either compendial or in-house. For in-house analytical procedures, a description of the method should be provided with information such as analytical conditions, system suitability tests, relative response factors, and method of quantification (external standard, normalization).

P.5.3 Validation of Analytical Procedures

Depending on the stage of product development, the validation results should be summarized and tabulated.

Analytical validation information for the analytical procedures used for testing the DP should be provided, including experimental data. These methods may include, but are not limited to, (i) identity assays for a DS, intermediates, and excipients; (ii) content assays for a DS, intermediates, and excipients; (iii) impurity profiling and quantification assays for a DS and proposed DP; (iv) dissolution assays for a proposed DP or DPs if more than one is included in the marketing application; and (v) stability-indicating assays for a DS and proposed DP. The report, data sheets, and typical chromatograms should be provided. A description of the data presented, evidence of supporting stability indicating the methods, and deviations, if any, should be provided.

P.5.4 Batch Analyses

A description of batches and results of batch analyses should be provided. A description and tabulation of the batches (e.g. strength, batch number and size, manufacturing site and date, use of the batch) and batch results, certificate of analyses for the batches, and typical spectrums and chromatograms should be included.

P.5.5 Characterization of Impurities

Information on the characterization of impurities should be provided, if not previously provided in S.3.2. Potential degradation products may include impurities identified in the DS section and degradation products resulting from the interaction of the DS with excipients or container closure system.

P.5.6 Justification of Specification(s)

Justification for the proposed DP specification(s) should be provided. The justification should include relevant tests and comment on the suitability, adequacy, and acceptability of the tests and the proposed acceptance criteria. The acceptance criteria should be based on the levels found in historic batches (e.g. used for nonclinical, clinical, and bioequivalence studies), relevant guidelines, and pharmacopoeia requirements.

P.6 Reference Standards or Materials

Information on the reference standards or reference materials used for testing the DP should be provided, including the description, source of reference standards, or reference materials for analysis of the DS and impurity(s). If the same reference standard(s) is/are used in the testing of the DP as in the DS, 3.2.S.5 may be referred to, where applicable.

P.7 Container Closure System

A description of the container closure system(s), including the identity and specification of materials of construction of each primary packaging component, should be provided. The specifications should include description, identification, and critical dimensions. A brief description of nonfunctional secondary packaging components and additional information for functional secondary packaging components should also be provided. Extractable and leachable data to support the container closure should be provided as needed.

P.8 Stability

This section is divided into three portions: a subsection indicating the stability summary and conclusions, a subsection on the post-approval stability protocol and commitment, and a subsection with the actual stability data/results in an appropriate format (e.g. tabulated form).

P.8.1 Stability Summary and Conclusion

The types of studies conducted, protocols used, and their results should be summarized. The summary should include, for example, conclusions with respect to storage conditions and shelf life, and, if applicable, in-use storage conditions and shelf-life. Information provided includes strength, batch numbers and sizes, manufacturing site, manufacturing date, container closure system(s), orientation for liquid dosage forms (e.g. upright or inverted), storage conditions (long-term, intermediate, and accelerated as applicable), and testing intervals.

P.8.2 Stability Protocol and Stability Commitment

Stability protocol and stability commitment should be provided when applicable. The stability protocols for the commitment batches (primary and commercial batches as necessary) and continuing (i.e. ongoing) batches should include, but are not limited to, the number of batches per strength and batch sizes, tests and acceptance criteria, container closure system(s), testing frequency, and storage conditions (and tolerances) of samples. Stability commitment (after approval) includes placement of the first three production batches (accelerated and long-term studies through the proposed shelf life) on stability studies, and thereafter, at least one batch per year of each strength of DP manufactured at each commercial site to confirm the stability unless otherwise justified.

P.8.3 Stability Data

Results of the stability studies should be presented in an appropriate format (e.g. tabular, graphical, and narrative). Information on the analytical procedures used to generate the data and validation of these procedures should be included. A discussion of stability data and trends should be provided, including observations noted for the various tests, ranges of analytical results, and any trends observed.

14.4 Synthetic Peptide CMC Development

Peptides are widely recognized as therapeutic agents in the treatment of various diseases such as metabolic diseases, tumors, cardiovascular, infectious diseases, and neurodegenerative disorders. Peptides with 40 or fewer amino acids, regardless of how they are made, are not biologics and are regulated as drugs under the FD&C Act in the United States (except for peptide vaccines). The main goals of CMC development for peptide APIs are to establish product characteristics, testing methods, and commercial manufacturing processes. Peptide manufacturing process development is a crucial element for the establishment of commercial processes and includes upstream and downstream processes, in-process controls, scale-up, and finally validation of the process.

14.4.1 Peptide Process Development

Peptides have been produced using both chemical and biological methods. Major techniques of peptide production include extraction, enzymatic hydrolysis, microbial fermentation, recombinant DNA technology, chemical synthesis, etc. [9]. These techniques can be applied individually or in combination depending on the complexity of peptides. In the past two decades, the two most popular methods for producing therapeutic peptides certainly are recombinant synthesis and chemical synthesis (Table 14.1).

14.4.1.1 Recombinant Production Method

In recombinant synthesis, a peptide is produced through recombinant DNA technology. This method involves inserting the DNA encoding the peptide into bacterial or mammalian cells, expressing the peptide in these cells and then isolating/purifying it. Human insulin and growth hormone are two early examples of peptides produced using this technology, with insulin being the first recombinant therapeutic polypeptide commercialized in 1982.

Recombinant process development involves the selection of an appropriate expression system, construction of expression vectors, and establishment of a bioprocess. Based on the expression system used, recombinant production is classified into *in vivo* and *in vitro* methods. The *in vitro* method is a more advanced technique in which the components for transcription and translation of a peptide gene are present *in vitro*, and synthesis takes place in the absence of the cell, and thus it is also called the cell-free system. The recombinant technique is primarily suited for the peptides containing only natural amino acids due to a lack of straightforward and effective methods for incorporating modifications (e.g. non-natural amino acids and novel structural elements).

Development of a recombinant manufacturing process is highly specific to the target peptide and is generally time-consuming and capital-intensive in the development stage. However, it has advantages in the feasibility of large-scale production, cost-effectiveness in commercialization, and environmental friendliness. FDA-approved top-selling peptide drugs produced by recombinant synthesis include teriparatide (2002), liraglutide (2010), teduglutide (2012), dulaglutide (2014), and semaglutide (2017).

14.4.1.2 Chemical Synthesis

Chemical synthesis is the preferred method to produce various-size peptides for fast deliveries of products. One huge advantage is that this production method allows the incorporation of unnatural or modified amino acids in the peptide chain. Chemical synthesis relies on organic chemistry using appropriately protected amino acid derivatives, organic solvents, and auxiliary chemicals. The chemical synthesis method is further divided into three subcategories: liquid-phase peptide synthesis (LPPS), SPPS, and hybrid synthesis. These chemical methods are well documented in the literature and have been demonstrated in the industry setting.

Table 14.1 Some peptide-based drugs (<50 AAs in length) approved by the FDA since 2000.

Approval	Peptide/API	Length/structure	Primary method of production
2000	Triptorelin	10 AAs	Chemical synthesis: liquid-phase peptide synthesis (LPPS)
2000	Cetrorelix	10 AAs	Chemical synthesis: LPPS
2000	Bivalirudin	20 AAs	Chemical synthesis: LPPS
2001	Nesiritide	32 AAs	Recombinant synthesis: rDNA
2002	Teriparatide	34 AAs	Recombinant synthesis: rDNA
2003	Abarelix	10 AAs	Chemical synthesis: solid-phase peptide synthesis (SPPS)
2003	Enfuvirtide	36 AAs	Chemical synthesis: Hybrid
2004	Ziconotide	25 AAs, 3 disulfide bridges	Chemical synthesis: SPPS
2005	Exenatide	39 AAs	Chemical synthesis: SPPS
2005	Pramlintide	37 AAs, 1 disulfide bridge	Chemical synthesis: SPPS
2007	Lanreotide	8 AAs	Chemical synthesis: SPPS
2008	Degarelix	8 AAs	Chemical synthesis: LPPS
2010	Tesamorelin	44 AAs, N-terminal modified by *trans*-3-hexenoic acid	Chemical synthesis: SPPS
2010	Liraglutide	31 AAs, Lys side-chain modification (C16 fatty acid and γGlu)	Recombinant synthesis: rDNA
2011	Icatibant	10 AAs	Chemical synthesis: SPPS
2012	Sinapultide	21 AAs	Chemical synthesis: SPPS
2012	Insulin Degludec	49 AAs, 3 disulfide bridges, Lys side-chain modification (C16 fatty acid and γGlu)	Recombinant synthesis: rDNA
2012	Peginesatide	21 AAs	Chemical synthesis: SPPS
2012	Carfilzomib	4 AAs, modification at C- and N-termini	Chemical synthesis: LPPS
2012	Pasireotide	6 AAs	Chemical synthesis: SPPS
2012	Teduglutide	33 AAs	Recombinant synthesis: rDNA
2016	Lixisenatide	44 AAs	Recombinant synthesis: rDNA

(Continued)

Table 14.1 (Continued)

Approval	Peptide/API	Length/structure	Primary method of production
2017	Linaclotide	14 AAs, 3 disulfide bridges	Chemical synthesis: SPPS
2017	Abaloparatide	34 AAs	Chemical synthesis: hybrid
2017	Angiotensin II	8 AA	Chemical synthesis: SPPS
2017	Etelcacetide	8 AAs, 1 disulfide bridge	Chemical synthesis: SPPS
2017	Macimorelin	3 AAs, *N*-formylated peusdotripeptide	Chemical synthesis: LPPS
2017	Plecanatide	16 AAs, two disulfide bridges	Chemical synthesis: SPPS
2017	Semaglutide	31 AAs, Lys side-chain modification (C18 fatty diacid, γGlu, and AEEAc)	Recombinant synthesis: rDNA (also partially SPPS)
2019	Afamelanotide	13 AAs	Chemical synthesis: LPPS
2019	Bremelanotide	7 AAs, cyclized through side chains (Asp and Lys)	Chemical synthesis: SPPS
2019	Ga-68 Edotreotide	8 AAs, 1 disulfide bridge	Chemical synthesis: SPPS
2020	Setmelanotide	8 AAs, 1 disulfide bridge	Chemical synthesis: LPPS
2021	Voclosporin	11 AAs, head-to-tail cyclized	Semi-synthesis
2021	Vosoritide	39 AAs, 1 disulfide bridge	Recombinant synthesis: rDNA
2021	Difelikefalin	4 AAs, D-amino acids	Chemical synthesis: LPPS, SPPS
2021	Pegcetacoplan	Two identical 13-AA peptides linked to a branched 40 kDa polyethylene glycol (PEG)	Chemical synthesis: SPPS
2021	Dasiglucagon	29 AAs	Chemical synthesis: SPPS
2022	Terlipressin	12 AAs, 1 disulfide bridge	Chemical synthesis: SPPS
2022	^{177}Lu Vipivotide Tetraxetan	4 AAs, with a urea linkage	Chemical synthesis: LPPS, SPPS
2022	Tirzepatide	39 AAs, Lys side-chain modification (C20 fatty acid, γGlu, and AEEAc)	Chemical synthesis: Hybrid
2022	Daxxify	35 AAs	Chemical synthesis: SPPS
2023	Trofinetide	3 AAs	Chemical synthesis: LPPS

Approval	Peptide/API	Length/structure	Primary method of production
2023	Rezafungin	6 AAs, cyclic lipopeptide	Semi-synthesis
2023	Flotufolastat F18	7 AAs, with a urea linkage	Chemical synthesis: SPPS
2023	Motixafortide	14 AAs, 1 disulfide bridge	Chemical synthesis: SPPS
2023	Zilucoplan	15 AAs, cyclized via cyclized via side chains (Asp and Lys), N-terminal Lys side-chain modification (C16 fatty acid, γGlu, and PEG24)	Chemical synthesis: SPPS

14.4.1.2.1 Liquid-phase Peptide Synthesis

Peptide synthesis was originally performed in solution, in which all reactions of amino acids occur in a solubilized state, known as solution-phase peptide synthesis or LPPS. The best-known LPPS is the synthesis of insulin (1953) [10]. An important advantage of this classic solution synthesis is the use of fewer raw materials benefiting from the preferred reaction kinetics and the opportunity for purification at each step of the synthesis. On the other hand, the need for isolation and purification of intermediates often makes the overall process lengthy and challenging. The major limitations of LPPS were and still are the length of the peptide and its time and labor intensity. Examples of peptide drugs initially produced by LPPS are desmopressin (1984), leuprolide (1985), buserelin (1985), goserelin (1987), lisinopril (1987), octreotide (1988), eptifibatide (1998), and others as listed in Table 14.1.

A subtype of LPPS that is different from classic solution synthesis is peptide synthesis using soluble tags. The soluble tags are used as a carrier for synthesis, which allows a reaction in solubilized conditions and ease in purification and isolation of intermediates at the end of each step (e.g. precipitate/filter or extract/phase-separate). In such LPPS, the properties of tags chosen must be significantly distinct from other reagents and byproducts to facilitate their simple removal. Some tags that have been evaluated include polydisperse PEG, non-cross-linked polystyrene, fluorous tags, ionic liquids, polycarbonate, and hydrophobic polymers. This method combines the advantages of solution-phase and solid-phase chemistry and is best suited for the synthesis of short peptides. Currently, there are no examples of approved drugs produced by this technology. However, interest in this method continues to grow due to its significantly reduced use of reagents and solvents and potential for scalability [11,12].

14.4.1.2.2 Solid-phase Peptide Synthesis

SPPS was invented in 1963 by Merrifield [13]. This technology utilizes an insoluble polymer (resin beads) as a solid support carrier for the synthesis. The peptide chain is assembled on the polymer, during which excessive reactants and reagents are used to quickly drive the reaction to completion. After each reaction, simple filtration and washing with solvent readily remove excessive reactants and separate them from the solid polymer-carrying peptide chains. Additionally, the SPPS process facilitates the entire peptide synthesis by combining amino acid coupling and deprotection in one single reactor. Presently, two SPPS synthesis approaches have been well established based on the type of $^{\alpha}$N-protection chemistry for amino acids: t-butyloxycarbonyl (Boc)-SPPS and fluorenylmethyloxycarbonyl (Fmoc)-SPPS. Both approaches have been well adopted in the pharmaceutical industry.

In Boc-SPPS, the alpha-amino group of the amino acid derivative is protected by the Boc group. During the synthesis, TFA solution is used to deprotect the Boc group. At the end of synthesis, hydrogen fluoride (HF) is used to deprotect side-chain protection groups and to release the peptide from the solid support. HF is highly corrosive and toxic, thus HF cleavage requires a special setup and is difficult to scale up.

In Fmoc-SPPS, the Fmoc group is used for the protection of the alpha-amino group of amino acid. The Fmoc group can be removed under milder conditions (e.g. piperidine solution). As opposed to Boc-SPPS, the cleavage of the peptide from the solid support after the Fmoc-SPPS is accomplished by a less hazardous acid, TFA. The TFA cleavage process is more scalable than the HF cleavage: for example, an industry-scale TFA cleavage allows to process quantities varying from kilograms to a hundred kilograms of peptide resin. Generally speaking, Fmoc-SPPS is preferred over Boc-SPPS.

The simplicity of SPPS has made it the most widely used method for producing peptides for research and industry. Nowadays, small-scale peptide synthesis is often carried out by automated peptide synthesizers, which may only take a few days to synthesize a peptide with ≤50 amino acids in length for research purposes. Roughly 50% of the peptide drugs approved by the FDA since 2000 are manufactured by SPPS (Table 14.1), undisputedly the leading method. A more detailed comparison between LPPS and SPPS is summarized in Table 14.2.

14.4.1.2.3 Hybrid Synthesis

In addition to the two primary methods of chemical synthesis, another synthesis strategy is called hybrid synthesis. The hybrid approach takes combined advantages of SPPS and LPPS, connecting protected peptide fragments made from SPPS into a full-length peptide through fragment condensations in solution (LPPS). For example, large-scale amounts of high-quality, protected peptide fragments (generally <20 amino acids) can readily be produced by SPPS and are then used in almost stoichiometric amounts for condensation in solution.

Table 14.2 Comparison between LPPS and SPPS.

Liquid-phase peptide synthesis (LPPS)	Solid-phase peptide synthesis (SPPS)
• More suitable for shorter peptides	• For small- to medium-length peptides
• Greener process due to the use of stoichiometric amounts of reagents and much less solvent, having lower raw material cost	• Use of excessive reagents and solvent, and high process mass intensity; raw material cost is generally higher
• Homogeneous reactions, easy in-process controls (IPCs), and more reliable IPC results	• Heterogeneous reaction, limited IPCs, and variation in sample preparation affect results
• Isolation of intermediates required at each synthesis step	• No isolation of intermediates is needed during the entire synthesis
• Workups and isolations of intermediates can be time-consuming and tedious in operations; full automation is not feasible	• Excessive reagents are simply removed by filtration, eases operations, and full automation is available
• Development generally takes a much longer time; scale-up needs to be done gradually	• Has a shorter development time (well-established technology) and is readily scalable
• Generally, no chromatographic purification is needed for the final product, and isolation of final API is often accomplished by precipitation and recrystallization in the common reactor (less restriction in scale)	• Chromatographic purification is mostly required. Purification can be intensive and challenging due to close impurities accumulated in SPPS; product isolation is often limited by the size of lyophilizer (equipment capacity)
• Once established, the production scale can be very large (more cost-effective for commercial manufacturing)	• Production scale is often limited by equipment capacity, such as SPPS reactor, prep-HPLC, and lyophilizer

The major advantages of a hybrid process over a stepwise SPPS approach lie in a simplified chromatographic purification and higher overall yield, having benefited from the use of high-quality peptide fragments. Uncommonly, peptide fragments from solution synthesis can also be used in fragment condensation in SPPS. For example, if a peptide contains several repeated sequence units, the repeat unit can be made by LPPS on a large scale (more cost-effective) and with high quality.

For the production of protected peptide acid fragments in Fmoc chemistry, 2-chlorotrityl chloride resin (2-CTC resin) is the most popular resin. Attachment of the C-terminal-protected amino acid to the resin is easy to control and does not involve a risk of racemization. The acid-labile property of the trityl linkage allows the liberation of the protected peptide from the resin in mild cleavage conditions by using diluted acids (≤5% TFA) or acidic solvents (trifluoroethyl alcohol [TFE] or hexafluoroisopropanol [HFIP]).

A well-known example of large-scale hybrid peptide synthesis is the manufacture of enfuvirtide (T-20), a peptide consisting of 36 all-natural amino acids for the treatment of human immunodeficiency virus. According to the report [14], the initial production process used the stepwise Fmoc-SPPS approach with an overall yield of about 6–8%. This yield is considered quite reasonable for a 36-mer peptide and was sufficient at the time for preclinical and Phase 1 clinical studies (a few kilograms in ~100 g batches). However, a hybrid synthesis process was needed in order to meet a much higher demand for peptide API in the commercial stage (metric tons annually were projected). In the developed hybrid process, three protected peptide fragments (each <20 amino acids) were synthesized using 2-CTC resin in large batch sizes (300–500 kg per batch) with high purity (>90%) and high yield (>80%). Liquid-phase fragment condensations were then employed to assemble them into the full-length sequence.

Indeed, the high quality of fragments contributed to a better crude quality over the crude produced by the stepwise Fmoc-SPPS method, as the manufacturing process with this hybrid strategy achieved a satisfactory overall yield of about 30%. A more recent example is the synthesis of tirzepaptide, a 39-mer dual glucagon-like peptide-1 and glucose-dependent insulinotropic peptide receptor agonist, for dual-targeted treatment of diabetes and obesity [15]. A four-fragment-based hybrid synthesis strategy was employed in conjunction with nanofiltration technology. Each fragment was individually synthesized by Fmoc-SPPS with a high purity (>97.0%). This hybrid synthesis reported an LPPS yield of >55% with a crude purity of >80%, thus facilitating a simplified chromatographic purification.

14.4.1.2.4 Synthesis Strategy

Different production methods have their own advantages, challenges, and limitations. The selection of a synthesis strategy for the production of a specific peptide is highly relevant to the length and complexity of the peptide and the production scale. Peptide length is the first factor to be considered when choosing a synthetic strategy. Figure 14.1 shows a schematic representation of the general applicability and practicability of synthetic methods in relation to peptide length.

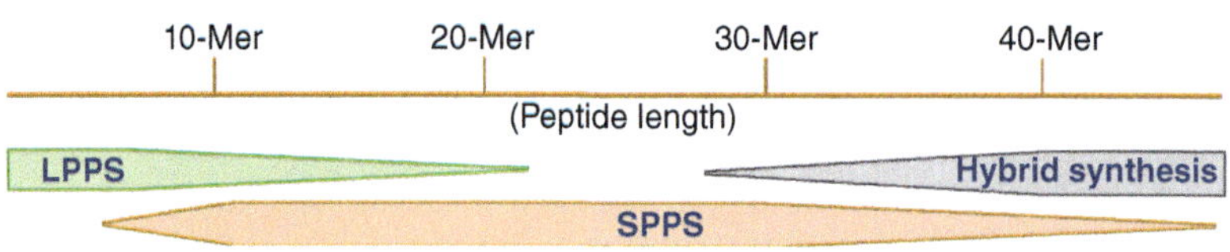

Figure 14.1 Synthetic methods versus peptide length.

Short peptides (2~10 amino acids) can easily be synthesized by either LPPS or SPPS. LPPS produces peptides of shorter sequences in high quality and does not require chromatographic purification. Generally speaking, the shorter the sequence and the larger the batch size, the more advantageous it will be to use LPPS.

Standard Fmoc-SPPS is the most widely used technique for producing peptides containing 5–50 amino acids. As peptide length increases in the upper portion of the range, stepwise SPPS is expected to generate more impurities that can also be more challenging to remove from the crude peptide. This will increase levels of complexity and difficulty in downstream purification and ultimately impact the efficiency of purification, yield, and cost of production. In small-scale production, such limitations may be acceptable, but in large-scale production, it will become a bottleneck or even an obstacle. Therefore, for a long-sequence peptide in need of a large batch size, a hybrid synthesis process should be developed.

The selection of appropriate fragments for SPPS is key to consider in the hybrid strategy. Protected fragments should be chosen to allow for high-quality, large-scale production with isolated intermediates. Subsequent fragment condensations in LPPS should not have major issues related to coupling efficiencies and racemization. The development of hybrid processes requires a significant amount of time and resources and should be carefully planned. This generally pays off, as the crude obtained from the hybrid synthesis enables a more efficient purification process and a higher recovery in production than from a stepwise SPPS process.

While the hybrid method can be more attractive and beneficial for complex and long peptide sequences, decisions should still be made on a case-by-case basis. Factors to consider include synthesis and purification difficulty, understanding of the peptide, near- and long-term API demand, regulatory strategy, and advantages and limitations of each approach.

The physical and chemical properties of a given peptide may also increase the complexity of the manufacturing process. Based on the sequence, amino acid composition, and type of modifications, one can predict potential difficulties in the synthesis and/or purification of a particular peptide. For example, a highly hydrophobic peptide may face chain aggregation during SPPS, leading to difficulties in couplings and deprotections. In such a situation, analytical development and downstream purification will also be challenging. Therefore, proper design of synthesis and necessary development are often required.

14.4.2 Phase-appropriate Considerations in CMC Development

14.4.2.1 The Graded Nature

In the early stages of development, limited knowledge about the peptide is still being accumulated. The regulatory agency reviews the Phase 1 CMC of an IND mainly from the patient safety perspective, generally relating to the identification,

Table 14.3 Typical CMC development activities for NCE peptide APIs.

Preclinical	Phase 1	Phase 2	Phase 3	Launch
Non-GMP	GMP production			Commercial
Process development		Scale-up	Process characterizations, PPQ	CMC lifecycle management
Analytical method development		Analytical method validation		
Stability studies: developmental, ICH-compliant				Follow-up

strength, quality, and purity of the API. CMC information for the initial IND submission will also depend on the novelty of the peptide, the formulation/dosage form, and relevant risks. Regardless, minimal CMC information is required in the Phase 1 stage.

As clinical development progresses to the late stages, the regulatory agency's review will expand to include the linkage of product quality between early-phase clinical uses and the one to be marketed. Increasing CMC development activities must be carried out to obtain sufficient information demonstrating control of safety risks and consistency of quality. Table 14.3 shows an abbreviated diagram of typical CMC development activities for an NCE peptide API. There is no doubt that, from the early phase to the second and third phase clinical stages, expectations and requirements for CMC content have increased significantly. This is known as the graded nature of requirements, which necessitates the adoption of a phase-appropriate, risk-based approach during drug development.

14.4.2.2 Preclinical and Phase 1 Stage CMC Development

The preclinical phase of drug development begins after the completion of discovery and lead optimization, when an NCE peptide is nominated for development [16,17]. Typically, a toxicology batch of peptide API needs to be produced to support *in vivo* studies to determine efficacy, toxicity, and pharmacokinetic information. Then, a clinical batch is produced under good manufacturing practice (GMP) conditions to provide materials for Phase 1 clinical trials.

Choosing a peptide synthesis strategy requires consideration of feasibility and scalability from a chemical synthesis and manufacturing perspective. The standard SPPS strategy allows for rapid access to API samples and is generally considered somewhat scalable, making it suitable for early-stage product development since peptides used in preclinical studies are often small in quantity and do not require high purity. Boc- or Fmoc-SPPS, chromatographic purification, and lyophilization are well-established technologies in the field of peptide manufacturing, which generally require only minimum development. Additionally, it is recommended that representative process impurities be considered for inclusion in toxicological materials so that toxicological studies also qualify the impurities.

Development of analytical methods, including raw material tests, in-process tests, and release tests, is an important part of CMC development and should start at this stage. To test peptide identity, high-resolution mass spectrometry, AAA, and NMR spectroscopy are typically used. To analyze purity, impurity, and content, a liquid chromatographic (LC) method is typically used, such as HPLC and UHPLC. A product-specific LC method must be developed to ensure the product meets its intended purpose. A suitable HPLC/UHPLC method for analysis of peptide API should be capable of assessing process-related impurities, degradation impurities, and also stability-indicating. For some peptides, a size-exclusion chromatography (SEC) method is developed to determine high molecular weight impurities such as dimers, polymers, and aggregates. Other common analytical methods for the determination of water content, elemental impurities, residual solvents, counterions, endotoxins, microbial limits, and so on may also be needed for testing the Phase 1 GMP material. Generic methods or compendial methods could be used. In the preclinical and Phase 1 stages, care needs to be paid to the peptide impurity profiles; the clinical batch should not have new impurities nor a higher corresponding impurity level than in the toxicology batch.

On the production side, generic or developmental batch records may be used during the production of GMP material for Phase 1 clinical trials, typically allowing for more flexibility because process knowledge is limited at this stage.

14.4.2.3 Late-phase CMC Development

At the Phase 2 stage, process development will focus on the establishment of a robust and scalable manufacturing process [18]. Based on the process established and experience gained from prior campaigns, a further assessment of basic strategy, including synthesis and downstream process, is still needed. Depending on the level of challenges, a modification of the strategy, or a change to a new strategy, may be considered. Evaluation and optimization of each process step are needed, including reaction conditions, amino acid derivative equivalents, coupling reagent amounts, deprotection conditions, washing cycles, solvent volumes, cleavage cocktail and conditions, purification media, column loading amount, mobile phases, purification conditions, and final product isolation. The developed or optimized process should be suitable for producing material used for Phase 2 clinical studies. A final scale-up production to the commercial batch size needs to be demonstrated, typically in the Phase 2 stage. Once the process and the final specifications are established, all analytical methods need to be validated. Since more process knowledge is obtained at this stage, the batch records are expected to have more details, with acceptance criteria or target values implemented. Instead of generic batch records, a product-specific master batch should be in place as a result of the development.

Process development at the Phase 3 stage focuses on process characterization for all process parameters. Process characterization is an essential step in the development of a commercial process, and it involves the identification and quantification of sources of variation in the manufacturing process that impact product quality and yield. Following the process characterization, process performance qualification (PPQ) (typically at least three batches) will be performed to confirm the process design and to scientifically prove that the process is reproducible and consistently delivers a quality product. The PPQ batches should be placed on stability. All these results are required for an NDA filing.

14.4.3 Control Strategies for Peptide APIs

A CQA is a physical, chemical, biological, or microbiological property or characteristic that should be within an appropriate limit, range, or distribution to ensure the desired product quality. CQAs are part of the overall target product profile based on required clinical performance. Identification of CQAs is critical in the drug development process [19,20]. To ensure that all CQAs are within their appropriate limits, ranges, or distributions, a control strategy should be developed for each peptide API. A strict control strategy should include controls on input material attributes, procedural control, in-process testing, and release testing. Impurities are an important class of drug substance CQAs because they may affect safety and efficacy. Therefore, they should be major concerns during development. Peptide-related impurities, which may originate from raw materials, manufacturing processes, and degradation during manufacturing or storage, must be adequately controlled. The major components of a DS control strategy are shown in Figure 14.2 and will be discussed in the subsequent sections.

14.4.3.1 Raw Materials Control

Raw materials used in the manufacture of peptides mainly include starting materials, reagents, solvents, and auxiliary materials.

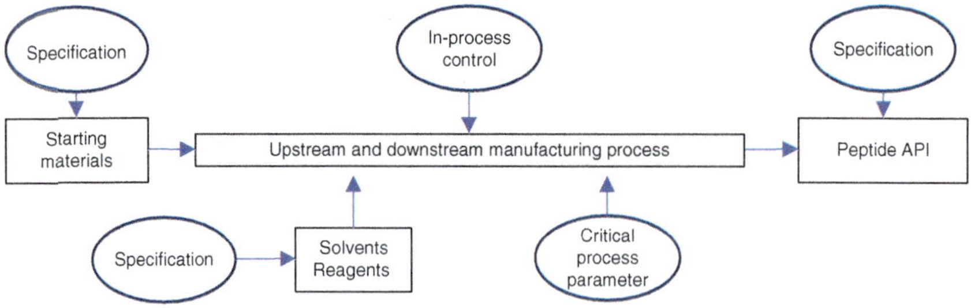

Figure 14.2 Control strategy components.

14.4.3.1.1 Starting Materials

Starting materials are the most important raw materials because they are incorporated into the final peptide sequence (i.e. becoming an integral part of the API) during the synthesis process [21]. Typical starting materials for peptide synthesis are protected amino acid derivatives, resins preloaded with amino acids (for SPPS), and protected peptide fragments.

14.4.3.1.1.1 Amino Acid Derivatives $^{\alpha}$N-protected amino acids are common starting materials for SPPS. For $^{\alpha}$N-Fmoc amino acid derivatives, side-chain protection employs TFA-liable protecting groups, such as tert-butyl (tBu) for hydroxyl and carboxyl groups, Boc for the amino group (Pbf for guanidino group), and Trt for thiol group [22]. For $^{\alpha}$N-Boc amino acids, benzyl-type groups are typically used for side-chain protection (Bzl for hydroxyl and carboxyl groups and carboxybenzyl (Cbz or Z) for amino groups) [23]. On the other hand, LPPS may use a wider variety of protecting groups, including $^{\alpha}$N-protected amino acids and $^{\alpha}$N-free but α-carboxyl protected amino acids.

Common impurities in amino acid derivatives can be, but are not limited to, (i) incorrect enantiomers of chiral amino acids (e.g. D-enantiomers in L-amino acid derivatives); (ii) incorrectly protected or unprotected amino acid impurities (e.g. $^{\alpha}$N-free amino acid or unprotected side chain or both); (iii) amino acid contaminants (e.g. incorrect amino acids); (iv) β-alanine-related contaminants in $^{\alpha}$N-Fmoc amino acids (e.g. Fmoc-β-Ala-Arg(Pbf)-OH in Fmoc-Arg(Pbf)-OH); (v) dipeptide impurities (e.g. Fmoc-Gly-Gly-OH in Fmoc-Gly-OH); (vi) isomeric amino acid impurities (such as Leu in Ile). Potential activation of these impurities during the coupling process may lead to the formation of peptide-related impurities, risking carryover to the final API. Strict impurity acceptance criteria should therefore be set for each starting material. In addition, the impurity control strategy for a specific amino acid derivative must consider the synthetic route, suppliers, and previous batch history. Impurity profiles across vendors can be evaluated via use tests during process development. Impurities can be classified as critical if they are likely to carry over to the final API, or noncritical if they are likely to be purged out during the process.

Unlike the critical impurities enumerated in the previous section, acceptance limits of non-amino acid-related impurities (residual solvents, elemental impurities, nitrosamine impurities, etc.) can often be set based on risk assessment as they generally do not interfere with synthesis. However, reactive solvents such as acetic acid, ethyl acetate (a potential source of acetic acid), or aldehydes can still cause unintended acetylation or chain termination during synthesis. Appropriate limits can be set based on knowledge gained from process development.

Additionally, any potential risk from materials containing TSEs should be assessed to confirm the absence of TSE. If the material is derived directly from a TSE-relevant animal species or is manufactured using materials derived from such species, a certificate of suitability and TSE compliance should be obtained.

In terms of quality control, analytical methods such as mass spectrometry/LC-MS (for identification), HPLC/UHPLC (for purity and related impurities), thin-layer chromatography (for free or unprotected amino acids), chiral HPLC or gas chromatography–mass spectrometry (GC-MS) (for enantiomer impurities), AAA (for incorrect amino acid impurities), and titration (for assay) are commonly used for testing amino acid derivatives.

14.4.3.1.1.2 Resins for SPPS Resins with a linker and preloaded with amino acids for SPPS are usually commercially available. Polystyrene is the most common core resin in SPPS and is typically cross-linked with 1–2% divinylbenzene. In general, the resin particles optimal for synthesis should be uniform, sufficiently mechanically strong, permeable to solvents and reagents, easy to filter, and chemically stable. Swelling characteristics (e.g. mL/g), degree of substitution (e.g. mmoL/g), and purity of loaded amino acids directly affect the final peptide purity and yield. Substitution of Fmoc amino acid-loaded resins can be easily determined by common color or UV tests. Too high a substitution of a preloaded resin can lead to synthetic difficulties, especially for long sequences. Too low a substitution can lead to inefficient batch throughput and higher production costs.

Common resins used for SPPS include hydroxymethylphenylacetamide-methyl resin and *p*-methoxybenzylamine resin for Boc-SPPS as well as 4-(2′,4′-dimethoxyphenylaminomethyl)-phenoxymethyl polystyrene (Rink) resin, [4-(hydroxymethyl)phenoxymethyl] polyethylene styrene (Wang) resin, and 2-CTC resin for Fmoc-SPPS.

Loading of the C-terminal chiral amino acid onto the resin through the formation of an ester bond should be carefully assessed. For example, Cys or His are particularly prone to enatiomerization if activation of their carboxyl groups is involved (e.g. loading onto Wang resin). This racemization problem can be avoided if 2-CTC resin is used for the synthesis instead of Wang resin, because this loading reaction does not involve the activation of carboxyl groups. 2-CTC resin is also preferred when the C-terminal amino acid is Gly or Pro. Compared with the 4-(hydroxymethyl)phenoxymethyl group of Wang resin, the bulky trityl group of 2-CTC resin suppresses the formation of diketopiperazine (DKP) and reduces the loss of dipeptides in the resin. Peptides synthesized on 2-CTC resin can be cleaved by diluted acids (TFA or AcOH) or acidic solvents (TFE or HFIP) and have been routinely used in industry to produce protected peptide fragments. Because of acid sensitivity, the use of 2-CTC resin is not recommended for long sequence peptides.

14.4.3.1.2 Reagents and Solvents

Unlike protected amino acids, solvents and reagents generally pose a low risk of affecting the quality of the final DS because they are less likely to be incorporated into the final peptide. Solvents and reagents used in upstream processes are typically removed to acceptable levels by precipitation/extraction in LPPS or washing in SPPS.

In peptide synthesis, commonly used reagents are *N,N′*-diisopropylcarbodiimide (DIC), hexafluorophosphate benzotriazole tetramethylurea (HBTU), hydroxybenzotriazole (HOBt), cyano (oxime) ethyl acetate (Oxyma), and *N,N*-diisopropylethylamine (DIEA). Commonly used solvents include piperidine, isopropyl alcohol, *N,N*-dimethylformamide (DMF), and methyl tert-butyl ether. For these raw materials, typical minimum quality control criteria are appearance, identity, and purity, with additional acceptance limits set for impurities that may react during synthesis (e.g. dimethylamine in DMF). Controls (i.e. specifications) should be based on the risks associated with the quality of the intermediate and final API.

The solvents and reagents used in the downstream process may impact the final peptide quality. The closer to the end of the downstream process, the more critical the quality of solvents and reagents is. In particular, the solvents and reagents used in the post-purification step must be carefully considered, since there are no purification mechanisms to remove any impurities introduced here. Relevant risk assessment should focus on potential reactive contaminants (e.g. formaldehyde, acetaldehyde in acetic acid) and nonvolatile residues (e.g. acetamide in acetonitrile). Acetonitrile, isopropanol, and acetic acid, often used in the final step of API isolation, should be tightly controlled.

14.4.3.2 Manufacturing Process Control

The ICH Q8 guideline states that quality cannot be tested into products. Therefore, quality must be built in by design. Quality by design is a systematic approach to drug development that starts with predefined objectives and emphasizes product and process understanding and process control based on sound science and quality risk management [19]. In the process design stage, every unit operation must be evaluated. The critical material attributes and CPPs that may affect the final product quality (i.e. CQAs) must be determined, and the ranges of operating and control of important variables must be established. After the process design is complete, process qualification is performed to validate the process through manufacturing commercial batches. The manufacturing process is maintained through continuous process verification in a state of control during commercial manufacture. Control of the peptide production process can be divided into upstream process control and downstream process control.

14.4.3.2.1 Upstream Process

14.4.3.2.1.1 Solid-phase Synthesis A standard SPPS process essentially consists of a number of repetitive synthesis cycles required to complete the assembly of a full-length sequence. When the C-terminal amino acid is attached to the resin, a typical synthesis cycle begins with removal of the $^{\alpha}$N-protection group (Boc or Fmoc) followed by appropriate washes and continuing with coupling of the next amino acid derivative.

In Fmoc chemistry, the removal of $^{\alpha}$N-Fmoc is accomplished by nucleophiles such as piperidine and morpholine solutions, typically 10–30% piperidine in DMF, while usually Boc-SPPS uses 20–50% TFA in dichloromethane (DCM) for $^{\alpha}$N-Boc deprotection. The deprotection step is usually performed twice to ensure complete removal of the temporary protecting group. Because large excesses of deprotection reagents are used in both chemistries, deprotection is typically very efficient and quantitative.

Incomplete deprotection may be encountered in the synthesis of long or difficult peptide sequences when peptide-chain aggregation (indicated by resin shrinking) occurs. Impurities resulting from incomplete deprotection are deletion peptides that pose challenges for downstream purification and may affect yields. In this case, the HPLC method can be employed to serve as a reliable in-process control, capable of providing a good assessment in Fmoc-SPPS. However, HPLC methods require cleavage of small amounts of peptidyl resin and subsequent analysis of the cleaved sample. An alternative approach is by measuring residual Fmoc, which involves treating a small amount of peptidyl resin with a solution of 1,8-diazabicyclo (5.4.0) undec-7-ene (DBU) followed by UV quantification. These in-process control methods are not applicable to Boc-SPPS.

Colorimetric tests such as ninhydrin (Kaiser) [24], trinitrobenzenesulfonate [25], chloranil [26] (mainly used for secondary amines), or bromphenol blue [27] are usually used to verify deprotection in Fmoc and Boc syntheses, but these tests do not confirm the completeness of deprotection. Measures to improve difficult deprotection encountered in Fmoc-SPPS include increasing the deprotection time, increasing the number of repetitions and temperature, and using solvent mixtures or stronger bases (e.g. DBU). However, it is important to note that intensive treatment with strong bases (such as DBU or piperidine) promotes DKP formation, which may result in aspartimide formation side reactions. In Boc-SPPS, a high concentration of TFA or prolonged duration of deprotection can also be applied.

After the deprotection step, there is a wash step before the next coupling step. In Fmoc-SPPS, incomplete piperidine removal may lead to premature deprotection of the next amino acid during coupling, creating insertion impurities or residual piperidine that unnecessarily consumes amino acids. To avoid these problems, an in-process control for residual piperidine should be in place. A common test is to evaluate piperidine content in the wash by pH measurement. In Boc-SPPS, after the deprotection with TFA solution, neutralization (as a separate step or *in situ*) is required to remove the TFA. Conventional washing with solvents does not remove the TFA bound to the amino groups. The neutralization step is typically performed twice using a solution of tertiary amine base in DCM (e.g. 10% triethyl amine or DIEA).

In the Fmoc- or Boc-SPPS process, the carbamate/urethane structure helps minimize epimerization of the activated amino acids in most cases during the coupling step. However, the risk of potential racemization is not eliminated and should still be evaluated during process development. The racemization mechanism occurs through either direct enolization or formation and epimerization of 5(4H) oxazolone [28–30]. Coupling reagents and solvents play roles in this side reaction and should be considered during process development. A variety of coupling reagents have been developed and are available for peptide synthesis [31]. Currently, three types of coupling reagents are commonly used in peptide synthesis: carbodiimides, uronium/aminium salts, and phosphonium salts. For carbodiimide, DIC is the most cost-effective in combination with an additive such as HOBt, 1-hydroxy-7-azabenzotriazole, 1-hydroxy-6-chloro-benzotriazole (6-Cl-HOBt), or Oxyma [32]. This coupling method suppresses the racemization of amino acid derivatives and avoids the formation of oxazolone by rapidly converting the very reactive *O*-acylisourea species into the corresponding reactive carboxyl ester. DIC/HOBt has been widely used in the commercial synthesis of many peptides. On the other hand, due to the potentially explosive properties of HOBt, Oxyma has now gained significant traction in the industry.

Uronium-based coupling reagents (HBTU, 2-(1H-Benzotriazole-1-yl)-1,1,3,3-tetramethylaminium tetrafluoroborate (TBTU), and hexafluorophosphate azabenzotriazole tetramethyl uronium (HATU)) as well as phosphonium-based coupling reagents (PyBOP and PyAOP) provide high coupling efficiencies, reduced racemization, and faster acylation rates. These reagents readily form OBt or OAt esters with the carboxyl groups *in situ* and pre-activation is not needed. Attention needs to be paid to amino acids that are prone to racemization (such as Cys and His), elevated temperatures, excessive base conditions, and unnecessary pre-activation.

In addition, uronium-based coupling reagents can react with amino groups, preventing further chain assembly. A slight excess of the carboxyl component relative to the coupling reagent is often used in practice. In contrast, phosphonium-based coupling reagents do not undergo such side reactions. During SPPS, colorimetric tests (Kaiser, 2,4,6-trinitrobenzenesulfonic acid [TNBS], Chloranil, etc.) are commonly used as in-process controls to detect the presence of residual amino groups.

In cases where coupling is difficult, recoupling can be performed using the same or a different method to drive the reaction to completion. If repeated coupling still cannot achieve completion, a capping step is required to mask unreacted residual amino functionality to avoid deletion sequences. In most cases, capped peptides (truncated sequences) are easier to resolve in chromatography than deletion sequences. When using a large excess of acetic anhydride for the acetylation step, adequate washing is required to avoid potential chain termination caused by residual anhydride in the following deprotection step.

14.4.3.2.1.2 Cleavage of Peptide Resin The global deprotection and cleavage process removes all side-chain protecting groups while simultaneously cleaving the peptide from the resin. For cleavage reactions, a TFA cocktail consisting of TFA and scavengers is used, in which various reactions occur and a number of reactive species (e.g. cations) are generated. To minimize or avoid attacks of deprotected peptides by reactive species, effective scavengers are required to adequately capture or quench these species. To cleave peptide resins obtained from Fmoc-SPPS, common scavengers such as triisopropylsilane, water, ethanedithiol (EDT), dithioerythritol (DTT), thioanisole, and phenol can be used. Among them, EDT, DTT, and thioanisole are particularly effective in preventing oxidation, such as disulfide bond formation, polymerization between two or more Cys residues, Met(O) formation, and S-alkylation (e.g. t-butylation) when the peptide contains Cys and Met residues. During development, the composition of the cleavage cocktail, the ratio of the cocktail relative to resin, cleavage temperature, and cleavage time need to be optimized to obtain optimal results. HPLC can be used as an in-process control method to monitor the cleavage reaction. LC-MS is a useful technique for identifying cleavage-related impurities.

To cleave peptide resins obtained from Boc-SPPS, anhydrous HF and appropriate scavengers (e.g. anisole, *p*-cresol, *p*-thiocresol, dimethyl sulfide [DMS]) are required. Temperature, time, and scavenger composition play key roles in minimizing side reactions during cleavage. HF cleavage is typically performed at 0–5 °C. A popular HF cleavage method is the low-high procedure developed by Tam [33], which consists first of a low HF concentration step followed by a high HF concentration step. Low-temperature conditions avoid the alkylation of Tyr, the acylation of the side-chain carboxyl group of Asp/Glu, and the formation of succinimide peptides. This can also effectively convert Met(O)/Trp(For) to Met/Trp with the presence of DMS. Subsequently, high-temperature conditions allow deprotection of Arg(Tos) and Arg(NO_2) and release of the peptide from the resin. Anisole is a universal scavenger in HF cleavage and effectively prevents the alkylation of Trp residues. For peptides containing sensitive residues, it is necessary to use a combination of scavenger mixtures. For example, anisole/DMS/*p*-thiocresol mixture can effectively prevent side-chain alkylation of Met and Cys.

14.4.3.2.1.3 Liquid-phase Synthesis In LPPS, all reactions occur in solution. In-process control of reactions is relatively easy using modern analytical tools. The HPLC method is most useful for monitoring reaction progress and quality control of isolated intermediates, as it usually provides reliable and quantitative results. Impurity characterization at each step is possible and can be accomplished conveniently by LC-MS.

In the case of fragment condensation, the length and the C-terminal amino acid of the fragments are two key factors to consider when selecting fragments and coupling sites. Typical fragment length should be between 5 and 15 amino acids, as high purity is often difficult to obtain with longer sequences. Unlike the activation and coupling of individual Fmoc or Boc amino acids, epimerization occurs more readily, so glycine or proline should be chosen as the C-terminal amino acid of the fragments. If this is not possible, sterically hindered amino acids (e.g. Ile and Val) and amino acids prone to racemization (e.g. His and Cys) should be avoided. When the C-terminal amino acid is chiral (except Pro), there is always a risk of racemization of the C-terminal amino acid during the activation of its carboxyl group. Development is needed to find optimal coupling methods with respect to racemization and coupling efficiency. Direct monitoring of epimerization by HPLC may not always be feasible (depending on the size of the condensed peptide). In this case, analysis of the deprotected product by HPLC, LC-MS, or analysis of the hydrolysis of the product for D/L form may be employed.

14.4.3.2.2 Downstream Process

Crude peptides produced by SPPS may contain deletions, insertions, truncated, modified sequences, and impurities related to the cleavage process. They are unlikely to meet the purity specifications set for the API and must therefore be purified. HPLC is the most commonly used separation technique in peptide production [34], particularly reverse phase (RP-HPLC), ion exchange (IEX), and SEC. Depending on the purity of the crude peptide, RP-HPLC and IEX chromatography are often used for crude purification.

14.4.3.2.2.1 Reverse-phase Chromatography RP-HPLC involves the separation of molecules based on hydrophobic interactions between solute molecules in the mobile phase and alkyl or aromatic ligands bound to the stationary phase (chromatography resin). RP-HPLC uses a polar aqueous solvent system (such as acetonitrile or alcohol in water) as the mobile phase. Elution and separation are generally according to the hydrophobicity order of the molecules.

In peptide purification, a variety of chromatography resins are used, such as C18, C8, and C4, of which C18 is the most widely used. These resins are based on silica gel beads, the surface of which is bonded with corresponding hydrocarbons. The particle spherical shape, particle size, pore size, uniformity/particle size distribution, and carbon loading of the resins will affect the separation efficiency and column loading capacity. The resin's particle size is the diameter of chromatography particles, with typical particle sizes ranging from 2 to 100 μm. For industry-preparative HPLC columns, 10–16 μm particles are generally used due to backpressure concerns and often achieve satisfactory separations. Pore size is another key parameter for chromatography resins. The pore size should be chosen to match the size of the peptide, i.e. low molecular weights are best suited to resins with small pore sizes, while larger molecules require resins with wider pore sizes. As a general rule, for analytes below 3000 Da, pore sizes of 100 *Å* or smaller should be used. For samples in the 3000–10,000 Da range, a pore size of 100–150 Å is recommended. For samples above 10,000 Da (e.g. pegylated peptides), 300 Å or larger provides better isolation efficiency and peak shape. Nevertheless, actual screening of purification resins is an essential task and is required during downstream process development.

Chromatographic conditions that need to be optimized also include acidifiers and pH of mobile phases, as well as gradients. The pH and ion strength of the mobile phases, which affect compound retention, separation performance, and potentially stability of stationary phase and product, also need to be evaluated. Optimized gradients should achieve effective removal of impurities, high recovery of products, and reasonable cycle times. For some complex peptides, multi-dimensional purification using different/orthogonal buffer/mobile phase systems may be required to achieve the required purity. In addition to its purification function, RP-HPLC is often used to convert the salt form of purified peptides into the desired counterion by washing adsorbed peptide on the column with a high concentration of the desired counterion, followed by eluting the peptide from the column.

The recent emergence of multicolumn chromatography techniques in peptide manufacturing [35–37] allows the integration of direct reprocessing of side fractions without needing to isolate them. This multicolumn technology represents a continuous chromatographic purification process that is expected to provide more reproducible high-purity peptides with higher yields and productivity compared to traditional single-column methods. Additionally, the implementation of advanced control and process analytical technology tools will make the technology more automated, robust, and promising.

14.4.3.2.2.2 Ion Exchange Chromatography IEX chromatography is considered a complement to RP-HPLC purification. Separation by IEX is based on charge differences of the peptides, most commonly performed using a gradient of increasing salt concentration. Uncharged structures (e.g. reagents and byproducts from the cleavage step), as well as smaller and less charged species, generally elute earlier than longer peptide sequences. In most cases, the net charge of the peptide can vary depending on the pH of the mobile phase. At acidic pH, carboxylate and basic residues (Arg, His, Lys) are mainly protonated. Therefore, through pH optimization, IEX is able to discriminate charge variants of peptides and thereby separate relevant peptide impurities. In addition to its purification function, IEX can also be used for salt conversion of purified peptides (e.g. TFA salts to HCl salts).

14.4.3.2.2.3 Size-exclusion Chromatography SEC methods separate molecules primarily based on their size. Higher molecular weights elute before lower molecular weights. The general two subcategories of SEC are gel filtration chromatography (using aqueous mobile phases) and gel permeation chromatography (using organic solvents). The application of SEC in peptide production is primarily to desalt peptide solutions or to remove polymeric forms of peptides or aggregates, typically using an aqueous mobile phase (e.g. aqueous acetic acid) for elution.

14.4.3.2.2.4 Fraction and Pool Analysis Analysis of fractions and pools is a critical control of intermediate product quality in downstream purification processes. Analytical HPLC techniques are often the primary IPC method used for process control, and their results are the basis of process decisions. One or more IPC methods may be required to ensure product purity and impurity criteria are met. In some cases, two RP-HPLC methods with orthogonal buffer systems are required to control for impurities. In addition to RP-HPLC, SEC is also used to control aggregates or high molecular weight impurities.

Compared with conventional analytical HPLC, UHPLC uses separation media with smaller particle sizes (e.g. 3.5 and $\leq 3\ \mu m$), which enables efficient, high-resolution separation and rapid turnaround of in-process analysis during the purification process. Its application in the field of peptides is increasingly common.

14.4.3.2.2.5 Final Product Isolation Generally, bulk solutions of peptide APIs are not stable for long-term storage, requiring isolation of the API in solid form. Methods for isolating peptides after purification include lyophilization, precipitation, and spray drying, with lyophilization being the most common but highly limited by equipment constraints. The latter two are more scalable.

Aqueous peptide solutions from the last step of downstream purification require a concentration step before lyophilization if the organic solvent (e.g. acetonitrile) concentration is high or the peptide concentration is low. High acetonitrile concentrations pose challenges to the lyophilization process and may raise concerns about high acetonitrile residues in the final peptide API. If the peptide concentration is too low, the freeze-drying volume will be too large. These concentration issues can be addressed by evaporation under reduced pressure, chromatography, or diafiltration. HPLC and GC can be used as IPC methods to control peptide and acetonitrile concentrations, respectively.

Lyophilization typically involves three process steps: (i) freezing the solution to convert water into ice (e.g. at −40 to −50 °C); (ii) primary drying to remove ice via sublimation under vacuum; and (iii) secondary drying to remove residual water by desorption. A typical lyophilization process using modern tray lyophilizers is automated, with strict monitoring and control of temperature and pressure. Data collected during the lyophilization process provide supporting interpretations of the progression of the cycle (e.g. endpoints of primary and secondary drying). For bulk lyophilization, product homogeneity is critical and should be ensured by appropriate homogeneity testing. A normal lyophilized product should exhibit self-supportive and uniform color and texture based on visual inspection of the cake appearance. Attention should be paid to potential meltback, which is a form of cake collapse caused by incomplete sublimation (i.e. insufficient primary drying) leading to a transition from solid to liquid.

Isolation of peptide APIs in solid form by precipitation is a more scalable and economical process compared to lyophilization. Depending on the physical properties of the peptide, antisolvent methods and pH adjustment methods have been successfully applied. Generally speaking, the precipitation process requires more development. During the process development, attention should be paid to factors, including, but not limited to, product loss in the mother liquor, solvent residue in the API, potential generation of degradation impurities, and particle size of the API. Spray drying has also been evaluated by many organizations as a potential routine application, but has not been widely adopted in part because of general concerns about the thermal stability of peptides.

14.5 Conclusions

Drug development is a lengthy process in which CMC development is a crucial element. This chapter aims to provide an overview for researchers interested in developing NCE peptide drugs relevant to CMC. The expectations and requirements for CMC content (DS and DP) in CTD format are introduced. Additionally, an aspect of peptide CMC development is illustrated, focusing on the chemical synthesis of peptides and related control strategies. Currently, guidelines and policies related to peptides are still being developed. A common risk-based approach should therefore be applied in quality considerations in the development and manufacturing of peptide APIs. Phase-appropriate focus should also be kept in mind during CMC development. Furthermore, CMC development and control strategies are highly specific to individual peptides and corresponding production process, and must be determined on a case-by-case basis.

References

1 Muttenthaler, M., King, G.F., Adams, D.J., and Alewood, P.F. Trends in peptide drug discovery. *Nat Rev Drug Discov*. 2021;20:309.
2 Wang, L., Wang, N., Zhang, W., et al. Therapeutic peptides: current applications and future directions. *Sig Transduct Target Ther*. 2022;7:48.
3 U.S. Department of Health and Human Services, Food and Drug Administration, Center for Drug Evaluation and Research (CDER). FDA Guidance for Industry: ANDAs for certain highly purified synthetic peptide drug products that refer to listed drugs of rDNA origin. May 2021.
4 Zane, D., Feldman, P.L., Sawyer, T., Sobol, Z., and Hawes, J. Development and regulatory challenges for peptide therapeutics. *Int J Toxicol*. 2021;40(2):108.
5 European Medicines Agency. EMA guideline: EMA/CHMP/QWP/735422/2022 concept paper on the establishment of a guideline on the development and manufacture of synthetic peptides. 15 September 2022.
6 European Medicines Agency. EMA guideline: EMA/CHMP/CVMP/QWP/387541/2023 development and manufacture of synthetic peptides. 18 October 2023.
7 ICH Tripartite Guideline M4 (R4) on common technical document (CTD) for the registration of pharmaceuticals for human use—Organization of CTD. Current step 4 version dated 15 June 2016.
8 ICH Harmonized Tripartite Guideline. M4Q(R1) on common technical document (CTD) for the registration of pharmaceuticals for human use—Quality. Current step 4 version dated 12 September 2002.
9 Akbarian, M., Khani, A., Eghbalpour, S., and Uversky, V.N. Bioactive peptides: synthesis, sources, applications, and proposed mechanisms of action. *Int J Mol Sci*. 2022;23:1445.
10 Katsoyannis, P. Synthesis of insulin. *Science*. 1966;154:1509.
11 Isidro-Llobet, A., Kenworthy, M.N., Mukherjee, S., et al. Sustainability challenges in peptide synthesis and purification: from R&D to production. *J Org Chem*. 2019;84(8):4615.
12 Sharma, A., Kumar, A., de la Torre, B.G., and Albericio, F. Liquid-phase peptide synthesis (LPPS): a third wave for the preparation of peptides. *Chem Rev*. 2022;122(16):13516.
13 Merrifield, R.B. Solid phase synthesis. I. The synthesis of a tetrapeptide. *J Am Chem Soc*. 1963;85:2149.
14 Bray, B.L. Large-scale manufacture of peptide therapeutics by chemical synthesis. *Nat Rev Drug Discov*. 2003;2:587.
15 Frederick, M.O., Boyse, R.A., Braden, T.M., et al. Kilogram-scale GMP manufacture of tirzepatide using a hybrid SPPS/LPPS approach with continuous manufacturing. *Org Process Res Dev*. 2021;25:1628.

16 U.S. Department of Health and Human Services, Food and Drug Administration, Center for Drug Evaluation and Research (CDER), Center for Biologics Evaluation and Research (CBER), Office of Regulatory Affairs (ORA). FDA Guidance for Industry: CGMP for Phase 1 investigational drugs. July 2008.
17 Center for Drug Evaluation and Research (CDER), Center for Biologics Evaluation and Rcscarch (CBER). FDA Guidance for Industry: content and format of investigational new drug applications (INDs) for Phase 1 studies of drugs, including well-characterized, therapeutic, biotechnology-derived products. November 1995.
18 U.S. Department of Health and Human Services, Food and Drug Administration, Center for Drug Evaluation and Research (CDER). FDA Guidance for Industry: INDs for Phase 2 and Phase 3 studies: chemistry, manufacturing, and controls information. May 2003.
19 ICH Harmonized Tripartite Guideline. Q8 (R2) pharmaceutical development. Current step 4 version dated August 2009.
20 ICH Harmonized Tripartite Guideline. Q11 development and manufacture of drug substances (chemical entities and biotechnological/biological entities). Current step 4 version dated 1 May 2012.
21 U.S. Department of Health and Human Services, Food and Drug Administration, Center for Drug Evaluation and Research (CDER), Center for Biologics Evaluation and Research (CBER). FDA Guidance for Industry: "The deemed to be a license" provision of the BPCI Act: Questions and Answers. March 2020.
22 Chan, W.C. and White, P.D. *Fmoc Solid Phase Peptide Synthesis: A Practical Approach*. Oxford; 2000.
23 Albericio, F. Orthogonal protecting groups for Nα-amino and C-terminal carboxyl functions in solid-phase peptide synthesis. *Biopoly*. 2000;55:123–139.
24 Bossinger, D. and Cook, P. Color test for detection of free terminal amino groups in the solid-phase synthesis of peptides. *Anal Biochem*. 1970;34:595.
25 Hancock, W.S. and Battersby, J.E. A new micro-test for the detection of incomplete coupling reactions in solid-phase peptide synthesis using 2, 4, 6-trinitrobenzene-sulphonic acid. *Anal Biochem*. 1976;71:260.
26 Vojkovsky, T. Detection of secondary amines on solid phase. *Pept Res*. 1995;8:236.
27 Krchnak, V. and Vagner, J. Color-monitored solid-phase multiple peptide synthesis under low-pressure continuous-flow conditions. *Pept Res*. 1990;3:182.
28 Antonovi, I. and Young, G.T. Amino-acids and peptides. Part XXV. Mechanism of base-catalysed racemisation of p-nitrophenyl esters of acylpeptides. *J Chem Soc C*. 1967;(0):595.
29 Bednarek, M.A., Springer, J.P., Cunningham, B.R., Bernick, A., and Bodanszky, M. Reactions of C-terminal omega-amino-acid residues in liquid-hydrogen fluoride. *Int J Pept Protein Res*. 1993;42(1):10.

30 Carpino, L.A. Methyl-ester nonequivalence in the H-1-NMR spectra of diastereomeric dipeptide esters incorporating N-terminal alpha-phenylglycine units. *J Org Chem*. 1988;53(4):875.
31 Albericio, F. and El-Faham, A. Choosing the right coupling reagent for peptides: a twenty-five-year journey. *Org Process Res Dev*. 2018;22:760.
32 Subirós-Funosas, R., Prohens, R., Barbas, R., El-Faham, A., and Albericio, F. Oxyma: an efficient additive for peptide synthesis to replace the benzotriazole-based HOBt and HOAt with a lower risk of explosion. *Chem Eur J*. 2009;15:9394.
33 Tam, J.P., Heath, W.F., and Merrifield, R.B. An SN2 deprotection of synthetic peptides with a low concentration of hydrofluoric acid in dimethyl sulfide: evidence and application in peptide synthesis. *J Am Chem Soc*. 1983;105:6442.
34 Pratap Singh Jadaun, G., Dixit, S., Saklani, V., Mendiratta, S., Jain, R., and Singh, S. HPLC for peptides and proteins: principles, methods and applications. *Pharm Methods*. 2017;8:1.
35 Luca, C.D., Felletti, S., Lievore, G., et al. From batch to continuous chromatographic purification of a therapeutic peptide through multicolumn countercurrent solvent gradient purification. *J Chromatogr A*. 2020; 1625:461304.
36 Silva, R.J.S., Rodrigues, R.C.R., Osuna-Sanchez, H., Bailly, M., Valéry, E., and Mota, J.P.B. A new multicolumn, open-loop process for center-cut separation by solvent-gradient chromatography. *J Chromatogr A*. 2010; 1217:8257.
37 Luca, C.D., Lievore, G., Bozza, D., et al. Downstream processing of therapeutic peptides by means of preparative liquid chromatography. *Molecules*. 2021;26(15):4688.

Index

Note: Page numbers in *italics* and **bold** refers to figures and tables respectively.

Peptide Science: Chemical Ligation, Lead Generation, and Therapeutic Advances, First Edition. Fa Liu.

t

Printed and bound by CPI Group (UK) Ltd, Croydon, CR0 4YY

17/09/2026

14964253-0007